Innovations and Obstacles: Microbial Communities in the Journey of Soil Remediation

Innovations and Obstacles: Microbial Communities in the Journey of Soil Remediation

Guest Editor

Dong Liu

MDPI

Basel • Beijing • Wuhan • Barcelona • Belgrade • Novi Sad • Cluj • Manchester

Guest Editor
Dong Liu
Kunming Institute of Botany
Chinese Academy of Sciences
Kunming
China

Editorial Office
MDPI AG
Grosspeteranlage 5
4052 Basel, Switzerland

This is a reprint of the Special Issue, published open access by the journal *Agronomy* (ISSN 2073-4395), freely accessible at: https://www.mdpi.com/journal/agronomy/special_issues/8SXZPP7R52.

For citation purposes, cite each article independently as indicated on the article page online and as indicated below:

Lastname, A.A.; Lastname, B.B. Article Title. *Journal Name* **Year**, *Volume Number*, Page Range.

ISBN 978-3-7258-6678-6 (Hbk)
ISBN 978-3-7258-6679-3 (PDF)
https://doi.org/10.3390/books978-3-7258-6679-3

Contents

About the Editor

Dong Liu

Dong Liu is an Associate Researcher at the Kunming Institute of Botany, Chinese Academy of Sciences, and Master's Supervisor. During his doctoral studies in Germany and Austria, he received systematic scientific training and was awarded the EMBO Scholarship and the Best Original Paper Award by the European Molecular Biology Organization. Since being selected for the Yunnan Provincial Thousand Talents Program in 2019, he has independently led and completed research on the microbiome and its interactions of precious wild edible fungi in Yunnan Province, supported by grants from the Yunnan Natural Science Foundation, the CAS Western Light Program, and local collaborative projects. Through years of research on the "plant–soil–microbe" system in forest ecosystems, the major achievements have been attained in the identification of fermentation microorganisms from agricultural, forestry, and livestock waste, supporting the preparation of saprophytic edible mushroom cultivation substrates; the implementation of the "forest–livestock–microbe" model has significantly enhanced soil fertility and ecological functions. To date, as first author he has published over 30 high-impact SCI papers in top-tier journals including *ACS Nano*, *Journal of Hazardous Materials*, *Geoderma*, and *Catena*, with 15 papers ranked in the top 1% (Q1) of the Chinese Academy of Sciences.

agronomy

MDPI

Editorial

Innovations and Obstacles: Microbial Communities in the Journey of Soil Remediation—Editorial for Closing Special Issue

Dong Liu

Yunnan Key Laboratory for Fungal Diversity and Green Development & Yunnan International Joint Laboratory of Fungal Sustainable Utilization in South and Southeast Asia, The Germplasm Bank of Wild Species, Kunming Institute of Botany, Chinese Academy of Sciences, Kunming 650201, China; liudongc@mail.kib.ac.cn

1. Introduction

The global imperative to transition towards sustainable agricultural practices has never been more pressing. As the world grapples with the dual challenges of ensuring food security and maintaining environmental integrity, agricultural soils have emerged as both a critical resource and a vulnerable ecosystem under threat. The excessive reliance on chemical inputs, improper application of pesticides, and intensive farming practices have collectively imposed unprecedented stresses on soil microbiomes, disrupting the delicate balance of microorganism-mediated carbon, nitrogen, and phosphorus cycles that underpin soil health and fertility.

This Special Issue, "Innovations and Obstacles: Microbial Communities in the Journey of Soil Remediation," was conceived to address these pressing challenges by exploring the multifaceted role of microbial communities in soil remediation processes. The central premise underlying this collection is the recognition that sustainable soil management must transcend traditional approaches focused solely on chemical amendments or physical interventions. Instead, we must embrace a more holistic paradigm that leverages the inherent potential of soil microbiomes to restore, maintain, and enhance soil health.

The contributions assembled in this issue reflect the diverse and dynamic nature of current research in soil microbiome engineering and application. From fundamental investigations into microbial community dynamics to applied studies demonstrating practical remediation strategies, the articles presented here collectively advance our understanding of how beneficial microorganisms can be harnessed to accelerate soil nutrient recovery, enhance disease resistance, and ultimately promote sustainable agricultural productivity.

2. Overview of Published Articles

This Special Issue comprises fourteen peer-reviewed articles that collectively represent the state of the art in soil remediation research through microbial interventions. The contributions span a broad spectrum of topics, methodologies, and geographical contexts, reflecting the global relevance and urgency of soil health restoration.

Several articles focus on the characterization and application of specific beneficial microbial strains. These studies demonstrate how targeted inoculation with carefully selected microorganisms can significantly enhance soil fertility, improve nutrient cycling, and promote plant growth under various stress conditions. The research encompasses both laboratory-based mechanistic studies and field-scale validation experiments, providing robust evidence for the efficacy of microbial-based remediation approaches.

A notable subset of contributions addresses the challenges associated with microbial community establishment and persistence in disturbed soils. These investigations reveal the complex interactions between introduced beneficial microorganisms and indigenous soil communities, highlighting both the opportunities and obstacles in achieving successful microbial colonization and long-term establishment. The studies employ advanced molecular techniques to track microbial dynamics and identify key factors influencing community stability and functionality.

Technological innovations in microbial preservation, formulation, and delivery systems constitute another important theme within this collection. Several articles present novel approaches for maintaining microbial viability during storage and application, including advanced encapsulation technologies, carrier material optimization, and delivery system design. These technological advances are crucial for translating laboratory discoveries into practical field applications.

The issue also includes comprehensive reviews that synthesize current knowledge on specific aspects of soil microbial remediation. These review articles provide valuable perspectives on emerging trends, identify knowledge gaps, and propose future research directions. They serve as essential resources for researchers entering the field and practitioners seeking to implement microbial-based soil remediation strategies.

Geographically, the contributions span diverse agroecological zones, from temperate agricultural regions to tropical farming systems, reflecting the universal applicability of microbial-based soil remediation approaches. This global perspective is particularly valuable as it demonstrates how local environmental conditions, farming practices, and soil characteristics influence the success of microbial interventions.

Methodologically, the collection showcases a diverse array of approaches, from traditional microbiological techniques to cutting-edge molecular biology tools, including high-throughput sequencing, metagenomics, and advanced bioinformatics analyses. This methodological diversity reflects the interdisciplinary nature of soil remediation research and the integration of multiple scientific disciplines in addressing complex soil health challenges.

3. Conclusions

The completion of this Special Issue marks an important milestone in our collective journey toward understanding and harnessing the power of microbial communities for soil remediation. The fourteen contributions presented here collectively demonstrate both the tremendous potential and the inherent complexities of microbial-based soil restoration approaches.

Several key insights emerge from this collection. First, the studies consistently affirm that beneficial microorganisms represent powerful tools for soil remediation, capable of significantly improving soil health, nutrient availability, and plant productivity. However, the success of microbial interventions is highly context-dependent, influenced by factors ranging from soil physicochemical properties to local climatic conditions and agricultural management practices.

Second, the research highlights the critical importance of understanding microbial community dynamics in achieving successful soil remediation. The interactions between introduced beneficial microorganisms and indigenous soil communities are complex and often unpredictable, necessitating a more nuanced approach to microbial application strategies. The studies suggest that future efforts should focus on developing microbial consortia rather than relying on single-strain inoculants, as this approach may better mimic natural microbial interactions and enhance resilience.

Third, technological innovations in microbial preservation, formulation, and delivery are essential for the practical implementation of microbial-based soil remediation strategies. The advances presented in this issue represent significant progress toward overcoming

https://doi.org/10.3390/agronomy16020202

key bottlenecks in the translation of laboratory discoveries to field applications. However, continued research is needed to develop cost-effective, scalable, and environmentally sustainable delivery systems.

Perhaps most importantly, this Special Issue underscores the need for continued interdisciplinary collaboration in addressing soil health challenges. The complexity of soil ecosystems demands integrated approaches that combine expertise from microbiology, soil science, agronomy, environmental engineering, and other disciplines. The success of microbial-based soil remediation will ultimately depend on our ability to develop holistic strategies that consider not only the biological components but also the physical, chemical, and socioeconomic factors that influence soil management decisions.

As we close this Special Issue, we hope that the research presented here will inspire continued innovation and collaboration in the field of soil remediation. The challenges we face in restoring and maintaining soil health are formidable, but the potential benefits—enhanced food security, improved environmental quality, and sustainable agricultural productivity—justify our continued efforts. We encourage readers to build upon the foundations laid by these contributions and to push the boundaries of our knowledge even further. The journey toward sustainable soil management through microbial interventions is far from complete, but the path forward is now clearer thanks to the dedicated work of the researchers whose contributions grace this collection. We express our sincere gratitude to all authors who contributed their outstanding research to this Special Issue, to the reviewers who provided invaluable feedback and ensured the high quality of the published works, and to the editorial team at MDPI for their professional support throughout this endeavor. It is our hope that this collection will serve as a valuable resource for the scientific community and contribute meaningfully to the global effort to achieve sustainable soil management and food security.

Conflicts of Interest: The author declares no conflict of interest.

List of Contributions:

1. Gao, J.; Ma, Z.; Liu, L.; Shi, Z.; Lv, J. Responses of Methane Emission and Bacterial Community to Fertilizer Reduction Plus Organic Materials over the Course of an 85-Day Leaching Experiment. *Agronomy* **2024**, *14*, 1972. https://doi.org/10.3390/agronomy14091972.
2. Gao, S.; Zhang, Y.; Shao, Y.; Li, B.; Liu, H.; Li, Y.; Ren, X.; Chen, Z. Combined Application of High-Throughput Sequencing and Metabolomics to Evaluate the Microbial Mechanisms of Plant-Growth-Promoting Bacteria in Enhancing the Remediation of Cd-Contaminated Soil by Hybrid Pennisetum. *Agronomy* **2024**, *14*, 2348. https://doi.org/10.3390/agronomy14102348.
3. Zhang, L.; Liu, T.; Yan, C.; Zhang, J.; Yu, R.; Luo, N.; Yu, Y. Isolation, Characterization, and Optimization of Culture Medium for Local Straw-Degrading Bacteria from Northeastern Black Soils of China. *Agronomy* **2024**, *14*, 2591. https://doi.org/10.3390/agronomy14112591.
4. Wang, Y.; Liu, J.; Deng, X.; Li, Y.; Gao, J.; Liu, L. Arbuscular Mycorrhizae Affect Soil Nitrogen Fertilizer Utilization, Denitrification Functional Genes, and N_2O Emissions During Biochar Amendment. *Agronomy* **2024**, *14*, 2627. https://doi.org/10.3390/agronomy14112627.
5. Li, Y.; Chen, W.; Xu, Y.; Yu, J.; Zuo, W.; Shan, Y.; Bai, Y. Environmental Constraints Mitigation Directly Drove the Diversifications of Fungal Community and Functional Profile in Amended Coastal Salt-Affected Soils. *Agronomy* **2024**, *14*, 2772. https://doi.org/10.3390/agronomy14122772.
6. Cruz, V.; Moreira, B.; Valério, T.; Frias, Y.; Silva, V.; Morais, E.; Vasconcelos, L.; Tropaldi, L.; Prado, E.; Montagnolli, R.; Lopes, P. Leguminous Plants and Microbial Inoculation: An Approach for Biocatalytic Phytoremediation of Tebuthiuron in Agricultural Soil. *Agronomy* **2024**, *14*, 2805. https://doi.org/10.3390/agronomy14122805.

https://doi.org/10.3390/agronomy16020202

7. Yan, Y.; Zhang, X.; Liu, Y.; Hou, L.; Geng, Z.; Hu, F.; Xu, C. Impact of Organic Fertilizer Substitution and Chemical Nitrogen Fertilizer Reduction on Soil Enzyme Activity and Microbial Communities in an Apple Orchard. *Agronomy* **2024**, *14*, 2917. https://doi.org/10.3390/agronomy14122917.
8. Dou, T.; Zhang, K.; Shi, X.; Liu, W.; Yu, F.; Liu, D. Crop–Mushroom Rotation: A Comprehensive Review of Its Multifaceted Impacts on Soil Quality, Agricultural Sustainability, and Ecosystem Health. *Agronomy* **2025**, *15*, 563. https://doi.org/10.3390/agronomy15030563.
9. Martirosyan, V.; Stavi, I.; Doniger, T.; Applebaum, I.; Sherman, C.; Levi, M.; Steinberger, Y. Bacterial Community Dynamics in Oil-Contaminated Soils in the Hyper-Arid Arava Valley. *Agronomy* **2025**, *15*, 1198. https://doi.org/10.3390/agronomy15051198.
10. Tan, M.; Feng, T.; Wang, C.; Hao, X.; Yu, H. Effects of Microbial Agents on Soil Improvement—A Review and Bibliometric Analysis. *Agronomy* **2025**, *15*, 1223. https://doi.org/10.3390/agronomy15051223.
11. Huang, L.; Wang, R.; Yu, F.; Liu, R.; He, C.; Huang, L.; Yang, S.; Liu, D.; Wan, S. Effects of N-P-K Ratio in Root Nutrient Solutions on Ectomycorrhizal Formation and Seedling Growth of *Pinus armandii* Inoculated with Tuber indicum. *Agronomy* **2025**, *15*, 1749. https://doi.org/10.3390/agronomy15071749.
12. Feng, L.; Liu, X.; Wang, N.; Shi, Z.; Wang, Y.; Jia, J.; Shi, Z.; Pu, T.; Yang, P. Genomic Analysis of Cadmium-Resistant and Plant Growth-Promoting *Burkholderia alba* Isolated from Plant Rhizosphere. *Agronomy* **2025**, *15*, 1780. https://doi.org/10.3390/agronomy15081780.
13. Nurzhanova, A.; Boulygina, E.; Sungurtseva, I.; Mamirova, A.; Berzhanova, R.; Muratova, A. Miscanthus × giganteus Rhizobacterial Community Responses to Zn and Oil Sludge Co-Contamination. *Agronomy* **2025**, *15*, 2232. https://doi.org/10.3390/agronomy15092232.
14. Khongchiu, P.; Murase, J.; Wongkaew, A.; Sajjaphan, K.; Kumdee, O.; Rakpenthai, A.; Nakasathien, S. Effect of Zinc Application on Maize Dry Matter, Zinc Uptake, and Soil Microbial Community Grown Under Different Paddy Soil pH. *Agronomy* **2026**, *16*, 78. https://doi.org/10.3390/agronomy16010078.

https://doi.org/10.3390/agronomy16020202

agronomy

MDPI

Article

Effect of Zinc Application on Maize Dry Matter, Zinc Uptake, and Soil Microbial Community Grown Under Different Paddy Soil pH

Phanuphong Khongchiu [1,2], Jun Murase [2], Arunee Wongkaew [3], Kannika Sajjaphan [4], Orawan Kumdee [5], Apidet Rakpenthai [3] and Sutkhet Nakasathien [3,*]

1 Faculty of Agriculture at Kamphaeng Saen, Kasetsart University, Kamphaeng Saen Campus, Nakhon Pathom 73140, Thailand; phanuphong@tistr.or.th
2 Graduate School of Bioagricultural Sciences, Nagoya University, Nagoya 464-8601, Japan; murase@agr.nagoya-u.ac.jp
3 Department of Agronomy, Faculty of Agriculture, Kasetsart University, Bangkok 10900, Thailand; arunee.wo@ku.ac.th (A.W.); fagradr@ku.ac.th (A.R.)
4 Department of Soil Science, Faculty of Agriculture, Kasetsart University, Bangkok 10900, Thailand; agrkks@ku.ac.th
5 Agricultural Research and Technology Transfer Center, Faculty of Agriculture, Kasetsart University, Bangkok 10900, Thailand; fagrowk@ku.ac.th
* Correspondence: agrskn@ku.ac.th

Abstract

Zinc (Zn) is often of deficient in paddy soils, and optimizing its application is crucial for improving maize productivity in intensive rice–maize cropping systems. This study aimed to develop practical Zn fertilizer strategies suitable for paddy soils with varying pH levels, thereby improving nutrient management and understanding of soil microbial responses. Field experiments were conducted during the 2020–2021 dry seasons at three sites: Chon Daen (pH 5.8), Noen Maprang (pH 6.7), and Lom Sak (pH 7.8). A two-factorial randomized complete block design with four replications was used, including four $ZnSO_4 \cdot H_2O$ rates (0, 1.5, 3, and 6 times the DTPA-extractable Zn in soil) and two hybrid maize varieties, Suwan 5731 and Suwan 5819. Results showed that at Chon Daen, Zn application significantly enhanced shoot Zn uptake and soil Zn concentration, with SW5819 exhibiting greater Zn efficiency and biomass production. At Noen Maprang, Zn application did not significantly affect dry matter, while, at Lom Sak, Zn responses were moderate, though SW5819 maintained better growth and Zn uptake. Across sites, maize Zn efficiency was highest under acidic conditions and in SW5819. Soil microbial communities remained largely unaffected by Zn fertilization and were primarily influenced by soil pH, with *Proteobacteria*, *Crenarchaeota*, and *Ascomycota* dominating bacterial, archaeal, and fungal groups, respectively. These findings support the feasibility of Zn fertilization strategies to enhance both crop productivity and nutritional quality without altering the microbial community composition.

Keywords: maize; soil pH; soil Zn fertilizer; Zn accumulation; soil microbial community

1. Introduction

Zinc is a micronutrient for many plant enzymes and proteins. It plays an important role in a wide range of processes, such as growth hormone production, seedling development, photosynthesis, and reproduction [1,2]. Zinc is transported via symplast pathways through a complex interaction of various mechanisms. Zinc is transported through the xylem to the phloem and is then remobilized from older to younger leaves [3]. Soil Zn fertilization facilitates Zn transport from roots to above-ground tissues and its redistribution to grains. This process enhances plant growth, increases yield, and promotes the biofortification of crops for better dietary nutrition [4]. Soil Zn deficiency occurs because Zn is poorly distributed and poorly soluble in soil solutions, resulting in widespread Zn deficiency in both crops and humans. The global Zn deficiency index indicates tropical regions still have widespread Zn deficiency in humans and soils, affecting crop yields and nutrient quality [1,4,5]. However, Zn solubility in soil solution depends on pH; in acidic soils, high Zn solubility often results from easy desorption and exchange processes on reactive soil mineral surfaces [6,7]. Meanwhile, increasing soil pH will decrease Zn solubility. The cause of high soil pH (>7) is due to a high content of $CaCO_3$ of pedogenic origin [6]. The quantified Zn concentration (DTPA-Zn extractable) with different solubility in 150 paddy soils was measured, with soil pH ranging from ultra-acidic to slightly alkaline. The median values for acid soil and neutral soil were 2.0 and 0.95 mg/kg, respectively. Meanwhile, the median Zn concentration across all samples was 1.5 mg/kg [7], which is the reported critical soil level of Zn [6].

Maize (*Zea mays* L.) is highly sensitive to Zn deficiency, which occurs when Zn levels in maize leaves are less than 15 mg/kg [8]. Maize requires about 500 mg/ha of Zn to achieve a grain yield of 12 t/ha [9]. Maize yield is particularly influenced by Zn availability during key growth stages. Zinc translocation from root to shoot after uptake is influenced by many factors, and Zn accumulation in the grain depends on both remobilization from shoots and continued uptake during grain filling [10,11]. Moderate Zn application rates have been shown to improve biomass accumulation, crop growth rate (CGR), and pollen viability, which in turn increase kernel number and weight—particularly in apical (inferior) kernels that are often poorly filled under Zn deficiency [12–14]. Lui et al. [15] reported that the response level was highly correlated with the amount of Zn added to the soil. The optimum soil Zn concentration for attaining high yields was 4.7 mg/kg, and the Zn application rate was 50 kg Zn/ha, which also improved yields [15]. In Thailand, a report of Zn application in high-pH or calcareous soils was noted. Maize responded positively to rates of Zn application between 2 and 4 mg/kg of soil to increase Zn uptake in corn and grain yield [16], while in the field under high soil pH (pH 7.8), a moderate Zn rate (5–10 kg Zn/ha) optimized maize performance, grain Zn uptake, and soil Zn availability [13]. However, for soil Zn fertilizer management in maize production on paddy soil, it is important to consider the long-term effects of Zn fertilizer application. After 9 years of Zn fertilizer application, the bacterial communities in Zn-treated soils were distinct from those in untreated soils. In the soil, the four most abundant phyla were *Actinobacteria*, *Proteobacteria*, *Acidobacteria*, and *Chloroflexi*. The environmental factor, soil Zn concentration, and total Zn concentration were significantly correlated with bacterial community composition [17]. Additionally, an increase in soil Zn concentration can alter the functional gene structure of microbial communities and the abundance of specific gene families, with a threshold in gene family abundance between 5 and 10 kg Zn/ha across the analyzed Zn concentrations [18].

Maize is the most significant crop grown in Thailand. Domestic maize production increased from 1.33 million tons in 2023 to 2.01 million tons in 2024 [19], but it still does not meet the rising local demand, leading to grain imports. Climate variability and pest infestations, such as the fall armyworm, have negatively impacted maize yields [19]. The planted area allocated mainly to maize is 75.5% in the early rainy season, 15.0% in the late rainy season, and only 9.5% in the dry season, which affects grain yield distribution, whereas demand is almost uniformly distributed year-round. Therefore, the Thai government proposed new planting-area ratios for maize: 30% in the early rainy season, 20% in the late rainy season, and 50% in the dry season. This ratio will increase the maize planting area during the dry season from 5% to 50% by converting a rice-rice cropping sequence to a rice–maize

https://doi.org/10.3390/agronomy16010078

sequence in paddy areas. The rice–maize cropping system is the most important emerging system, capable of producing yields similar to those of water-saving relative systems [20].

Despite considerable research on Zn fertilizer management in maize production, there remains a limited understanding of how Zn fertilization performs across different soil pH levels in flooded or rice-based cropping systems. Such soils may exhibit distinct chemical conditions that influence Zn availability and plant uptake differently from those in typical upland systems. These uncertainties highlight the need for further investigation to inform fertilizer strategies suited to similar transitional agroecosystems. To address this gap, the present study aimed to evaluate the effects of varying Zn application rates on the dry matter accumulation, Zn uptake, and grain yield responses of hybrid maize grown in different pH paddy soils and to provide information on the impact of agricultural practices related to Zn fertilizer management on soil microbial communities.

2. Materials and Methods

2.1. Sites Description

The three-field experiment was conducted in the dry season after rice cultivation from December 2020 to April 2021. The study sites were located in Chon Daen (CD) and Lom Sak (LS), Phetchabun province, and Noen Maprang (NM), Phitsanulok province, Thailand, where a rice–maize rotation system has been maintained consistently for a period exceeding five years: rice is cultivated during the rainy season, and maize during the dry season. Prior to maize planting, rice residues are incorporated into the soil following harvest plowing [13]. Meteorological data, including mean temperature and total precipitation, were recorded throughout the experimental period using a weather station installed at each of the three field sites (Table 1 and Figure S1).

Table 1. Geographical and climatic characteristics in the three study sites.

Characteristics		Sites		
		Chon Daen	Noen Maprang	Lom Sak
Geographic	Latitude (°N)	16°12′28.0″	16°26′10.2″	16°45′00.4″
	Longitude (°E)	100°51′17.4″	100°41′12.7″	101°10′17.9″
Climatic	Mean temperature (°C)	28.4	28.8	27.6
	Total precipitation (mm)	243.2	128.7	272.2

Bulk soil samples were collected at 0–20 cm depth from the plot area before the experiment and analyzed for soil texture and chemical properties (Table 2).

Table 2. Soil characteristics information of soil texture, soil organic matter, soil pH, available phosphorus, exchangeable potassium, and extractable Zn for each experimental site.

Characteristics	Experimental Sites		
	Chon Daen	Noen Maprang	Lom Sak
Soil type	clay	clay loam	clay
Sand (%)	30	42	34
Silt (%)	26	28	24
Clay (%)	44	30	42
pH (1:1)	5.8	6.7	7.8
Soil organic matter (%)	0.35	0.91	1.63
Available phosphorus (mg/kg)	1.50	20.75	26.84
Exchangeable potassium (mg/kg)	27.50	62.18	60.45
DTPA-Zn (mg/kg)	0.55	0.87	0.53

Note: DTPA-Zn, diethylenetriaminepentaacetic acid-extractable zinc.

https://doi.org/10.3390/agronomy16010078

2.2. Experimental Management and Design

The experimental field was laid out in a 4 × 2 factorial in a randomized complete block design (RCBD) with four replications. Zinc fertilizer was applied in the form of Zn sulfate monohydrate ($ZnSO_4 \cdot H_2O$), containing 36% elemental Zn. These application rates correspond to 0, 1.5, 3, and 6 times the baseline soil Zn concentration (0.55, 0.87, and 0.53 mg/kg in CD, NM, and LS, respectively), calculated to represent a gradient of Zn availability relative to the existing soil Zn status. The four rates equivalent to 0, 5, 10, and 20.6 kg Zn/ha per crop season in CD and LS, and 0, 8.8, 16.8, and 33.8 kg Zn/ha in NM. Zn fertilizer was applied at the V6 stage by banding to the soil surface at all experimental sites. Two varieties were selected: Suwan 5731 (SW 5731) and Suwan 5819 (SW 5819), new hybrid maize varieties recently developed and introduced by the National Corn and Sorghum Research Center, Thailand. The experiment plot size was 28 m^2 (5.6 m × 5.0 m) with eight rows per plot. The seeds of each maize variety were sown using a manual planting machine, with two seeds per hill in eight rows per plot and thinned at two weeks after planting (WAP) to maintain one plant per hill. The density was 71,500 plants/ha, with a row spacing of 0.7 m and a plant spacing of 0.2 m. The N-P-K fertilizer was applied as follows, based on soil chemical analysis recommended by the Department of Agriculture, Thailand [21]. All plots were applied as follows: 125 kg N/ha (as urea), 65 kg P_2O_5/ha (as diammonium phosphate), and 65 kg K_2O/ha (as potassium chloride) in CD, 125 kg N/ha, 35 kg P_2O_5/ha, and 65 kg K_2O/ha in NM, and 105 kg N/ha, 35 kg P_2O_5/ha, and 65 kg K_2O/ha in LS (Table S1). All field operations included immediate sprinkler irrigation after sowing, which was maintained weekly until maturity. Herbicide control at 21 WAP involved the use of Acetochlor and Atrazine. Emamectin benzoate and Spinetoram were applied to control fall armyworm outbreaks detected at 30 (V6), 40 (V10), and 63 (VT) days after planting [13].

2.3. Plant Sampling and Measuring

At the milking (R3) stage, leaves, stalks, and ear samples were collected, prepared, and analyzed for plant dry weight as described by Khongchiu et al. [13]. Plant samples were ground with a stainless-steel grinder. The 0.2 g of sample was digested with HNO_3-$HClO_4$ (2:1), and the solution was heated at 200 °C until dense white fumes appeared. The mixture was allowed to cool, and deionized water was added to bring the volume to 25 mL. The Zn concentration was determined by an atomic absorption spectrometer (iCE 3000; Thermo Scientific, Waltham, MA, USA). All results are expressed on a dry weight basis. The formula for Zn uptake was calculated as the content per area following Liu et al. [22].

2.4. Soil Sampling and Measuring

Soil samples were kept in all treatments at the milking (R3) stage of maize growth. Soil samples were air-dried and crushed to pass a 0.5 mm sieve. Soil pH was measured using a pH meter with a soil-to-water ratio of 1:1. Soil Zn concentration was determined by end-to-end shaking (180 rpm) for 2 h of 5 g of soil with 20 mL of 0.005 mol/L DTPA (diethylene triamine pentacetic acid), 0.01 mol/L and 0.1 mol/L TEA (tri-ethanol amine) buffered at pH 7.0 [23]. Zn concentrations in the extractants were determined by an atomic absorption spectrometer (iCE 3500; Thermo Scientific, USA). For soil microbial community analysis, the fresh soil samples were separately collected, sieved (<2 mm), transported to the laboratory in an ice box, and stored at −20 °C until use.

2.5. DNA Extraction, PCR Amplification, and Sequencing

Total DNA extraction was performed using a FastDNA SPIN Kit (MP Biomedicals, Cleveland, OH) from 0.5 g fresh soil following the manufacturer's instructions. Extracted

https://doi.org/10.3390/agronomy16010078

DNA was quantified using a PicoGreen dsDNA assay kit (Invitrogen, Thermo Scientific, Waltham, MA, USA) and Fluoroskan™ Microplate Fluorometer (Thermo Scientific, USA). The primer pairs 515F (5′-TGC TCG GCA GCG TCA GAT GTG TAT AAG AGA CAG GTG CCA GCM GCC GCG GTA A-3′) and 806R (5′-GTC TCG TGG GCT CGG AGA TGT GTA TAA GAG ACA GGG ACT ACH VGG GTW TCT AAT-3′) were used to amplify the V3-V4 region of the 16s rRNA gene. PCR conditions were 94 °C for 2 min, followed by 28 cycles of 94 °C for 30 s, 55 °C for 30 s, and 72 °C for 30 s, with final extension at 72 °C for 5 min.

The fungal primer pairs gITS7 5′-GTGAATCATCGAATCTTTG-3′ and ITS4 5′-TCCTCCGCTTATTGATATGC-3′ were used to amplify the ITS2 region. The PCR conditions were 94 °C for 4 min, followed by 30 cycles of 94 °C for 1 min 58 °C for 1 min, and 72 °C for 1 min, with final extension at 72 °C for 7 min. The PCR products were purified using AmPure XP bead (Beckman coulter, Brea, CA, USA) and quantified using a PicoGreen dsDNA assay Kit (Invitrogen). Dual indexed using the Nextera XT Index Kit v2 (Illumina, San Diego, CA, USA) according to the manufacturer's instructions. Indexed amplicons were again purified using magnetic beads and quantified using a PicoGreen dsDNA Quantification Kit (Invitrogen). Samples were pooled in equal concentrations and sequenced on an Illumina MiSeq reagent Kit v3 (600 cycles, Illumina) according to the manufacturer's instructions.

2.6. Statistical Analysis

The data were analyzed to Analysis of variance (ANOVA) of the measured parameters using STAR (Statistical Tool for Agricultural Research) software, version 2.0.1 [24]. The least significant difference (LSD) test was used to test the significant differences between treatments. The sequencing was conducted using an Illumina MiSeq platform and sequences were submitted to the NCBI database under the accession numbers PRJNA1371364 for bacteria and archaea and PRJNA1371415 for fungi. Paired-end raw sequencing reads were processed using QIIME 2 (version 2024.10) according to best-practice workflows. Denoising and amplicon sequence variant (ASV) inference were performed with the DADA2 plugin. ASVs were taxonomically classified using a naïve Bayes classifier trained on the relevant regions of reference databases: the SILVA database release 138 for the 16S rRNA gene and the UNITE database version 10 for the fungal ITS region. Non-target ASVs and singletons were filtered prior to downstream analysis. Microbial community composition and alpha- and beta-diversity metrics were calculated using standard QIIME 2 plugins. The sequence data were compared using R software (version 4.1.3). Alpha-diversity was evaluated using several major indices, including Chao1 and Shannon, and the indices were compared using a *t*-test. The beta-diversity was assessed by the non-metric multidimensional scaling (NMDS) based on the Bray–Curtis algorithm to analyze overall structural changes in the microbial communities. The relative abundance values of bacteria, archaea, and fungal taxa at phylum levels are more than 0.1% and at class levels are more than 1%. Spearman's correlations were conducted between major microbial genera, maize growth parameters, and soil chemical properties to identify relationships with environmental variables, with significance set at $p < 0.05$.

3. Results

3.1. Dry Matter Accumulation and Shoot Zn Uptake at the Milking (R3) Stage

Dry matter accumulation at the R3 stage, under the interaction between Zn application rates and maize varieties in 2020, is presented in Figure 1. At the Chon Daen site, the interaction between Zn application rate and maize variety significantly affected leaf dry matter. In contrast, stalk and ear dry matter did not show significant differences (Figure 1A). Leaf dry matter of SW 5819 was significantly higher under Zn application rates of 5 and 10 kg Zn/ha compared with SW 5731 at 5 and 20.6 kg Zn/ha and the control (no Zn application), representing a 21–25% increase relative to SW 5731 under those treatments. Stalk dry matter constituted the largest portion of total biomass, followed by ear and leaf dry matter. Stalk dry matter ranged from approximately 7.5 to 9.2 t/ha, while ear dry matter varied from 6.3 to 7.6 t/ha with no significant differences among treatments. Total dry matter accumulation at the R3 stage ranged between approximately 17.7 and 19.5 t/ha across Zn application rates and maize varieties (Figure 1A). The main effect (Figure 2A) confirmed this varietal difference, with SW 5819 showing significantly greater leaf dry matter accumulation. Increasing Zn rates from 0 to 20.6 kg Zn/ha did not result in significant changes in total dry matter. Overall, these results suggest that maize dry matter production was more influenced by genetic variation between varieties than by Zn fertilization levels.

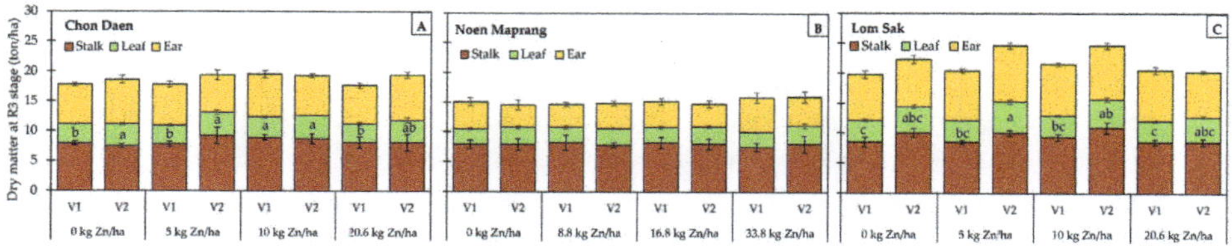

Figure 1. Dry matter accumulation in plant parts at the milking (R3) stage as influenced by the interaction effect of Zn application rates and maize varieties (V1: SW 5731, V2: SW 5819) at Chon Daen (**A**), Noen Maprang (**B**), Lom Sak (**C**) experimental sites in 2020. Data are mean values ± SE (n = 4). Different letters above the bars indicate significant differences in Zn application rates and maize varieties, as determined by the LSD test at $p < 0.05$.

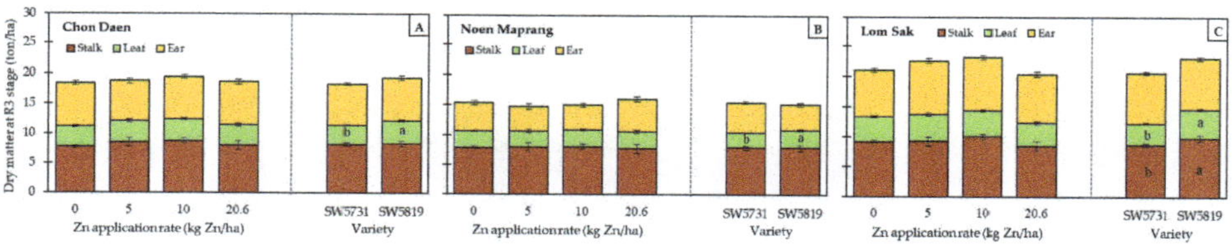

Figure 2. Dry matter accumulation in plant parts at the milking (R3) stage as influenced by the main effect of Zn application rates and maize varieties at Chon Daen (**A**), Noen Maprang (**B**), and Lom Sak (**C**) experimental sites in 2020. Data are mean values ± SE (n = 4). Different letters above the bars indicate significant differences in Zn application rates and maize varieties, as determined by the LSD test at $p < 0.05$.

At Noen Maprang, the dry matter in different plant parts showed no significant interaction effect between Zn application rates and maize varieties (Figure 1B). Stalk dry matter ranged between 7.5 and 8.8 t/ha, leaf dry matter between 2.5 and 3.1 t/ha, and ear dry matter between 3.8 and 5.9 t/ha. The proportions of dry matter among plant parts remained consistent across all Zn application rates and maize varieties, suggesting that Zn supply under the prevailing soil conditions did not notably affect dry matter production or distribution in maize. Total dry matter at the R3 stage ranged from 14.6 to 16.0 t/ha. Zn application rates and maize varieties had no significant effect on partitioned dry matter (Figure 2B). While the leaf dry matter of variety SW5819 showed a slightly significantly higher dry matter compared with SW5731. These findings indicate that under the environmental and soil conditions at Noen Maprang, Zn application had minimal impact on dry matter accumulation and distribution, and maize growth was largely stable across treatments.

At Lom Sak, a significant Zn application rate × variety interaction was detected for leaf dry matter ($p < 0.05$; Figure 1C). Consequently, Zn effects were examined within each variety. SW5819 supplied with 5 kg Zn/ha produced significantly higher leaf dry matter (5.3 t/ha) than SW5731 at all Zn application rates, whereas leaf dry matter within SW5819 did not differ

https://doi.org/10.3390/agronomy16010078

significantly among Zn treatments. Stalk dry matter ranged from 8.5 to 10.9 t/ha and ear dry matter from 7.5 to 9.3 t/ha. Compared with the control, ear dry matter increased by 7.8–20.8% with Zn application; however, at 20.6 kg Zn/ha, ear dry matter declined by 10.4% in SW5731 and 3.8% in SW5819. Total dry matter accumulation ranged from 19.8 to 24.6 t/ha. The main effect analysis confirmed that SW5819 accumulated significantly more total dry matter than SW5731, primarily due to greater leaf and stalk biomass, while Zn application rate had no significant main effect on total dry matter (Figure 2C).

Shoot Zn uptake at the R3 stage varied among sites and was significantly influenced by Zn application rates, varieties, and their interactions (Figures 3 and 4). At Chon Daen, a significant Zn application rate × variety interaction was observed (Figure 3A). The highest Zn uptake was recorded in SW5819 at 20.6 kg Zn/ha (approximately 481.7 g/ha) and SW 5819 at 5 kg Zn/ha (approximately 437.4 g/ha), while those without Zn application show lower than in all Zn treatments (333.3 and 343.6 g/ha in SW 5731 and SW 5819, respectively). Consistent with this interaction, the main effect analysis showed that shoot Zn uptake increased with Zn application rate and was significantly higher in SW5819 than in SW5731 (Figure 4A), indicating a strong Zn fertilization response and greater Zn accumulation capacity of SW5819 under acidic soil conditions.

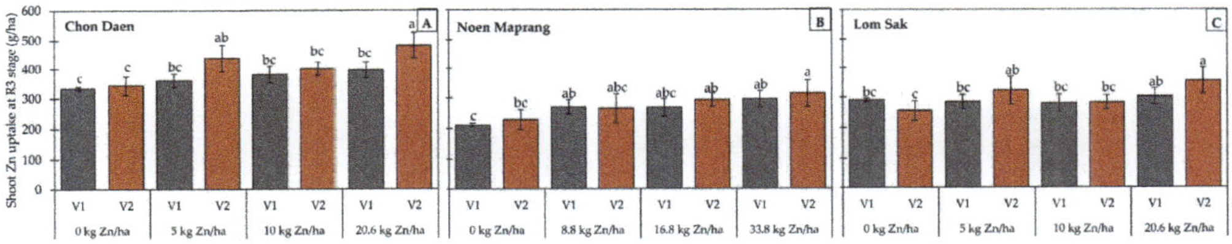

Figure 3. Zn uptake in maize at the milking (R3) stage as influenced by the interaction effect of Zn application rates and maize varieties (V1: SW 5731, V2: SW 5819) at Chon Daen (**A**), Noen Maprang (**B**), Lom Sak (**C**) experimental sites in 2020. Data are mean values ± SE (n = 4). Different letters above the bars indicate significant differences in Zn application rates and maize varieties, as determined by the LSD test at $p < 0.05$

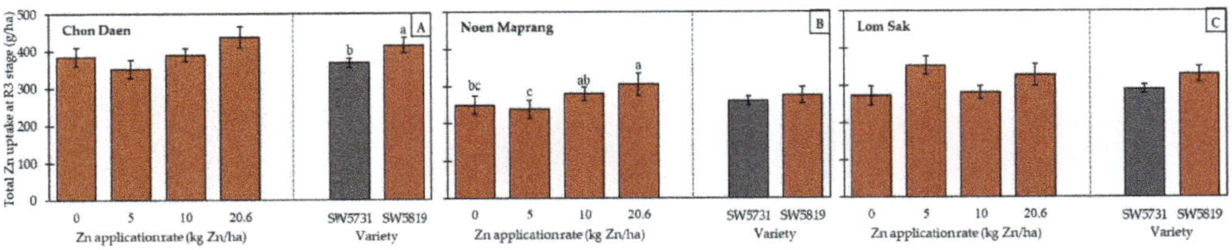

Figure 4. Zn uptake in maize at the milking (R3) stage as influenced by the main effect of Zn application rates and maize varieties at Chon Daen (**A**), Noen Maprang (**B**), Lom Sak (**C**) experimental sites in 2020. Data are mean values ± SE (n = 4). Different letters above the bars indicate significant differences in Zn application rates and maize varieties, as determined by the LSD test at $p < 0.05$.

At Noen Maprang, no significant interaction between Zn application rate and maize variety was detected for shoot Zn uptake (Figure 3B); therefore, only main effects are presented. Shoot Zn uptake increased progressively with increasing Zn application rate, ranging from approximately 210.9 g/ha in SW5731 without Zn to 314.9 g/ha in SW5819 at 33.8 kg Zn/ha. The main effect of Zn application rate showed a similar trend (Figure 4B), where total Zn uptake was lowest at 0 and 5 kg Zn/ha (250.0 and 238.4 g/ha, respectively) and highest at 10 and 20.6 kg Zn/ha (280.7 and 305.4 g/ha, respectively), while the main effect of maize variety was not significant, indicating comparable Zn uptake capacity between varieties at this site.

At Lom Sak, Zn application rate × maize variety interaction was significantly different in shoot Zn uptake at the R3 stage (Figure 3C). SW 5819 at 5 and 20.6 kg Zn/ha (322.8 and 355.0 g/ha, respectively) showed higher Zn uptake than SW 5819 at 10 kg Zn/ha and without Zn application (282.6 and 255.9 g/ha, respectively). SW 5731 at 20.6 kg Zn/ha was higher than Zn application rates at 5 and 10 kg Zn/ha, increasing approximately 6–8%. Despite these interaction effects, the main effects of Zn application rate and maize variety were not significant (Figure 4C); however, SW5819 consistently exhibited higher shoot Zn uptake than SW5731, with an average increase of 13.7% across Zn treatments.

3.2. *Soil Zn Concentration and Soil pH*

Soil Zn concentration at the R3 stage was reported in Table 3. At Chon Daen, application of 20.6 kg Zn ha^{-1} resulted in significantly higher soil Zn concentrations than the control across both maize varieties, reaching 4.50 mg/kg in SW5731 and 3.41 mg/kg in SW5819. The main effect of Zn application rate showed the highest soil Zn concentration at 20.6 kg Zn/ha (3.96 mg/kg), representing a 2.65-fold increase compared with no Zn application, whereas rates of 5 and 10 kg Zn/ha did not differ significantly from the control. The main effect of maize variety was not significant.

Table 3. Soil Zn concentration at the milking stage (R3) as affected by Zn application rates and maize varieties in 2020.

Experimental Sites	Zn Application Rate (kg Zn/ha)	Variety		Mean
		SW 5731	SW 5819	
Chon Daen	0	1.52 ± 0.28 ^{bc}	1.46 ± 0.20 ^{c}	1.49 ± 0.22 B
	5	2.23 ± 0.62 ^{abc}	1.60 ± 0.05 ^{bc}	1.91 ± 0.52 B
	10	1.63 ± 0.47 ^{bc}	1.67 ± 0.34 ^{bc}	1.65 ± 0.37 B
	20.6	4.50 ± 2.03 ^{a}	3.41 ± 1.94 ^{ab}	3.96 ± 1.88 A
	Mean	2.47 ± 1.57	2.03 ± 1.19	
Source of variation	Zn application rate	**	Variety (Var)	ns
	Zn × Var	**		
Noen Maprang	0	1.54 ± 0.35 ^{bc}	1.37 ± 0.05 ^{c}	1.46 ± 0.24 C
	8.8	2.85 ± 1.11 ^{abc}	2.06 ± 0.52 ^{bc}	2.46 ± 0.89 B
	16.8	2.12 ± 0.43 ^{abc}	2.64 ± 0.89 ^{abc}	2.38 ± 0.69 BC
	33.8	4.50 ± 1.01 ^{a}	3.89 ± 1.40 ^{ab}	4.20 ± 1.14 A
	Mean	2.75 ± 1.35	2.49 ± 1.2	
Source of variation	Zn application rate	**	Variety	ns
	Zn × Var	**		
Lom Sak	0	0.52 ± 0.11 ^{d}	1.46 ± 0.90 ^{bcd}	0.99 ± 0.77 B
	5	0.80 ± 0.12 ^{cd}	1.21 ± 0.58 ^{bcd}	1.00 ± 0.44 B
	10	1.63 ± 0.45 ^{abc}	1.70 ± 0.73 ^{abc}	1.67 ± 0.75 AB
	20.6	2.43 ± 0.78 ^{a}	1.89 ± 0.61 ^{ab}	2.16 ± 0.69 A
	Mean	1.35 ± 1.06	1.56 ± 0.67	
Source of variation	Zn application rate	**	Variety	ns
	Zn × Var	**		

Data represent the mean ± standard errors (n = 4). Different lowercase letters indicate significant differences in the interaction effect of treatments, and uppercase letters indicate significant differences in the main effect of Zn application rate and maize variety ($p < 0.05$). ns indicates no significant differences. ** indicate significant difference at $p < 0.01$, respectively.

At Noen Maprang, soil Zn concentration also increased significantly with Zn application rate. The highest rate (33.8 kg Zn/ha) produced significantly greater soil Zn concentrations than the control in both varieties (4.50 and 3.89 mg/kg in SW5731 and SW5819, respectively). Main effect analysis confirmed that soil Zn concentration was greatest at 33.8 kg Zn/ha (4.20 mg/kg), while no significant varietal differences were observed.

At Lom Sak, soil Zn concentration increased with increasing Zn application rate. The application rates of Zn at 10 and 20.6 kg/ha were significantly higher than the rate without Zn application under SW 5731. The range of soil Zn concentration from 0.80 to 2.43 mg/kg under applied Zn fertilizer. The main effect of Zn application rate was significant at 20.6 kg

https://doi.org/10.3390/agronomy16010078

Zn/ha relative to 0 and 5 kg Zn/ha, while no significant differences between maize varieties were detected.

The soil pH values varied considerably among the three experimental sites (Figure 5). At Chon Daen, the soils were distinctly acidic, with pH values ranging from 4.98 to 6.1 and an average of 5.5. In contrast, Noen Maprang soils showed moderately acidic conditions, with pH values clustered between 5.6 and 6.4, and a median of about 6.0. Meanwhile, soil pH at Lom Sak was markedly more alkaline compared with the other sites, with pH values consistently ranging from 7.1 to 7.7 and a median of about 7.5. These results indicate site-specific differences in soil reaction, with Chon Daen representing strongly acidic soils, Noen Maprang moderately acidic soils, and Lom Sak slightly alkaline soils. Such differences are likely to influence soil Zn availability and the overall response of maize to Zn fertilization across sites.

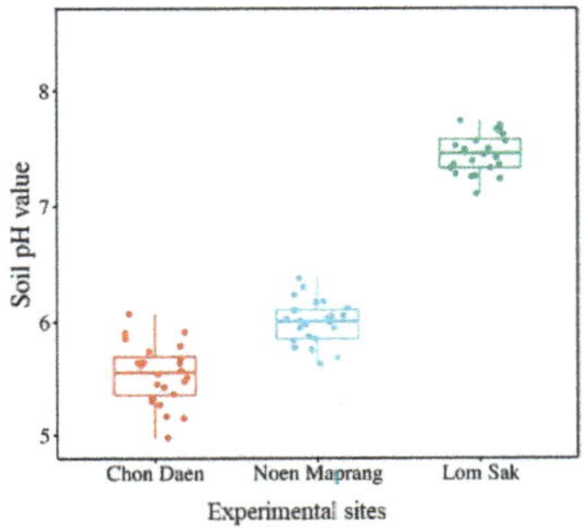

Figure 5. Soil pH in bulk soil at silking at the Chon Daen, Noen Maprang, and Lom Sak experimental sites.

3.3. Soil Bacterial and Archaeal Community

There were no significant differences in α- and β-diversity of bacterial and archaeal communities among the various Zn application rates and two maize varieties (Figure S2). However, soil bacterial and archaeal community structures differed significantly among the three experimental sites as a function of their contrasting soil pH level (Figure 6). The Chao1 and Shannon indices were used to indicate the α-diversity of the bacterial and archaeal communities, as shown in Figure 6A,B. The average Chao1 index values for bacterial and archaeal communities across the three experimental sites were 787.1, 642.3, and 767.4 in Chon Daen, Noen Maprang, and Lom Sak, respectively. The Shannon index values of bacterial and archaeal communities were 5.9, 5.5, and 5.9, respectively. The β-diversity analysis showed that the difference in soil pH influenced the composition of soil microbial communities. The nonmetric multidimensional scaling (NMDS) analysis demonstrated distinct clustering patterns among the three experimental sites (p = 0.02), indicating that location strongly influenced bacterial and archaeal community structure (Figure 6C).

In Figures 7 and 8, the relative abundance of the dominant phylum and class at ChonDaen, it was found that *Proteobacteria*, *Chloroflexi*, and *Actinobacteria* were the three most abundant phyla, with relative abundances of 16.4–21.5%, 18.1–22.7%, and 13.3–19.3%, respectively. *Crenarchaeota* was the dominant archaea in this soil, with a relative abundance range of 2.3–3.6% (Figure 7). In addition, the five most abundant bacteria in the class were *Actinobacteria*, *Gammaproteobacteria*, *Alphaproteobacteria*, *Ktedonobacteria*, and *Anaerolineae*, respectively. The most abundant archaeon at the class level was *Nitrososphaeria* (Figure 8A). At Noen Maprang, *Proteobacteria*, *Chloroflexi*, and *Actinobacteria* were the dominant bacterial phyla. The relative abundances were 22.5–28.2%, 13.9–17.3% and 14.1–21.5%, respectively. *Crenarchaeota* occupied 2.3–2.9% of community (Figure 7). The five most abundant classes at Noen Maprang were *Actinobacteria*, *Gammaprotebacteria*, Bacilli, Alphaproteobacteria, and *Ktedonobacteria*, respectively. *Nitrososphaeria* was the dominant archaeal class in the soil

https://doi.org/10.3390/agronomy16010078

(Figure 8B). At Lom Sak, the four most abundant bacterial phyla were *Proteobacteria, Chloroflexi, Actinobacteria, and Acidobacteria.* The relative abundance was 16.6–20.8%, 11.1–13.8%, 9.5–10.8%, and 14.5–16.5%, respectively. These experimental fields had the highest Crenarchaeota abundance, exceeding those of other fields, with relative abundances ranging from 12.4% to 14.7% (Figure 7). The most abundant bacterial classes were *Alphaproteobacteria, Anaerolineae, Gammaproteobacteria, Vicinamibacteria*, and *Bacteroidia*. *Nitrososphaeria* had a higher relative abundance than the bacterial class (Figure 8C).

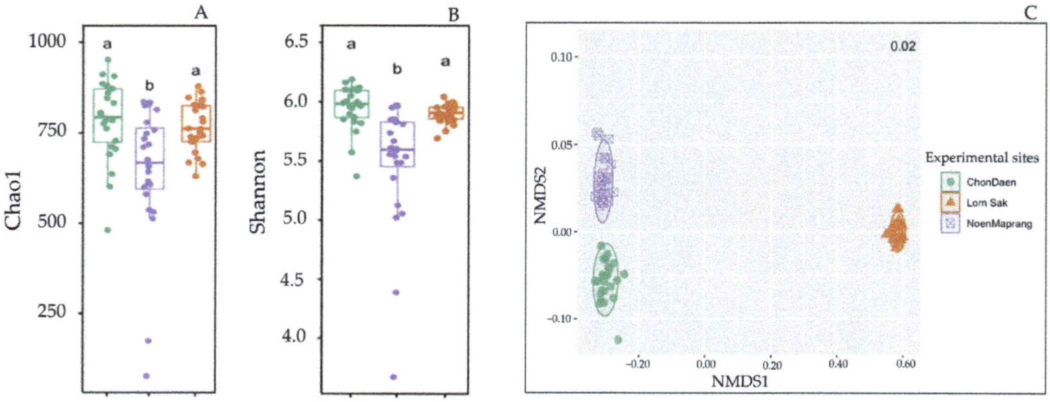

Figure 6. Alpha diversity of the soil bacterial and archaeal community: Chao1 index (**A**) and Shannon index (**B**) in different soil pH after applying Zn fertilizer. The nonmetric multidimensional scaling (NMDS) based on the Bray–Curtis algorithm to analyze overall structural changes in the bacterial and archaeal communities (**C**). Different letters above the bars indicate significant differences between experimental sites ($p < 0.05$).

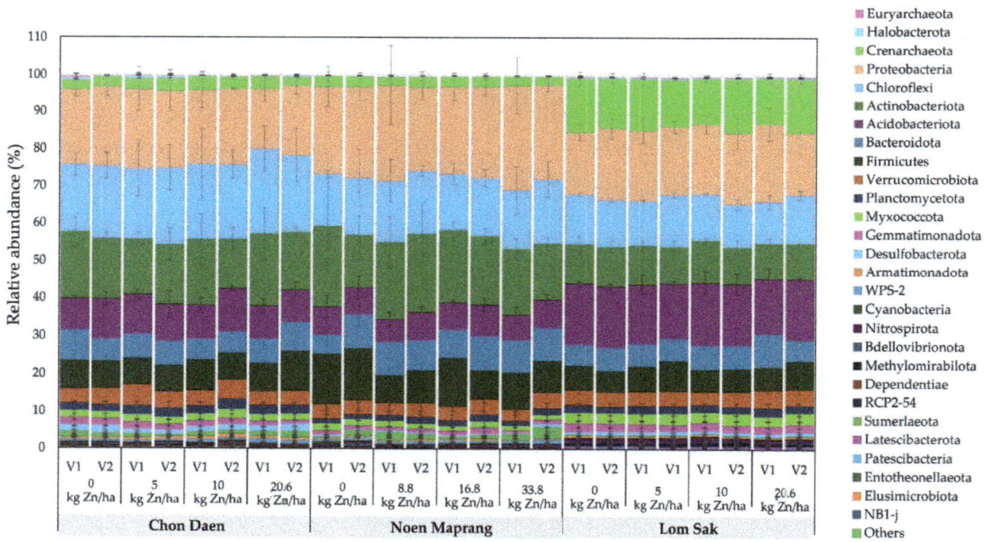

Figure 7. Soil bacterial and archaeal community analysis in response to different interactions between Zn application rates and maize varieties (V1: SW 5731; V2: SW 5819). Relative abundance of the dominant phyla (relative abundance > 0.1%) of different Zn application rates and maize varieties in Chone Daen, Noen Maprang, and Lom Sak experimental fields.

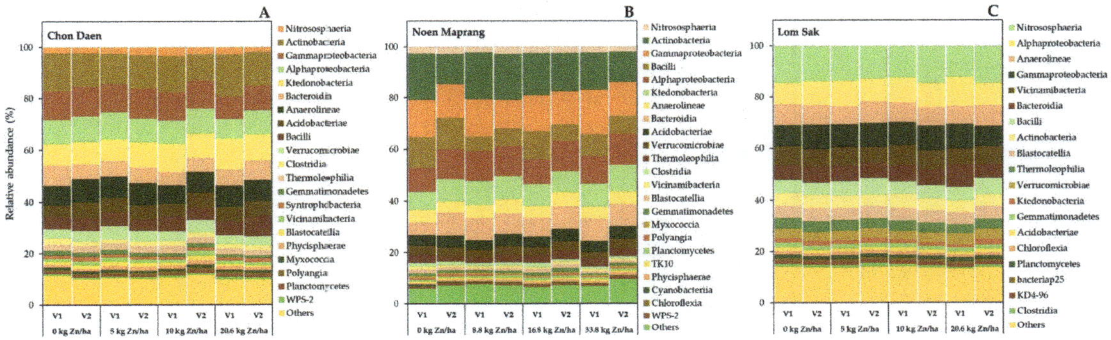

Figure 8. Soil bacterial and archaeal community analysis in response to different interactions between Zn application rates and maize varieties (V1: SW 5731; V2: SW 5819). Relative abundance of the dominant class (relative abundance > 1.0%) at Chon Daen (**A**), Noen Maprang (**B**), and Lom Sak (**C**).

3.4. Soil Fungal Community

No significant differences in α- and β-diversity of fungal communities were observed across different Zn application rates and maize varieties (Figure S3). However, the three experimental sites, characterized by distinct soil pH levels, exhibited notable differences in their soil fungal communities (Figure 9). The Chao1 and Shannon indices were used to evaluate the α-diversity of soil fungal communities in soils with different pH levels. The Chao1 and Shannon indices were significant across different soil pH levels. The average Chao1 index values for the fungal community across the three experimental sites were 137.4, 112.5, and 124.1 in Chon Daen, Noen Maprang, and Lom Sak, respectively. The Shannon index values of bacterial and archaeal communities were 2.8, 2.7, and 2.3, respectively. Changes in soil microbial β-diversity under different Zn application rates and maize varieties in three different soil pH levels in paddy soils. The NMDS based on the Bray–Curtis algorithm was used to analyze overall structural changes in the fungal microbial communities. The soil pH showed significantly altered soil fungal community structure in paddy soil (Figure 9C).

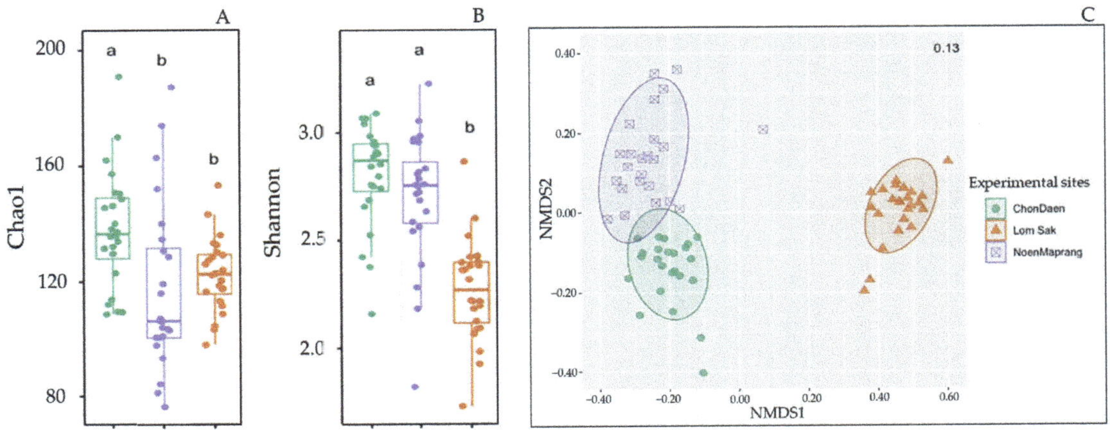

Figure 9. Alpha diversity of the fungal community: Chao1 index (**A**) and Shannon index (**B**) in different soil pH after applying Zn fertilizer. Nonmetric multidimensional scaling (NMDS) analysis of fungal communities based on the Bray–Curtis algorithm (**C**). Different letters above the bars indicate significant differences between experimental sites ($p < 0.05$).

https://doi.org/10.3390/agronomy16010078

The relative abundance results for three experimental sites are shown in Figure 10. At Chon Daen, the main fungal phyla were *Ascomycota*, *Mucoromycota*, and *Basidiomycota*. The relative abundance ranges were 89.2–98.6%, 0.1–9.8%, and 0.4–4.8%, respectively. The fungal class was *Dothideomycetes* (51.1–72.4%), followed by *Sordariomycetes* (24.8–38.0%), *Mucoromycetes* (0.1–9.8%), *Tremellomycetes* (0.4–4.8%), and *Eurotiomycetes* (0.6–6.3%), respectively (Figure 11A). At Noen Maprang, the most abundant fungal phyla were *Ascomycota* (57.4–88.6%), *Basidiomycota* (6.5–29.2%), and *Mucoromycota* (1.0–28.8%), and classes were *Soridiomycetes* (24.4–51.6%), *Dothideomycetes* (21.9–42.5%), *Tremellomycetes* (6.5–29.2%), *Mucoromycetes* (1.0–28.8%), and *Eurotiomycetes* (0.9–10.0%) (Figure 11B). At Lom Sak, the most abundant fungal phyla were *Ascomycota* (99.5–99.7%) and *Basidiomycota* (0.1–0.3%). The main fungal classes were *Eurotiomycetes* (43.5–61.9%), *Soridiomycetes* (28.5–48.6%), *Dothideomycetes* (7.3–22.9%), *Tremellomycetes* (0.1–0.2%), and *Leotiomycetes* (0.03–0.2%) (Figure 11C).

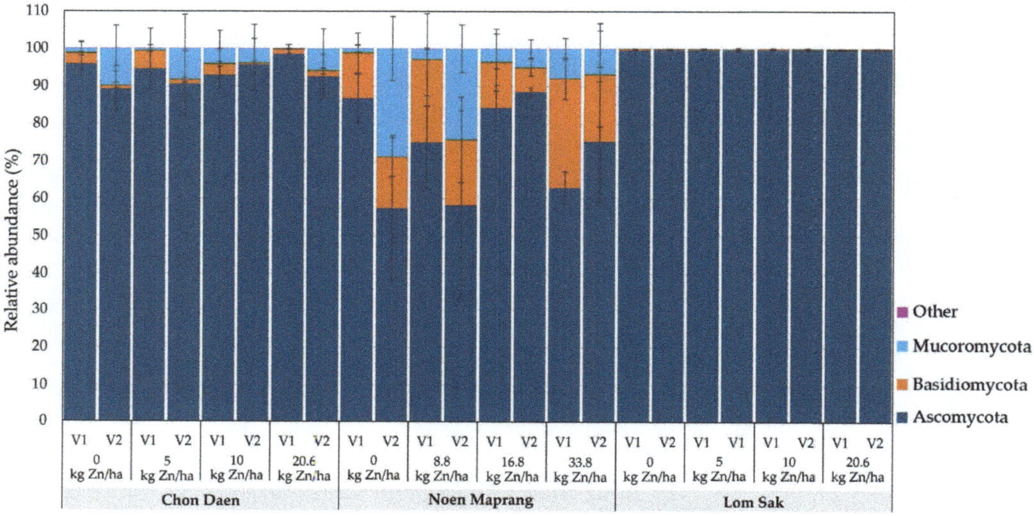

Figure 10. Soil fungal community analysis in response to different interactions between Zn application rates and maize varieties (V1: SW 5731; V2: SW 5819). Relative abundance of the dominant phyla (relative abundance > 0.1%) of different Zn application rates and maize varieties in Chone Daen, Noen Maprang, and Lom Sak experimental fields.

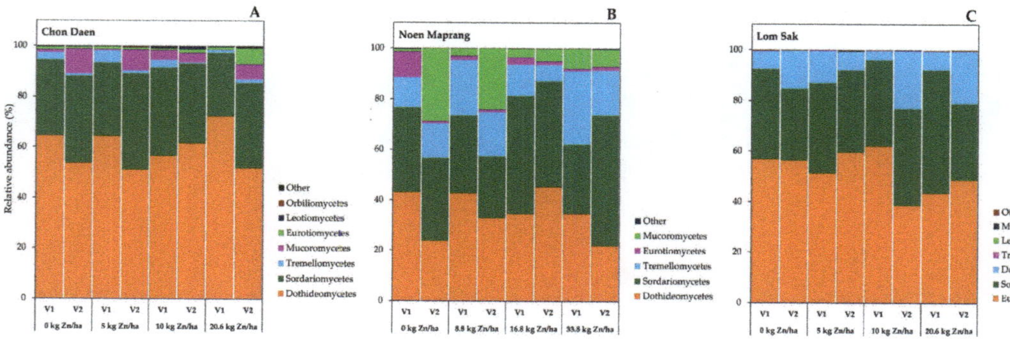

Figure 11. Soil fungal community analysis in response to different interactions between Zn application rates and maize varieties (V1: SW 5731; V2: SW 5819). Relative abundance of the dominant class (relative abundance > 0.1%) at Chon Daen (**A**), Noen Maprang (**B**), and Lom Sak (**C**).

3.5. Correlation Between Soil Microbial Community and Environmental Factors

The correlation analysis revealed relationships among bacterial and archaeal genera, plant growth parameters, and environmental factors. The bacterial community, *Marinobacter*, FCPU453, *Chalmydiaceae*, *Candidatus*_Jorgensenbacteria, and B122 had a significant positive correlation with stalk dry weight. In addition, soil pH had a significant positive correlation with *Vicinamibacteria*, TRA3-20, RB41, bacteriap25, and *Anaerolineae*. In contrast, *Acidothermus* had a significant negative correlation with soil pH. *Singulisphaera* and *Abditibacterium* showed a significant positive correlation with soil Zn availability (Figure 12A). In Noen Maprang, the genus of *Arenimonas* showed a significant positive correlation with soil pH. In addition, leaf dry weight was positively correlated with genus *Solitalea* (Figure 12B). No factors were significantly correlated with the bacterial and archaeal community in the Lom Sak field.

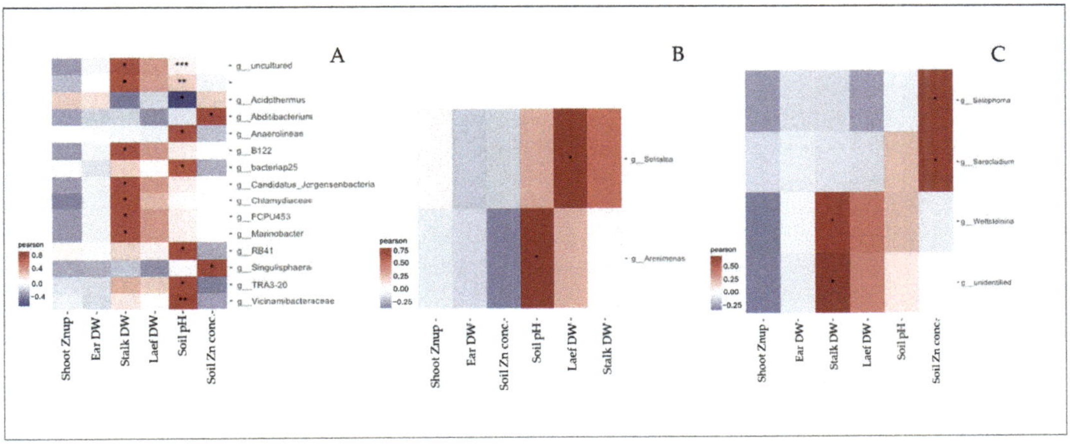

Figure 12. Correlation analyses of the genus of soil bacteria and archaea community with growth parameters and environmental factors (* $p < 0.05$, ** $p < 0.01$, *** $p < 0.001$) at Chon Daen (**A**), Noen Maprang (**B**), and correlation analyses of the genus of soil fungal community with growth parameters and environmental factors at ChonDaen (**C**). Note: Shoot Znup: shoot Zn uptake, DW: dry weight, Soil Zn conc.: soil Zn concentration.

Correlation between the soil fungal community and growth parameters and environmental factors was observed only in the Chon Daen experimental field (Figure 12C). The genus *Setophoma* in class *Dothideomycetes* and the genus *Sarocladium* in class *Sordariomycetes* were significantly related to the soil Zn concentration. Moreover, the genus *Welttsteinina* in the class *Dothideomycetes* was associated considerably with stalk dry weight accumulation. On the other hand, no factors were significantly correlated with the fungal community at the Noen Maprang and Lom Sak experimental sites.

4. Discussion

4.1. Effect of Zn Application on Biomass, Zn Accumulation, and Soil Zn Concentration

The effects of Zn fertilization on biomass production at the R3 stage varied significantly across the three different soil pH types, reflecting both environmental differences and site-specific soil conditions. At Chon Daen (pH 4.9–6.1), Zn fertilization significantly increased shoot Zn uptake (333.3 to 481.7 g/ha) (Figure 3A) and soil Zn concentration (up to 3.96 mg/kg) (Table 1). Variety SW 5819 showed superior Zn use efficiency and growth, indicating good adaptation to low pH (Figures 2A and 4A). At Noen Maprang (pH 5.6–6.4), Zn availability was moderate, resulting in a limited response from maize. Zn

https://doi.org/10.3390/agronomy16010078

uptake increased slightly (210.9 to 314.9 g/ha) (Figure 3B), but dry matter accumulation was stable (14.6–16.0 t/ha) (Figure 1B), suggesting the soil already supplied sufficient Zn for growth. At Lom Sak (pH 7.1–7.7), Zn fertilization improved ear dry matter by 7.8–20.8% (Figure 1C), and Zn uptake increased modestly (273.2 to 353.7 g/ha) (Figure 3C). The variety SW 5819 performed better than SW 5731, showing higher biomass and Zn uptake (Figures 2C and 4C), yet overall benefits were limited by reduced Zn solubility under alkaline conditions.

Soil pH is the key factor influencing Zn distribution in soil, as it determines Zn solubility. Higher soil pH increases Zn adsorption onto soil particle cation exchange sites, thereby decreasing Zn availability in the soil solution [25]. The results showed that soil Zn concentration at Lom Sak was lower than at Chaon Daen and Noen Maprang (Table 3), where the soil pH was high (Figure 5). The difference in soil pH across the three experimental sites results in maize Zn concentrations exceeding the upper critical values reported in previous studies: 0.99 mg/kg [26] and 1.02 mg/kg [27].

The soil Zn concentration, extracted by the DTPA method, can be used to predict maize absolute yield response: high Zn response at <0.9 mg/kg, medium Zn response at 0.9–1.3 mg/kg, and low Zn response at >1.30 mg/kg [26]. A previous study showed that applying 10.0 kg of Zn/ha annually resulted in higher grain size, straw, and total Zn concentrations in maize and wheat than in the no-Zn treatment [28]. Additionally, applying biofertilizer inoculated with arbuscular mycorrhizal fungi (AMF) and supplemented with 20 mg Zn/kg Zn fertilizer enhanced photosynthesis, transpiration, and stomatal conductance. This treatment effectively promotes better growth and maize yield while preventing Zn deficiency or toxicity [29].

4.2. Effect of Zn Application on Soil Microbial Community Under Different Soil pH

The results of this study demonstrate that Zn application rates ranging from 5 to 33.8 kg Zn/ha had minimal impact on soil microbial community structure in paddy soils. This observation is consistent with a previous study that reported that moderate Zn application rates (5.7 kg Zn/ha) did not significantly affect microbial diversity, whereas higher rates (34.1 kg Zn/ha) had adverse effects on enzyme activities [17]. The stability of microbial communities observed in our study suggests that the applied Zn rates were within the tolerance range of soil microorganisms. Location emerged as the primary driver of microbial community variation, rather than Zn application, as supported by research showing that soil properties such as total nitrogen and organic carbon are significant factors influencing microbial community structure [30]. This finding suggests that inherent soil characteristics and environmental conditions have stronger effects on microbial communities than moderate fertilizer applications. Some studies have found that Zn fertilizer application can reduce microbial community diversity but increase the connectivity and metabolic function of bacterial networks, potentially improving soil function and crop productivity [30]. However, these changes were not observed in this study, likely due to differences in Zn application rates, soil types, or cropping systems.

The previous research suggests that microbial communities can adapt over time to fertilizer inputs, but this adaptation is typically slow and is more influenced by seasonal and environmental changes than by moderate fertilizer application [31,32]. The minimal impact of Zn application on microbial communities in this study contrasts with research showing that excessive Zn concentrations can adversely affect microbial growth and enzyme activities [33]. However, our applied rates were considerably lower than the toxic thresholds identified in greenhouse studies, explaining the lack of adverse effects observed. The dominance of *Proteobacteria*, *Actinobacteria*, and *Bacteroidota* across all locations corresponds with findings from other rice-based systems. Previous studies have identified these phyla as critical bacterial taxa in agricultural soils, particularly *Actinobacteria, Bacteroidetes, Proteobacteria,* and *Chloroflexi,* which play important topological roles in soil microbial networks [30]. The presence of these groups benefits soil health by promoting nutrient cycling and plant growth. In other rice-paddy experiments with higher Zn application rates (e.g., 30 kg Zn/ha), soil Zn application reduced bacterial community diversity but increased network connectivity and metabolic function [30]. *Actinobacteria* often increased in abundance, while *Bacteroidetes* were sometimes suppressed. Metabolic pathways such as lipid, amino acid, and carbohydrate metabolism and the biodegradation of xenobiotics were upregulated after Zn application [30].

Both *Proteobacteria* and *Actinobacteria* were dominant and stable across soil treatments and locations in this study, and their abundance was not significantly altered by Zn fertilizer application [Figure 7]. In comparable experiments, increased abundance of *Actinobacteria* following Zn application was linked to improved soil metabolic functions and enhanced plant nutrient uptake. In studies where *Actinobacteria* and *Proteobacteria* were enriched by Zn application, there were measurable increases in grain yield and Zn content in crops; for example, rice yields increased by 17–19% and Zn enrichment in grains. These phyla are recognized for their critical roles in soil health, including organic matter recycling, nutrient cycling, and plant growth promotion [34,35]. However, in this experiment, while these phyla remained abundant and stable, the primary drivers of yield improvement were likely the direct effects of Zn fertilization and site-specific factors, rather than dramatic shifts in these microbial groups.

The fungal community showed greater location-specific variation than bacterial communities, with *Ascomycota* dominating across all sites. This pattern is typical in agricultural soils, where *Ascomycota* often represents the most abundant fungal phylum. The variation in fungal community composition among locations may reflect differences in soil pH, organic matter content, and other physicochemical properties that influence fungal establishment and survival. Fungi generally showed less pronounced changes compared to bacteria, with community shifts more influenced by soil properties and moisture than by fertilizer application [36].

Correlation analyses showed that microbial-plant interactions are location-specific, emphasizing the need to consider local soil conditions in microbial-based agriculture. These findings highlight the importance of Zn management to optimize maize productivity and maintain soil microbial stability.

5. Conclusions

Maize dry matter accumulation and Zn uptake at the R3 stage were significantly affected by Zn application rate, maize variety, and their interaction, with SW 5819 consistently outperforming SW 5731 under Zn fertilization, particularly at Chon Daen and Lom Sak. No yield or uptake response to Zn was observed at Noen Maprang, indicating adequate native Zn supply at that site. Soil Zn concentrations increased with Zn application across all sites, confirming effective enhancement of Zn availability. The range of Zn applications (5–33.8 kg Zn/ha) in this study did not alter bacterial, archaeal, or fungal diversity, with microbial community composition driven mainly by soil pH and site-specific conditions rather than Zn inputs. These results emphasize that Zn fertilization can improve maize performance without disrupting soil microbial communities, but its effectiveness depends strongly on soil pH. This highlights the site-specific Zn management strategies to optimize maize grain yield and grain Zn biofortification while maintaining soil microbial stability.

https://doi.org/10.3390/agronomy16010078

Supplementary Materials: The following supporting information can be downloaded at https://www.mdpi.com/article/10.3390/agronomy16010078/s1. Figure S1: The daily precipitation, maximum and minimum temperature during the maize growing in the dry season from January to April 2021 at Chon Daen, Noen Maprang, and Lom Sak experimental sites. The red arrow indicates the time of planting, V6, tasseling (VT), milking (R3), and physiological maturity stage (PM) of maize growth; Figure S2: Alpha diversity of the soil bacterial and archaeal community: Chao1 index and Shannon index under the interaction effect between Zn application rates and maize varieties at Chon Daen (CD), Noen Maprang (NM), and Lom Sak (LS); Figure S3: Alpha diversity of the soil fungal community: Chao1 index and Shannon index under the interaction effect between Zn application rates and maize varieties at Chon Daen (CD), Noen Maprang (NM), and Lom Sak (LS); Table S1: The fertilizer rate for maize production in Thailand was analyzed based on soil chemical properties.

Author Contributions: Conceptualization, S.N., A.W., J.M., K.S. and P.K.; methodology, S.N., J.M., A.W., K.S. and P.K.; software, J.M. and P.K.; validation, P.K., J.M., S.N., A.W. and A.R.; formal analysis, J.M. and P.K.; investigation, S.N., A.W., J.M., K.S., O.K. and P.K.; resources, S.N., J.M., A.W., K.S., O.K. and P.K.; data curation, P.K., J.M., A.R. and S.N.; writing—original draft preparation, P.K.; writing—review and editing, S.N., A.W., J.M., K.S., O.K. and A.R.; visualization, J.M. and P.K.; supervision, S.N. and J.M.; funding acquisition, J.M. and S.N. All authors have read and agreed to the published version of the manuscript.

Funding: The Graduate School of Bioagricultural Sciences, Nagoya University, the Development of Smart Thai Agriculture Using Big Data (2018) project of the Office of the Ministry of Higher Education, Science, Research and Innovation (grant no. 2561NRCT71013), and the Agricultural Research Development Agency grant 2021 (grant no. PRP6405032480).

Data Availability Statement: The data presented in this study are available upon request from the first and corresponding author.

Acknowledgments: This research was supported in part by the Graduate Program Scholarship from The Graduate School, Kasetsart University, and the Graduate School of Bioagricultural Sciences, Nagoya University. We gratefully acknowledge the Crop Physio-Molecular Biology Laboratory, Department of Agronomy, Kasetsart University, and Laboratory of Information Science in Agricultural Lands, Graduate School of Bioagricultural Sciences, Nagoya University, for providing experimental facilities and technical assistance. All authors agreed to acknowledge this.

Conflicts of Interest: The authors declare no conflicts of interest.

References

1. Alloway, B.J. *Zinc in Soils and Crop Nutrition*, 2nd ed.; International Zinc Association and International Fertilizer Industry Association: Paris, France, 2008.
2. Stanton, C.; Sanders, D.; Krämer, U.; Podar, D. Zinc in Plants: Integrating Homeostasis and Biofortification. *Mol. Plant* **2022**, *15*, 65–85. [CrossRef]
3. Jalal, A.; da Silva Oliveira, C.E.; Fernandes, G.C.; Horschut, B.; Furlani, E.; Henrique, P.; Galindo, F.S.; Martins, I.; Carvalho, M. Nanozinc and Plant Growth-Promoting Bacteria Improve Biochemical and Metabolic Attributes of Maize in Tropical Cerrado. *Front. Plant Sci.* **2023**, *13*, 1046642. [CrossRef] [PubMed]
4. Jalal, A.; Júnior, E.F.; Teixeira Filho, M.C.M. Interaction of Zinc Mineral Nutrition and Plant Growth-Promoting Bacteria in Tropical Agricultural Systems: A Review. *Plants* **2024**, *13*, 571. [CrossRef] [PubMed]
5. Mutambu, D.; Kihara, J.; Mucheru-Muna, M.; Bolo, P.; Kinyua, M. Maize Grain Yield and Grain Zinc Concentration Response to Zinc Fertilization: A Meta-Analysis. *Heliyon* **2023**, *9*, e16040. [CrossRef] [PubMed]
6. Alloway, B.J. Soil Factors Associated with Zinc Deficiency in Crops and Humans. *Environ. Geochem. Health* **2009**, *31*, 537–548. [CrossRef]
7. Wisawapipat, W.; Janlaksana, Y.; Christl, I. Zinc Solubility in Tropical Paddy Soils: A Multi-Chemical Extraction Technique Study. *Geoderma* **2017**, *301*, 1–10. [CrossRef]
8. Schulte, E. *Understanding Plant Nutrients Soil and Applied Zinc*; Cooperative Extension Publications; University of Wisconsin: Madison, WI, USA, 2002; Available online: https://corn.aae.wisc.edu/Management/pdfs/a2528.pdf (accessed on 29 April 2020).

9. Bender, R.R.; Haegele, J.W.; Ruffo, M.L.; Below, F.E. Nutrient Uptake, Partitioning, and Remobilization in Modern, Transgenic Insect-Protected Maize Hybrids. *Agron. J.* **2013**, *105*, 161–170. [CrossRef]
10. Impa, S.M.; Morete, M.J.; Ismail, A.M.; Schulin, R.; Johnson-Beebout, S.E. Zn Uptake, Translocation and Grain Zn Loading in Rice (*Oryza sativa* L.) Genotypes Selected for Zn Deficiency Tolerance and High Grain Zn. *J. Exp. Bot.* **2013**, *64*, 2739–2751. [CrossRef]
11. Waters, B.M.; Uauy, C.; Dubcovsky, J.; Grusak, M.A. Wheat (*Triticum aestivum*) NAM Proteins Regulate the Translocation of Iron, Zinc, and Nitrogen Compounds from Vegetative Tissues to Grain. *J. Exp. Bot.* **2009**, *60*, 4263–4274. [CrossRef]
12. Singh, A.; Singh, A.; Gill, H.K. Zinc Biofortification through Biofertilizers, Foliar and Soil Application in Maize (*Zea mays* L.). *J. Exp. Agric. Int.* **2024**, *46*, 119–127. [CrossRef]
13. Khongchiu, P.; Wongkaew, A.; Murase, J.; Sajjaphan, K.; Rakpenthai, A.; Kumdee, O.; Nakasathien, S. Zinc Application Enhances Biomass Production, Grain Yield, and Zinc Uptake in Hybrid Maize Cultivated in Paddy Soil. *Agronomy* **2025**, *15*, 1501. [CrossRef]
14. Liu, D.-Y.; Zhang, W.; Liu, Y.-M.; Chen, X.-P.; Zou, C.-Q. Soil Application of Zinc Fertilizer Increases Maize Yield by Enhancing the Kernel Number and Kernel Weight of Inferior Grains. *Front. Plant Sci.* **2020**, *11*, 188. [CrossRef] [PubMed]
15. Liu, D.-Y.; Zhang, W.; Yan, P.; Chen, X.-P.; Zhang, F.-S.; Zou, C.-Q. Soil Application of Zinc Fertilizer Could Achieve High Yield and High Grain Zinc Concentration in Maize. *Plant Soil* **2017**, *411*, 47–55. [CrossRef]
16. Takrattanasaran, N.; Chanchareonsook, J.; Johnson, P.G.; Thongpae, S.; Sarobol, E. Amelioration of Zinc Deficiency of Corn in Calcareous Soils of Thailand: Zinc Sources and Application Methods. *J. Plant Nutr.* **2013**, *36*, 1275–1286. [CrossRef]
17. Liu, Y.; Cao, W.; Chen, X.; Yu, B.-G.; Lang, M.; Chen, X.; Zou, C. The Responses of Soil Enzyme Activities, Microbial Biomass and Microbial Community Structure to Nine Years of Varied Zinc Application Rates. *Sci. Total Environ.* **2020**, *737*, 140245. [CrossRef]
18. Navarrete, A.A.; Mellis, E.V.; Escalas, A.; Lemos, L.B.; Henrique, F.; Quaggio, J.A.; Zhou, J.; Tsai, S.M. Zinc Concentration Affects the Functional Groups of Microbial Communities in Sugarcane-Cultivated Soil. *Agric. Ecosyst. Environ.* **2017**, *236*, 187–197. [CrossRef]
19. Office of Agricultural Economics. *Agricultural Statistics of Thailand 2024*; Office of Agricultural Economics, Ministry of Agriculture and Cooperatives: Bangkok, Thailand, 2025; pp. 29–33. Available online: https://oae.go.th/uploads/files/2025/05/06/13e808 9a3f69ea96.pdf (accessed on 10 October 2025).
20. Kadiyala, M.D.M.; Mylavarapu, R.S.; Li, Y.C.; Reddy, G.B.; Reddy, M.D. Impact of Aerobic Rice Cultivation on Growth, Yield, and Water Productivity of Rice-Maize Rotation in Semiarid Tropics. *Agron. J.* **2012**, *104*, 1757–1765. [CrossRef]
21. Department of Agriculture. Fertilizer Recommendation Based on Soil Chemical Analysis for Economic Crops. Department of Agriculture, Ministry of Agriculture and Cooperatives. Available online: https://www.doa.go.th/apsrdo/?page_id=3241 (accessed on 2 November 2021).
22. Liu, H.; Gan, W.; Rengel, Z.; Zhao, P. Effects of Zinc Fertilizer Rate and Application Method on Photosynthetic Characteristics and Grain Yield of Summer Maize. *J. Soil Sci. Plant Nutr.* **2016**, *16*, 95162016005000045. [CrossRef]
23. Lindsay, W.L.; Norvell, W.A. Development of a DTPA Soil Test for Zinc, Iron, Manganese, and Copper. *Soil Sci. Soc. Am. J.* **1978**, *42*, 421–428. [CrossRef]
24. IRRI. Statistical Tool for Agricultural Research (STAR). Biometrics and Breeding Informatics, PBGB Division. International Rice Research Institute: Los Baños, Philippines. Available online: https://news.irri.org/2013/08/irri-biometrics-group-releases.html (accessed on 1 March 2020).
25. Gupta, N.; Ram, H.; Kumar, B. Mechanism of Zinc Absorption in Plants: Uptake, Transport, Translocation and Accumulation. *Rev. Environ. Sci. Bio/Technol.* **2016**, *15*, 89–109. [CrossRef]
26. Barbieri, P.A.; Sainz, R.; Wyngaard, N.; Eyherabide, M.; Calvo, R.; Salvagiotti, F.; Correndo, A.A.; Barbagelata, P.A.; Espósito, G.P.; Colazo, J.C.; et al. Can Edaphic Variables Improve DTPA-Based Zinc Diagnosis in Corn? *Soil Sci. Soc. Am. J.* **2017**, *81*, 556–563. [CrossRef]
27. Cuesta, N.M.; Wyngaard, N.; Rozas, H.S.; Calvo, N.R.; Carciochi, W.; Eyherabide, M.; Colazo, J.C.; Barraco, M.; Guertal, E.A.; Barbieri, P. Determining Mehlich-3 and DTPA Extractable Soil Zinc Optimum Economic Threshold for Maize. *Soil Use Manag.* **2020**, *37*, 736–748. [CrossRef]
28. Kumar, D.; Patel, K.C.; Ramani, V.P.; Shukla, A.K.; Behera, S.K.; Patel, R.A. Influence of Different Rates and Frequencies of Zn Application to Maize–Wheat Cropping on Crop Productivity and Zn Use Efficiency. *Sustainability* **2022**, *14*, 15091. [CrossRef]
29. Saboor, A.; Ali, M.A.; Ahmed, N.; Skalicky, M.; Danish, S.; Fahad, S.; Hassan, F.; Hassan, M.M.; Brestic, M.; EL Sabagh, A.; et al. Biofertilizer-Based Zinc Application Enhances Maize Growth, Gas Exchange Attributes, and Yield in Zinc-Deficient Soil. *Agriculture* **2021**, *11*, 310. [CrossRef]
30. Xiao, Y.S.; Zhou, B.; Han, Z.; Liu, S.; Ding, C.; Jia, F.; Zeng, W. Microbial Mechanism of Zinc Fertilizer Input on Rice Grain Yield and Zinc Content of Polished Rice. *Front. Plant Sci.* **2022**, *13*, 962246. [CrossRef]
31. Upadhyay, P.K.; Dey, A.; Singh, V.K.; Dwivedi, B.S.; Singh, R.K.; Rajanna, G.A.; Babu, S.; Rathore, S.S.; Shekhawat, K.; Rai, P.K.; et al. Changes in Microbial Community Structure and Yield Responses with the Use of Nano-Fertilizers of Nitrogen and Zinc in Wheat–Maize System. *Sci. Rep.* **2024**, *14*, 1100. [CrossRef]

32. Yang, R.; Zhou, S.; Zhang, L.; Qin, C. Pronounced Temporal Changes in Soil Microbial Community and Nitrogen Transformation Caused by Benzalkonium Chloride. *J. Environ. Sci.* **2023**, *126*, 827–835. [CrossRef]
33. Wyszkowska, J.; Boros-Lajszner, E.; Borowik, A.; Baćmaga, M.; Kucharski, J.; Tomkiel, M. Implication of Zinc Excess on Soil Health. *J. Environ. Sci. Health Part B* **2016**, *51*, 261–270. [CrossRef]
34. Ahlawat, V.; Ahalawat, N.; Boora, N.; Dadarwal, R.S.; Dhanda, D.; Sangwan, P.; Kumar, V.; Kumar, R.; Yadav, P.K. Impact of Long Term Integrated Nutrient Management Practices on Soil Bacterial Diversity: A 16S RRNA-Based Metagenomic Approach. *bioRxiv* **2025**, 639006. [CrossRef]
35. Boukhatem, Z.F.; Merabet, C.; Tsaki, H. Plant Growth Promoting Actinobacteria, the Most Promising Candidates as Bioinoculants? *Front. Agron.* **2022**, *4*, 849911. [CrossRef]
36. Han, L.; Qin, H.; Wang, J.; Yao, D.; Zhang, L.; Guo, J.; Zhu, B. Immediate Response of Paddy Soil Microbial Community and Structure to Moisture Changes and Nitrogen Fertilizer Application. *Front. Microbiol.* **2023**, *14*, 1130298. [CrossRef]

https://doi.org/10.3390/agronomy16010078

agronomy

MDPI

Article

Miscanthus × *giganteus* Rhizobacterial Community Responses to Zn and Oil Sludge Co-Contamination

Asil Nurzhanova [1], Eugenia Boulygina [2], Irina Sungurtseva [3], Aigerim Mamirova [4,*], Ramza Berzhanova [4] and Anna Muratova [3]

1 Institute of Plant Biology and Biotechnology, Timiryazev 45, Almaty 050040, Kazakhstan; gen_asil@mail.ru
2 Institute of Fundamental Medicine and Biology, Kazan Federal University, Marx, 76, 420012 Kazan, Russia; boulygina@gmail.com
3 Institute of Biochemistry and Physiology of Plants and Microorganisms, Saratov Scientific Centre of the Russian Academy of Sciences (IBPPM RAS), Entuziastov 13, 410049 Saratov, Russia; airinmind@yandex.ru (I.S.); muratova_a@ibppm.ru (A.M.)
4 Department of Biotechnology, Faculty of Biology and Biotechnology, Al-Farabi Kazakh National University, Al-Farabi 71, Almaty 050040, Kazakhstan; ramza05@mail.ru
* Correspondence: aigerim.mamirova@mail.com; Tel.: +7-702-348-8009

Abstract: Soil contamination in industrial areas often involves complex mixtures of contaminants, making remediation a significant challenge. Microbe-assisted phytoremediation offers a promising solution, yet its success depends on understanding interaction between plants, microorganisms, and contaminants in rhizosphere. This study examined the effects of organic (oil sludge) and inorganic (Zn) contaminants, applied individually and in combination, on the rhizosphere bacterial community of *Miscanthus* × *giganteus* Greef et Deu (*M*×*g*), with emphasis on strains exhibiting plant growth-promoting, hydrocarbon-degrading, and metal-tolerant traits. A one-season greenhouse experiment included soils spiked with Zn (1650 mg kg^{-1}) and/or oil sludge (15 mL kg^{-1}). Oil sludge exerted a stronger influence on the taxonomic structure of rhizobacterial communities than Zn, largely shaping the patterns observed under co-contamination. Zn exposure increased the relative abundance of Actinobacteriota, whereas oil sludge favoured Proteobacteriota. Both contaminants, individually and together, enhanced the proportion of Sphingomonadaceae. Across all treatments, taxa with potential plant-growth-promoting traits were present, while co-contaminated soil harboured microorganisms capable of hydrocarbon degradation, heavy metal tolerance, and plant growth promotion. These findings highlight the adaptive capacity of the *M*×*g* rhizobiome and support its application in phytoremediation. The isolation and characterisation of rhizosphere-associated strains provide basis for developing microbial bioagents to enhance biomass production and remediation efficiency in multi-contaminated environments.

Keywords: *Miscanthus* × *giganteus*; rhizobacterial community; taxonomic structure; Zn; oil sludge

1. Introduction

Under real conditions, technogenic soil pollution typically involves complex mixtures of contaminants, with petroleum hydrocarbons and heavy metals being the most prevalent [1–6]. A recent study estimated that 14–17% of cropland soils already exceed agricultural thresholds for at least one heavy metal [7]. Hydrocarbons are classified as priority pollutants because of their persistence, high toxicity, mutagenic and carcinogenic potential, and teratogenic effects in humans [8]. Combined contamination commonly occurs in industrial zones of oil extraction and processing, roadside soils, and landfill areas [4,5,9–11]. The coexistence of pollutants with distinct physicochemical properties poses a significant challenge for remediation, as they respond differently to conventional rehabilitation strategies [9]. In this context, phytoremediation, which relies on vegetation to restore contaminated soils, is regarded as one of the most effective modern technologies, as it enables the simultaneous removal of both organic and inorganic pollutants through diverse mechanisms [12]. The principal mechanism for the remediation of organic pollutants, such as petroleum hydrocarbons, is rhizosphere-mediated microbial degradation [13,14], whereas heavy metal remediation primarily relies on phytoaccumulation and phytostabilisation [15–17]. The efficiency of these processes is determined by both the remediating plant and its associated rhizosphere microbiome.

In recent years, plants of the *Miscanthus* genus have attracted considerable interest owing to their economic value and broad applicability across multiple sectors [18,19]. Among these, *Miscanthus* × *giganteus* Greef et Deu (*M*×*g*) is particularly recognised as a promising bioenergy crop capable of establishing in marginal and abandoned lands. This species is characterised by high biomass productivity and has shown potential for the remediation of soils contaminated with both organic and inorganic pollutants [20–25]. The ability of plants to diverse environmental conditions is strongly influenced by ecological services provided by their microbial symbionts, including biofertilisation, disease suppression, and enhanced tolerance to abiotic stress [26]. Soil functions as a reservoir of microbial diversity from which plants selectively assemble their microbiomes according to the requirements of their specific habitats [26]. Understanding the formation of rhizosphere microbial communities that support plant growth under adverse conditions and facilitate soil remediation is essential for evaluating the feasibility of cultivating *M*×*g* in contaminated environments.

The microbial communities associated with *M*×*g* have been extensively studied using both culture-independent and culture-based approaches. Since the majority of microorganisms are non-culturable (>99%), even when diverse media are employed [27,28], modern culture-independent molecular techniques are indispensable for assessing microbial diversity. A comprehensive study of 26 plant species identified *Clavibacter*, *Chryseobacterium*, and *Acidovorax* as taxa uniquely associated with *M*×*g* [29]. Increasing evidence indicates that *Miscanthus* species actively shape their rhizosphere microbiomes to support growth in disturbed soils [30–34]. Predictive functional profiling using Phylogenetic Investigation of Communities by Reconstruction of Unobserved States (PICRUSt) revealed that *Miscanthus* cultivation modifies the metabolic capabilities of bacterial communities, including pathways related to carbon and nitrogen metabolism and the phosphorus cycle [35]. Furthermore, FUNGuild analysis demonstrated that *Miscanthus* cultivation alters the trophic modes of fungal communities [35]. *Miscanthus*-associated diazotrophs supply plants with bioavailable nitrogen [36–38], and nitrogen fertilisation has been shown to influence the composition of the *Miscanthus* microbiome [39]. Several factors affecting the assembly of microbial communities in the *Miscanthus* rhizosphere are currently under investigation, including plant species identity [40], genetic variability [41], plant age [39], rooting depth [42], and exposure to different pollutants [22,43–46].

Several studies have documented shifts in the composition and structure of the *M*×*g* rhizosphere microbial community in response to heavy metal contamination [22,44,46,47]. In contrast, information on the effects of hydrocarbons on the *M*×*g* rhizosphere microbiome remains scarce [48]. These observations provide a basis for further investigations into the ecological functions of microorganisms in *Miscanthus* growth and offer valuable insights for the cultivation of this species in marginal soils, particularly those contaminated with heavy metals or other pollutants.

Increasing attention has recently been directed toward understanding plant responses and remediation efficiency in soils contaminated with multiple pollutants, particularly heavy metals and hydrocarbons [49–52]. However, only three studies published in 2025 have investigated the effects of complex pollution on the root zone microbiome. Conlon et al. [53] examined rhizodegradation by *Sinapis alba*, *Lolium perenne*, *L. perenne* + *Trifolium repens*, and *Cichorium intybus* in petroleum hydrocarbon- and heavy metal-co-contaminated soil in a microcosm-scale pot trial, reporting rhizosphere enrichment with hydrocarbon-degrading bacteria, including KCM-B-112, C1-B045, *Hydrogenophaga*, unclassified *Saccharimonadales* sp., and *Pedobacter*. Another microcosm experiment demonstrated that elevated levels of Cu, Ni, and Zn (100 mg kg^{-1} each) negatively affected the growth of *Lolium multiflorum* and reduced the abundance of hydrocarbon-degraders, while also altering the structure of hydrocarbon-degrading bacterial communities [54]. In a field study evidencing the co-contaminated nature of soils in industrial areas, the rhizosphere of *Carmona microphylla*, a native plant growing at an operating ink factory site polluted with polycyclic aromatic hydrocarbons (~500–800 mg kg^{-1}) and heavy metals (As, Cd, Cr, Pb and Zn), was analysed [55]. High pollution levels markedly reduced the number of specific operational taxonomic units (OTUs), with heavy metals identified as the primary factor shaping bacterial community structure [55]. Soil bacterial functions were influenced mainly by pollution levels rather than rhizosphere effects, with high contamination altering α diversity as well as the structure and composition of C- and N-cycling bacteria. Compared to hydrocarbons, heavy metals played a more significant role in shaping bacterial communities by affecting metabolite profiles [55]. Our previous research described the physiological and biochemical responses of $M \times g$ to soils contaminated with heavy metals and petroleum hydrocarbons, as well as its effective application in remediating soils affected by mixed pollution [56]. Zinc (Zn) and oil sludge, both widespread anthropogenic contaminants [8,57], were selected as model pollutants in the present study. Elevated concentrations of these pollutants in soils often result from industrial activities such as mining and smelting [10].

This study investigated the rhizobacterial community of $M \times g$ cultivated in soils subjected to distinct contamination scenarios: Zn, oil sludge, or their combination, with uncontaminated soil serving as a control. The research was grounded in the premise that soil contamination adversely affects rhizosphere microbial diversity, which is essential for both plant health and soil functioning [58,59]. It was hypothesized that complex contamination involving both inorganic (Zn) and organic (oil sludge) pollutants would exert a more pronounced negative effect on rhizosphere microbial diversity than single-pollutant exposure. In particular, a decline was expected in the abundance and diversity of cultivable microbial taxa with plant growth-promoting traits and contaminant tolerance [54]. Moreover, it was anticipated that organic contamination (oil sludge) would be less detrimental, or potentially more favourable, to rhizobacterial community composition than inorganic contamination (Zn), owing to differences in bioavailability, toxicity mechanisms, and the potential for microbial degradation. Within this framework, the study aimed to elucidate how contamination type and complexity shape rhizosphere microbial communities and their functional potential in the context of phytoremediation with $M \times g$.

2. Materials and Methods

2.1. Soil Origin and Characteristics

The experimental design was previously described by Muratova et al. [56]. Briefly, the soil used in the pot experiments was loamy leached chernozem (World Reference Base [60]: Luvic Chernozems) collected near Novye Burasy village, Saratov region, Russian Federation. Prior to the experiment, the soil was analysed for agrochemical parameters. Soil pH was measured using a Metler Toledo Delta 320 pH meter (Mettler-Toledo Instrument Co., Ltd.; Shanghai, China) [61]. The contents of mobile (available) phosphorus (P_2O_5) [62], nitrate (NO_3-N) [63], and water-soluble ammonium (NH_4-N) [64] were quantified by photocolorimetry (Table 1).

The principal characteristics of the soil were as follows: particle-size distributions—1–3 mm, 0.036%; 1–0.5 mm, 0.23%; 0.5–0.25 mm, 31.1%; and <0.25 mm, 68.5%. The total organic carbon content was 4.4% of the air-dried soil.

Table 1. Soil agrochemical profiles.

Parameter	Unit	Control Soil	Zn-Spiked Soil	Zn- & Oil Sludge-Spiked Soil	Oil Sludge-Spiked Soil
pH	–	5.64	5.37	5.54	5.53
NO_3-N	mg kg^{-1}	37.8 ± 6.29	18.6 ± 0.71	17.0 ± 3.24	20.6 ± 0.20
NH_4-N	mg kg^{-1}	31.7 ± 1.55	39.6 ± 2.14	31.7 ± 1.87	31.8 ± 6.60
P_2O_5	mg kg^{-1}	63.8 ± 0.98	51.0 ± 6.53	53.2 ± 5.55	58.3 ± 4.57

2.2. Experimental Layout

The soil for the pot experiment was sieved through a 7 mm mesh and divided into four portions. The first portion was spiked with a Zn solution obtained by dissolving $ZnSO_4 \cdot 7H_2O$ (>99% purity, Reakhim, Russia) in deionised water, resulting in a final soil concentration of 1650 mg Zn kg^{-1}, corresponding to a 15 × MPC [65]. The second portion was spiked with oil sludge (15 mL kg^{-1}), yielding a final concentration of total petroleum hydrocarbons (TPHs) of 8810 mg kg^{-1}. The oil sludge was collected from a pit located on the grounds of a petroleum refinery in Saratov, Russian Federation. The oil sludge was collected from an oil sludge pit located on the grounds of a petroleum refinery in Saratov, Russian Federation. Its characteristics were as follows: pH 6.4, suspended solids ≤ 2%, water content 14.2%. The petroleum fraction comprised paraffins (18.8%), naphthene (18.1%), mono- and bicyclic aromatic hydrocarbons (MBCA, 7.1%), polycyclic aromatic hydrocarbons (PAHs, 11.2%), and alcohol-benzene tars (ABTs) plus asphaltenes (44.7%). The third portion was spiked with both the Zn solution and oil sludge at the same final concentrations. The fourth portion served as the control and was treated with an equivalent amount of distilled water. Thus, the experimental treatments were: (i) control soil, (ii) Zn-spiked soil, (iii) oil sludge-spiked soil, and (iv) Zn- and oil sludge-spiked soil.

Pots were lined with black plastic bags to prevent leaching of soil solution, including water-soluble contaminants and nutrients. A 1 kg layer of expanded clay (granule diameter: 15–20 mm) was placed at the bottom of each pot as drainage and covered with gauze. This was overlaid with 1 kg of purified river sand, which was again covered with gauze. Finally, 3 kg dry weight (DW) of either control or spiked soil was added. Two one-year-old *M×g* rhizomes (sourced from the plantation of the Institute of Plant Biology and Biotechnology, Almaty, Kazakhstan) were planted in each pot, and the soil surface was covered with 200 g of purified river sand.

All treatments were established in triplicate. Soil moisture was maintained at 50% of field moisture capacity by daily watering, determined gravimetrically. Plants were cultivated in a greenhouse under natural illumination at 24–28 °C during the day and 20–22 °C at night for six months (March–September 2018).

2.3. Isolation of Strains Capable of Tolerating Contaminants and Promoting Plant Growth

Heavy-metal-resistant microorganisms were enumerated and isolated using Lysogeny Broth (LB) agar supplemented with either $ZnSO_4 \cdot 7H_2O$ or $Pb(CH_3COO)_2$ at a final metal-ion concentration of 0.5 mmol L^{-1}. The isolates were subsequently tested for metal resistance by culturing on LB agar containing 0.5–5 mM Zn^{2+} or Pb^{2+}.

All isolates were assessed for plant growth–promoting traits, including the ability to fix atmospheric nitrogen (N_2) on nitrogen-free agar medium [66,67], produce the phytohormone indole-acetic acid (IAA) [68], and solubilise phosphates [69]. The paraffin-degrading ability of isolates was tested using the membrane-filter technique and Bushnell–Haas medium [67]. PAH degradation was evaluated by a plate assay based on the formation of clearance zones around bacterial colonies [70].

2.4. 16S rRNA Gene Sequencing Analysis of Rhizosphere Soil

At the end of the experiment, plant roots were removed from the pots and soil loosely adhering to the roots was shaken off. Young roots with tightly attached rhizosphere soil were collected (~1.0 g) for metagenomic analysis. Prior to analysis, rhizosphere soil was carefully separated from the roots using a laboratory spatula. Two replicate subsamples (0.3 g each) were taken for DNA extraction.

Soil DNA was extracted and purified using the Fast DNA SPIN Kit for Soil (MP Biomedicals, Solon, OH, USA) and homogenised with a FastPrep-24 instrument (MP Biomedicals, Solon, OH, USA) according to the manufacturer's instructions. A 16S rRNA gene sequencing library was constructed following the Illumina 16S Metagenomics Sequencing Library Preparation protocol (Illumina, San Diego, CA, USA), targeting the V3–V4 hypervariable regions. The initial PCR amplification was carried out with region-specific primers containing Illumina adapter overhang sequences (forward primer: 5′- TCGTCGGCAGCGTCAGATGTGTATAAGAGACAGCCTACGGGNGGCWGCAG-3′; reverse primer: 5′- GTCTCGTGGGCTCGGAGATGTGTATAAGAGACAGGACTACHVGGG TATCT-AATCC-3′). PCR products were visualised by gel electrophoresis and purified with AMPure XT magnetic beads. A second PCR was then performed using the Nextera XT Index Kit (Illumina, San Diego, CA, USA).

The DNA concentration of purified PCR products was quantified with a Qubit dsDNA HS Assay Kit (Thermo Scientific, Waltham, MA, USA) on a Qubit 2.0 fluorometer. The pooled library (4 nM) was denatured with 0.2 M NaOH, diluted to 10 pM, and combined with 20% (*v*/*v*) denatured 4 pM PhiX control, according to Illumina's guidelines. Sequencing of the V3–V4 regions was performed on an Illumina MiSeq platform using a 2 × 300 bp paired-end run.

Raw sequence reads were deposited in the NCBI Sequence Read Archive (SRA) under BioProject accession number PRJNA1119175 in FASTQ format (https://www.ncbi.nlm.nih.gov/bioproject/PRJNA1119175, accessed on 14 August 2025).

2.5. Bioinformatics and Statistical Analysis

Sequence data were processed using QIIME2 software, version 2022.8 (http://qiime2.org/; accessed on 1 August 2025) [71]. On average, 118,339 read pairs per sample were obtained before filtering. Raw reads were denoised with the DADA2 algorithm implemented in QIIME2 [72]. After quality filtering and removal of chimeric and phiX sequences, an average of 17,870 joined read pairs per sample remained. Taxonomic classification was performed using the SILVA 138 database (99% OTUs) and a Naive Bayes classifier [73]. The number of observed features ranged from 601 to 1030.

Alpha (α) diversity, reflecting richness and evenness of the bacterial community, was estimated by using Chao1, Shannon, and Simpson indices. Beta (β) diversity, representing differences in microbial community composition among samples, was assessed using weighted and unweighted UniFrac distances and visualised by non-metric multidimensional scaling (nMDS). Venn diagrams were constructed with Creately software (https://creately.com/lp/venn-diagram-maker/; accessed on 1 August 2025).

Statistical analyses were performed with Statistica 13 (TIBCO Software Inc. 2017, Palo Alto, CA, USA). Fisher's least significant difference (LSD) test was used to evaluate treatment effects. Mean comparisons were conducted with Tukey's HSD, and Spearman's rank correlation coefficients were also calculated.

3. Results

3.1. Diversity of Rhizosphere Communities

Sequencing of the 16S rRNA gene from 10 rhizosphere samples generated 1,122,256 raw reads. After denoising and chimera removal, 703,550 sequences were retained for downstream analyses. Sequences with >97% similarity were clustered into OTUs, which were assigned to 32 phyla, 98 classes, 229 orders, 346 families, and 587 genera. Rarefaction curves based on normalised OTU counts nearly reached saturation for all samples, indicating sufficient sequencing depth and comprehensive coverage of the bacterial communities (Figure S1).

Bacterial diversity and richness were evaluated using α- and β-diversity metrics (Table 2, Figure 1).

Table 2. The α-diversity indices for $M \times g$ rhizospheric microbial communities.

Treatment	Observed Features	Chao1	Shannon Index	Simpson Index	Faith PD
Control soil	1054 ± 148	1062 ± 147	8.74 ± 0.18	0.994 ± 0.001	82.5 ± 8.96
Zn-spiked soil	709 ± 130 *	718 ± 135 *	8.46 ± 0.39	0.995 ± 0.002	64.3 ± 9.64
Oil sludge-spiked soil	575 ± 202 *	582 ± 208 *	8.26 ± 0.43	0.995 ± 0.001	49.3 ± 14.8 *
Zn- & oil sludge-spiked soil	814 ± 207	822 ± 211	8.77 ± 0.35	0.997 ± 0.001 *	71.9 ± 15.8

Notes: PD—phylogenetic diversity; *—significantly different compared control at $p < 0.05$.

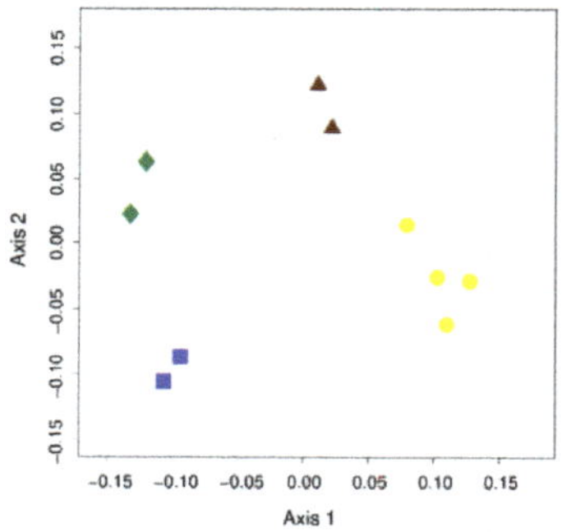

Figure 1. NMDS for $M \times g$ rhizosphere bacterial communities: •—control soil; ▲—Zn-spiked soil; ■—oil sludge-spiked soil; ♦—Zn- & oil sludge-spiked soil.

The α-diversity of rhizosphere bacterial communities was assessed using species richness indices (Chao1, Shannon, and Simpson) and Faith's phylogenetic diversity (Table 2). The rhizobacterial community of $M \times g$ grown in control soil exhibited the greatest richness, as indicated by the observed features and Chao1 indices. Rhizobacterial community of plants grown in Zn-spiked and oil sludge-spiked soils showed significantly reduced richness (observed features: $p = 0.04$; Chao1: $p = 0.007$). The lowest overall richness, including phylogenetic diversity (Faith PD, $p = 0.008$), was recorded in oil sludge-spiked soil. By comparison, the highest diversity was detected in rhizobacterial communities from plants grown in Zn- and oil sludge-co-contaminated soil, where both Shannon and Simpson indices reached their maximum values.

Non-metric multidimensional scaling (nMDS) visualisation further demonstrated that soil treatment significantly shaped the composition of $M \times g$ rhizosphere communities, with clear clustering of samples according to treatment (Figure 1).

3.2. Taxonomic Structure of Rhizosphere Communities

MiSeq sequencing revealed that the $M{\times}g$ rhizosphere communities comprised 587 bacterial genera belonging to 346 families across 32 phyla. The predominant bacterial phyla under different contamination conditions are shown in Figure 2.

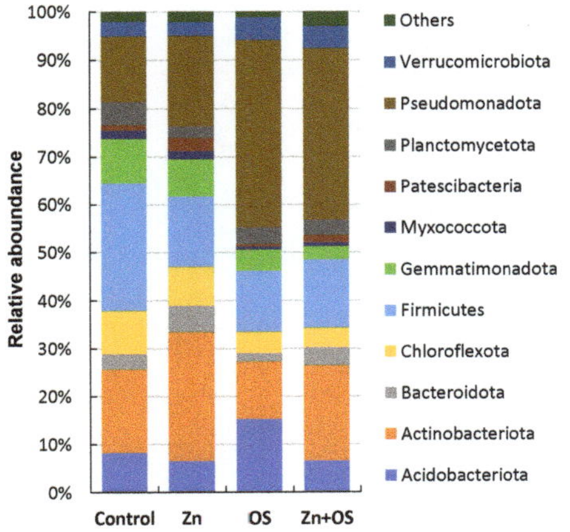

Figure 2. Relative abundance of bacterial communities at the phylum level in the rhizosphere of $M{\times}g$ grown in control and contaminated soils. Note: OS—oil sludge.

The rhizosphere bacterial community of $M{\times}g$ grown in control loamy leached chernozem was dominated by bacteria rather than archaea. Across all samples, the dominant phyla were Pseudomonadota (13 ± 2.5% to 39 ± 4.1%), Actinobacteriota (12 ± 0.1% to 27 ± 5.5%), and Firmicutes (13 ± 0.8% to 27 ± 0.6%). However, their relative abundances varied depending on the type of contaminant. A high proportion of Firmicutes in the control soil was associated with the use of material stored without moistening for one year. Contamination with both heavy metal and hydrocarbons reduced the abundance of Firmicutes, although this group remained dominant.

Statistical analysis confirmed that soil treatments significantly influenced the composition of $M{\times}g$ rhizosphere communities (Figure 1). The addition of contaminants decreased the proportion of Firmicutes, with oil sludge exerting a stronger effect than Zn (Figure 2). Zn contamination significantly ($p < 0.05$) increased the proportions of Actinobacteriota (by 54%), Pseudomonadota (by 37%), and Bacteroidota (by 69%), while decreasing those of Firmicutes (by 45%), Planctomycetota (by 51%), and Acidobacteriota (by 20%). Conversely, oil sludge contamination promoted the growth of Pseudomonadota (by 186%), Acidobacteriota (by 84%), and Verrucomicrobia (by 56%), while reducing the abundance of Actinobacteriota (by 41%), Bacteroidota (by 46%), Chloroflexota (by 51%), Firmicutes (by 52%), and Gemmatimonadetes (by 54%). The taxonomic profile of communities under mixed contamination was intermediate, reflecting the combined but distinct impacts of each contaminant.

Family-level taxa identified in control, Zn-spiked, oil sludge-spiked, and Zn- and oil sludge-co-spiked soils are listed in Table S1. Statistically significant differences were observed in the community composition of oil sludge-spiked soil compared with Zn-spiked soil (LSD = 0.025299), control soil (LSD = 0.021431), and co-contaminated soil (LSD = 0.030404).

Quantitative characteristics of predominant taxa (≥0.5%) identified at the genus level, including members with potential roles as Zn and oil sludge detoxifiers and/or plant growth promoters, are shown in Figure 3.

Phyllum	Family and genus	Control	Zn	Oil sludge	n+Oil sludg
Acidobacteriota	o__Acidobacteriales;__;__	0.01 ± 0.00	0.06 ± 0.02	0.71 ± 0.46	0.28 ± 0.08
	f__Acidobacteriaceae_(Subgroup_1);__	0.18 ± 0.08	0.00 ± 0.00	0.55 ± 0.06	0.15 ± 0.04
	g__Candidatus_Koribacter	0.24 ± 0.06	0.34 ± 0.18	2.25 ± 0.99	0.45 ± 0.00
	f__uncultured;g__uncultured	1.81 ± 0.18	2.04 ± 0.24	4.59 ± 1.97	0.86 ± 0.14
	f__Bryobacteraceae;g__Bryobacter	1.79 ± 0.02	2.14 ± 0.68	2.61 ± 0.15	2.30 ± 0.21
	g__Candidatus_Solibacter	1.33 ± 0.68	0.90 ± 0.68	1.58 ± 0.66	1.28 ± 0.09
	f__Blastocatellaceae;g__uncultured	0.40 ± 0.09	0.06 ± 0.05	0.61 ± 0.24	0.02 ± 0.01
	f__Pyrinomonadaceae;g__RB41	0.20 ± 0.09	0.03 ± 0.02	0.81 ± 0.60	0.00 ± 0.00
	o__Vicinamibacterales;f__uncultured;g__ucultur	0.61 ± 0.04	0.54 ± 0.20	0.41 ± 0.01	0.14 ± 0.03
	g__Vicinamibacteraceae	0.71 ± 0.26	0.10 ± 0.09	0.15 ± 0.08	0.04 ± 0.02
Actinobacteriota	g__Mycobacterium	0.03 ± 0.01	0.00 ± 0.00	1.40 ± 0.70	1.19 ± 0.67
	g__Williamsia	0.00 ± 0.00	0.00 ± 0.00	0.00 ± 0.00	0.52 ± 0.09
	g__Blastococcus	0.62 ± 0.13	0.39 ± 0.19	0.56 ± 0.03	0.48 ± 0.23
	f__Intrasporangiaceae;__	0.60 ± 0.18	1.25 ± 0.14	0.28 ± 0.01	0.54 ± 0.18
	f__Microbacteriaceae;__	0.13 ± 0.09	0.34 ± 0.02	0.47 ± 0.47	0.67 ± 0.32
	g__Microbacterium	0.00 ± 0.00	0.00 ± 0.00	0.52 ± 0.10	0.30 ± 0.07
	f__Micrococcaceae;__	0.04 ± 0.03	1.08 ± 0.88	0.04 ± 0.02	0.21 ± 0.04
	g__Pseudarthrobacter	1.07 ± 0.23	5.96 ± 1.81	0.53 ± 0.07	2.15 ± 0.88
	f__Micromonosporaceae;__	0.52 ± 0.26	0.38 ± 0.21	0.25 ± 0.12	0.28 ± 0.08
	g__Dactylosporangium	0.54 ± 0.19	0.11 ± 0.06	0.00 ± 0.00	0.00 ± 0.00
	g__Micromonospora	0.39 ± 0.22	2.14 ± 0.45	0.14 ± 0.09	0.32 ± 0.26
	g__Nocardioides	0.26 ± 0.17	1.12 ± 0.95	0.00 ± 0.00	0.53 ± 0.29
	g__Microlunatus	1.21 ± 0.28	1.21 ± 0.01	0.19 ± 0.07	0.25 ± 0.14
	g__Pseudonocardia	0.08 ± 0.06	0.44 ± 0.35	0.68 ± 0.22	0.65 ± 0.01
	g__Streptomyces	0.71 ± 0.60	0.62 ± 0.41	0.63 ± 0.26	0.07 ± 0.01
	g__MB-A2-108	0.62 ± 0.10	0.90 ± 0.24	1.03 ± 0.46	1.32 ± 0.53
	g__Rubrobacter	2.14 ± 0.18	1.62 ± 1.22	0.55 ± 0.11	0.92 ± 0.16
	g__Gaiella	1.31 ± 0.08	1.29 ± 0.25	1.18 ± 0.45	1.05 ± 0.10
	o__Gaiellales;f__uncultured;g__uncultured	3.34 ± 0.53	2.86 ± 0.16	2.02 ± 0.93	3.98 ± 0.67
	o__Solirubrobacterales;f__67-14;g__67-14	0.7 5± 0.38	1.25 ± 0.24	0.12 ± 0.04	0.80 ± 0.09
Bacteroidota	g__Chitinophaga	0.10 ± 0.03	0.52 ± 0.17	0.03 ± 0.02	0.25 ± 0.17
	g__Flavisolibacter	0.72 ± 0.11	0.58 ± 0.07	0.07 ± 0.02	0.00 ± 0.00
	g__Segetibacter	0.54 ± 0.06	0.41 ± 0.32	0.14 ± 0.09	0.23 ± 0.01
	f__Chitinophagaceae;g__uncultured	0.48 ± 0.03	0.81 ± 0.04	0.19 ± 0.09	0.35 ± 0.18
	g__Mucilaginibacter	0.03 ± 0.02	0.09 ± 0.07	0.76 ± 0.02	0.88 ± 0.15
Chloroflexota	f__Roseiflexaceae;g__uncultured	1.58 ± 1.28	0.01 ± 0.01	0.97 ± 0.18	0.04 ± 0.02
	f__JG30-KF-CM45;g__JG30-KF-CM45	2.09 ± 0.23	2.72 ± 0.01	1.64 ± 0.09	1.96 ± 0.14
	f__KD4-96;g__KD4-96	1.16 ± 0.41	1.05 ± 0.56	0.35 ± 0.04	1.03 ± 0.11
	f__C0119;g__C0119	1.49 ± 0.43	2.13 ± 0.31	0.37 ± 0.12	0.25 ± 0.18
	f__TK10;g__TK10	0.90 ± 0.35	0.28 ± 0.22	0.19 ± 0.09	0.01 ± 0.00
Cyanobacteriota	g__Obscuribacter	0.01 ± 0.00	0.09 ± 0.01	0.16 ± 0.07	0.65 ± 0.19
Firmicutes	g__Bacillus	24.64 ± 0.46	12.9 6 ±4.15	12.14 ± 0.34	12.48 ± 0.35
Gemmatimonadota	f__Gemmatimonadaceae;__	1.24 ± 0.04	0.41 ± 0.04	0.99 ± 0.27	0.07 ± 0.04
	g__Gemmatimonas	4.39 ± 0.02	4.19 ± 0.24	1.20 ± 0.16	0.87 ± 0.49
	f__Gemmatimonadaceae;g__uncultured	3.30 ± 0.49	3.03 ± 1.30	2.07 ± 0.28	1.80 ± 0.49
Myxococcota	g__Haliangium	0.99 ± 0.74	0.61 ± 0.14	0.16 ± 0.05	0.41 ± 0.12
	g__Pajaroellobacter	0.14 ± 0.09	0.62 ± 0.60	0.03 ± 0.02	0.22 ± 0.10
Nitrospirota	g__Nitrospira	0.50 ± 0.08	0.69 ± 0.18	0.14 ± 0.04	0.08 ± 0.03
Patescibacteria	f__LWQ8;g__LWQ8	0.14 ± 0.02	0.69 ± 0.02	0.07 ± 0.04	0.46 ± 0.12
	g__Saccharimonadales	0.52 ± 0.07	0.58 ± 0.24	0.04 ± 0.02	0.20 ± 0.07
Planctomycetota	g__WD2101_soil_group	3.31 ± 0.94	0.82 ± 0.15	2.88 ± 1.01	1.30 ± 0.38
Pesudomonadota	c__Alphaproteobacteria;__;__;__	0.05 ± 0.01	0.66 ± 0.40	0.29 ± 0.11	0.77 ± 0.08
	f__Acetobacteraceae;__	0.00 ± 0.00	0.00 ± 0.00	0.69 ± 0.33	0.42 ± 0.07
	g__Phenylobacterium	0.12 ± 0.00	0.57 ± 0.18	0.18 ± 0.09	0.07 ± 0.03
	g__Dongia	0.05 ± 0.06	0.08 ± 0.04	0.13 ± 0.03	0.81 ± 0.29
	f__Micropepsaceae;g__uncultured	0.37 ± 0.13	0.81 ± 0.62	1.46 ± 0.04	1.02 ± 0.12
	g__Parvibaculum	0.00 ± 0.00	0.00 ± 0.00	1.75 ± 0.36	1.06 ± 0.58
	f__Beijerinckiaceae;__	0.52 ± 0.28	0.20 ± 0.02	0.43 ± 0.15	0.01 ± 0.00
	g__Methylobacterium-Methylorubrum	0.00 ± 0.00	0.36 ± 0,11	0.03 ± 0.01	0.60 ± 0.08
	g__Mesorhizobium	0.30 ± 0.14	0.57 ± 0.06	0.69 ± 0.00	1.00 ± 0.39
	f__Xanthobacteraceae;__	0.76 ± 0.02	0.31 ± 0.07	1.29 ± 0.33	0.77 ± 0.04
	g__Pseudolabrys	0.53 ± 0.17	0.68 ± 0.24	0.48 ± 0.02	0.22 ± 0.04
	g__Sphingomonas	1.73 ± 1.03	3.39 ± 1.50	5.97 ± 0.66	5.34 ± 0.98
	g__KCM-B-112	0.00 ± 0.00	0.00 ± 0.00	6.65 ± 2.61	3.93 ± 3.06
	g__Burkholderia-Caballeronia-Paraburkholderia	0.04 ± 0.02	0.30 ± 0.00	0.53 ± 0.15	1.20 ± 0.20
	f__Comamonadaceae;__	0.31 ± 0.06	0.06 ± 0.02	0.73 ± 0.09	0.12 ± 0.02
	g__Ellin6067	0.36 ± 0.21	0.23 ± 0.13	0.97 ± 0.00	0.13 ± 0.07
	g__MND1	0.61 ± 0.44	0.61 ± 0.16	0.40 ± 0.09	0.00 ± 0.00
	f__Oxalobacteraceae;__	0.77 ± 0.08	0.29 ± 0.20	1.18 ± 0.07	0.60 ± 0.02
	g__Massilia	0.42 ± 0.09	0.44 ± 0.10	0.68 ± 0.26	0.09 ± 0.04
	g__Noviherbaspirillum	0.20 ± 0.09	0.30 ± 0.04	0.51 ± 0.16	0.05 ± 0.02
	f__SC-I-84;g__SC-I-84	0.78 ± 0.03	0.97 ± 0.18	1.19 ± 0.45	0.40 ± 0.05
	f__Porticoccaceae;g__C1-B045	0.00 ± 0.00	0.00 ± 0.00	0.49 ± 0.08	0.90 ± 0.01
	g__Aquicella	0.18 ± 0.08	1.10 ± 0.58	0.54 ± 0.06	0.41 ± 0.03
	g__Acidibacter	0.11 ± 0.05	0.01 ± 0.01	1.11 ± 0.84	0.26 ± 0.04
	g__Immundisolibacter	0.00 ± 0.00	0.00 ± 0.00	0.82 ± 0.38	0.29 ± 0.08
	g__Legionella	0.02 ± 0.01	0.07 ± 0.04	0.08 ± 0.01	0.83 ± 0.55
	g__Cavicella	0.00 ± 0.00	0.01 ± 0.00	0.26 ± 0.16	0.50 ± 0.28
	g__Pseudomonas	0.04 ± 0.02	0.00 ± 0.00	0.07 ± 0.05	0.63 ± 0.06
	f__Rhodanobacteraceae;__	0.02 ± 0.01	0.36 ± 0.16	0.19 ± 0.07	0.61 ± 0.06
	g__Rhodanobacter	0.17 ± 0.09	0.59 ± 0.30	0.00 ± 0.00	2.38 ± 0.21
	g__Lysobacter	0.22 ± 0.09	0.23 ± 0.09	0.53 ± 0.19	0.00 ± 0.00
Verrucomicrobiota	g__Neochlamydia	0.03 ± 0.01	0.32 ± 0.18	0.00 ± 0.00	0.65 ± 0.44
	f__Chthoniobacteraceae;g__Candidatus_Udaeob	1.46 ± 0.28	1.13 ± 0.27	2.23 ± 0.31	0.80 ± 0.32
	f__Xiphinematobacteraceae;g__Candidatus_Xiph	0.61 ± 0.08	0.06 ± 0.04	0.42 ± 0.25	0.15 ± 0.02
	g__Pedosphaeraceae	0.26 ± 0.11	0.58 ± 0.03	0.73 ± 0.27	0.44 ± 0.23

Figure 3. Effect of Zn and oil sludge on the relative abundance of predominant genus-level OTUs in the rhizosphere of $M{\times}g$. Values significantly different from the control are indicated in red (higher) or blue (lower) at $p \leq 0.05$.

Among the Actinobacteriota, the dominant orders in the $M \times g$ rhizosphere were Micrococcales and Gaiellales, each maintaining a relative abundance of ~4% across all treatments. The predominant families included Micrococcaceae, Micromonosporaceae, Rubrobacteriaceae, Gaiellaceae, and an uncultivated family within Gaiellales. Zn contamination promoted Actinobacteriota growth, showing strong correlations with increased proportions of Intrasporangiaceae (2.4-fold; $R^2 = 0.830$, $p < 0.05$) and Micrococcaceae (6.4-fold; $R^2 = 0.976$, $p < 0.05$). Enrichment of the rhizosphere with *Pseudarthrobacter* (Micrococcaceae), *Micromonospora* (Micromonosporaceae), and *Nocardioides* (Nocardioidaceae) was also observed, with increases of 5.6-, 5.5-, and 4.3-fold, respectively (Figure 3).

In contrast, oil sludge reduced the abundance of several Actinobacteriota families. In oil sludge-spiked soil, Intrasporangiaceae, Micrococcaceae, Rubrobacteriaceae, Micromonosporaceae, Nocardioidaceae, and Propionibacteriaceae decreased by 2.0-, 2.0-, 3.9-, 5.8-, 2.8-, and 6.5-fold, respectively, compared with the control (Table S1). Strong correlations were detected between oil sludge presence and increased proportions of OTUs affiliated with Nocardiaceae ($R^2 = 0.995$, $p < 0.05$) and Microbacteriaceae ($R^2 = 0.959$, $p < 0.05$). Oil sludge also promoted the growth of *Microbacterium* and *Mycobacterium* (Figure 3).

The effects of oil sludge, both alone and in combination with Zn, significantly increased the proportions of families within the phylum Proteobacteriota, particularly those belonging to Alphaproteobacteria. In oil sludge-spiked soils, Alphaproteobacteria families were enriched as follows: Sphingomonadaceae (3.4-fold), Xanthobacteraceae (1.3-fold), Micropepsaceae (3.8-fold), Rhizobiaceae (2.2-fold), and Parvibaculaceae (2.2-fold) ($R^2 = 0.990$, $p < 0.05$). Betaproteobacteria also increased, with Burkholderiaceae (13-fold), Oxalobacteraceae (1.7-fold), and Comamonadaceae (1.4-fold) showing enrichment ($R^2 = 0.955$, $p < 0.05$). Within Gammaproteobacteria, increases were observed in Acidithiobacillaceae, Porticoccaceae, and Xanthomonadaceae ($R^2 = 0.950$, $p < 0.05$), compared to the control.

At the genus level, the most notable increases under oil sludge contamination included *KCM-B-112* (Acidithiobacillaceae), *Sphingomonas* (Sphingomonadaceae), uncultured Micropepsaceae, *Parvibaculum* (Parvibaculaceae), *Acidibacter*, *Noviherbaspirillum* (Oxalobacteraceae), *SC-I-84*, *C1-B045* (Porticoccaceae), *Immundisolibacter* (Immundisolibacteraceae), and *Lysobacter* (Xanthomonadaceae) (Figure 3). Zn contamination also significantly increased the abundance of Micropepsaceae (1.3-fold), Sphingomonadaceae (1.9-fold), Burkholderiaceae (7.5-fold), and Rhodanobacteraceae (2.0-fold) compared to the control.

Genera enriched under Zn treatment included *Sphingomonas* (Sphingomonadaceae), *Aquicella* (Diplorickettsiaceae), *SC-I-84* (Burkholderiales order), *Methylobacterium-Methylorubrum* (Beijerinckiaceae), and *Phenylobacterium* (Caulobacteraceae) (Figure 3). Under mixed contamination, the proportions of Rhizobiaceae, Sphingomonadaceae, Burkholderiaceae, and Rhodanobacteraceae were significantly higher ($R^2 = 0.999$, $p < 0.05$), while Nitrosomonadaceae (*SC-I-84* genus) was reduced compared with the other treatments. Genera enriched in Zn- and oil sludge-co-contaminated soils included *Sphingomonas* (Sphingomonadaceae), *KCM-B-112* (Acidithiobacillaceae), *Burkholderia-Caballeronia-Paraburkholderia* (Burkholderiaceae), *Rhodanobacter* (Rhodanobacteraceae), *Mesorhizobium* (Rhizobiaceae), *Pseudomonas* (Pseudomonadaceae), and *Legionella* (Legionellaceae) (Figure 3).

Oil sludge also strongly promoted Acidobacteriota. The abundance of Koribacteraceae increased by 838% ($R^2 = 0.984$), while uncultivated Acidobacteriales increased by 154% ($R^2 = 0.969$). In contrast, Zn, either alone or in combination with oil sludge, did not significantly affect Acidobacteriota (Table S1). Within this phylum, the genera *Bryobacter*, Candidatus_*Koribacter*, and Candidatus_*Solibacter* were significantly enriched in oil sludge-spiked soil (Figure 3).

Within the Bacteroidota phylum, oil sludge altered community structure by greatly increasing Sphingobacteriaceae (19-fold under oil sludge, 25-fold under Zn + oil sludge), with

Mucilaginibacter emerging only under oil sludge (Figure 3). In contrast, Chitinophagaceae decreased (4.3-fold under oil sludge, 1.4-fold under Zn + oil sludge), with *Flavisolibacter* disappearing under oil sludge exposure (Figure 3). Zn contamination alone did not significantly affect Bacteroidota composition.

The Chloroflexota phylum showed marked reductions in predominant families under oil sludge contamination, while Zn contamination had no significant effect. Oil sludge also reduced Bacillaceae abundance in the *M*×*g* rhizobacterial community by twofold, with similar reductions under Zn and mixed contamination. Bacillaceae abundance was closely correlated with the presence of oil sludge ($R^2 = 0.953$, $p < 0.05$). In addition, oil sludge reduced Gemmatimonadaceae abundance by 2.2-fold, with a further 3.4-fold decrease under Zn + oil sludge co-contamination. Both Zn and mixed contamination supressed members of the Planctomycetota phylum, significantly reducing the predominant unspecified family of the WD2101 order by 4.0- and 2.5-fold, respectively (Figure 3).

3.3. Common and Unique Taxa Among Rhizosphere Bacterial Communities

Comparative analyses were performed to identify OTUs common or specific to the rhizosphere of *M*×*g* under control contamination treatments (Figure 4). A total of 126 genera were shared across all conditions, representing approximately one-third of the rhizosphere communities.

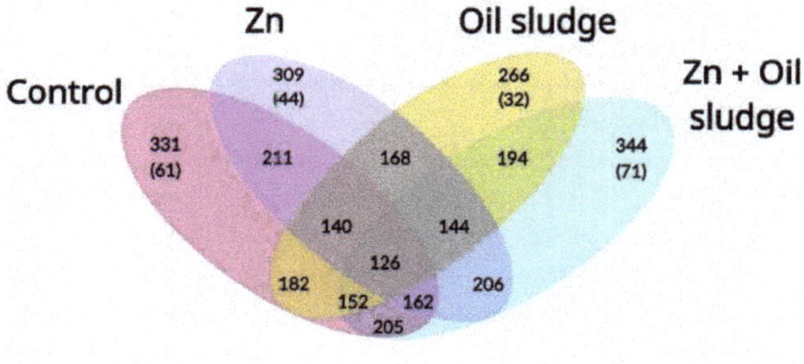

Figure 4. Venn diagram of common and unique (in parentheses) OTUs at the genus level in the rhizosphere bacterial communities of *M*×*g* grown in control and contaminated soils.

Control and mixed-contaminated soils exhibited the highest total number of OTUs at the genus level, as well as the largest number of unique genera (Figure 4). In control soil, only 38.1% of genera were shared with other treatments, while 18.4% (61 genera) were unique. These taxa were mainly affiliated with Firmicutes (9 genera), Gammaproteobacteria (8), Verrucomicrobiota (7), Myxococcota (6), Alphaproteobacteria (6), and Patescibacteria (5) (Table S2).

The Zn-spiked rhizosphere soil shared 40.8% of genera with other treatments, with 14.2% (44 genera) unique. Unique taxa in this soil were primarily affiliated with Firmicutes (9 genera), Bacteroidota (6), Alphaproteobacteria (6), Actinobacteriota (5), and Patescibacteria (5) (Table S3).

Oil sludge-spiked rhizosphere soil contained 47.4% of genera in common with other treatments, with the lowest proportion of unique taxa (12.0%; 32 genera). Unique genera were mainly *Alphaproteobacteria* (11 genera), with no unique Firmicutes detected (Table S4).

Rhizosphere soil co-contaminated with Zn and oil sludge had the highest total number of genera (344), of which only 36.6% were shared with other treatments. This community also contained the largest proportion of unique taxa (20.6%; 71 genera; Table S5). These were predominantly Gammaproteobacteria (20 genera), Actinobacteriota (12), Firmicutes (12), and Bacteroidota (6).

3.4. Plant-Growth Promoting and Contaminant-Tolerant Microbial Strain Isolation

From polluted rhizosphere samples, 76 microbial strains were isolated and screened for heavy metal resistance and plant growth-promoting traits, including nitrogen fixation, phosphorous solubilisation, and production of siderophores and phytohormones. Of these, 12 isolates were resistant to Pb^{2+} or Zn^{2+} at concentrations >5 mM, 10 isolates degraded paraffins, and 2 degraded phenanthrene. In addition, 33 isolates synthesised the phytohormone IAA, 10 fixed atmospheric N_2, 8 produced siderophores, and 2 solubilised phosphorus. Isolates combining plant-growth-promoting and pollutant-resistance traits were selected, identified, and are listed in Table 3.

Table 3. Microbial isolates with potential applications in bioremediation.

Isolate	$N_{2.5}$ Fixation	P Mobilization	Siderophore Production	IAA Synthesis	Resistance, mM		HC Degradation
					Zn^{2+}	Pb^{2+}	
Chitinophaga sp. Zn19	+	−	+	+	4.0	−	−
Mycolicibacterium sp. Pb113	+	−	+	+	−	≥5.0	+ (Phe)
Stenotrophomonas sp. Zn210	−	−	−	±	≥5.0	−	+ (alk)
Mesorhizobium sp. Pb210	+	++	±	+	−	2.5	+ (Phe)
Mycolicibacterium sp. Pb216	+	−	+	+	−	3.0	+ (Phe)
Microbacterium sp. Pb24	+	+	−	+	−	3.0	−

Notes: HC—hydrocarbons; Phe—degradation of phenanthrene in Kiyohara's test; alk—degradation of paraffines on Bushnell–Haas's medium.

4. Discussion

The ability of $M{\times}g$ to colonise and grow on marginal, disturbed, and polluted lands without compromising biomass accumulation has attracted considerable attention [74]. Such adaptability is determined not only by plant genetics but also by the microorganisms associated with its roots. Soil microorganisms are known to induce plant resistance to adverse environmental conditions, including drought and anthropogenic pollution [75,76]. Plants selectively recruit microorganisms from surrounding soil that enhance their survival, making root-associated microbiota crucial for successful colonisation and establishment [26]. Among these microorganisms, bacteria represent the most abundant group [77]. Accordingly, this study investigated how bacterial community composition in the rhizosphere of $M{\times}g$ is shaped by different types of soil contamination.

Our comparative analyses revealed that the rhizosphere microbiome of $M{\times}g$ was reorganised under heavy metal, hydrocarbon, and mixed contamination, with patterns of change depending on the type of pollutant. In Zn-spiked soil, overall microbial diversity and the abundance of unique taxa decreased, consistent with previous reports [78]. The most pronounced decreases were observed in *Bacillus* (Firmicutes), *Rubrobacter* (Actinobacteriota), *Vicinamibacter* (Acidobacteriota), and the WD2101_soil_group (Planctomycetota). Conversely, Zn contamination stimulated the enrichment of *Pseudarthrobacter*, *Micromonospora*, *Nocardioides* (Actinobacteriota), *Sphingomonas*, *Methylobacterium-Methylorubrum*, *Aquicella* (Alphaproteobacteria), JG30-KF-CM45, C0119 (Chloroflexota), and *Bryobacter* (Acidobacteriota) (Figure 3). Similar metal-induced shifts in microbial community structure have been documented [22,43,44,46]. For instance, Bourgeois et al. [43] found that $M{\times}g$ cultivation on soils contaminated with persistent organic pollutants and heavy metals (Zn, Pb, Cu, and Cd) promoted the growth of *Rhizobiales*, *Nitrospira*, *Azospira*, and *Gemmatimonas*, as well as fungi of the phyla Glomeromycota and Mortierella. Pham et al. [44] reported that heavy metal contamination shifted rhizosphere bacterial communities towards opportunistic genera such as *Pseudomonas* and *Stenotrophomonas*, though with only minor effects on overall diversity and richness. Other studies have shown that bacterial diversity in the rhizosphere, rhizoplane, and endosphere of $M \times g$ remains relatively stable under metal exposure, despite strong uptake by roots [46]. Enrichment of *Luteolibacter* and *Micromonospora* in the endosphere under heavy metal stress has been linked to improved plant performance [46]. Our findings align with these observations, particularly the enrichment of *Micromonospora* in the $M \times g$ rhizosphere under Zn contamination (Figure 3). Although resistance mechanisms of *Micromonospora* to the metalloid TeO_3^{2-} have been described [79], their bioremediation potential remains largely unexplored.

Several taxa stimulated by Zn in our study are known to play dual roles in remediation and plant growth promotion. These include *Bacillus* [80], *Sphingomonas* [35,81], *Pseudarthrobacter* [82,83], and *Micromonospora* [84], all of which exhibit metal resistance and contribute to phytostabilisatioin or phytoaccumulation processes.

Soil contaminated with oil sludge exhibited the lowest diversity and the fewest unique taxa. The abundance of *Bacillus* (Firmicutes), *Micromonospora*, *Pseudarthrobacter*, *Microlunatus*, and *Rubrobacter* (Actinobacteriota), JG30-KF-CM45, KD4-96, and C0119 (Chloroflexota), and *Gemmatimonas* (Gemmatimonadota) declined most notably under oil sludge contamination. In contrast, taxa stimulated by oil sludge included Candidatus_Koribacter, uncultured Acidobacteriales, *Bryobacter* (Acidobacteriota), *Mycobacterium* (Actinobacteriota), *Sphingomonas*, KCM-B-112, *Parvibaculum*, *Acidibacter*, *Immundisolibacter*, *Mesorhizobium*, as well as representatives of Xanthobacteraceae and Oxalobacteraceae (Pseudomonadota) and Candidatus_Udaeobacter (Verrucomicrobiota). Notably, *Sphingomonas* [85,86] and *Mycobacterium* [87,88] are established hydrocarbon degraders.

Our results on changes in the rhizosphere bacterial community of $M \times g$ are consistent with previous findings. Nebeská et al. [48] reported a decline in Actinobacteriota, a predominant phylum, and an increase in Verrucomicrobia, Bacterioides, and Acidobacteriota in the rhizosphere community of $M \times g$ grown in oil-contaminated soil. Hydrocarbon degraders such as *Pseudomonas*, *Shinella*, *Altererythrobacter*, *Azospirillum*, *Mesorhizobium*, and *Dyella* were also enriched in Miscanthus-planted soils compared with unplanted controls [48].

Mixed contamination with Zn and oil sludge supported the highest diversity and the largest number of unique taxa. Besides a reduction in *Bacillus*, the most prominent feature was the enrichment of Proteobacteria members including *Sphingomonas*, *KCM-B-112*, *Rhodanobacter*, *Burkholderia-Caballeronia-Paraburkholderia*, *Mesorhizobium*, *Dongia*, and *Pseudomonas*. Comparative analysis of treatments suggested that taxa such as *Rhodanobacter* and *Pseudarthrobacter* were primarily stimulated by Zn contamination, whereas oil sludge promoted *KCM-B-112*, *Sphingomonas*, and *Mycobacterium*. Unique taxa in co-contaminated soils = included *Burkholderia-Caballeronia-Paraburkholderia*, *Mesorhizobium*, *Dongia*, and *Pseudomonas*.

Although only 2–5% of rhizosphere microorganisms directly promote plant growth [89], plants selectively recruit beneficial taxa under stress [90]. Our metagenomic analysis revealed several plant growth–promoting taxa in the $M \times g$ rhizosphere, including *Bacillus* (Bacillaceae) [35,91], *Sphingomonas* (Sphingomonadaceae) [92], *Mesorhizobium* (Phyllobacteriaceae) [93], *Pseudarthrobacter* [82,83], and *Micromonospora* [84].

In addition to community-level analysis, we isolated strains from various taxa with strong remediation potential. These isolates combined heavy metal resistance and/or hydrocarbon-degrading capacity with plant-growth-promoting activity, making them promising candidates for enhancing $M \times g$ performance on contaminated soils. Nonetheless, validation in field conditions is required before practical applications can be confirmed.

Overall, this study characterised the taxonomic composition of the $M\times g$ rhizosphere microbiome and its response to soil contamination. Future research should extend beyond taxonomic profiling to include fungal communities and, importantly, functional characterisation using metagenome-assembled genomes (MAGs) and metatranscriptomics. These approaches will enable the linking of microbial taxa to ecologically relevant functions, including plant growth promotion and pollutant resistance or degradation.

5. Conclusions

This comparative study of the $M\times g$ rhizosphere microbiome in soils spiked with Zn, oil sludge, or both revealed pollutant-specific reorganisation of the rhizosphere bacterial community. Both contaminants reduced microbial diversity when applied individually, selectively stimulating or inhibiting particular taxa. Oil sludge induced more pronounced taxonomic shifts than Zn and largely determined in the structure of communities under mixed contamination. Specifically, Zn contamination increased the proportions of Actinobacteriota, Pseudomonadota, and Bacteroidota, while decreasing Firmicutes, Planctomycetota, and Acidobacteriota. In contrast, oil sludge promoted Pseudomonadota, Acidobacteriota, and Verrucomicrobiota, but suppressed Actinobacteriota, Bacteroidota, Chloroflexota, Firmicutes, and Gemmatimonadetes. The co-contaminated soil supported the highest diversity and contained the greatest number of unique genera, particularly Gammaproteobacteria, Actinobacteriota, Firmicutes and Bacteroidota.

The rhizosphere community of $M\times g$ under co-contamination harboured taxa with hydrocarbon-degrading, heavy metal–tolerant, and plant-growth-promoting traits. Notably, Sphingomonadaceae increased under all contamination scenarios, suggesting their involvement in oil sludge degradation, Zn tolerance, and adaptation to the $M\times g$ rhizosphere.

In total, 76 strains were isolated from $M\times g$ rhizosphere and screened, of which several exhibited combined metal resistance, hydrocarbon-degrading capacity, and plant growth–promoting activities such as nitrogen fixation, phosphorus mobilisation, siderophore production, and phytohormone synthesis.

Thus, $M\times g$–associated rhizosphere microorganisms represent promising candidates for the development of microbial bioagents to enhance plant growth and improve the efficiency of soil remediation in contaminated environments.

Supplementary Materials: The following supporting information can be downloaded at: https://www.mdpi.com/article/10.3390/agronomy15092232/s1. Figure S1: Relationship between the number of detected OTUs and sequencing depth in $M\times g$ rhizosphere soil samples; Table S1: Predominant taxonomic composition of rhizosphere bacterial communities of $M\times g$ grown in control and differently spiked soils; Table S2: Unique genus-level OTUs in the rhizobiome of $M\times g$ grown in control soil; Table S3: Unique genus-level OTUs in the rhizobiome of $M\times g$ grown in Zn-spiked soil; Table S4: Unique genus-level OTUs in the rhizobiome of $M\times g$ grown in oil sludge-spiked soil; Table S5: Unique genus-level OTUs in the rhizobiome of $M\times g$ grown in Zn- & oil sludge-spiked soil.

Author Contributions: Conceptualization, A.M. (Anna Muratova); methodology, I.S.; validation, E.B. and A.M. (Aigerim Mamirova); formal analysis, E.B., I.S., R.B. and A.M. (Anna Muratova); investigation, E.B. and I.S.; resources, A.N.; data curation, E.B., I.S., R.B. and A.M. (Anna Muratova); writing—original draft preparation, A.M. (Aigerim Mamirova) and A.M. (Anna Muratova); writing—review and editing, A.N., E.B., I.S., A.M. (Aigerim Mamirova), R.B. and A.M. (Anna Muratova); supervision, A.N.; project administration, A.N.; funding acquisition, A.N. All authors have read and agreed to the published version of the manuscript.

Funding: This research was funded by the Committee of Science, the Ministry of Science and Higher Education, the Republic of Kazakhstan, grant number AP23487419.

Data Availability Statement: The original contributions presented in the study are included in the article, further inquiries can be directed to the corresponding author/s.

Acknowledgments: We are very grateful to Sergey Golubev (IBPPM RAS) for his help in depositing raw reds in the NCBI Sequence Read Archive.

Conflicts of Interest: The authors declare no conflicts of interest.

Abbreviations

The following abbreviations are used in this manuscript:

ABTs	Alcohol-benzene tars
DW	Dry weight
DNA	Deoxyribonucleic acid
IAA	Indole-acetic acid
LB	Luria–Bertani medium
LSD	Least significant difference
M×g	*Miscanthus × giganteus*
MAGs	Metagenome-assembled genomes
MBCA	Mono- and bicyclic aromatic hydrocarbons
MPC	Maximum permissible concentration
NCBI	National Centre for Biotechnology Information
nMDS	Nonmetric multidimensional scaling
OTU	Operational taxonomic unit
PAHs	Polycyclic aromatic hydrocarbons
PCA	Principal component analysis
PCR	Polymerase chain reaction
rRNA	Ribosomal ribonucleic acids

References

1. Das, S.; Raj, R.; Mangwani, N.; Dash, H.R.; Chakraborty, J. 2—Heavy Metals and Hydrocarbons: Adverse Effects and Mechanism of Toxicity. In *Microbial Biodegradation and Bioremediation*; Das, S., Ed.; Elsevier: Oxford, UK, 2014; pp. 23–54; ISBN 978-0-12-800021-2.
2. Mukherjee, A.G.; Wanjari, U.R.; Eladl, M.A.; El-Sherbiny, M.; Elsherbini, D.M.A.; Sukumar, A.; Kannampuzha, S.; Ravichandran, M.; Renu, K.; Vellingiri, B.; et al. Mixed Contaminants: Occurrence, Interactions, Toxicity, Detection, and Remediation. *Molecules* **2022**, *27*, 2577. [CrossRef] [PubMed]
3. National Academies of Sciences, Engineering, and Medicine. Interactions of Soil Chemical Contaminants, Soil Health, and Human Health. In *Exploring Linkages Between Soil Health and Human Health*; National Academies Press (US): Washington, DC, USA, 2024.
4. Zhang, X.; Zhang, X.; Wang, S.; Zhao, S. Improved Remediation of Co-Contaminated Soils by Heavy Metals and PAHs with Biosurfactant-Enhanced Soil Washing. *Sci. Rep.* **2022**, *12*, 3801. [CrossRef]
5. Wu, J.; Jiang, J.; Xu, C.; Cai, Y.; Li, M.; Yang, Y.; Yang, G.; Meng, X.-Z.; Leib, J.; Zhangb, H.; et al. A Comprehensive Assessment of Heavy Metals, VOCs and Petroleum Hydrocarbon in Different Soil Layers and Groundwater at an Abandoned Al/Cu Industrial Site. *Ecotoxicol. Environ. Saf.* **2024**, *284*, 116927. [CrossRef] [PubMed]
6. Omo-Okoro, P.; Ofori, P.; Amalapridman, V.; Dadrasnia, A.; Abbey, L.; Emenike, C. Soil Pollution and Its Interrelation with Interfacial Chemistry. *Molecules* **2025**, *30*, 2636. [CrossRef] [PubMed]
7. Hou, D.; Jia, X.; Wang, L.; McGrath, S.P.; Zhu, Y.-G.; Hu, Q.; Zhao, F.-J.; Bank, M.S.; O'Connor, D.; Nriagu, J. Global Soil Pollution by Toxic Metals Threatens Agriculture and Human Health. *Science* **2025**, *388*, 316–321. [CrossRef]
8. Daâssi, D.; Qabil Almaghribi, F. Petroleum-Contaminated Soil: Environmental Occurrence and Remediation Strategies. *3 Biotech* **2022**, *12*, 139. [CrossRef]
9. Choden, D.; Pokethitiyook, P.; Poolpak, T.; Kruatrachue, M. Phytoremediation of Soil Co-Contaminated with Zinc and Crude Oil Using *Ocimum gratissimum* (L.) in Association with Pseudomonas Putida MU02. *Int. J. Phytoremediat.* **2021**, *23*, 181–189. [CrossRef]
10. Gorlenko, N.V.; Timofeeva, S.S. Assessment of Environmental Damage from Oil Sludge to Land Resources in the Irkutsk Region. *IOP Conf. Ser. Earth Environ. Sci.* **2019**, *229*, 012019. [CrossRef]
11. Sindireva, A.V.; Terentyev, G.R. Ecological Assessment of Heavy Metals and Petroleum Hydrocarbons Concentrations in Roadside Soils of Tyumen City. *Soc. Ecol. Technol.* **2023**, *13*, 57–76. [CrossRef]
12. EPA US. *Phytoremediation Resource Guide*; Solid Waste and Emergency Response (5102G); U.S. Environmental Protection Agency, Office of Solid Waste and Emergency Response, Technology Innovation Office: Washington, DC, USA, 1999; p. 56.

13. Gaskin, S.E.; Bentham, R.H. Rhizoremediation of Hydrocarbon Contaminated Soil Using Australian Native Grasses. *Sci. Total Environ.* **2010**, *408*, 3683–3688. [CrossRef]
14. Khan, S.; Afzal, M.; Iqbal, S.; Khan, Q.M. Plant–Bacteria Partnerships for the Remediation of Hydrocarbon Contaminated Soils. *Chemosphere* **2013**, *90*, 1317–1332. [CrossRef]
15. Awa, S.H.; Hadibarata, T. Removal of Heavy Metals in Contaminated Soil by Phytoremediation Mechanism: A Review. *Water Air Soil Pollut.* **2020**, *231*, 47. [CrossRef]
16. Nnaji, N.D.; Onyeaka, H.; Miri, T.; Ugwa, C. Bioaccumulation for Heavy Metal Removal: A Review. *SN Appl. Sci.* **2023**, *5*, 125. [CrossRef]
17. Yaqoob, S.; Zahoor, I.; Kazerooni, E.A. Phytostabilization of Contaminated Soils. *Int. J. Chem. Biochem. Sci. (IJCBS)* **2024**, *25*, 136–153.
18. Heaton, E.A.; Dohleman, F.G.; Miguez, A.F.; Juvik, J.A.; Lozovaya, V.; Widholm, J.; Zabotina, O.A.; McIsaac, G.F.; David, M.B.; Voigt, T.B.; et al. Chapter 3—Miscanthus: A Promising Biomass Crop. In *Advances in Botanical Research*; Kader, J.-C., Delseny, M., Eds.; Academic Press: Cambridge, MA, USA, 2010; Volume 56, pp. 75–137.
19. Arnoult, S.; Obeuf, A.; Béthencourt, L.; Mansard, M.-C.; Brancourt-Hulmel, M. Miscanthus Clones for Cellulosic Bioethanol Production: Relationships between Biomass Production, Biomass Production Components, and Biomass Chemical Composition. *Ind. Crops Prod.* **2015**, *63*, 316–328. [CrossRef]
20. Figala, J.; Vranová, V.; Rejšek, K.; Formánek, P. Giant Miscanthus (*Miscanthus* × *giganteus* Greef et Deu.)—A Promising Plant for Soil Remediation: A Mini Review. *Acta Univ. Agric. Silvic. Mendel. Brun.* **2016**, *63*, 2241–2246. [CrossRef]
21. Krzyżak, J.; Pogrzeba, M.; Rusinowski, S.; Clifton-Brown, J.; McCALMONT, J.P.; Kiesel, A.; Mangold, A.; Mos, M. Heavy Metal Uptake by Novel Miscanthus Seed-Based Hybrids Cultivated in Heavy Metal Contaminated Soil. *Civ. Environ. Eng. Rep.* **2017**, *26*, 121–132. [CrossRef]
22. Nebeská, D.; Trögl, J.; Pidlisnyuk, V.; Popelka, J.; Veronesi Dáňová, P.; Ust'ak, S.; Honzík, R. Effect of Growing *Miscanthus* × *giganteus* on Soil Microbial Communities in Post-Military Soil. *Sustainability* **2018**, *10*, 4021. [CrossRef]
23. Nurzhanova, A.; Pidlisnyuk, V.; Abit, K.; Nurzhanov, C.; Kenessov, B.; Stefanovska, T.; Erickson, L. Comparative Assessment of Using *Miscanthus* × *giganteus* for Remediation of Soils Contaminated by Heavy Metals: A Case of Military and Mining Sites. *Environ. Sci. Pollut. Res.* **2019**, *26*, 13320–13333. [CrossRef]
24. Pidlisnyuk, V.; Erickson, L.; Stefanovska, T.; Popelka, J.; Hettiarachchi, G.; Davis, L.; Trögl, J. Potential Phytomanagement of Military Polluted Sites and Biomass Production Using Biofuel Crop *Miscanthus* × *giganteus*. *Environ. Pollut.* **2019**, *249*, 330–347. [CrossRef] [PubMed]
25. Pidlisnyuk, V.; Herts, A.; Khomenchuk, V.; Mamirova, A.; Kononchuk, O.; Ust'ak, S. Dynamic of Morphological and Physiological Parameters and Variation of Soil Characteristics during *Miscanthus* × *giganteus* Cultivation in the Diesel-Contaminated Land. *Agronomy* **2021**, *11*, 798. [CrossRef]
26. Gopal, M.; Gupta, A. Microbiome Selection Could Spur Next-Generation Plant Breeding Strategies. *Front. Microbiol.* **2016**, *7*, 1971. [CrossRef] [PubMed]
27. Bakken, L. Culturable and Nonculturable Bacteria in Soil. In *Modern Soil Microbiology*; van Elsas, J.D., Trevors, J.T., Wellington, E.M.H., Eds.; Marcel Dekker Inc.: New York, NY, USA, 1997; pp. 47–61. ISBN 978-0-8247-9436-1.
28. Vieira, F.C.S.; Nahas, E. Comparison of Microbial Numbers in Soils by Using Various Culture Media and Temperatures. *Microbiol. Res.* **2005**, *160*, 197–202. [CrossRef]
29. Fan, D.; Schwinghamer, T.; Smith, D.L. Isolation and Diversity of Culturable Rhizobacteria Associated with Economically Important Crops and Uncultivated Plants in Québec, Canada. *Syst. Appl. Microbiol.* **2018**, *41*, 629–640. [CrossRef]
30. Xu, Y.; Zheng, C.; Liang, L.; Yi, Z.; Xue, S. Quantitative Assessment of the Potential for Soil Improvement by Planting Miscanthus on Saline-Alkaline Soil and the Underlying Microbial Mechanism. *GCB Bioenergy* **2021**, *13*, 1191–1205. [CrossRef]
31. Guo, Y.; Nishizawa, T.; Sakagami, N.; Fujimura, R.; Kamijo, T.; Ohta, H. Root Bacteriome of a Pioneer Grass *Miscanthus Condensatus* along Restored Vegetation on Recent Miyake-Jima Volcanic Deposits. *Rhizosphere* **2021**, *19*, 100422. [CrossRef]
32. Zhao, C.; Yue, Y.; Guo, Q.; Wu, J.; Song, J.; Wang, Q.; Li, C.; Hu, Y.; Wang, S.; Yuan, N.; et al. Metagenomic Insights into the Alteration of Soil N-Cycling-Related Microbiome and Functions under Long-Term Conversion of Cropland to Miscanthus. *GCB Bioenergy* **2023**, *15*, 1105–1118. [CrossRef]
33. Kane, J.L.; Schartiger, R.G.; Daniels, N.K.; Freedman, Z.B.; McDonald, L.M.; Skousen, J.G.; Morrissey, E.M. Bioenergy Crop *Miscanthus* × *giganteus* Acts as an Ecosystem Engineer to Increase Bacterial Diversity and Soil Organic Matter on Marginal Land. *Soil Biol. Biochem.* **2023**, *186*, 109178. [CrossRef]
34. Kane, J.L.; Liseski, K.B.; Dang, C.; Freedman, Z.B.; Morrissey, E.M. Trade or Scavenge? *Miscanthus*-Microbiome Interactions Depend upon Soil Fertility. *Appl. Soil Ecol.* **2024**, *196*, 105289. [CrossRef]
35. Chen, Z.-J.; Tian, W.; Li, Y.-J.; Sun, L.-N.; Chen, Y.; Zhang, H.; Li, Y.-Y.; Han, H. Responses of Rhizosphere Bacterial Communities, Their Functions and Their Network Interactions to Cd Stress under Phytostabilization by *Miscanthus* spp. *Environ. Pollut.* **2021**, *287*, 117663. [CrossRef]

36. Li, D.; Voigt, T.B.; Kent, A.D. Plant and Soil Effects on Bacterial Communities Associated with *Miscanthus* × *giganteus* Rhizosphere and Rhizomes. *GCB Bioenergy* **2016**, *8*, 183–193. [CrossRef]
37. Liu, Y.; Ludewig, U. Nitrogen-Dependent Bacterial Community Shifts in Root, Rhizome and Rhizosphere of Nutrient-Efficient *Miscanthus* × *giganteus* from Long-Term Field Trials. *GCB Bioenergy* **2019**, *11*, 1334–1347. [CrossRef]
38. Zhao, C.; Li, X.; Yue, Y.; Hou, X.; Guo, Q.; Song, J.; Li, C.; Zhang, W.; Wang, C.; Hou, Y.; et al. Cropland-to-Miscanthus Conversion Alters Soil Bacterial and Archaeal Communities Influencing N-Cycle in Northern China. *GCB Bioenergy* **2021**, *13*, 1528–1544. [CrossRef]
39. Ma, L.; Rocha, F.I.; Lee, J.; Choi, J.; Tejera, M.; Sooksa-Nguan, T.; Boersma, N.; VanLoocke, A.; Heaton, E.; Howe, A. The Impact of Stand Age and Fertilization on the Soil Microbiome of *Miscanthus* × *giganteus*. *Phytobiomes J.* **2021**, *5*, 51–59. [CrossRef]
40. Chen, Y.; Tian, W.; Shao, Y.; Li, Y.-J.; Lin, L.-A.; Zhang, Y.-J.; Han, H.; Chen, Z.-J. Miscanthus Cultivation Shapes Rhizosphere Microbial Community Structure and Function as Assessed by Illumina MiSeq Sequencing Combined with PICRUSt and FUNGUIld Analyses. *Arch. Microbiol.* **2020**, *202*, 1157–1171. [CrossRef]
41. Ji, N.; Liang, D.; Clark, L.V.; Sacks, E.J.; Kent, A.D. Host Genetic Variation Drives the Differentiation in the Ecological Role of the Native Miscanthus Root-Associated Microbiome. *Microbiome* **2023**, *11*, 216. [CrossRef]
42. Zhang, B.; Penton, C.R.; Xue, C.; Quensen, J.F.; Roley, S.S.; Guo, J.; Garoutte, A.; Zheng, T.; Tiedje, J.M. Soil Depth and Crop Determinants of Bacterial Communities under Ten Biofuel Cropping Systems. *Soil Biol. Biochem.* **2017**, *112*, 140–152. [CrossRef]
43. Bourgeois, E.; Dequiedt, S.; Lelièvre, M.; van Oort, F.; Lamy, I.; Ranjard, L.; Maron, P.A. Miscanthus Bioenergy Crop Stimulates Nutrient-Cycler Bacteria and Fungi in Wastewater-Contaminated Agricultural Soil. *Environ. Chem. Lett.* **2015**, *13*, 503–511. [CrossRef]
44. Pham, H.N.; Pham, P.A.; Nguyen, T.T.H.; Meiffren, G.; Brothier, E.; Lamy, I.; Michalet, S.; Dijoux-Franca, M.-G.; Nazaret, S. Influence of Metal Contamination in Soil on Metabolic Profiles of *Miscanthus* × *giganteus* Belowground Parts and Associated Bacterial Communities. *Appl. Soil Ecol.* **2018**, *125*, 240–249. [CrossRef]
45. Sauvêtre, A.; Węgrzyn, A.; Yang, L.; Vestergaard, G.; Miksch, K.; Schröder, P.; Radl, V. Enrichment of Endophytic Actinobacteria in Roots and Rhizomes of *Miscanthus* × *giganteus* Plants Exposed to Diclofenac and Sulfamethoxazole. *Environ. Sci. Pollut. Res.* **2020**, *27*, 11892–11904. [CrossRef] [PubMed]
46. Zadel, U.; Nesme, J.; Michalke, B.; Vestergaard, G.; Płaza, G.A.; Schröder, P.; Radl, V.; Schloter, M. Changes Induced by Heavy Metals in the Plant-Associated Microbiome of *Miscanthus* × *giganteus*. *Sci. Total Environ.* **2020**, *711*, 134433. [CrossRef]
47. Muratova, A.; Golubev, S.; Romanova, V.; Sungurtseva, I.; Nurzhanova, A. Effect of Heavy-Metal-Resistant PGPR Inoculants on Growth, Rhizosphere Microbiome and Remediation Potential of *Miscanthus* × *giganteus* in Zinc-Contaminated Soil. *Microorganisms* **2023**, *11*, 1516. [CrossRef]
48. Nebeská, D.; Trögl, J.; Ševců, A.; Špánek, R.; Marková, K.; Davis, L.; Burdová, H.; Pidlisnyuk, V. *Miscanthus* × *giganteus* Role in Phytodegradation and Changes in Bacterial Community of Soil Contaminated by Petroleum Industry. *Ecotoxicol. Environ. Saf.* **2021**, *224*, 112630. [CrossRef] [PubMed]
49. Lu, M.; Zhang, Z.-Z.; Wang, J.-X.; Zhang, M.; Xu, Y.-X.; Wu, X.-J. Interaction of Heavy Metals and Pyrene on Their Fates in Soil and Tall Fescue (*Festuca arundinacea*). *Environ. Sci. Technol.* **2014**, *48*, 1158–1165. [CrossRef] [PubMed]
50. Somtrakoon, K.; Chouychai, W. Removal of Phenanthrene and Cadmium from Co-Contaminated Alkaline Soil by Carpet Grass, Iam Weed and Winged Bean. *J. Environ. Biol.* **2018**, *39*, 558–564. [CrossRef]
51. Jeelani, N.; Yang, W.; Qiao, Y.; Li, J.; An, S.; Leng, X. Individual and Combined Effects of Cadmium and Polycyclic Aromatic Hydrocarbons on the Phytoremediation Potential of *Xanthium sibiricum* in Co-Contaminated Soil. *Int. J. Phytoremediat.* **2018**, *20*, 773–779. [CrossRef] [PubMed]
52. Lyubun, Y.; Muratova, A.; Dubrovskaya, E.; Sungurtseva, I.; Turkovskaya, O. Combined Effects of Cadmium and Oil Sludge on Sorghum: Growth, Physiology, and Contaminant Removal. *Environ. Sci. Pollut. Res.* **2020**, *27*, 22720–22734. [CrossRef]
53. Conlon, R.; Dowling, D.N.; Germaine, K.J. Assessing Microbial Activity and Rhizoremediation in Hydrocarbon and Heavy Metal-Impacted Soil. *Microorganisms* **2025**, *13*, 848. [CrossRef]
54. Odion, E.E.; Ma, J.; Brau, L.; Lim, J.W.; Lin, C. Interactive Effects of Heavy Metals and Soil Conditioners on Rhizodegradation of Long-Chain Petroleum Hydrocarbons in Soils. *Environ. Pollut.* **2025**, *384*, 126995. [CrossRef]
55. Zhang, Y.; Qian, F.; Bao, Y. Variations of Microbiota and Metabolites in Rhizosphere Soil of *Carmona microphylla* at the Co-Contaminated Site with Polycyclic Aromatic Hydrocarbons and Heavy Metals. *Ecotoxicol. Environ. Saf.* **2025**, *290*, 117734. [CrossRef]
56. Muratova, A.; Lyubun, Y.; Sungurtseva, I.; Turkovskaya, O.; Nurzhanova, A. Physiological and Biochemical Characteristic of *Miscanthus* × *giganteus* Grown in Heavy Metal—Oil Sludge Co-Contaminated Soil. *J. Environ. Sci.* **2022**, *115*, 114–125. [CrossRef] [PubMed]
57. Van, H.-T.; Hoang, V.H.; Nga, L.T.Q.; Nguyen, V.Q. Effects of Zn Pollution on Soil: Pollution Sources, Impacts and Solutions. *Results Surf. Interfaces* **2024**, *17*, 100360. [CrossRef]

58. Berg, G.; Köberl, M.; Rybakova, D.; Müller, H.; Grosch, R.; Smalla, K. Plant Microbial Diversity Is Suggested as the Key to Future Biocontrol and Health Trends. *FEMS Microbiol. Ecol.* **2017**, *93*, fix050. [CrossRef]
59. Chauhan, P.; Sharma, N.; Tapwal, A.; Kumar, A.; Verma, G.S.; Meena, M.; Seth, C.S.; Swapnil, P. Soil Microbiome: Diversity, Benefits and Interactions with Plants. *Sustainability* **2023**, *15*, 14643. [CrossRef]
60. FAO. *World Reference Base for Soil Resources 2014: International Soil Classification Systems for Naming Soils and Creating Legends for Soil Maps (Update 2015)*; World Soil Resources: Rome, Italy, 2014.
61. *GOST 26483-85*; Soils. Preparation of Salt Extract and Determination of Its pH by CINAO Method. CINAO: Minsk, Belarus, 1986.
62. *GOST 26207-91*; Soils. Determination of Mobile Compounds of Phosphorus and Potassium by Kirsanov Method Modified by CINAO. StandardInform: Moscow, Russia, 1993.
63. *GOST 26488-85*; Soils. Determination of Nitrates by CINAO Methods. Russian Institute of Standardization: Moscow, Russia, 2021.
64. *GOST 26489-85*; Soils. Determination of Exchangeable Ammonium by CINAO Method. Russian Institute of Standardization: Moscow, Russia, 2021.
65. *GN 2.1.7.2511-09*; Tentative Allowable Concentrations (TAC) of Chemicals in Soils: Hygienic Standards. Russian Federal Service on Customers' Rights Protection and Human Well-being Surveillance: Moscow, Russia, 2009.
66. Golubev, S.N.; Schelud'ko, A.V.; Muratova, A.Y.U.; Makarov, O.E.; Turkovskaya, O.V. Assessing the Potential of Rhizobacteria to Survive under Phenanthrene Pollution. *Water Air Soil Pollut.* **2009**, *198*, 5–16. [CrossRef]
67. Muratova, A.; Hübner, T.; Narula, N.; Wand, H.; Turkovskaya, O.; Kuschk, P.; Jahn, R.; Merbach, W. Rhizosphere Microflora of Plants Used for the Phytoremediation of Bitumen-Contaminated Soil. *Microbiol. Res.* **2003**, *158*, 151–161. [CrossRef] [PubMed]
68. Belimov, A.A.; Hontzeas, N.; Safronova, V.I.; Demchinskaya, S.V.; Piluzza, G.; Bullitta, S.; Glick, B.R. Cadmium-Tolerant Plant Growth-Promoting Bacteria Associated with the Roots of Indian Mustard (*Brassica juncea* L. Czern.). *Soil Biol. Biochem.* **2005**, *37*, 241–250. [CrossRef]
69. Muromtsev, G.S. Some Methods for Studying the Dissolution of Calcium Phosphates by Microorganisms. *Microbiologiya* **1957**, *26*, 172–178.
70. Kiyohara, H.; Nagao, K.; Yana, K. Rapid Screen for Bacteria Degrading Water-Insoluble, Solid Hydrocarbons on Agar Plates. *Appl. Environ. Microbiol.* **1982**, *43*, 454–457. [CrossRef] [PubMed]
71. Bolyen, E.; Rideout, J.R.; Dillon, M.R.; Bokulich, N.A.; Abnet, C.C.; Al-Ghalith, G.A.; Alexander, H.; Alm, E.J.; Arumugam, M.; Asnicar, F.; et al. Author Correction: Reproducible, Interactive, Scalable and Extensible Microbiome Data Science Using QIIME 2. *Nat. Biotechnol.* **2019**, *37*, 1091. [CrossRef]
72. Callahan, B.J.; McMurdie, P.J.; Rosen, M.J.; Han, A.W.; Johnson, A.J.A.; Holmes, S.P. DADA2: High-Resolution Sample Inference from Illumina Amplicon Data. *Nat. Methods* **2016**, *13*, 581–583. [CrossRef]
73. Yilmaz, P.; Parfrey, L.W.; Yarza, P.; Gerken, J.; Pruesse, E.; Quast, C.; Schweer, T.; Peplies, J.; Ludwig, W.; Glöckner, F.O. The SILVA and "All-Species Living Tree Project (LTP)" Taxonomic Frameworks. *Nucleic Acids Res.* **2014**, *42*, D643–D648. [CrossRef]
74. Lewin, E.; Clifton-Brown, J.; Jensen, E.; Lewandowski, I.; Krzyżak, J.; Pogrzeba, M.; Hartung, J.; Wolfmüller, C.; Kiesel, A. Testing Agronomic Treatments to Improve the Establishment of Novel Miscanthus Hybrids on Marginal Land. *Agronomy* **2025**, *15*, 1297. [CrossRef]
75. Aroca, R.; Ruiz-Lozano, J.M. Induction of Plant Tolerance to Semi-Arid Environments by Beneficial Soil Microorganisms—A Review. In *Climate Change, Intercropping, Pest Control and Beneficial Microorganisms*; Lichtfouse, E., Ed.; Springer Netherlands: Dordrecht, The Netherlands, 2009; pp. 121–135. ISBN 978-90-481-2716-0.
76. Gureeva, M.V.; Gureev, A.P. Molecular Mechanisms Determining the Role of Bacteria from the Genus Azospirillum in Plant Adaptation to Damaging Environmental Factors. *Int. J. Mol. Sci.* **2023**, *24*, 9122. [CrossRef] [PubMed]
77. Khan, A.; Rao, T.S. Chapter 36—Molecular Evolution of Xenobiotic Degrading Genes and Mobile DNA Elements in Soil Bacteria. In *Microbial Diversity in the Genomic Era*; Das, S., Dash, H.R., Eds.; Academic Press: Cambridge, MA, USA, 2019; pp. 657–678; ISBN 978-0-12-814849-5.
78. Staninska-Pięta, J.; Czarny, J.; Piotrowska-Cyplik, A.; Juzwa, W.; Wolko, Ł.; Nowak, J.; Cyplik, P. Heavy Metals as a Factor Increasing the Functional Genetic Potential of Bacterial Community for Polycyclic Aromatic Hydrocarbon Biodegradation. *Molecules* **2020**, *25*, 319. [CrossRef]
79. Piacenza, E.; Campora, S.; Carfì Pavia, F.; Chillura Martino, D.F.; Laudicina, V.A.; Alduina, R.; Turner, R.J.; Zannoni, D.; Presentato, A. Tolerance, Adaptation, and Cell Response Elicited by *Micromonospora* Sp. Facing Tellurite Toxicity: A Biological and Physical-Chemical Characterization. *Int. J. Mol. Sci.* **2022**, *23*, 12631. [CrossRef]
80. Barra Caracciolo, A.; Terenzi, V. Rhizosphere Microbial Communities and Heavy Metals. *Microorganisms* **2021**, *9*, 1462. [CrossRef]
81. Bilal, S.; Khan, A.L.; Shahzad, R.; Kim, Y.-H.; Imran, M.; Khan, M.J.; Al-Harrasi, A.; Kim, T.H.; Lee, I.-J. Mechanisms of Cr(VI) Resistance by Endophytic *Sphingomonas* Sp. LK11 and Its Cr(VI) Phytotoxic Mitigating Effects in Soybean (*Glycine max* L.). *Ecotoxicol. Environ. Saf.* **2018**, *164*, 648–658. [CrossRef]

82. Oubohssaine, M.; Sbabou, L.; Aurag, J. Native Heavy Metal-Tolerant Plant Growth Promoting Rhizobacteria Improves *Sulla spinosissima* (L.) Growth in Post-Mining Contaminated Soils. *Microorganisms* **2022**, *10*, 838. [CrossRef] [PubMed]
83. Shi, A.; Hu, Y.; Zhang, X.; Zhou, D.; Xu, J.; Rensing, C.; Zhang, L.; Xing, S.; Ni, W.; Yang, W. Biochar Loaded with Bacteria Enhanced Cd/Zn Phytoextraction by Facilitating Plant Growth and Shaping Rhizospheric Microbial Community. *Environ. Pollut.* **2023**, *327*, 121559. [CrossRef] [PubMed]
84. Ortúzar, M.; Trujillo, M.E.; Román-Ponce, B.; Carro, L. *Micromonospora* Metallophores: A Plant Growth Promotion Trait Useful for Bacterial-Assisted Phytoremediation? *Sci. Total Environ.* **2020**, *739*, 139850. [CrossRef]
85. Chaudhary, D.K.; Kim, J. *Sphingomonas olei* Sp. Nov., with the Ability to Degrade Aliphatic Hydrocarbons, Isolated from Oil-Contaminated Soil. *Int. J. Syst. Evol. Microbiol.* **2017**, *67*, 2731–2738. [CrossRef] [PubMed]
86. Asaf, S.; Numan, M.; Khan, A.L.; Al-Harrasi, A. Sphingomonas: From Diversity and Genomics to Functional Role in Environmental Remediation and Plant Growth. *Crit. Rev. Biotechnol.* **2020**, *40*, 138–152. [CrossRef]
87. Cerniglia, C.E. Recent Advances in the Biodegradation of Polycyclic Aromatic Hydrocarbons by Mycobacterium Species. In *The Utilization of Bioremediation to Reduce Soil Contamination: Problems and Solutions*; Šašek, V., Glaser, J.A., Baveye, P., Eds.; Springer: Dordrecht, The Netherlands, 2003; pp. 51–73. ISBN 978-94-010-0131-1.
88. Deng, Y.; Mou, T.; Wang, J.; Su, J.; Yan, Y.; Zhang, Y.-Q. Characterization of Three Rapidly Growing Novel Mycobacterium Species with Significant Polycyclic Aromatic Hydrocarbon Bioremediation Potential. *Front. Microbiol.* **2023**, *14*, 1225746. [CrossRef]
89. Antoun, H.; Prévost, D. Ecology of Plant Growth Promoting Rhizobacteria. In *PGPR: Biocontrol and Biofertilization*; Siddiqui, Z.A., Ed.; Springer: Dordrecht, The Netherlands, 2006; pp. 1–38. ISBN 978-1-4020-4152-5.
90. Mohanram, S.; Kumar, P. Rhizosphere Microbiome: Revisiting the Synergy of Plant-Microbe Interactions. *Ann. Microbiol.* **2019**, *69*, 307–320. [CrossRef]
91. Babu, A.G.; Kim, J.-D.; Oh, B.-T. Enhancement of Heavy Metal Phytoremediation by *Alnus firma* with Endophytic *Bacillus thuringiensis* GDB-1. *J. Hazard. Mater.* **2013**, *250–251*, 477–483. [CrossRef] [PubMed]
92. Chen, B.; Zhang, Y.; Rafiq, M.T.; Khan, K.Y.; Pan, F.; Yang, X.; Feng, Y. Improvement of Cadmium Uptake and Accumulation in *Sedum alfredii* by Endophytic Bacteria *Sphingomonas SaMR12*: Effects on Plant Growth and Root Exudates. *Chemosphere* **2014**, *117*, 367–373. [CrossRef]
93. Hao, X.; Xie, P.; Zhu, Y.-G.; Taghavi, S.; Wei, G.; Rensing, C. Copper Tolerance Mechanisms of *Mesorhizobium amorphae* and Its Role in Aiding Phytostabilization by *Robinia pseudoacacia* in Copper Contaminated Soil. *Environ. Sci. Technol.* **2015**, *49*, 2328–2340. [CrossRef] [PubMed]

agronomy

MDPI

Article

Genomic Analysis of Cadmium-Resistant and Plant Growth-Promoting *Burkholderia alba* Isolated from Plant Rhizosphere

Luyao Feng [1,2,3,†], Xin Liu [1,2,†], Nan Wang [2], Zhuli Shi [1,2], Yu Wang [1,2], Jianpeng Jia [1,2], Zhufeng Shi [2], Te Pu [2] and Peiwen Yang [2,*]

1 School of Agriculture, Yunnan University, Kunming 650500, China; huan2004212@163.com (L.F.); liuxin7@stu.ynu.edu.cn (X.L.); 15969576820@139.com (Z.S.); yuwang@stu.ynu.edu.cn (Y.W.); jiajianpeng0715@163.com (J.J.)
2 Institute of Agricultural Environment and Resources, Yunnan Academy of Agricultural Sciences, Kunming 650204, China; 2024110087@stu.ynau.edu.cn (N.W.); shizhfe@163.com (Z.S.); pt806232385@outlook.com (T.P.)
3 Chongqing Three Gorges Academy of Agricultural Sciences, Chongqing 404155, China
* Correspondence: pwyang2000@126.com
† These authors contributed equally to this work.

Abstract: Reducing the application of chemical fertilizers and remediating heavy metal pollution in soil are important directions in current agricultural research. Utilizing the plant-growth-promoting and remediation capabilities of bacteria can provide more environmentally friendly assistance to agricultural production. In this study, the *Burkholderia alba* YIM B08401 strain was isolated and identified from rhizospheric soil, subjected to whole-genome sequencing and analysis, and its Cd^{2+} adsorption efficiency and characteristics were confirmed using multiple experimental methods, including atomic absorption spectrometry (AAS), Fourier transform infrared spectroscopy (FTIR), and scanning electron microscopy with energy-dispersive X-ray spectroscopy (SEM-EDS). The results showed that the genome of strain YIM B08401 has a total length of 7,322,157 bp, a GC content of 66.39%, and predicts 6504 protein-coding sequences. It contains abundant functional genes related to nutrient conversion (phosphate solubilization, sulfur metabolism, zinc solubilization, siderophore production), plant hormone regulation (indole-3-acetic acid secretion, ACC deaminase production), phenolic acid degradation, root colonization, heavy metal tolerance, pathogen antagonism, and the production of antagonistic secondary metabolites. Additionally, strain YIM B08401 can specifically bind to Cd^{2+} through various functional groups on the cell surface, such as C-O-C, P=O, and O-H, enabling biosorption. In conclusion, strain YIM B08401 is an excellent strain with plant-growth-promoting, disease-resistant, and bioremediation capabilities, warranting further development as a biofertilizer for agricultural applications to promote green and sustainable agricultural development.

Keywords: *Burkholderia alba*; genome analysis; plant-growth promotion; heavy metals; bioremediation

1. Introduction

With the rapid advancement of intensive modern agricultural production models, soil ecosystems are facing unprecedented multiple stresses. Excessive application of chemical fertilizers has increasingly exacerbated soil acidification and nutrient imbalance [1], while continuous cropping has led to phenolic acid accumulation and cropping obstacles, significantly reducing arable land productivity [2,3]. Industrial activities have also caused heavy metal pollution, such as cadmium (Cd) with high bioavailability, posing severe threats to agricultural product safety, agricultural productivity, and human health [4–6]. Traditional chemical remediation methods (e.g., lime amendment, chemical passivators), although capable of short-term pollution mitigation, often suffer from high costs and risks of secondary contamination [7,8]. Against this backdrop, rhizospheric microorganisms have emerged as a research hotspot in soil remediation due to their environmental friendliness and multifunctionality. As core mediators of "plant-soil" interactions, rhizospheric microorganisms enhance soil fertility and reduce heavy metal bioavailability through diverse mechanisms, including nitrogen fixation, phosphate solubilization, plant hormone synthesis, and heavy metal passivation [9–11]. These characteristics enable rhizospheric microorganisms to not only boost crop yields but also act as "biological filters" in contaminated soil remediation, achieving the dual goals of agricultural production and ecological security.

The *Burkholderia* represents a group of prominent plant-growth-promoting rhizobacteria (PGPR) with remarkable genetic and metabolic diversity. Certain strains within this genus exhibit multifunctional capabilities, including phosphate solubilization, nitrogen fixation, indole-3-acetic acid (IAA) production, pathogen antagonism, and heavy metal adsorption, achieved through mechanisms such as quorum sensing, nutrient conversion, and secondary metabolite synthesis [12–14]. These attributes enable *Burkholderia* strains to facilitate plant growth and enhance tolerance to abiotic stresses. For instance, *Burkholderia* sp. J62 has been shown to both absorb lead (Pb) and Cd from contaminated soils and significantly promote plant growth [15]. *Burkholderia* sp. LD-11 demonstrates dual functionality in copper- or lead-contaminated soils, combining heavy metal adsorption with the secretion of IAA, 1-aminocyclopropane-1-carboxylic acid (ACC) deaminase, and siderophores. This strain not only increases plant dry weight in polluted soils, but also enhances soil urease activity and rhizobacterial community diversity [16]. Additionally, *Burkholderia* sp. Y4 can colonize rice root surfaces, enriching Cd ions through direct biosorption mediated by -C=O and -NH/-C=O groups in amino acids and proteins. Concurrently, it promotes the accumulation of essential elements (e.g., Fe, Mn) in rice while reducing Cd content [14].

However, the precise mining of rhizomicrobial functional potential still faces challenges. On one hand, the interaction mechanisms between microorganisms and plants are complex, involving multi-dimensional networks such as signal recognition, metabolic pathway regulation, and environmental adaptability. On the other hand, traditional phenotypic screening methods are insufficient to comprehensively decode functional genes and metabolic pathways, limiting the directional optimization and application of strains [17–19]. In recent years, the rapid development of genomics, metabolomics, and synthetic biology has provided new insights for addressing this challenge. Through whole-genome sequencing and functional annotation, it is possible to systematically uncover gene clusters related to plant growth promotion, stress resistance, and biocontrol, screen key functional modules, and evaluate their application potential, thereby laying a foundation for the development of efficient multifunctional microbial inoculants [20–23].

This study focuses on *Burkholderia alba* YIM B08401, a strain previously isolated from the rhizospheric soil of *Pinus yunnanensis* in Wuliang Mountain. Preliminary experiments have confirmed its multifunctional capabilities, including phosphate solubilization, nitrogen fixation, siderophore secretion, ACC deaminase production, and pathogen antagonism [24]. To deeply dissect its molecular mechanisms, whole-genome sequencing was performed, followed by comprehensive functional annotation using multi-database integration (COG (Clusters of Orthologous Groups of proteins), KEGG (Kyoto Encyclopedia of Genes and Genomes), antiSMASH (Antibiotics and Secondary Metabolites Analysis

Shell)) to systematically identify gene clusters associated with plant growth-promoting functions. Meanwhile, phenotypic assays were conducted to validate the expression and reliability of selected functional genes, ensuring their practical applicability. The findings enrich the rhizomicrobial resource library with a high-value strain and provide a theoretical foundation for designing "microbe-plant" synergistic remediation systems and developing stress-resistant bioinoculants for yield improvement. Additionally, this research contributes to driving the sustainable transition of agriculture from "chemical dependency" to "biological-driven" models.

2. Materials and Methods

2.1. DNA Extraction, Identification, and Sequencing of Strain YIM B08401

The strain YIM B08401 was inoculated into NB medium and cultured at 30 °C with shaking at 180 r/min for 24 h. The bacterial cells were harvested by centrifugation of the culture at 10,000 rpm for 15 min. Genomic DNA was extracted using a bacterial genome DNA extraction kit following the manufacturer's protocol (Thermo Fisher Scientific, Waltham, MA, USA). The DNA samples were then fragmented into approximately 400 bp segments using a Covaris M220 Focused-ultrasonicator (Thermo, Waltham, MA, USA) to construct the library. Paired-end (PE150) sequencing of the prepared library was performed on an Illumina sequencing platform (MajorBio, Shanghai, China).

2.2. Genome Assembly and Annotation of Strain YIM B08401

Genome assembly and annotation were performed following procedures described previously [25]. The optimized sequencing reads were assembled using SOAPdenovo v2.04. Plasmid identification in the bacterial genome assembly was conducted using PLASMe v1.1. Plasmid sequences were further annotated via BLAST+ v2.3.0 and the PLSDB Database v202106. For coding sequence (CDS) prediction, Glimmer v3.02, GeneMarkS v4.3, and Prodigal v2.6.3 were employed. tRNA and rRNA were predicted using tRNAscan-SE v2.0 and Barrnap v0.9, respectively. The predicted CDS were functionally annotated against databases including NR (Non-Redundant Protein Database), Swiss-Prot (Swiss-Prot Protein Knowledgebase), Pfam (Protein Families Database), COG, GO (Gene Ontology), and KEGG. Additionally, carbohydrate-active enzymes (CAZymes) were predicted using the Carbohydrate-Active Enzymes database (CAZy Database v8), and biosynthetic gene clusters (BGCs) responsible for secondary metabolite production were identified via AntiSMASH v7.1.0.

2.3. Experiment on Root Colonization Ability of Strain YIM B08401

Tomato seeds with no adhesion, plump particles, and uniform size were selected. The seeds were soaked in room-temperature water for 12 h, then disinfected in a sterile workbench with 75% ethanol for 5 min, followed by immersion in 15% sodium hypochlorite for 15 min. After disinfection, the seeds were transferred to a sterilized 50 mL centrifuge tube, rinsed with sterile water, shaken for 5 s, and the supernatant was discarded. This rinsing procedure was repeated three times before replacing it with a new centrifuge tube, and the rinsing steps were continued. After a total of five centrifuge tube replacements, the seeds were spread on sterile filter paper, transferred to Murashige and Skoog (MS) Medium using sterile tweezers, and 50 seeds were evenly distributed per bottle of medium. The cultures were placed in a light incubator with a 16 h light/8 h dark cycle at 26 °C, and aseptic seedlings were obtained after 5–6 days [26]. The strain YIM B08401 was inoculated into Luria-Bertani (LB) liquid medium and cultured with constant shaking at a constant temperature for 24 h. After centrifugation, the bacterial cells were resuspended in sterile ddH_2O to prepare a bacterial suspension of strain YIM B08401 with an OD_{600} of 0.5. For inoculation, 100 mL of this bacterial suspension was used to immerse the roots of aseptic tomato seedlings for 2 h, with 10 replicates performed. Subsequently, the inoculated tomato seedlings were transferred to fresh MS medium for static culture for 2 days. Prior to observation, the roots were rinsed with sterile ddH_2O, and at least 20 root tips were examined under a scanning electron microscope (SEM) to assess the colonization of strain YIM B08401 [27].

2.4. *Qualitative Assays for Enzyme Production by Strain YIM B08401*

Chitinase screening medium formulation: 20 mL of 1% colloidal chitin solution, 3 g yeast extract, 3 g peptone, 0.5 g $MgSO_4 \cdot 7H_2O$, 0.7 g KH_2PO_4, 0.3 g K_2HPO_4, 0.01 g $FeSO_4 \cdot 7H_2O$, 1000 mL distilled water, 20 g agar powder, pH 7. Preparation of colloidal chitin: 2 g of chitin powder was dissolved in 45 mL pre-cooled concentrated hydrochloric acid, followed by standing at 4 °C for 24 h. Then 300 mL of 50% ethanol was added to obtain a flocculent mixture. The mixture was centrifuged at 4000 rpm for 20 min to collect the precipitate, which was washed with distilled water until neutral and resuspended in 200 mL of water to prepare 1% colloidal chitin. Strain YIM B08401 was activated and inoculated onto the chitin medium; the formation of clear zones around colonies indicated its ability to produce chitinase [28]. Xylanase screening medium formulation: 2 g xylan, 1 g KH_2PO_4, 2 g $(NH_4)_2SO_4$, 0.5 g $MgSO_4 \cdot 7H_2O$, 0.5 g NaCl, 18 g agar, 1000 mL distilled water, natural pH. Preparation of Congo red solution (1 mg/mL): 1 g of Congo red was weighed and dissolved in 1000 mL of distilled water, and the solution was stored at 4 °C for later use. Activated strain YIM B08401 was inoculated onto the xylan screening medium and cultured at 30 °C for 3 days. The plate was covered with Congo red solution and left to stand for 15 min, then rinsed with 1 mol/L sodium chloride solution for 15 min. The presence of clear zones around colonies was used to determine the strain's ability to secrete xylanase [29].

2.5. *Cd Adsorption Assays of Strain YIM B08401*

2.5.1. Cd^{2+} Adsorption by Strain YIM B08401

Strain YIM B08401 was inoculated into sterile NB medium and cultured at 30 °C with shaking at 180 rpm until the optical density at 600 nm (OD_{600}) reached 0.6, serving as the seed solution. The seed solution was inoculated at a 1% inoculum ratio into NB liquid media containing 0 mg/L, 100 mg/L, and 200 mg/L Cd^{2+}, with uninoculated medium as the blank control. Cultures were incubated at 30 °C and 180 rpm for 2 days, with three replicates per treatment. After cultivation, the culture broth was centrifuged at 4500 rpm for 15 min. The supernatant was collected, filtered through a 0.22 μm membrane, and the Cd^{2+} concentration was measured using atomic absorption spectroscopy (AAS) to calculate the strain's adsorption efficiency. Meanwhile, the centrifuged bacterial cells were washed three times with distilled water via centrifugation at 4500 rpm for 15 min, followed by discarding the supernatant. The cell pellets were frozen at −80 °C for at least 5 h and then lyophilized using a freeze-dryer to obtain lyophilized biomass. The experiment was conducted from November 2023 to July 2024.

2.5.2. Fourier Transform Infrared Spectroscopy (FTIR) Analysis

Lyophilized bacterial cells from different Cd^{2+} concentrations were mixed with dry powdered potassium bromide (KBr) at a 1:100 ratio. The mixture was uniformly spread into a suitable disc, pressed at 800 MPa using a Fourier transform infrared (FTIR) spectrometer, and spectral data were recorded. The FTIR peaks were scanned over a wavenumber range of 4000 to 400 cm^{-1}. The sample treated with 0 mg/L Cd^{2+} (lyophilized cells + KBr) served as the reference, and the treated samples were compared based on peaks indicating various bond stretching and bending vibrations [30].

2.5.3. SEM-EDS Analysis

Bacterial cells and sediments were collected from the cultures described in Section 2.5.1. (after incubation with different Cd^{2+} concentrations), washed three times with phosphate buffer solution, and then fixed with 2.5% (v/v) glutaraldehyde solution for 12 h. The samples were then dehydrated using an ethanol gradient series for purification, freeze-dried into powder, and subjected to scanning electron microscopy with energy-dispersive X-ray spectroscopy (SEM-EDS) to analyze morphological and elemental composition changes before and after bacterial adsorption [31].

3. Results and Discussion

3.1. General Genome Characteristics and Strain Identification of YIM B08401

Strain YIM B08401 was isolated from the rhizosphere soil of *Pinus armandii* Franch (100°49′25″ N, 24°27′00″ E) in Wuliangshan Mountain International Nature Reserve, Yunnan Province, China. Strain YIM B08401 showed the highest 16S rRNA gene sequence similarity of 100.00% with the closest type strain *Burkholderia alba* AD18^T. Furthermore, it also showed high 16S rRNA similarities with *B. oklahomensis* C6786^T (98.76%), *B. rinojensis* A396^T (98.69%) and *B. plantarii* ATCC 43733^T (98.63%) by Ez-biocloud. The NJ evolutionary tree based on 16S rRNA was constructed using MEGA 7.0, and the ML and MP evolutionary tree was used for auxiliary identification. Combined with physiological and biochemical results, strain YIM B08401 was identified as *B. alba* (Figure 1). However, *B. alba* AD18^T had been effectively published, and the taxonomic name has not been verified.

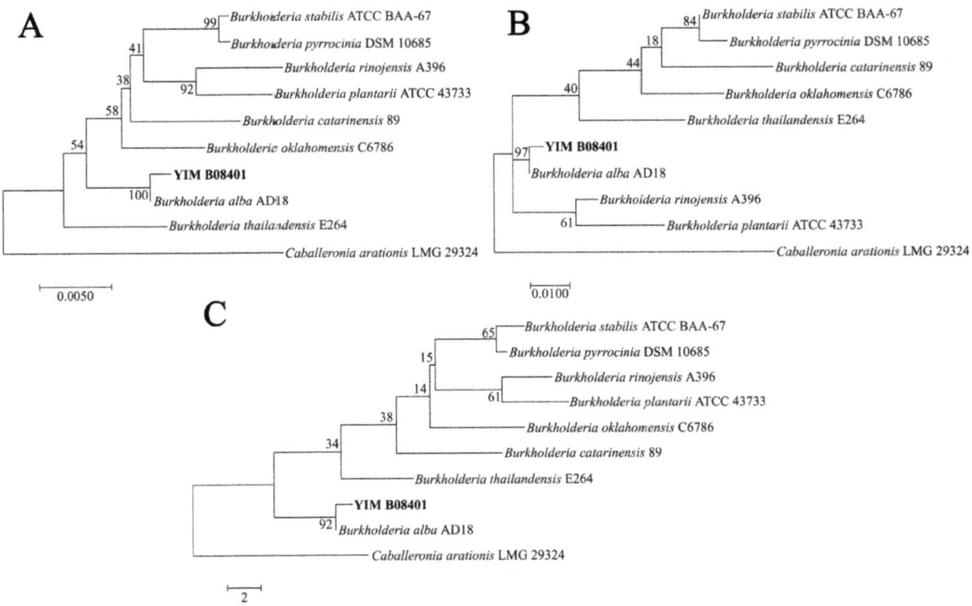

Figure 1. Phylogenetic trees of strain YIM B08401 based on 16S rDNA gene sequences. (**A**) Neighbor-joining (NJ) method; (**B**) maximum likelihood (ML) method; (**C**) maximum parsimony (MP) method.

The genome of YIM B08401 contained 204 scaffolds with the total length of 7,322,157 bp and GC content of 66.39%. A total of 6504 protein-coding sequences (CDSs) were annotated, and 3 rRNAs and 63 tRNA were also predicted. In addition, 15 gene islands and 2 prophage regions were predicted by the IslandViewer and Phage Finder predictions of the strains. The general genomic characteristics of strain YIM B08401 are shown in Table 1. The round genome representing the entire genome assembly is shown in Figure 2.

Table 1. The genome characterization of YIM B08401.

The Characterization	Value
Assembly information	
The total length (bp)	7,322,157
Total scaffold No	204
Scaffold N50(bp)	201,844
Contigs	193
Conting N50 (bp)	143,587
G + C content (%)	66.39
Annotation information	
CDSs (total)	6504
rRNA	3
tRNA	63
Mobile Genetic Element (MGE)	
Gene island	15
Prophage regions	2
CRISPR-Cas	22
Insertion sequence	9
Tns	62
NCBI	
BioProject ID	PRJNA1027102
BioSample ID	SAMN37776913
GeneBank accession number	JAWJWL000000000

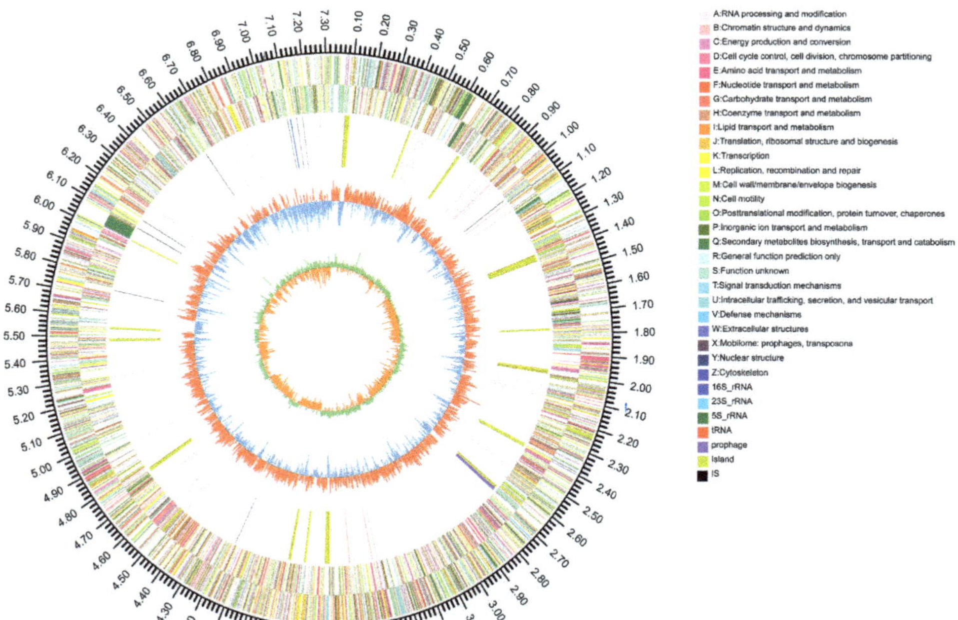

Figure 2. Circos genome map of strain YIM B08401. Note: The outermost circle indicates the genome size. The second and third circles represent the coding sequences (CDS) on the positive and negative strands, respectively, with different colors corresponding to different COG (Clusters of Orthologous Groups) functional categories of the CDS. The fourth circle shows rRNA and tRNA. The fifth circle denotes the GC content. The innermost circle represents the GC-Skew values.

3.2. Genome Functional Annotation

The genome of strain YIM B08401 was functionally annotated using the NR, Swiss-Prot, Pfam, COG, GO and KEGG databases. A total of 6504 genes were annotated, with 6284 genes annotated in the NR database, 4592 in Swiss-Prot, 5358 in Pfam, 5353 in COG, 3990 in GO, and 3196 in KEGG.

Annotation with the COG database identified 5353 genes in strain YIM B08401, representing 82.30% of all annotated genes (Figure 3A). These genes were hierarchically classified into 4 major categories and 22 subcategories [32]. Given that some genes were assigned to multiple categories, genes related to "Information Storage and Processing", "Cellular Processes and Signaling", "Metabolism", and "Poorly Characterized" accounted for 17.60%, 17.78%, 36.69%, and 27.93% of the total COG annotations, respectively. These results highlight the metabolic activity of strain YIM B08401, indicating its capacity to support bacterial growth by facilitating material synthesis, energy conversion, and metabolite production. Among the 22 subcategories, excluding genes with unknown functions (27.93%), genes involved in transcription (8.91%), amino acid transport and metabolism (8.83%), inorganic ion transport and metabolism (5.83%), cell wall/membrane/envelope biogenesis (5.81%), and carbohydrate transport and metabolism (5.54%) were highly abundant. Notably, the high proportions of genes involved in amino acid transport and metabolism, as well as inorganic ion transport and metabolism, suggest that strain YIM B08401 may possess inherent capabilities to compete with other microorganisms [33,34]. Furthermore, previous studies have demonstrated that in plant growth-promoting bacteria such as *Bacillus velezensis* SH-1471 and *Pseudomonas qingdaonensis* ZCR6, genes related to transcription and amino acid transport/metabolism also constitute the most abundant gene groups [11,26]. Collectively, these characteristics indicate that strain YIM B08401 has potential advantages in functional expression, environmental survival, and competitive nutrient acquisition.

The KEGG database annotated 3196 genes across six categories: metabolism, genetic information processing, environmental information processing, human diseases, organismal systems, and cellular processes [35], representing 49.14% of all annotated genes (Figure 3B). Despite overlapping annotations, the "Metabolism" category (2732 genes, 65.74%) harbored the largest gene set, consistent with COG annotation results. Among the six categories, excluding 1074 genes assigned to "Global and overview maps", genes associated with membrane transport (345 genes, 8.30%), amino acid metabolism (344 genes, 8.28%), and carbohydrate metabolism (313 genes, 7.53%) were the most prevalent. These findings align with GO and COG annotations, further validating the strain's high abundance of genes related to energy transport and biosynthesis.

Using the GO database, 3990 genes were annotated in strain YIM B08401, accounting for 61.35% of the total annotated genes (Figure 3C). The database classified all annotated genes into three main categories, with partial genes assigned to multiple categories: "Cellular Component" (1682 genes, 24.57%), "Molecular Function" (3349 genes, 48.92%), and "Biological Process" (1815 genes, 26.51%) [36]. Notably, nearly half of the annotated genes fell under "Molecular Function", indicating the strain's rich functional expression and strong potential for growth promotion. Among all annotated subcategories, the most abundant were "integral component of membrane" (14.78%), "DNA binding" (7.21%), "ATP binding" (5.57%), "cytoplasm" (4.78%), "plasma membrane" (4.5%), and "DNA-binding transcription factor activity" (4.47%). A comparison with known plant growth-promoting rhizobacteria revealed that strain YIM B08401 exhibits significantly higher abundances of genes associated with integral components of membrane, ATP binding, and DNA-binding transcription factor activity compared to strains such as *Bacillus velezensis* SH-1471 and *Bacillus subtilis* MG-1 [26,37]. These characteristics further confirm that strain YIM B08401 possesses remarkable functional versatility and metabolic activity, with distinct advantages particularly in membrane-related functions, molecular interactions, and transcriptional regulation.

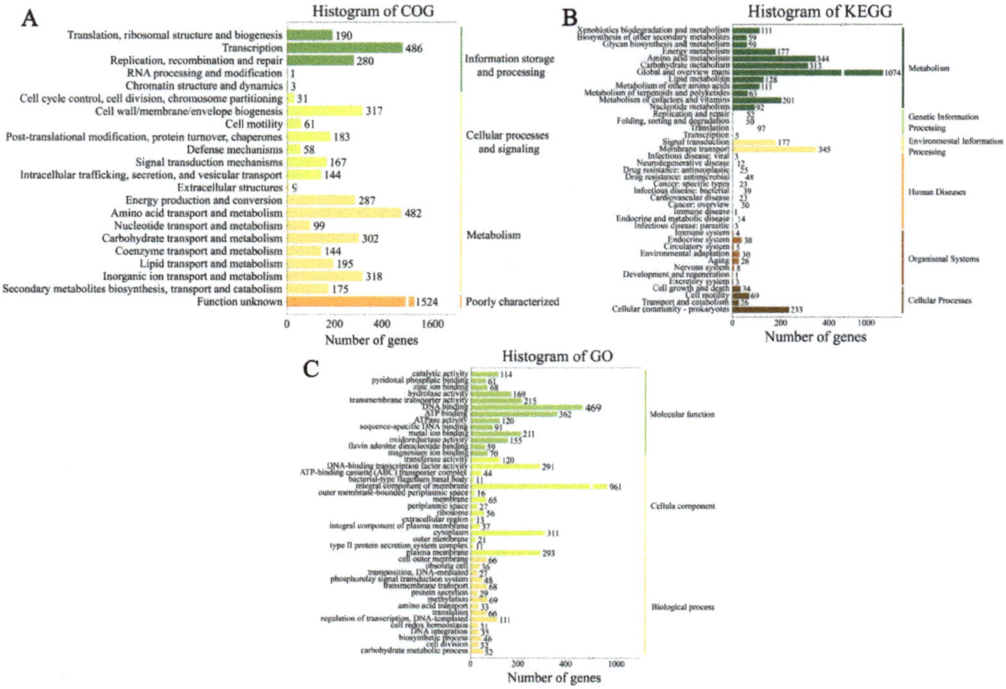

Figure 3. Functional annotation results of the YIM B08401 strain genome across databases. (**A**) COG database; (**B**) KEGG database; (**C**) GO database (showing highly abundant categories among the three classifications).

3.3. *CAZyme Database Analysis*

A total of 196 CAZyme-related enzymes were identified in the genome of YIM B08401, including 64 Glycoside Hydrolases (GH), 54 GlycosylTransferases (GT), 1 Polysaccharide Lyase (PL), 46 Carbohydrate Esterases (CE), 30 Auxiliary Activities (AA), and 1 Carbohydrate-Binding Module (CBM). Among these, families associated with chitin degradation, starch hydrolysis, and cellulose degradation were detected, including GH18, GH15, GH54, GH39, and GH3. Genomic mining further revealed the presence of the chitinase synthesis gene *chiA*, which enables the conversion of exogenous chitin into N-acetylglucosamine (GlcNAc) and GlcNAc monomer polymers, subsequently transported into bacterial cells [38]. GlcNAc is phosphorylated to GlcNAc-6P by NagK (N-acetylglucosamine kinase encoded by *nagK*), then converted to glucosamine-6P and acetate by NagA (N-acetylglucosamine-6-phosphate deacetylase encoded by *nagA*), where acetate serves as a carbon source for bacterial growth. Additionally, glucosamine-6P is transformed into fructose-6-phosphate by NagB (glucosamine-6-phosphate deaminase encoded by *nagB*), with released NH_3 utilized as a nitrogen source by the bacteria [39]. Qualitative assays confirmed the production of chitinase and xylanase by strain YIM B08401 (Figure S2), highlighting its potential for carbohydrate metabolism and utilization in ecological niches.

3.4. *antiSMASH Database Analysis*

Burkholderia species typically possess large and complex genomes with numerous secondary metabolite biosynthesis clusters [40]. Using the antiSMASH database (with the MIBiG database as the reference dataset and stringent cluster detection), strain YIM B08401 was analyzed for potential biosynthetic gene clusters (BGCs) (Table S1). KnownClusterBlast identified 21 secondary metabolite biosynthesis gene clusters, of which only 5 showed >85% similarity to known compound clusters, 10 exhibited <85% similarity, and 6 had no homology [41]. Studies indicate that BGCs with >85% similarity likely encode the same compound as their closest known cluster, while those with <85% similarity may harbor novel metabolites [42]. The five BGCs with >85% similarity include fragin (with antifungal activity) [43]; ornibactin C8/C4/C6 (siderophores with pathogen-antagonistic capabilities) [44]; icosalide A/B (antibiotics inhibiting insect pathogens) [45]; and rhizomide A/B/C (with potential antibacterial activity) [46,47]. These predicted secondary metabolites are all reported to inhibit pathogen growth, consistent with the strain's antibacterial results in previous studies [24], indicating a strong potential for YIM B08401 to synthesize these compounds. Additionally, the clusters were predicted to contain homoserine lactones (hserlactone), signaling molecules of the quorum-sensing system, aligning with the universal quorum-sensing capability of *Burkholderia* [48]. BGCs with <85% similarity represent promising targets for discovering novel natural products. Heterologous expression could be employed to further characterize its functional expression and uncover novel natural secondary metabolites.

3.5. Plant Growth Promotion Potential

3.5.1. Phosphorus Transformation

Phosphorus is an essential element for plant growth, but large amounts of insoluble phosphorus in soil cannot be absorbed by plant roots. Bacteria can leverage their microbial characteristics to convert insoluble phosphorus into forms absorbable by plants. Phosphorus is also a vital nutrient for microbes, participating in the synthesis of cellular structures (e.g., nucleic acids, lipids) and energy-supplying processes (e.g., the tricarboxylic acid cycle). To ensure long-term phosphorus acquisition, microbes have evolved diverse regulatory systems for phosphate uptake and utilization, with three well-characterized mechanisms: inorganic phosphorus solubilization and organic phosphorus mineralization, phosphate uptake and transport, and phosphorus starvation response regulation. A primary mechanism of bacterial inorganic phosphorus solubilization involves secreting low-molecular-weight organic acids to dissolve insoluble phosphorus by reducing pH, chelating cations bound to phosphate, and competing for phosphorus absorption sites [49]. Examples of such organic acids include gluconic acid (GA), oxalic acid, malic acid, and citric acid [50]. Goldstein et al. proposed that gluconic acid is the main agent for inorganic phosphorus solubilization in Gram-negative bacteria [51]. Gluconic acid synthesis requires glucose dehydrogenase (GDH encoded by *gcd*) and the redox cofactor pyrroloquinoline quinone (PQQ encoded by *pqqABCDEF*), which together convert glucose to gluconic acid. Genomic annotation of YIM B08401 identified genes involved in gluconic acid synthesis (*pqqBCDF*, *gcd*, *gdh*), indicating the strain's potential to secrete gluconic acid-driven organic acids for insoluble phosphate solubilization. Utilization of insoluble organic phosphorus relies on bacterial phosphatases to mineralize organic phosphates. YIM B08401 harbors *phoA* and *phnCDEFGHIJKLMN* genes, conferring the potential to secrete alkaline phosphomonoesterase (EC: 3.1.3.1) and phosphonatases/carbon-phosphorus (C-P) lyases. C–P lyases enable microbes to utilize phosphonates as the sole phosphorus source under phosphorus starvation [52]. Additionally, YIM B08401 possesses phosphate uptake-related genes (*pit*, *pitA*, *PhoR-PhoB*, *pstSCAB*, *PhoU*), forming a low-affinity constitutive Pit system and a high-affinity phosphate starvation-induced Pst system. Details of phosphate-solubilizing genes are listed in Table S2. Previous experiments showed that YIM B08401 solubilized 455.63 ± 59.65 mg/L of phosphorus from tricalcium phosphate in a phosphorus-solubilizing medium after

2 days and 18.46 ± 1.71 mg/L from lecithin after 3 days [24], confirming its phosphorus solubilization capacity.

3.5.2. Nitrogen Fixation

In studies on whether the strain can participate in the nitrogen cycle, preliminary experiments verified that strain YIM B08401 could grow on Ashby nitrogen-fixing bacteria medium, demonstrating biological nitrogen fixation capacity [24]. Gene annotation revealed genes related to the nitrogen cycle in the strain (Table S3), including *nirB*, which encodes nitrite reductase involved in nitrate reduction. However, the genome of YIM B08401 lacks genes directly associated with nitrogen fixation, though it possesses a gene (*iscU*) with primary structural similarity to *nifU*. This gene may be involved in synthesizing iron-sulfur (Fe-S) clusters required for nitrogenase [53,54].

3.5.3. Sulfur Metabolism

Multiple studies have demonstrated that hydrogen sulfide (H_2S) acts as a gaseous signaling molecule in various plant physiological processes, including promoting seed germination, alleviating heavy metal toxicity, and mitigating effects of abiotic stresses (e.g., high salinity, drought, and extreme temperatures) [55–58]. However, direct exogenous supplementation of H_2S in agricultural practice is challenging, making microbial-mediated H_2S production a promising research direction. Experimental validation confirmed that strain YIM B08401 can utilize thiosulfate and convert it into H_2S.

Genomic annotation of YIM B08401 revealed a gene pathway for assimilatory sulfate reduction (Figure 4). Sulfate is first converted to adenosine 5′-phosphosulfate (APS) by sulfate adenylyltransferase (encoded by *cysN*), followed by transformation to 3′-phosphoadenosine-5′-phosphosulfate (PAPS) via adenylylsulfate kinase (encoded by *cysC*). PAPS is then reduced to sulfite by phosphoadenosine phosphosulfate reductase (encoded by *cysH*), and finally converted to H_2S by sulfite reductase (NADPH) flavoprotein alpha-component (EC: 1.8.1.2, encoded by *cysJ*) [34]. Additionally, the *sqr* gene, a marker for sulfide oxidation, encodes sulfide: quinone oxidoreductase, which catalyzes the oxidation of sulfide and reduction of ubiquinone [59]. YIM B08401 harbors a highly homologous *sqr* gene, suggesting its potential for sulfur oxidation. Details of sulfur metabolism-related genes are listed in Table S4.

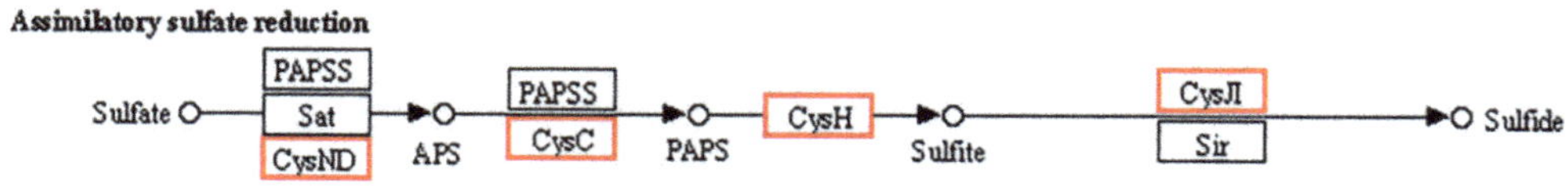

Figure 4. Schematic diagram of the assimilatory sulfate reduction gene pathway. Red boxes indicate genes possessed by strain YIM B08401.

3.5.4. Zinc Solubilization

Zinc, a trace element essential for living organisms, is the second most important transition metal ion after iron. Its divalent ion form (Zn^{2+}) serves as a cofactor for metalloproteins, participating in structural, catalytic, and regulatory functions. Zinc in soil predominantly exists in insoluble forms, making it inaccessible to plants. Microbial zinc solubilization can convert insoluble zinc into plant-available Zn^{2+}. *Burkholderia* species are Gram-negative bacteria that utilize multiple systems for zinc uptake, including the ZnuABC transporter, ZIP protein family, and TonB system. *zur*, a key regulator of zinc homeostasis, controls zinc uptake by activating high-affinity transporters under zinc-limiting conditions. When zinc is abundant, low-affinity ZIP (ZRT, IRT-like proteins) family members like *ZupT* mediate zinc uptake [60]. Under zinc deficiency, *zur* senses low zinc levels and activates the high-affinity ABC transporter *znuABC*, which specifically imports extracellular zinc for bacterial utilization. Genomic annotation of YIM B08401 identified *znuA* and *znuC* genes involved in forming the zinc-uptake ABC transporter, but *znuB* was absent, suggesting this may not be the primary zinc uptake pathway for the strain. Additionally, the genome harbors TonB system genes (*tonB, exbB, exbD*) associated with zinc uptake. Collectively, YIM B08401 possesses a suite of genes related to zinc uptake and utilization (Table S5), indicating its zinc solubilization potential. Experimental validation showed that the strain solubilized 18.60 ± 4.19 mg/kg of zinc after 7 days [24], confirming its capacity to convert insoluble zinc into bioavailable forms.

3.5.5. Siderophore Production

Iron, the most critical transition metal ion for living organisms, is an essential nutrient for microbial growth. Siderophores, with their high affinity for Fe^{3+}, chelate insoluble iron into plant-available soluble forms. They also compete with pathogens for Fe^{3+}, indirectly reducing pathogen abundance. Previous studies have confirmed that strain YIM B08401 produces siderophores [24], though the specific type remains uncharacterized. Genomic annotation revealed genes involved in synthesizing catechol-type siderophores Bacillibactin and Enterochelin. These siderophores, which are encoded by *entABCEF* and *dhbF* genes, function not only as potential antimicrobial agents but also as iron-chelating agents to supply iron for bacterial growth. Details of the relevant genes are listed in Table S6.

3.5.6. Indole-3-Acetic Acid (IAA) Production

The genome of strain YIM B08401 harbors a complete pathway for tryptophan synthesis encoded by *trpABCDEFG* genes (Figure 5), indicating the strain's potential to produce IAA without exogenous tryptophan supplementation. Genes involved in the second and third steps of the indole pyruvic acid (IPA) pathway were identified, including *ipdC* (encoding indolepyruvate decarboxylase [EC: 4.1.1.74]) and *aldh* (encoding aldehyde dehydrogenase [NAD+] [EC: 1.2.1.3]). Additionally, genes governing the final step of the indoleacetamide (IAM) pathway, specifically *amiE* and *iaaH*, were annotated. These genes encode amidase [EC: 3.5.1.4] and indoleacetamide hydrolase [EC: 3.5.1.-], respectively. Both enzymes specifically hydrolyze indole-3-acetamide (IAM) into IAA and ammonia. Details of IAA biosynthesis-related genes annotated in YIM B08401 are listed in Table S7.

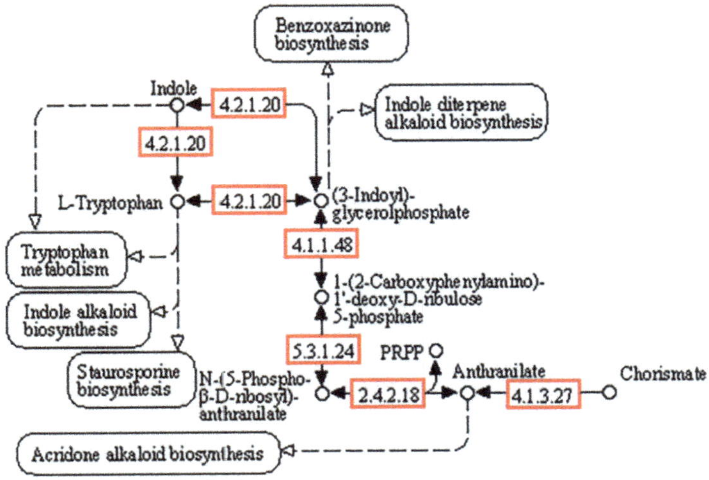

Figure 5. Schematic diagram of the tryptophan biosynthesis gene pathway. Red boxes indicate genes possessed by strain YIM B08401.

3.5.7. Ethylene Regulation

Ethylene is a key phytohormone regulating plant growth and development, orchestrating physiological processes and coordinating with other signaling pathways to mediate responses including fruit ripening, organ senescence, flowering, seed dormancy and germination, root hair development, and biotic/abiotic stress responses. However, excessive ethylene accelerates plant senescence, detrimentally affecting crop yield and storage. One strategy to mitigate ethylene production involves the *acdS*-encoded enzyme 1-aminocyclopropane-1-carboxylate deaminase (ACCD), which catalyzes the conversion of 1-aminocyclopropane-1-carboxylate (ACC, the immediate precursor of ethylene) into ammonia and α-ketobutyrate. This not only reduces ethylene biosynthesis but also provides ammonia as a nitrogen source for plant growth [61,62].

Genomic annotation of strain YIM B08401 using the Swiss-Prot database identified the presence of the *acdS* gene (Table S8). Blastp analysis of the translated protein revealed highest homology (89.35%) to ACCD from *Paraburkholderia guartelaensis*. Structural alignment with functionally verified ACCD sequences highlighted conserved residues E295 (glutamic acid) and L322 (leucine), which are critical for catalytic activity. These findings strongly suggest that YIM B08401 harbors a functional ACCD enzyme. Experimental validation demonstrated that YIM B08401 maintained viability when repeatedly subcultured on ADF (ACC Deaminase Activity Assay Medium) [24], confirming its ability to utilize ACC as a nitrogen source and further supporting its role in ethylene modulation. This mechanism likely contributes to plant growth promotion and stress tolerance under adverse conditions.

3.5.8. Degradation of Phenolic Acids

Continuous monocropping in agriculture often leads to the accumulation of phenolic acids (e.g., benzoic acid, p-hydroxybenzoic acid, salicylic acid, ferulic acid) secreted by plant roots, which inhibit both plant growth and soil microbial activity, thereby contributing to replant disease. Microbial degradation of these compounds represents a promising strategy to alleviate such obstacles. The genome of strain YIM B08401 harbors a complete benzoate degradation pathway (Figure S1). Benzoic acid is sequentially converted to catechol by the actions of benzoate 1,2-dioxygenase (encoded by *benA-xylX*) and dihydrodiol dehydrogenase (encoded by *benD-xylL*). Through the ortho-cleavage pathway, catechol is further metabolized via enzymes encoded by *catB*, *catC*, *pcaD*, *pcaI*, and *fadA*, ultimately generating succinyl-CoA and acetyl-CoA, which enter the citric acid cycle to provide energy for microbial growth. Additional genes involved in this degradation process that are present in the genome of strain YIM B08401 are listed in Table S9.

3.5.9. Root Colonization

The ability of plant growth-promoting microorganisms to colonize the rhizosphere is a critical prerequisite for their effectiveness in promoting plant growth. Bacterial chemotaxis serves as the primary driver of colonization; bacteria must first recognize chemical signals from root exudates and exhibit chemotactic responses to migrate toward the rhizosphere and aggregate for colonization. This process relies on bacterial motility, primarily mediated by flagella, which enable movement and adhesion in response to chemotactic cues [63,64]. Genomic analysis of strain YIM B08401 identified a suite of genes associated with chemotaxis, including *tar*, *tsr*, *trg*, and *aer* (encoding chemotactic signal sensory proteins) and *cheABCDRWYZ* (encoding regulatory proteins). Genes involved in flagellar assembly, including *flgABCDEEEFGHIKLMN*, *flhABCD*, *fliACDDEFGHIJKMNOPQRST*, and *motAB*, were also annotated. Collectively, these genes suggest that YIM B08401 possesses the genetic machinery for rhizosphere colonization (Figure 6, Table S10). To validate this potential, colonization assays were conducted, demonstrating that YIM B08401 successfully colonized the roots of tomato seedlings (Figure S3). This phenotypic result aligns with genomic predictions, providing robust evidence for the strain's chemotactic and motile capabilities essential for root association.

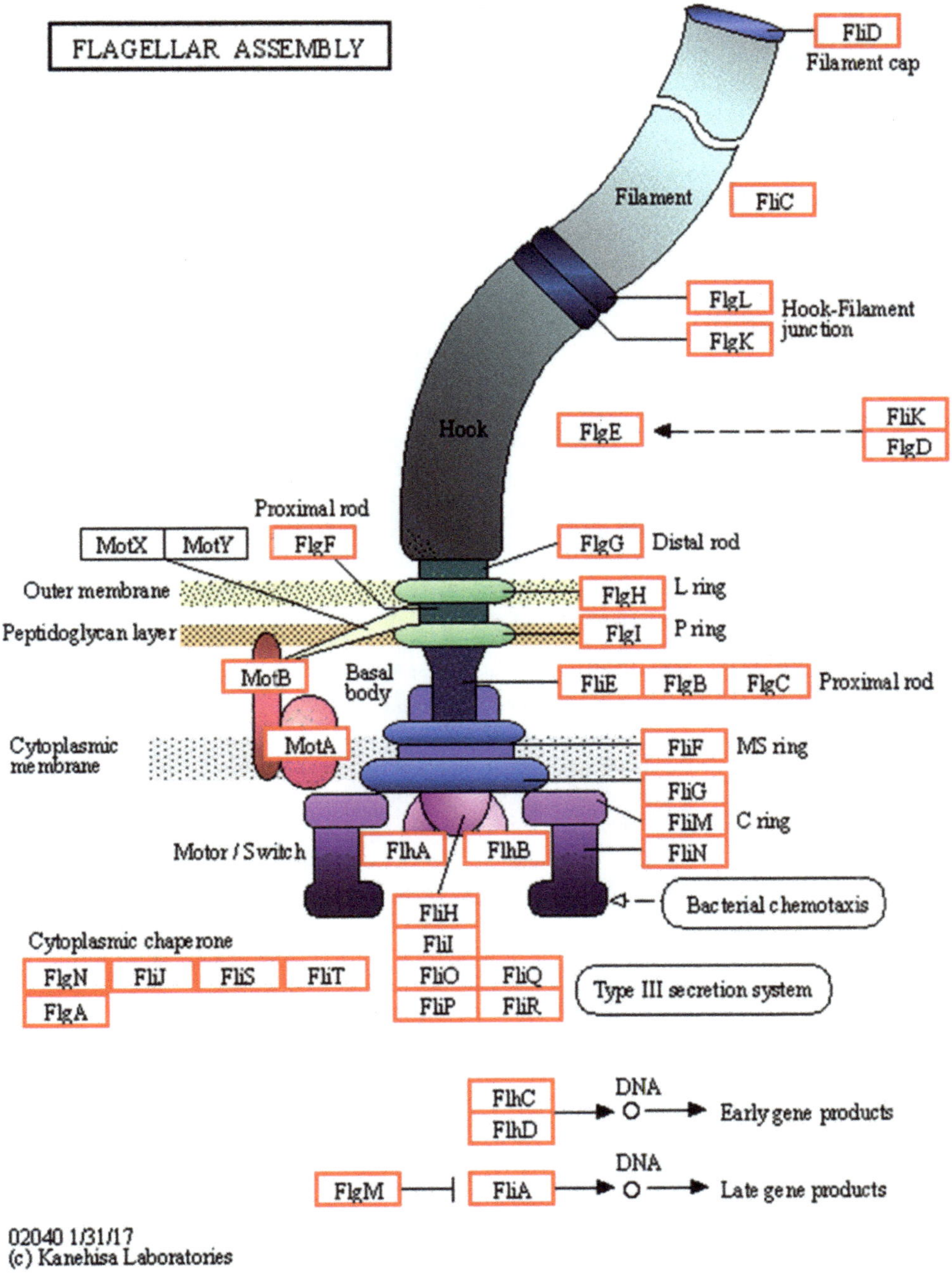

Figure 6. Schematic diagram of flagellar assembly genes. Red boxes indicate genes present in strain YIM B08401.

3.5.10. Heavy Metal Adsorption

Heavy metal ions are typical inorganic pollutants, with cadmium (Cd) recognized as one of the most toxic, posing lethal threats to humans and ecosystems [65,66]. Microbial-based remediation leveraging metal tolerance and adsorption capabilities has emerged as a research hotspot in environmental restoration due to its cost-effectiveness, minimal ecological disturbance, and sustainability. Microorganisms employ diverse strategies to counteract heavy metal stress, including biotransformation (e.g., valence alteration), complexation-precipitation (e.g., extracellular polymeric substance binding), transmembrane transport (e.g., efflux pumps), and intracellular detoxification (e.g., thiol compound chelation) [67,68]. Genomic annotation of strain YIM B08401 revealed numerous genes associated with tolerance, resistance, and transport of multiple heavy metals (e.g., Cd, Cu, Cr, Hg) (Table S11). Key genetic determinants include glutathione (GSH) biosynthesis genes *gshA* (encoding glutamate-cysteine ligase) and *gshB* (glutathione synthetase), which mediate intracellular detoxification by forming low-toxicity complexes with heavy metal ions (e.g., Cd^{2+}, Hg^{2+}) via thiol (-SH) groups [69,70]. Phytochelatin (PCs) synthesis gene *pcs-1* encodes enzymes that chelate heavy metals through multiple thiol groups, thereby reducing their bioavailability and toxicity [71]. The copper resistance system involves the two-component regulatory system *cusBR* (sensing Cu^{2+} and activating downstream responses) and *copA* (encoding a Cu-CPx type ATPase for active copper efflux), which collaboratively maintain intracellular copper homeostasis [72,73]. The RND (Resistance–Nodulation–Division) transporter complex *czcCBA* mediates the efflux of divalent cations (Cd^{2+}, Zn^{2+}, Co^{2+}) and represents a critical resistance mechanism in Gram-negative bacteria [74,75]. Heavy metal adsorption assays confirmed that live cells of strain YIM B08401 can reduce Cd^{2+} concentrations. The underlying mechanisms may involve not only direct cellular adsorption but also the gene-mediated transport and detoxification processes described above.

3.6. *Cd^{2+} Adsorption Efficiency and Characteristics of Strain YIM B08401*

Heavy metal tolerance tests revealed that strain YIM B08401 could grow on NA medium supplemented with 500 mg/L of Cd^{2+}, Zn^{2+}, or Cu^{2+}, and adsorption experiments further demonstrated the strain effectively reduced Cd^{2+} concentrations in solution within two days, with adsorption efficiencies of 20.76% (100 mg/L initial Cd^{2+}) and 38.49% (200 mg/L) as measured by atomic absorption spectroscopy (AAS).

Scanning electron microscopy–energy dispersive spectroscopy (SEM-EDS) analysis of cells exposed to 0, 100, and 200 mg/L Cd^{2+} showed that precipitates in the 0 mg/L group appeared as large, smooth aggregates (Figure 7A) likely formed by extracellular polymeric substances (EPS) secreted during normal growth [76], while those in the 100 mg/L group fragmented into smaller particles (Figure 7C), possibly due to interactions between Cd^{2+} and cell surface functional groups (e.g., carboxyl, hydroxyl, amino) disrupting EPS structure via complexation or ion exchange [77,78]. When the Cd^{2+} concentration reaches 200 mg/L, the precipitate exhibits a rough and porous surface (Figure 7E), potentially resulting from high-concentration Cd^{2+}-induced cell lysis. In this scenario, the released intracellular substances react with Cd^{2+} and extracellular polymeric substances (EPS) to form irregular complexes [79–81].

EDS detected characteristic Cd peaks accounting for 1.78% and 14.75% of total elemental composition in the 100 and 200 mg/L groups, respectively (Figure 7D,F), indicating multiple Cd^{2+} removal mechanisms including surface adsorption and biomineralization [77,81]. The absence of distinct bacterial structures in SEM images could be attributed to surface masking by EPS-Cd^{2+} complexes or Cd precipitates, or cell damage and lysis from high Cd^{2+} toxicity mixing cellular debris with other components [80,82].

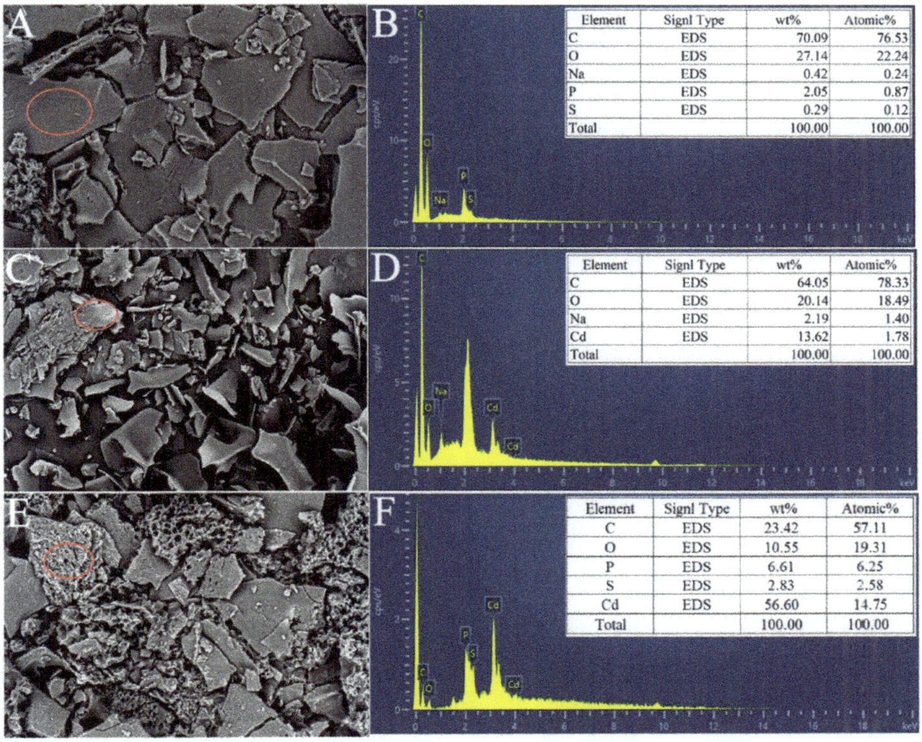

Figure 7. SEM images and EDS elemental analysis of YIM B08401 under non-Cd and Cd stress conditions. (**A**,**B**) SEM image and EDS spectrum of YIM B08401 without Cd stress; (**C**,**D**) SEM image and EDS spectrum of YIM B08401 under 100 mg/L Cd^{2+} stress; (**E**,**F**) SEM image and EDS spectrum of YIM B08401 under 200 mg/L Cd^{2+} stress. Red circles indicate the selected points for spectral analysis.

3.7. FTIR Spectroscopy of Strain YIM B08401

To further clarify the adsorption mechanism of Cd^{2+} by strain YIM B08401, Fourier transform infrared spectroscopy (FTIR) was used to characterize the functional groups on the cell surface of YIM B08401 before and after cadmium adsorption, revealing multiple characteristic absorption peaks within the wavenumber range of 397–4003 cm^{-1} (Figure 8). Specifically, a peak near 1055 cm^{-1} corresponded to stretching vibrations of C-O, C-C, C-O-C, and C-O-P bonds in polysaccharides [83], while the peak at 1231 cm^{-1} represented P=O vibrations of phosphodiester bonds in nucleic acids and phospholipids [84]. The absorption peak at 1390 cm^{-1} reflected C-H vibrations of CH_3/CH_2 groups and C-O vibrations of carboxylate (COO^-), confirming the presence of extracellular polymeric substances (EPS) secreted by bacteria that can effectively chelate metal ions through cation exchange to reduce toxicity [85,86]. A characteristic absorption peak at 1534 cm^{-1} was likely attributed to bending vibrations of N-H bonds and stretching vibrations of C-N bonds in amide groups, potential structural components of the bacterial cell wall [87], and the broad absorption peak at 3281 cm^{-1} was assigned to stretching vibrations of hydroxyl (O-H) groups [81]. These results indicate that under Cd^{2+} stress, strain YIM B08401 interacts with Cd^{2+} through multiple cell surface functional groups (e.g., C-O-C, P=O, O-H), with groups associated with proteins, polysaccharides, and lipids (such as phospholipids) potentially involved in Cd assimilation and metabolic processes.

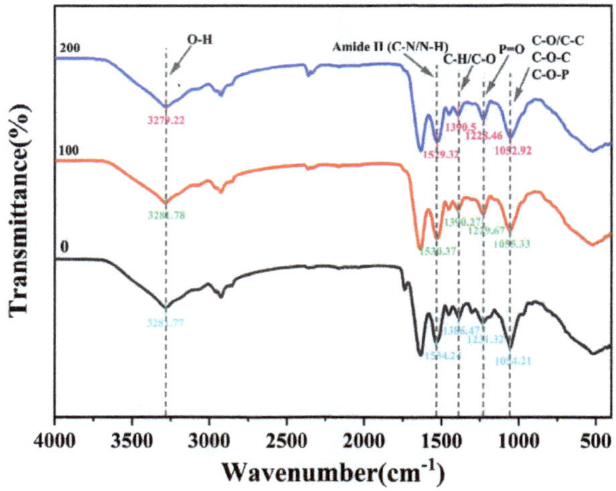

Figure 8. FTIR spectra of YIM B08401 before and after Cd^{2+} adsorption.

4. Conclusions

In this study, whole-genome sequencing and annotation of Burkholderia alba YIM B08401 were performed to identify genes and metabolic pathways associated with plant growth promotion, biocontrol, and abiotic stress tolerance. The strain exhibited a G+C content of 66.39% and harbored 6504 annotated genes, including those involved in nutrient conversion, indole-3-acetic acid (IAA) secretion, and siderophore production. KEGG pathway annotation revealed metabolic pathways related to nutrient utilization, such as ABC transporter genes (*pstABCS*, *phnDEC*, *cysPUWA*, *znuABC*), two-component systems (*PhoR-PhoB*, *cheWARYB*), and complete pathways for tryptophan synthesis and benzoic acid degradation. AntiSMASH annotation identified 21 secondary metabolism gene clusters, including those for terpenoids, non-ribosomal peptides (NRPS), polyketides (PKS), and homoserine lactones (hserlactone). CAZy annotation further confirmed the presence of enzymes involved in chitin, starch, cellulose, and xylan degradation. FTIR and SEM-EDS analyses demonstrated that YIM B08401 specifically binds Cd^{2+} through cell surface functional groups (e.g., C-O-C, P=O, O-H), enabling biosorption of heavy metal ions. Collectively, *Burkholderia alba* YIM B08401 harbors genetically encoded potentials for plant growth promotion, biocontrol of plant diseases, and abiotic stress resistance, warranting further molecular investigations to unlock its full genetic repertoire and expand its application potential in agricultural and environmental biotechnology.

Supplementary Materials: The following supporting information can be downloaded at: https://www.mdpi.com/article/10.3390/agronomy15081780/s1, Figure S1: Schematic diagram of the benzoic acid degradation gene pathway (blue-bordered region). Red boxes indicate genes present in strain YIM B08401; Figure S2. Photographs of qualitative enzyme activity assays for strain YIM B08401. (A) Chitinase; (B) Xylanase. Figure S3. Photographs of tomato seedling root colonization by strain YIM B08401. Table S1: Results of secondary metabolite prediction for YIM B08401 by antiSMASH; Table S2: Phosphate solubilization related genes; Table S3: Nitrogen cycle related genes; Table S4: Sulfur metabolism related genes; Table S5: Zinc solubilization related genes; Table S6: Iron carrier production related genes; Table S7: Indole acetic acid (IAA) secretion related genes; Table S8: Ethylene regulation related genes; Table S9: Benzoic acid degradation related genes; Table S10: Root colonization related genes; Table S11: Heavy metal related genes.

Author Contributions: Conceptualization, L.F., X.L., N.W., Z.S. (Zhuli Shi), Y.W., J.J., Z.S. (Zhufeng Shi), T.P. and P.Y.; methodology, P.Y.; software, N.W.; formal analysis, X.L.; investigation, Z.S. (Zhuli

Shi), Y.W. and J.J.; resources, Z.S. (Zhufeng Shi) and T.P.; writing—original draft, L.F. and X.L.; writing—review and editing, L.F., X.L. and P.Y.; funding acquisition, P.Y. All authors have read and agreed to the published version of the manuscript.

Funding: This research was supported by the Yunnan Province "Xingdian Talent" Support Program (XDYC-CYCX-2002-0071) and the Central Government Guided Local Science and Technology Development Fund (202407AC110006).

Data Availability Statement: Genome sequences were deposited in the GenBank database under BioProject accession number PRJNA1027102 under accession number JAWJWL000000000 (https://www.ncbi.nlm.nih.gov/nuccore/JAWJWL000000000, accessed on 16 October 2023).

Conflicts of Interest: The authors declare that they have no known competing financial interests or personal relationships that could have appeared to influence the work reported in this paper.

References

1. Ning, C.; Gao, P.; Wang, B.; Wang, B.; Lin, W.; Jiang, N.; Cai, K. Impacts of chemical fertilizer reduction and organic amendments supplementation on soil nutrient, enzyme activity and heavy metal content. *J. Integr. Agric.* **2017**, *16*, 1819–1831. [CrossRef]
2. Bao, L.; Liu, Y.; Ding, Y.; Shang, J.; Wei, Y.; Tan, Y.; Zi, F. Interactions Between Phenolic Acids and Microorganisms in Rhizospheric Soil from Continuous Cropping of *Panax notoginseng*. *Front. Microbiol.* **2022**, *13*, 791603. [CrossRef] [PubMed]
3. Jia, M.; Wang, X.; Zhu, X.; Du, Y.; Zhou, P.; Wang, G.; Wang, N.; Bai, Y. Accumulation of coumaric acid is a key factor in tobacco continuous cropping obstacles. *Front. Plant. Sci.* **2024**, *15*, 1477324. [CrossRef]
4. Li, X.; Yu, Y.; He, R.; Zhen, Q.; She, D. Synergistic effects of aged lignin-based biochar and selenium fertilization on heavy metal remediation in agricultural soils. *Ind. Crops. Prod.* **2025**, *225*, 120464. [CrossRef]
5. Ouyang, X.; Ma, J.; Zhang, R.; Li, P.; Gao, M.; Sun, C.; Weng, L.; Chen, Y.; Yan, S.; Li, Y. Uptake of atmospherically deposited cadmium by leaves of vegetables: Subcellular localization by NanoSIMS and potential risks. *J. Hazard. Mater.* **2022**, *431*, 128624. [CrossRef]
6. Roberts, T. Cadmium and phosphorous fertilizers: The issues and the science. *Procedia. Eng.* **2014**, *83*, 52–59. [CrossRef]
7. Mandal, R.; Bashir, Z.; Raj, D. Microbe-assisted phytoremediation for sustainable management of heavy metal in wastewater—A green approach to escalate the remediation of heavy metals. *J. Environ. Manag.* **2025**, *375*, 124199. [CrossRef]
8. Sahu, O.; Chaudhari, P. Review on chemical treatment of industrial waste water. *J. Appl. Sci. Environ. Manag.* **2013**, *17*, 241–257. [CrossRef]
9. Wang, Y.; Zhao, Q.; Sun, Z.; Li, Y.; He, H.; Zhang, Y.; Yang, X.; Wang, D.; Dong, B.; Zhou, H.; et al. Whole-genome analysis revealed the growth-promoting mechanism of endophytic bacterial strain Q2H1 in potato plants. *Front. Microbiol.* **2022**, *13*, 1035901. [CrossRef]
10. Luo, Y.; Chen, L.; Lu, Z.; Zhang, W.; Liu, W.; Chen, Y.; Wang, X.; Du, W.; Luo, J.; Wu, H. Genome sequencing of biocontrol strain *Bacillus amyloliquefaciens* Bam1 and further analysis of its heavy metal resistance mechanism. *Bioresour. Bioprocess.* **2022**, *9*, 74. [CrossRef]
11. Chlebek, D.; Płociniczak, T.; Gobetti, S.; Kumor, A.; Hupert-Kocurek, K.; Pacwa-Płociniczak, M. Analysis of the Genome of the Heavy Metal Resistant and Hydrocarbon-Degrading Rhizospheric *Pseudomonas qingdaonensis* ZCR6 Strain and Assessment of Its Plant-Growth-Promoting Traits. *Int. J. Mol. Sci.* **2021**, *23*, 214. [CrossRef] [PubMed]
12. Janaki, M.; Kirupanantha-Rajan, P.; Senthil-Nathan, S.; Stanley-Raja, V.; Al Farraj, D.; Aljeidi, R.; Arokiyaraj, S. Beneficial role of *Burkholderia cepacia* in heavy metal bioremediation in metal-polluted soils and enhance the tomato plant growth. *Biocatal. Agric. Biotechnol.* **2024**, *57*, 103032. [CrossRef]
13. Lan, Y.; Liu, M.; Song, Y.; Cao, Y.; Li, F.; Luo, D.; Qiao, D. Distribution, characterization, and evolution of heavy metal resistance genes and Tn7-like associated heavy metal resistance Gene Island of *Burkholderia*. *Front. Microbiol.* **2023**, *14*, 1252127. [CrossRef] [PubMed]
14. Wang, C.; Huang, Y.; Yang, X.; Xue, W.; Zhang, X.; Zhang, Y.; Pang, J.; Liu, Y.; Liu, Z. *Burkholderia* sp. Y4 inhibits cadmium accumulation in rice by increasing essential nutrient uptake and preferentially absorbing cadmium. *Chemosphere* **2020**, *252*, 126603. [CrossRef]
15. Jiang, C.; Sheng, X.; Qian, M.; Wang, Q. Isolation and characterization of a heavy metal-resistant *Burkholderia* sp. from heavy metal-contaminated Paddy field soil and its potential in promoting plant growth and heavy metal accumulation in metal-polluted soil. *Chemosphere* **2008**, *72*, 157–164. [CrossRef]
16. Huang, G.; Tian, H.; Liu, H.; Fan, X.; Liang, Y.; Li, Y. Characterization of plant-growth-promoting effects and concurrent promotion of heavy metal accumulation in the tissues of the plants grown in the polluted soil by *Burkholderia* strain LD-11. *Int. J. Phytoremediation* **2013**, *15*, 991–1009. [CrossRef]

17. Chaudhry, V.; Runge, P.; Sengupta, P.; Doehlemann, G.; Parker, J.; Kemen, E. Shaping the leaf microbiota: Plant-microbe-microbe interactions. *J. Exp. Bot.* **2021**, *72*, 36–56. [CrossRef]
18. Prashar, D.; Sharma, M.; Sharma, D.; Mir, Z.; Sood, H. Exploring the complexity of plant stress response: Role of metabolomics and signalling pathways in metabolic diversification. *J. Plant Biochem. Biotechnol.* **2025**, 1–14. [CrossRef]
19. Pan, H.; Ren, Y.; Sun, P.; Jing, X. Toward better understanding of denitrifying phosphorus accumulating organisms (DPAOs) via Single-cell Raman spectroscopy: A review. *Environ. Technol. Innov.* **2025**, *38*, 104205. [CrossRef]
20. Chavan, A.; Khardenavis, A. Annotating multiple prebiotic synthesizing capabilities through whole genome sequencing of *Fusarium* strain HFK-74. *Appl. Biochem. Biotechnol.* **2023**, *196*, 4993–5012. [CrossRef]
21. Bhanse, P.; Singh, L.; Qureshi, A. Functional and Genomic Potential of *Burkholderia contaminans* PB_AQ24 Isolate for Boosting the Growth of Bamboo Seedlings in Heavy Metal Contaminated Soils. *Appl. Biochem. Biotechnol.* **2025**, *197*, 2437–2456. [CrossRef] [PubMed]
22. Singh, A.; Gupta, R.; Purohit, H.; Khardenavis, A. Genomic characterization of denitrifying methylotrophic *Pseudomonas aeruginosa* strain AAK/M5 isolated from municipal solid waste landfill soil. *World J. Microbiol. Biotechnol.* **2022**, *38*, 140. [CrossRef]
23. Shen, W.; Zhang, J.; Geng, B.; Qiu, M.; Hu, M.; Yang, Q.; Bao, W.; Xiao, Y.; Zheng, Y.; Peng, W.; et al. Establishment and application of a CRISPR-Cas12a assisted genome-editing system in *Zymomonas mobilis*. *Microb. Cell Fact.* **2019**, *18*, 162. [CrossRef] [PubMed]
24. Feng, L.; Zhao, J.; Shi, Z.; Mo, Y.; Yang, T.; Shen, Y.; He, F.; Li, M.; Yang, P. Isolation and Identification of Bacteria in Forest Rhizosphere Soil and Their Biological Activity Screening. *Biotechnol. Bull.* **2024**, *40*, 294–307.
25. Qin, G.; Wang, D.; Luo, K.; Liu, Y.; Xie, Y.; Wang, M.; Li, C.; Fan, R.; Tian, X. Evaluation of probiotic properties and safety of a *Bacillus* strain for shrimp farming: Integrating in vitro testing, genomic analysis and in vivo validation. *Microbiol. Res.* **2025**, *297*, 128179. [CrossRef]
26. Shen, Y.; Shi, Z.; Zhao, J.; Ling, M.; Tang, J.; Wang, N.; Mo, Y.; Yang, T.; Zhou, X.; Chen, Q.; et al. Whole genome sequencing provides evidence for *Bacillus velezensis* SH-1471 as a beneficial rhizosphere bacterium in plants. *Sci. Rep.* **2023**, *13*, 20929. [CrossRef]
27. Ding, H.; Mo, W.; Yu, S.; Cheng, H.; Peng, L.; Liu, Z. Whole genome sequence of *Bacillus velezensis* strain GUMT319: A potential biocontrol agent against tobacco black shank disease. *Front. Microbiol.* **2021**, *12*, 658113. [CrossRef]
28. Rishad, K.; Rebello, S.; Shabanamol, P.; Jisha, M. Biocontrol potential of Halotolerant bacterial chitinase from high yielding novel *Bacillus pumilus* MCB-7 autochthonous to mangrove ecosystem. *Pestic. Biochem. Physiol.* **2017**, *137*, 36–41. [CrossRef]
29. Ninawe, S.; Lal, R.; Kuhad, R. Isolation of three xylanase-producing strains of actinomycetes and their identification using molecular methods. *Curr. Microbiol.* **2006**, *53*, 178–182. [CrossRef]
30. Iqbal, M.; Saeed, A.; Zafar, S. FTIR spectrophotometry, kinetics and adsorption isotherms modeling, ion exchange, and EDX analysis for understanding the mechanism of Cd^{2+} and Pb^{2+} removal by mango peel waste. *J. Hazard. Mater.* **2009**, *164*, 161–171. [CrossRef]
31. Liu, S.; Huang, Y.; Zheng, Q.; Zhan, M.; Hu, Z.; Ji, H.; Zhu, D.; Zhao, X. Cd-Resistant Plant Growth-Promoting Rhizobacteria *Bacillus siamensis* R27 Absorbed Cd and Reduced Cd Accumulation in Lettuce (*Lactuca sativa* L.). *Microorganisms* **2024**, *12*, 2321. [CrossRef]
32. Tatusov, R.; Koonin, E.; Lipman, D. A genomic perspective on protein families. *Science* **1997**, *278*, 631–637. [CrossRef]
33. Szczerba, H.; Komoń-Janczara, E.; Krawczyk, M.; Dudziak, K.; Nowak, A.; Kuzdraliński, A.; Waśko, A.; Targoński, Z. Genome analysis of a wild rumen bacterium *Enterobacter aerogenes* LU2-a novel bio-based succinic acid producer. *Sci. Rep.* **2020**, *10*, 1986 [CrossRef]
34. Eida, A.; Bougouffa, S.; L'Haridon, F.; Alam, I.; Weisskopf, L.; Bajic, V.; Saad, M.; Hirt, H. Genome Insights of the Plant-Growth Promoting Bacterium *Cronobacter muytjensii* JZ38 With Volatile-Mediated Antagonistic Activity Against *Phytophthora infestans* *Front. Microbiol.* **2020**, *11*, 369. [CrossRef] [PubMed]
35. Kanehisa, M.; Sato, Y.; Kawashima, M.; Furumichi, M.; Tanabe, M. KEGG as a reference resource for gene and protein annotation *Nucleic Acids Res.* **2016**, *44*, D457–D462. [CrossRef] [PubMed]
36. Ashburner, M.; Ball, C.; Blake, J.; Botstein, D.; Butler, H.; Cherry, J.; Davis, A.; Dolinski, K.; Dwight, S.; Eppig, J.; et al. Gene Ontology: Tool for the unification of biology. *Nat. Genet.* **2000**, *25*, 25–29. [CrossRef] [PubMed]
37. Ren, J.; Yu, D.; Li, N.; Liu, S.; Xu, H.; Li, J.; He, F.; Zou, L.; Cao, Z.; Wen, J. Biological Characterization and Whole-Genome Analysis of *Bacillus subtilis* MG-1 Isolated from Mink Fecal Samples. *Microorganisms* **2023**, *11*, 2821. [CrossRef]
38. Mannaa, M.; Han, G.; Jeong, T.; Kang, M.; Lee, D.; Jung, H.; Seo, Y. Taxonomy-guided selection of *Paraburkholderia busanensis* sp nov.: A versatile biocontrol agent with mycophagy against *Colletotrichum scovillei* causing pepper anthracnose. *Microbiol. Spectr* **2023**, *11*, e0242623. [CrossRef]
39. Meena, B.; Rajan, L.; Vinithkumar, N.; Kirubagaran, R. Novel marine actinobacteria from emerald Andaman & Nicobar Islands A prospective source for industrial and pharmaceutical byproducts. *BMC Microbiol.* **2013**, *13*, 145. [CrossRef]
40. Bach, E.; Passaglia, L.; Jiao, J.; Gross, H. *Burkholderia* in the genomic era: From taxonomy to the discovery of new antimicrobial secondary metabolites. *Crit. Rev. Microbiol.* **2022**, *48*, 121–160. [CrossRef]

41. Blin, K.; Shaw, S.; Kloosterman, A.; Charlop-Powers, Z.; van Wezel, G.; Medema, M.; Weber, T. antiSMASH 6.0: Improving cluster detection and comparison capabilities. *Nucleic Acids Res.* **2021**, *49*, W29–W35. [CrossRef]
42. Naughton, L.; Romano, S.; O'Gara, F.; Dobson, A. Identification of Secondary Metabolite Gene Clusters in the *Pseudovibrio* Genus Reveals Encouraging Biosynthetic Potential toward the Production of Novel Bioactive Compounds. *Front. Microbiol.* **2017**, *8*, 1494. [CrossRef]
43. Jenul, C.; Sieber, S.; Daeppen, C.; Mathew, A.; Lardi, M.; Pessi, G.; Hoepfner, D.; Neuburger, M.; Linden, A.; Gademann, K.; et al. Biosynthesis of fragin is controlled by a novel quorum sensing signal. *Nat. Commun.* **2018**, *9*, 1297. [CrossRef]
44. Deng, P.; Foxfire, A.; Xu, J.; Baird, S.; Jia, J.; Delgado, K.; Shin, R.; Smith, L.; Lu, S. The Siderophore Product Ornibactin Is Required for the Bactericidal Activity of *Burkholderia contaminans* MS14. *Appl. Environ. Microbiol.* **2017**, *83*, e00051-17. [CrossRef] [PubMed]
45. Dose, B.; Niehs, S.; Scherlach, K.; Flórez, L.; Kaltenpoth, M.; Hertweck, C. Unexpected Bacterial Origin of the Antibiotic Icosalide: Two-Tailed Depsipeptide Assembly in Multifarious *Burkholderia* Symbionts. *ACS Chem. Biol.* **2018**, *13*, 2414–2420. [CrossRef] [PubMed]
46. Adeniji, A.; Ayangbenro, A.; Loots, D. Genome Sequence Resource of *Pseudomonas fulva* HARBPS9.1-Candidate Biocontrol Agent. *Phytopathology* **2021**, *111*, 896–898. [CrossRef] [PubMed]
47. Wang, X.; Zhou, H.; Chen, H.; Jing, X.; Zheng, W.; Li, R.; Sun, T.; Liu, J.; Fu, J.; Huo, L.; et al. Discovery of recombinases enables genome mining of cryptic biosynthetic gene clusters in *Burkholderiales* species. *Proc. Natl. Acad. Sci. USA* **2018**, *115*, E4255–E4263. [CrossRef]
48. Suarez-Moreno, Z.; Caballero-Mellado, J.; Coutinho, B.; Mendonca-Previato, L.; James, E.; Venturi, V. Common Features of Environmental and Potentially Beneficial Plant-Associated *Burkholderia*. *Microb. Ecol.* **2012**, *63*, 249–266. [CrossRef]
49. Rawat, P.; Das, S.; Shankhdhar, D.; Shankhdhar, S. Phosphate-Solubilizing Microorganisms: Mechanism and Their Role in Phosphate Solubilization and Uptake. *J. Soil Sci. Plant Nutr.* **2021**, *21*, 49–68. [CrossRef]
50. Nacoon, S.; Jogloy, S.; Riddech, N.; Mongkolthanaruk, W.; Kuyper, T.; Boonlue, S. Interaction between Phosphate Solubilizing Bacteria and Arbuscular Mycorrhizal Fungi on Growth Promotion and Tuber Inulin Content of *Helianthus tuberosus* L. *Sci. Rep.* **2020**, *10*, 4916. [CrossRef]
51. Goldstein, A. Recent progress in understanding the molecular genetics and biochemistry of calcium phosphate solubilization by gram negative bacteria. *Biol. Agric. Horti.* **1995**, *12*, 185–193. [CrossRef]
52. Stosiek, N.; Talma, M.; Klimek-Ochab, M. Carbon-Phosphorus Lyase-the State of the Art. *Appl. Biochem. Biotechnol.* **2020**, *190*, 1525–1552. [CrossRef] [PubMed]
53. Dos Santos, P.; Johnson, D.; Ragle, B.; Unciuleac, M.; Dean, D. Controlled expression of nif and isc iron-sulfur protein maturation components reveals target specificity and limited functional replacement between the two systems. *J. Bacteriol.* **2007**, *189*, 2854–2862. [CrossRef]
54. Tokumoto, U.; Kitamura, S.; Fukuyama, K.; Takahashi, Y. Interchangeability and distinct properties of bacterial Fe-S cluster assembly systems: Functional replacement of the isc and suf operons in *Escherichia coli* with the nifSU-like operon from Helicobacter pylori. *J. Biochem.* **2004**, *136*, 199–209. [CrossRef]
55. Yusuf, M.; Saeed, T.; Almarri, H.; Khan, T.; Faizan, M.; Elsayed, N. Hydrogen sulfide counteract copper induced inhibition of photosynthetic performance through altered proline metabolism and enhanced antioxidants in Cucumis sativus. *Plant Stress* **2023**, *10*, 100222. [CrossRef]
56. Pourebrahimi, M.; Eshghi, S.; Ramezanian, A.; Faghih, S. Effect of combined application of selenium and hydrogen sulfide under salinity stress on yield, physiological traits and biofortification of strawberries in hydroponic cultivation. *Sci. Hortic.* **2023**, *315*, 111982. [CrossRef]
57. Wang, L.; Huang, X.; Liu, C.; Zhang, C.; Shi, K.; Wang, M.; Wang, Y.; Song, Q.; Chen, X.; Jin, P.; et al. Hydrogen sulfide alleviates chilling injury by modulating respiration and energy metabolisms in cold-stored peach fruit. *Postharvest Biol. Technol.* **2023**, *199*, 112291. [CrossRef]
58. Wang, L.; Chen, S.; Shao, J.; Zhang, C.; Mei, L.; Wang, K.; Jin, P.; Zheng, Y. Hydrogen sulfide alleviates chilling injury in peach fruit by maintaining cell structure integrity via regulating endogenous H_2S, antioxidant and cell wall metabolisms. *Food Chem.* **2022**, *391*, 133283. [CrossRef]
59. Hu, L.; Wang, Y.; Ci, M.; Long, Y. Unravelling microbial drivers of the sulfate-reduction process inside landfill using metagenomics. *Chemosphere* **2023**, *313*, 137537. [CrossRef]
60. Roberts, C.; Ni, F.; Mitra, B. The Zinc and Iron Binuclear Transport Center of ZupT, a ZIP Transporter from *Escherichia coli*. *Biochem.* **2021**, *60*, 3738–3752. [CrossRef]
61. Dubois, M.; Van den Broeck, L.; Inzé, D. The Pivotal Role of Ethylene in Plant Growth. *Trends Plant Sci.* **2018**, *23*, 311–323. [CrossRef]
62. Yang, S. Biosynthesis and action of ethylene. *HortScience* **1985**, *20*, 41–45. [CrossRef]
63. Karmakar, R. State of the art of bacterial chemotaxis. *J. Basic Microbiol.* **2021**, *61*, 366–379. [CrossRef] [PubMed]

64. Eisenbach, M.; Caplan, S. Bacterial chemotaxis: Unsolved mystery of the flagellar switch. *Curr. Biol.* **1998**, *8*, R444–R446. [CrossRef] [PubMed]
65. Huang, H.; Jia, Q.; Jing, W.; Dahms, H.; Wang, L. Screening strains for microbial biosorption technology of cadmium. *Chemosphere* **2020**, *251*, 126428. [CrossRef] [PubMed]
66. Cai, X.; Zheng, X.; Zhang, D.; Iqbal, W.; Liu, C.; Yang, B.; Zhao, X.; Lu, X.; Mao, Y. Microbial characterization of heavy metal resistant bacterial strains isolated from an electroplating wastewater treatment plant. *Ecotoxicol. Environ. Saf.* **2019**, *181*, 472–480. [CrossRef]
67. Khanna, K.; Kohli, S.; Kaur, R.; Handa, N.; Bakshi, P.; Sharma, P.; Ohri, P.; Bhardwaj, R. Reconnoitering the efficacy of plant growth promoting rhizobacteria in expediting phytoremediation potential of heavy metals. *J. Plant Growth. Regul.* **2023**, *42*, 6474–6502. [CrossRef]
68. Margaryan, A.; Panosyan, H.; Birkeland, N. Heavy metal resistance in prokaryotes: Mechanism and application. In *Microbial Communities and Their Interactions in the Extreme Environment*; Springer: London, UK, 2021; pp. 273–313.
69. Rauser, W. Phytochelatins and related peptides. Structure, biosynthesis, and function. *Plant Physiol.* **1995**, *109*, 1141. [CrossRef]
70. Lu, S. Glutathione synthesis. *Biochim. Biophys. Acta.* **2013**, *1830*, 3143–3153. [CrossRef]
71. Li, M.; Stragliati, L.; Bellini, E.; Ricci, A.; Saba, A.; Sanità di Toppi, L.; Varotto, C. Evolution and functional differentiation of recently diverged phytochelatin synthase genes from *Arundo donax* L. *J. Exp. Bot.* **2019**, *70*, 5391–5405. [CrossRef]
72. Nies, D. Efflux-mediated heavy metal resistance in prokaryotes. *FEMS Microbio. Rev.* **2003**, *27*, 313–339. [CrossRef]
73. Gaballa, A.; Helmann, J. *Bacillus subtilis* CPx-type ATPases: Characterization of Cd, Zn, Co and Cu efflux systems. *Biometals* **2003**, *16*, 497–505. [CrossRef]
74. Nies, D. Resistance to cadmium, cobalt, zinc, and nickel in microbes. *Plasmid* **1992**, *27*, 17–28. [CrossRef] [PubMed]
75. Nies, D. The cobalt, zinc, and cadmium efflux system CzcABC from *Alcaligenes eutrophus* functions as a cation-proton antiporter in *Escherichia coli*. *J. Bacteriol.* **1995**, *177*, 2707–2712. [CrossRef] [PubMed]
76. Decho, A.; Gutierrez, T. Microbial Extracellular Polymeric Substances (EPSs) in Ocean Systems. *Front. Microbiol.* **2017**, *8*, 922. [CrossRef]
77. Sreedevi, P.; Suresh, K.; Jiang, G. Bacterial bioremediation of heavy metals in wastewater: A review of processes and applications. *J. Water Pro. Engin.* **2022**, *48*, 102884. [CrossRef]
78. Abbas, S.; Ismail, I.; Mostafa, T.; Sulaymon, A. Biosorption of heavy metals: A review. *J. Chem. Sci. Technol.* **2014**, *3*, 74–102.
79. Zeng, Q.; Zhu, T.; Wen, Y.; Li, F.; Cheng, Y.; Chen, S.; Lan, T.; Yang, Y.; Liao, J.; Sun, Q.; et al. The dynamic behavior and mechanism of uranium (VI) biomineralization in *Enterobacter* sp. X57. *Chemosphere* **2022**, *298*, 134196. [CrossRef]
80. Alidoust, L.; Zahiri, H.; Maleki, H.; Soltani, N.; Vali, H.; Noghabi, K. *Nostoc entophytum* cell response to cadmium exposure: A possible role of chaperon proteins GroEl and HtpG in cadmium-induced stress. *Ecotoxicol. Environ. Saf.* **2019**, *169*, 40–49. [CrossRef]
81. Xie, Y.; He, N.; Wei, M.; Wen, T.; Wang, X.; Liu, H.; Zhong, S.; Xu, H. Cadmium biosorption and mechanism investigation using a novel *Bacillus subtilis* KC6 isolated from pyrite mine. *J. Clean. Prod.* **2021**, *312*, 127749. [CrossRef]
82. Qiao, W.; Zhang, Y.; Xia, H.; Luo, Y.; Liu, S.; Wang, S.; Wang, W. Bioimmobilization of lead by *Bacillus subtilis* X3 biomass isolated from lead mine soil under promotion of multiple adsorption mechanisms. *R. Soc. Open Sci.* **2019**, *6*, 181701. [CrossRef]
83. D'Souza, L.; Devi, P.; Shridhar, D.; Naik, C. Use of Fourier Transform Infrared (FTIR) spectroscopy to study cadmium-induced changes in *Padina tetrastromatica* (Hauck). *Anal. Chem. Insights* **2008**, *3*, 135–143. [CrossRef]
84. Yee, N.; Benning, L.; Phoenix, V.; Ferris, F. Characterization of metal-cyanobacteria sorption reactions: A combined macroscopic and infrared spectroscopic investigation. *Environ. Sci. Technol.* **2004**, *38*, 775–782. [CrossRef] [PubMed]
85. Naik, M.; Pandey, A.; Dubey, S. Biological characterization of lead-enhanced exopolysaccharide produced by a lead resistant *Enterobacter cloacae* strain P2B. *Biodegradation* **2012**, *23*, 775–783. [CrossRef] [PubMed]
86. Kepenek, E.; Gozen, A.; Severcan, F. Molecular characterization of acutely and gradually heavy metal acclimated aquatic bacteria by FTIR spectroscopy. *J. Biophot.* **2019**, *12*, e201800301. [CrossRef] [PubMed]
87. Wei, X.; Fang, L.; Cai, P.; Huang, Q.; Chen, H.; Liang, W.; Rong, X. Influence of extracellular polymeric substances (EPS) on Cd adsorption by bacteria. *Environ. Pollut.* **2011**, *159*, 1369–1374. [CrossRef]

agronomy

MDPI

Article

Effects of N-P-K Ratio in Root Nutrient Solutions on Ectomycorrhizal Formation and Seedling Growth of *Pinus armandii* Inoculated with *Tuber indicum*

Li Huang [1,†], Rui Wang [1,†], Fuqiang Yu [2], Ruilong Liu [1], Chenxin He [3], Lanlan Huang [4], Shimei Yang [1,2], Dong Liu [2,*] and Shanping Wan [1,*]

1 College of Resources and Environment, Yunnan Agricultural University, Kunming 650201, China; 18869443604@163.com (L.H.); nyyy730@163.com (R.W.); lrl17056810@163.com (R.L.)

2 Germplasm Bank of Wild Species & Yunnan Key Laboratory for Fungal Diversity and Green Development & Yunnan International Joint Laboratory of Fungal Sustainable Utilization in South and Southeast Asia, Kunming Institute of Botany, Chinese Academy of Sciences, Kunming 650201, China; fqyu@mail.kib.ac.cn

3 Meteorological Institute of Shaanxi Province, Xi'an 710016, China

4 College of Yunnan Rural Revitalizing Education, Yunnan Open University, Kunming 650221, China

* Correspondence: liudongc@mail.kib.ac.cn (D.L.); wsp871117@163.com (S.W.)

† These authors contributed equally to this work.

Abstract: Ectomycorrhizal symbiosis is a cornerstone of ecosystem health, facilitating nutrient uptake, stress tolerance, and biodiversity maintenance in trees. Optimizing *Pinus armandii–Tuber indicum* mycorrhizal synthesis enhances the ecological stability of coniferous forests while supporting high-value truffle cultivation. This study conducted a pot experiment to compare the effects of three root nutrient regulations—Aolu 318S (containing N-P_2O_5-K_2O in a ratio of 15-9-11 (w/w%)), Aolu 328S (11-11-18), and Youguduo (19-19-19)—on the mycorrhizal synthesis of *P. armandii–T. indicum*. The results showed that root nutrient supplementation significantly improved the seedling crown, plant height, ground diameter, biomass dry weight, and mycorrhizal infection rate of both the control and mycorrhizal seedlings, with the slow-release fertilizers Aolu 318S and 328S outperforming the quick-release fertilizer Youguduo. The suitable substrate composition in this experiment was as follows: pH 6.53–6.86, organic matter content 43.25–43.49 g/kg, alkali-hydrolyzable nitrogen 89.25–90.3 mg/kg, available phosphorus 83.69–87.32 mg/kg, available potassium 361.5–364.65 mg/kg, exchangeable magnesium 1.17–1.57 mg/kg, and available iron 33.06–37.3 mg/kg. It is recommended to mix the Aolu 318S and 328S solid fertilizers evenly into the substrate, with a recommended dosage of 2 g per plant. These results shed light on the pivotal role of a precise N-P-K ratio regulation in fostering sustainable ectomycorrhizal symbiosis, offering a novel paradigm for integrating nutrient management with mycorrhizal biotechnology to enhance forest restoration efficiency in arid ecosystems.

Keywords: black truffle; *pine*; ectomycorrhizal synthesis; fertilization

1. Introduction

Truffles (*Tuber*) refer to a group of subterranean fungi with important economic and ecological values in the *Tuberaceae*, *Pezizales*, and *Ascomycota* [1,2]. Recent taxonomic and molecular phylogenetic studies of the *Tuber* have shown that at least 82 species exist in China [3]. As the value of truffles continues to rise, over-harvesting has led to the severe destruction of truffle habitats, causing a yearly decline in production and an increasingly serious threat to their existence [4]. In recent years, in order to protect and stabilize the natural resources and ecological environment, as well as to meet the market demand for edible ectomycorrhizal fungi, the artificial cultivation and resource conservation of ectomycorrhizal fungi have attracted much attention [5].

Truffles are among the world's most valuable edible ectomycorrhizal fungi (EMF), capable of forming mutually beneficial mycorrhizal symbioses with dominant tree species in northern temperate forests, such as those in the genera *Pinus* (pines), *Picea* (spruces), *Populus* (poplars), *Salix* (willows), *Corylus* (hazels), and *Quercus* (oaks) [2]. EMF serve as a crucial interface between the soil matrix and host plants, facilitating bidirectional material exchange and signal transmission.

The formation of mycorrhizae not only makes the artificial cultivation of EMF possible but also enhances the ability of plants to absorb carbon, nitrogen, phosphorus, water, and mineral elements [6]. Specifically, in terms of improving resistance to abiotic stresses, EMF can expand the absorption range of host plant roots through their extensive mycelial networks, enabling more efficient uptake of soil moisture and nutrients under drought or nutrient-deficient conditions and alleviating stress [7]. For example, under drought stress, the hyphae of EMF can penetrate into tiny soil pores that root hairs cannot reach, absorbing and transporting water to the host plant while also regulating the expression of plant drought-responsive genes to reduce water loss [8]. In the case of heavy metal stress, EMF can sequester heavy metal ions in their mycelia through chelation and compartmentalization, preventing these toxic substances from entering the host plant's vascular system, thus reducing damage [9]. Additionally, EMF can enhance the stability of plant cell membranes and the activity of antioxidant enzymes, helping plants mitigate oxidative damage caused by high temperatures, salinization, and other stresses [10–14]. Regarding carbon elimination and cycling, EMF play a vital role in the forest carbon cycle. Host plants transfer a portion of their photosynthetically fixed carbon to EMF, which is used for the growth and metabolism of the fungi. A significant portion of this carbon is allocated to the mycelial network in the soil, where some is decomposed into organic matter and stored in the soil, contributing to long-term carbon sequestration [15,16]. This process not only reduces the concentration of carbon dioxide in the atmosphere but also improves the soil's organic carbon content, enhancing soil fertility and stability [17].

Due to the high profitability of truffles, their production has promoted the use of agricultural land in low-productivity areas through private forestation, as well as important associated ecosystem services such as carbon fixation and soil quality improvement [18]. Thus, EMF play a significant role in constructing forest ecological networks and maintaining the health and stability of ecosystems [19].

The soil matrix is a critical medium for the growth of ectomycorrhizal seedlings. To ensure the normal growth of both the seedlings and mycorrhizae, the soil must continuously and appropriately provide and balance essential factors for plant growth, such as water, nutrients, air, heat, and other necessary conditions [20]. Soil properties, including pH and nutrients (such as nitrogen, phosphorus, potassium, magnesium, calcium, and organic carbon), significantly influence the growth of EMF [2]. These soil environmental variables, in turn, interact with EMF, helping to regulate the carbon and nutrient balance at the interface between the soil matrix and the host plant [21,22].

Previous studies have found that EMF yields are higher in fertile soils compared to barren ones. The reasons for these differential results may lie in the effectiveness of soil background nutrients and the impact of exogenous nutrient additions, which influence the degree of nutrient limitation and alleviation in EMF fungi, thereby affecting the direction and extent of the mycelial response to nutrients. Throughout this extended life cycle, soil properties and climatic conditions are considered to be critical for the development of black truffle sporocarps [23,24]. Research indicates that phosphorus (P) is one of the key factors influencing truffle mycelial dynamics in natural forest plantations during black truffle production [25]. Meanwhile, the availability of nitrogen (N) significantly affects the growth characteristics (such as biomass, yield, density, and turnover rate) and functional traits of EMF mycelia [26,27]. Appropriate nitrogen supplementation increased EMF mycelial yield, biomass, and density by 79%, 39%, and 73%, respectively. Additionally, proper nitrogen application can enhance EMF phosphatase activity, thereby promoting phosphorus absorption in plants [28–30]. Recent evidence suggests that this relationship follows an optimum curve rather than a linear trend. In particular, for truffles, Barou et al. [25] revealed that phosphorus (P) availability shows a non-monotonic relationship with *T. melanosporum* mycelial biomass; it is beneficial up to 20 mg kg^{-1} but inhibitory at higher concentrations in plantation soils. Similarly, while nitrogen (N) availability significantly influences EMF's growth characteristics (biomass, yield, and density) and functional traits [26,27], excessive N can disrupt the mutualistic balance, as demonstrated by the increases in mycelial parameters being contingent on maintaining appropriate C: N ratios [24].

Research has shown that inoculating EMF and the addition of a low amount of nitrogen (15 kg/(hm^2·a)) exhibit a synergistic effect on the root morphology and nutrient content of *P. sylvestris* var. *mongolica* seedlings [31]. Through EMF inoculation and exponential fertilization, the growth indicators, nutrient accumulation (N and P), and mycorrhizal colonization rate of Scots pine seedlings were significantly higher than those under conventional fertilization [32]. In addition to using external additions to regulate the nutrient composition of the substrate [33], a reasonable substrate composition is also a key factor for the growth of mycorrhizal seedlings. Dupre et al. [34] studied the effects of different substrates, such as soil, vermiculite, and perlite, as well as factors such as soil N, P, and K content, on the mycorrhizal synthesis of *T. melanosporum*. The results showed that vermiculite as a growth substrate was the most conducive to the formation of truffle mycorrhizae, followed by perlite, while forest topsoil performed the worst. Chen et al. [35] used a mixed substrate of vermiculite, peat, and river sand in a certain volume ratio to inoculate *T. melanosporum* during the seedling stage of *Castanopsis hystrix*, achieving a mycorrhizal infection rate as high as 90% after 26 weeks, indicating that the mixed soil nutrients and structure are very suitable for the growth of *T. melanosporum* mycorrhizal seedlings. The growth and development of *T. indicum* mycorrhizae are also affected by soil nutrients and structure; sufficient nutrients and favorable soil environmental conditions increase the number of *T. indicum* mycorrhizae [36].

In addition, vegetation type is a fundamental factor influencing soil fungal communities and controlling the ecological functions of mycorrhizal systems [37–39]. Vegetation type can alter the physical and chemical properties of the soil (such as nutrient availability and pH) due to differences in growth rates, litter, and root exudates, thereby affecting the formation and growth of mycorrhizal fungi [40,41]. Therefore, the type of host plant can lead to significant differences in mycelial biomass due to its utilization of available soil nutrients. Research has shown that nitrogen application can increase mycelial biomass in *P. armandii* forests by 39%; specifically, nitrogen application can significantly increase the relative abundance of truffle mycelia in these forests [26]. Moreover, the strong selectivity and specificity of truffles for host tree species are evident even at the seedling stage, where the root morphology of the host plant can influence the appearance of the formed mycorrhizae. Thus, selecting suitable symbiotic plants for inoculation is not only a prerequisite for cultivating mycorrhizal seedlings but also a key factor in the success of truffle cultivation [42,43].

Pinus armandii, one of China's native pine species, plays a significant role in ecological protection and regional socio-economic development. Proper fertilization has a marked effect on improving seedling productivity and expanding the spread of the root system. In particular, the use of nitrogen (N) and combinations of nitrogen with other plant nutrients can effectively shorten the harvesting period of trees and enhance the yield and quality of timber [44–47]. It has been pointed out that appropriate fertilization methods, substrate combinations, and fertilizer ratios can significantly promote the growth of container seedlings of *P. armandii* [48–50]. *Pinus armandii* is an ideal host for the synthesis of truffle mycorrhizae [51]. For truffle mycorrhizal seedlings, proper fertilization can promote the growth of both *P. armandii* and mycelia, thereby increasing the mycorrhizal colonization rate, which is crucial for the subsequent growth of truffles and the improvement of their yield.

In the field of mycorrhizology, research on the effects of fertilization on ectomycorrhizae is far less extensive than on other types of mycorrhizae. In particular, studies on the impact of fertilization on specific economically valuable mycorrhizal fungi and host combinations are rarely reported. While previous studies have examined ectomycorrhizal relationships [10,23] and fertilization effects on tree growth [34,46], few have systematically investigated how specific N-P-K ratios in root nutrients affect the economically important *T. indicum–P. armandii* symbiosis. Notably, no studies have compared slow-release versus quick-release fertilizers for this specific host–fungus combination. Thus, in this study, we hypothesized that slow-release fertilizers (Aolu 318S and 328S) with balanced N-P-K ratios would promote superior ectomycorrhizal formation and seedling growth compared to a quick-release fertilizer (Youguduo) due to sustained nutrient availability that matches fungal and plant requirements. An optimal substrate nutrient range exists that simultaneously maximizes both *T. indicum* mycorrhization and *P. armandii* growth. The interaction between specific N-P-K ratios and mycorrhization would create synergistic effects on plant performance beyond additive individual effects. Therefore, this study employed a pot experiment to comparatively analyze the effects of three types of compound fertilizers, Aolu 318S, Aolu 328S, and Yuguoduo, on the mycorrhizal synthesis of *T. indicum* with *P. armandii* and the growth of mycorrhizal seedlings. By analyzing changes in the seedling crown, plant height, ground diameter, aboveground and belowground biomass, substrate nutrient content, and the number of mycorrhizae of the seedlings, this study aims to explore the optimal nutrient substrate ratio and fertilization effect, providing valuable insights for the artificial cultivation of truffles, mycorrhizology research, and the efficient cultivation of robust *P. armandii* seedlings.

2. Materials and Methods

2.1. Sterile Seedling Cultivation and Inoculation

The experimental design used a mixed substrate with a peat/vermiculite/perlite volume ratio of 2:2:1. The moisture content was adjusted to 30%, and the substrate was sterilized at 121–126 °C. The pH was adjusted to 7.5 using calcium carbonate and magnesium carbonate. Mature ascomata of *T. indicum* were used as the inoculum, with an inoculation concentration of 5×10^6 spores/mL and an inoculation amount of 0.5 g per seedling. The inoculation time was 5 March 2023. *Pinus armandii* was fertilized on 5 June 2023 after inoculation. The seedlings used for inoculation were cultivated from germinated seeds of *P. armandii*. The spore suspension was inoculated onto the seedling roots in two stages, followed by covering with the substrate. Mycorrhizal formation was examined at 90 days (5 September 2023), 180 days (5 December 2023), and 270 days (5 March 2024) after fertilization. Each time, 5 seedlings were randomly selected from each of the 8 treatment groups for inspection. All seedlings were cultivated in an outdoor environment with good ventilation and sufficient sunlight and managed under conventional practices with alternating dry and wet cultivation conditions throughout the growth period.

2.2. Experimental Design for the Fertilization and Cultivation of Mycorrhizal Seedlings

This experiment used three types of compound fertilizers (Table 1) and designed 8 experimental treatments: CK (control), T (inoculation with *T. indicum*), 318S (application of Aolu 318S alone), 328S (application of Aolu 328S alone), P19 (application of Yuguoduo alone), T318S (inoculation with *T. indicum* + Aolu 318S), T328S (inoculation with *T. indicum* + Aolu 328S), and TP19 (inoculation with *T. indicum* + Yuguoduo). Fertilization was carried out three months after inoculation. Aolu 318S and Aolu 328S were evenly mixed into the substrate as solid fertilizers, with an application rate of 2 g per pot; Yuguoduo was diluted 600 times with water and applied to the roots at a rate of 0.66 g per pot. Each treatment group contained 15 seedlings, with a total of 120 seedlings across the 8 treatment groups. Mycorrhizal inspections were conducted at 90 days, 180 days, and 270 days after fertilization, with 5 seedlings randomly selected from each of the 8 treatment groups for inspection each time.

Table 1. Application methods, fertility duration, and nutrient content of the three fertilizers.

Fertilizer	Fertilizer Application	Fertile Period	NPK Content and Ratio	Trace Element
Aolu 318S	roots	9 months	15-9-11 + TE	Fe0.45, Mg0.06, Cu0.05, Zn0.015, B0.03
Aolu 328S	roots	9 months	11-11-18 + TE	Fe0.25, Mg0.03, Cu0.055, Zn0.01, B0.01
Youguduo	roots	1 week	19-19-19 + TE	Fe0.05, B0.12, Zn0.06, Mo0.015

Note: Data are expressed as the mean ± standard deviation, and the significance level is set at 0.05.

2.3. Methods for the Inspection, Identification, and Statistical Analysis of Mycorrhizae

Mycorrhizal inspection is conducted by observing morphological changes in plant roots or examining root tissues under a microscope to determine whether mycorrhizae have formed. For mycorrhizal identification, molecular-level identification is performed on the formed mycorrhizae. Before extracting DNA from mycorrhizal root tips using the modified CTAB method [52], the root tips must first be cleaned by rinsing gently to remove any attached substrate particles, then be soaking the roots repeatedly in a beaker filled with water to wash away impurities. After DNA extraction, PCR amplification and sequencing are carried out. The sequencing chromatograms are proofread and assembled using the sequence assembly software SeqMan v7.0. After removing low-quality sequences and primer regions, the sequences are uploaded to NCBI for NCBI Basic Local Alignment Search Tool (BLAST; https://blast.ncbi.nlm.nih.gov/Blast.cgi (accessed on 1 July 2025)) to confirm that the colonized mycorrhizae are the target mycorrhizae.

After completing mycorrhizal verification, statistical analysis is conducted using a stereomicroscope and a root scanner. First, root parameters are obtained by scanning with the root scanner. Then, the roots are observed under the stereomicroscope to count the mycorrhizal infection rate. The calculation formula is: mycorrhizal infection rate (%) = number of mycorrhizae/number of nutrient roots × 100%. Experimental data were analyzed using Microsoft Excel 2010 and SPSS 20.0.

2.4. Measurement of the Seedling Crown, Plant Height, Ground Diameter, and Aboveground and Belowground Biomass of Mycorrhizal Seedlings

For morphological index measurements, the dry weight of the seedlings in each treatment container was determined using an analytical balance. The ground diameter was measured with a vernier caliper, and the plant height and seedling crown were measured with a standard measuring tape. The determination of the dry weight of *P. armandii* seedlings requires drying treatment, with the specific procedure as follows. First, the seedlings are blanched at 105 °C for 30 min to quickly terminate cellular physiological activities and prevent substance decomposition. Then, they are placed in an environment at 80 °C to dry to constant weight. Among these steps, continuous drying at a low temperature of 80 °C can not only effectively remove moisture but also reduce the loss of volatile components in the seedlings. "Drying to constant weight" means that the weight difference between two weighings (with an interval of more than 2 h) does not exceed 0.001 g, which ensures the complete removal of moisture and guarantees the stability and accuracy of the determination results.

2.5. Measurement Indicators and Analytical Methods for Nutrient Substrates

The measurement methods for pH, organic matter, alkali-hydrolyzable nitrogen, available phosphorus, available potassium, exchangeable calcium and magnesium, and available iron in the nutrient substrate refer to those outlined in "Soil Agrochemical Analysis", edited by Bao Shidan [53].

2.6. Statistical Analysis

To explore the significant differences in variables among different treatments, one-way analysis of variance (ANOVA) was conducted at a 95% confidence level ($p < 0.05$), followed by Tukey's honest significant difference test (Tukey's HSD). When the data showed a normal distribution and homogeneous variance, Tukey's HSD test was used; if these assumptions were not met, the Kruskal–Wallis H non-parametric test was adopted instead.

3. Results

3.1. Effects of Different Treatments on the Seedling Crown, Plant Height, and Ground Diameter of T. indicum–P. armandii Seedlings

At the same growth stage, the growth of the seedling crown, plant height, and ground diameter of *P. armandii* varied under different fertilization treatments (Figure 1A–C, $p < 0.05$). Among the non-inoculated treatments, *P. armandii* seedlings treated with Aolu 318S and 328S had significantly higher seedling crowns, plant heights, and ground diameters compared to those treated with Yuguoduo and the control group ($p < 0.05$), while no significant difference was found between Aolu 318S and 328S ($p > 0.05$). The difference between Yuguoduo and the control group was not significant ($p > 0.05$).

Similarly, among the inoculated treatments, seedlings treated with Aolu 318S and 328S also exhibited significantly higher seedling crowns, heights, and ground diameters compared to those treated with Yuguoduo and the group inoculated with *T. indicum* alone ($p < 0.05$), with no significant difference between Yuguoduo and the group inoculated with *T. indicum* alone ($p > 0.05$).

3.2. Effects of Different Treatments on the Aboveground and Belowground Biomass Dry Weight of T. indicum–P. armandii Seedlings

At different growth stages, the aboveground and belowground biomass dry weights of *P. armandii* seedlings showed significant differences under different treatments during the same time period. On day 90, the aboveground and belowground biomass dry weights of the T328S treatment were higher than those of the T318S treatment, but the difference was not significant. However, both T318S and T328S treatments showed significant differences in biomass compared to other treatments. On days 180 and 270, the aboveground and belowground biomass dry weights of the T328S treatment were significantly higher than those of other treatments (Table 2). Throughout the entire growth period, inoculation combined with fertilization significantly increased the aboveground and belowground biomass dry weights of *P. armandii* seedlings. The effectiveness of fertilization on the *P. armandii* seedlings inoculated with *T. indicum* was as follows: Aolu 328S > Aolu 318S > Yuguoduo.

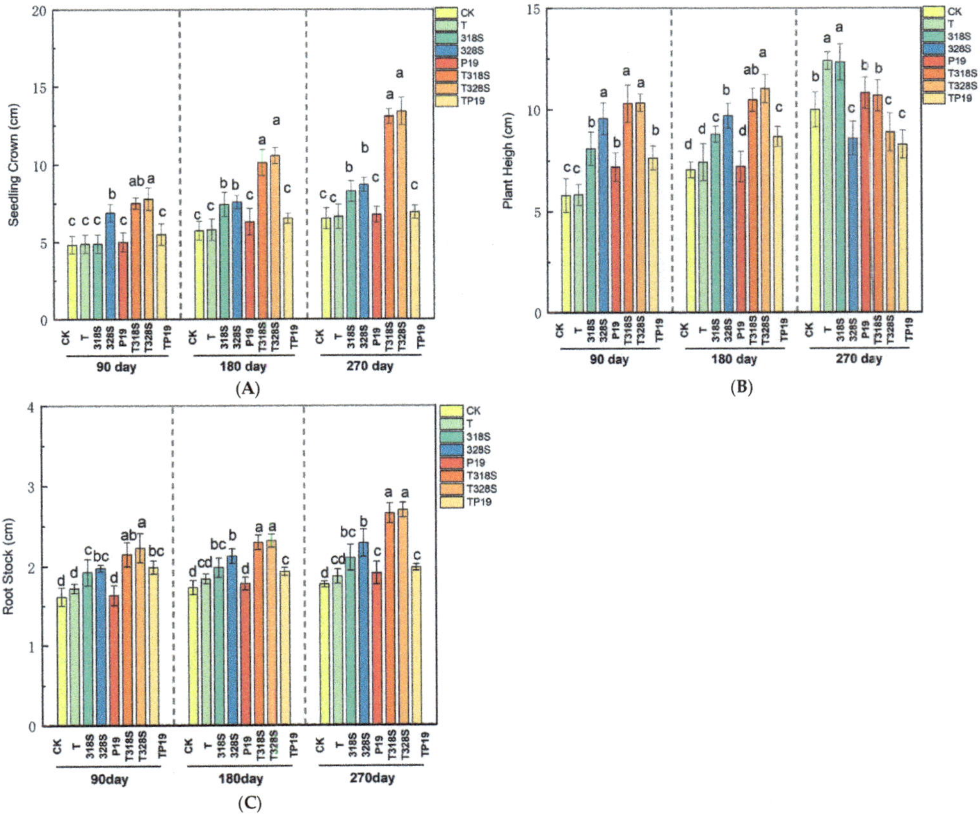

Figure 1. Effects of different treatments on the seedling crown (**A**), plant height (**B**), and ground diameter (**C**) of *P. armandii* at different growth stages. Note: CK: control treatment; T: inoculation with *T. indicum* only; 318S: application of Aolu 318S only; 328S: application of Aolu 328S only; P19: application of Youguduo only; T318S: combined inoculation with *T. indicum* and application of Aolu 318S; T328S: combined inoculation with *T. indicum* and application of Aolu 328S; TP19: combined inoculation with *T. indicum* and application of Youguduo. Data represent the mean of five replicates. Different letters indicate significant differences at the 0.05 level, while the same letter indicates no significant difference.

Table 2. Aboveground and belowground biomass dry weights (g) of *P. armandii* seedlings under different treatments at different growth stages.

Treatments	Aboveground Dry Weight			Belowground Dry Weight		
	90 Day	180 Day	270 Day	90 Day	180 Day	270 Day
CK	2.25 ± 0.04 d	2.41 ± 0.06 d	2.44 ± 0.04 d	2.14 ± 0.03 d	2.27 ± 0.08 d	2.33 ± 0.07 e
T	2.29 ± 0.08 cd	2.42 ± 0.05 d	2.46 ± 0.08 d	2.17 ± 0.04 d	2.30 ± 0.07 d	2.35 ± 0.08 de
318S	2.35 ± 0.08 bc	2.49 ± 0.14 cd	2.67 ± 0.06 c	2.30 ± 0.12 c	2.37 ± 0.07 cd	2.44 ± 0.08 cd
328S	2.39 ± 0.06 bc	2.64 ± 0.05 b	2.70 ± 0.03 c	2.36 ± 0.08 bc	2.43 ± 0.07 c	2.47 ± 0.08 c
P19	2.30 ± 0.07 cd	2.47 ± 0.08 cd	2.49 ± 0.04 d	2.20 ± 0.04 d	2.31 ± 0.07 d	2.34 ± 0.06 de
TF318S	2.41 ± 0.05 ab	2.71 ± 0.13 ab	3.06 ± 0.17 b	2.40 ± 0.04 ab	2.67 ± 0.06 b	2.71 ± 0.06 b
TF328S	2.49 ± 0.08 a	2.84 ± 0.15 a	3.19 ± 0.12 a	2.46 ± 0.04 a	2.82 ± 0.14 a	2.86 ± 0.09 a
TFP19	2.32 ± 0.08 cd	2.55 ± 0.06 bc	2.68 ± 0.08 c	2.21 ± 0.04 d	2.34 ± 0.05 cd	2.38 ± 0.06 cde

Note: Data are expressed as the mean ± standard deviation, and the significance level is set at 0.05. Different letters (e.g., a, b, c, d, e) indicate significant differences between treatments at the 0.05 level based on ANOVA; treatments sharing the same letter are not significantly different from each other.

3.3. *Effects of Different Fertilization Treatments on the Mycorrhizal Infection Rate of T. indicum–P. armandii Seedlings*

The mycorrhizal infection rate showed significant differences under different treatments during the same growth period. In the T treatment, the mycorrhizal infection rate remained relatively stable between 40% and 50% across the three growth stages, with no significant changes. In the three treatments with both inoculation and fertilization, the mycorrhizal infection rate continuously increased throughout the growth period, with the rates in the T318S and T328S treatments being significantly higher than in the TP19 treatment. However, there was no significant difference in the mycorrhizal infection rate between the T318S and T328S treatments. Overall, the T318S and T328S treatments performed better than the TP19 and T treatments, indicating that the fertilizer effect of Osmocote 318S and 328S was superior to that of Yuguoduo. This suggests that fertilization is more beneficial than no fertilization in enhancing the mycorrhizal infection rate of *P. armandii* inoculated with *T. indicum* (Figure 2).

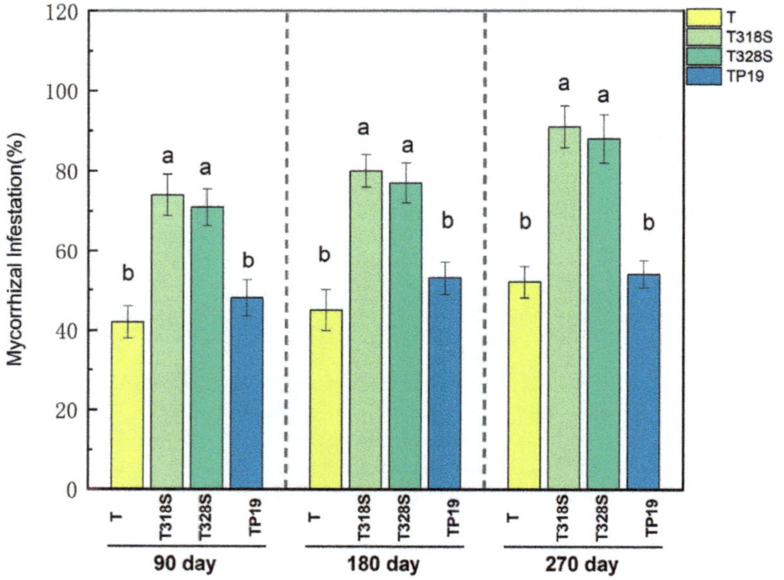

Figure 2. Effects of different treatments on the mycorrhizal infection rate of *P. armandii* at different growth stages. Note: CK: control treatment; T: inoculation with *T. indicum* only; 318S: application of Aolu 318S only; 328S: application of Aolu 328S only; P19: application of Youguduo only; T318S: combined inoculation with *T. indicum* and application of Aolu 318S; T328S: combined inoculation with *T. indicum* and application of Aolu 328S; TP19: combined inoculation with *T. indicum* and application of Youguduo. Data represent the mean of five replicates. Different letters indicate significant differences at the 0.05 level, while the same letter indicates no significant difference.

Under a stereomicroscope, the ectomycorrhizae formed by *T. indicum* and *P. armandii* displayed various morphological appearances (Figure 3). The ectomycorrhizal system was mostly bifurcated or irregularly pinnate, sometimes appearing coralloid or intermediate between these types. The young mycorrhizae were light yellow in color with numerous white emanating hyphae. As the mycorrhizae aged, their color deepened to dark brown. These are consistent with well-documented *T. indicum–P. armandii* morphotypes in peer-reviewed studies [49].

Figure 3. Morphological appearance of ectomycorrhizae formed by *T. indicum* and *P. armandii*. (**A**) bifurcated mycorrhiza; (**B**) coral-like mycorrhiza; (**C**) simple unbranched mycorrhiza.

3.4. Nutrient Status Changes in the Substrate and Establishing the Optimal Nutrient Substrate Ratio

The changes in substrate pH under different treatments at different growth stages ranged between 6.18 and 7.63 (Figure 4A). According to the Chinese soil pH classification standards, the substrate pH values under different treatments and at different growth stages are within the weakly alkaline range. During the same growth period, there were significant differences in the organic matter content of the substrate among different treatments. Throughout the entire growth period, the organic matter content in each treatment showed a decreasing trend. However, the organic matter content in the T318S, T328S, and T treatments was slightly higher than in the CK, 318S, 328S, and P19 treatments. After 90 days of fertilization, the organic matter content in the T328S treatment reached 46.34 g/kg. In the T318S treatment, it reached 45.20 g/kg, and in the T treatment, it reached 45.93 g/kg. These three treatments were at relatively higher levels compared to the other treatments (Figure 4B).

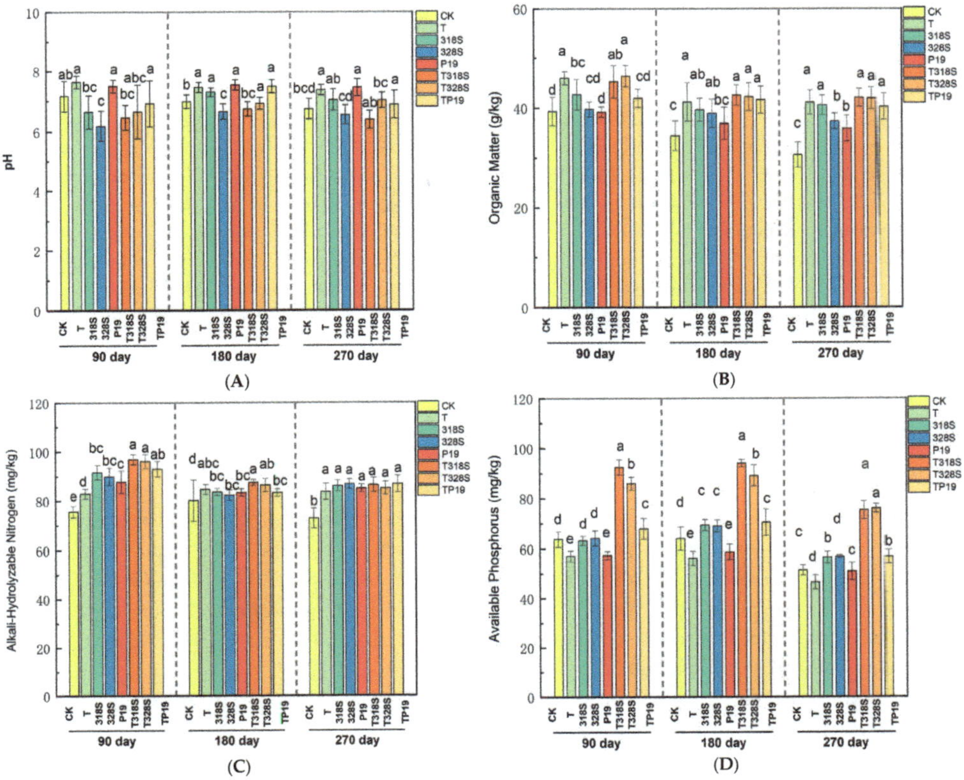

Figure 4. *Cont.*

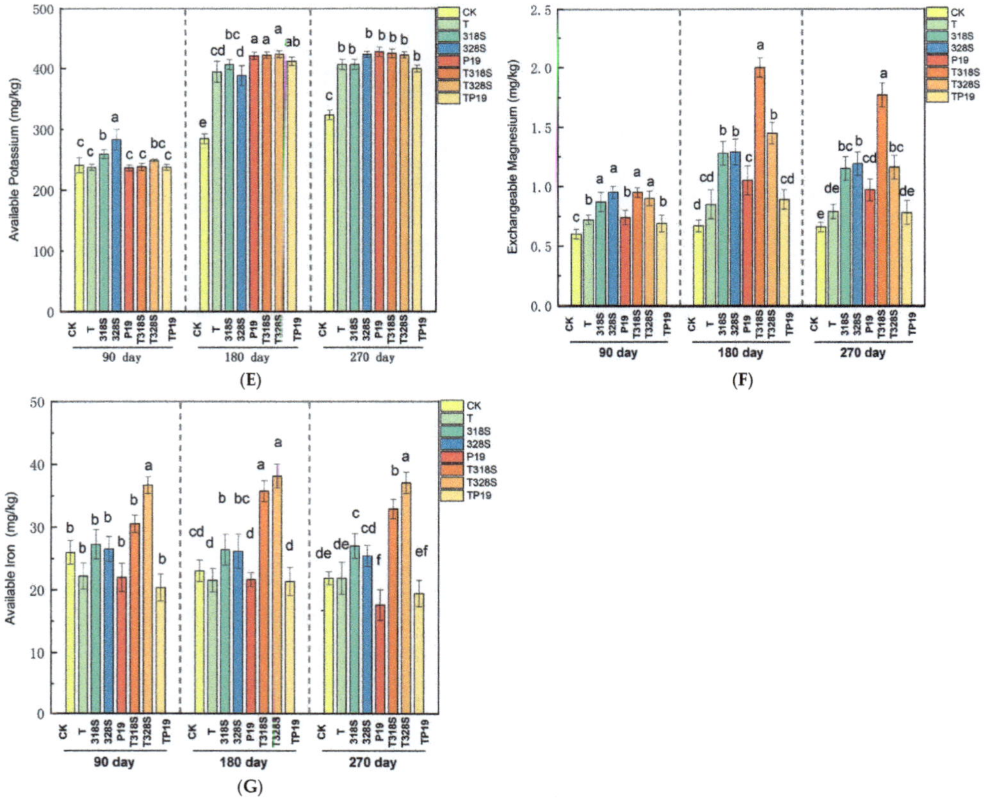

Figure 4. Changes in substrate pH (**A**), organic matter (**B**), alkali-hydrolyzable nitrogen (**C**), available phosphorus (**D**), available potassium (**E**), exchangeable magnesium (**F**), and available iron (**G**) under different treatments at different growth stages. Note: CK: control treatment; T: inoculation with *T. indicum* only; 318S: application of Aolu 318S only; 328S: application of Aolu 328S only; P19: application of Youguduo only; T318S: combined inoculation with *T. indicum* and application of Aolu 318S; T328S: combined inoculation with *T. indicum* and application of Aolu 328S; TP19: combined inoculation with *T. indicum* and application of Youguduo. Data represent the mean of five replicates. Different letters indicate significant differences at the 0.05 level, while the same letter indicates no significant difference.

During the same growth period, there were significant differences in the alkali-hydrolyzable nitrogen content of the substrate among different treatments. In the CK and T treatments, the alkali-hydrolyzable nitrogen content initially increased and then decreased, while in the other treatments, it showed a decreasing trend with no significant changes. The alkali-hydrolyzable nitrogen content in the substrate of all treatments was at a deficient level. After 90 and 180 days of fertilization, the alkali-hydrolyzable nitrogen content in the TF318S treatment was the highest among all treatments, whereas the CK treatment had the lowest content. From 180 to 270 days after fertilization, the alkali-hydrolyzable nitrogen content in the substrate showed no significant changes in treatments other than CK and T (Figure 4C).

The changes in available phosphorus in the substrate under different treatments during various growth stages showed an overall trend of first increasing and then decreasing. This may be related to the decomposition of applied fertilizers into the substrate and the subsequent consumption of a large amount of phosphorus in the later stages of growth.

Among them, the available phosphorus content in the substrate for the TF318S and TF328S treatments was significantly higher than in other treatments, while the other treatments showed no significant changes throughout the growth period (Figure 4D).

The trend in available potassium in the substrate under different treatments during various growth stages, except for the CK treatment, was an increase followed by no significant changes. The available potassium content in the substrate of the 328S treatment showed a slight upward trend in the later stages. The available potassium content in the CK treatment was lower compared to other treatments. The maximum available potassium content in the substrate was 423.56 mg/kg, and the minimum was 236.49 mg/kg, indicating an extremely rich level (Figure 4E).

The changes in exchangeable magnesium content in the substrate across all treatments during different growth stages showed an initial increase followed by a decrease. In the T318S and T328S treatments, the exchangeable magnesium content in the substrate was significantly higher than in the other treatments (Figure 4F).

For available iron, the content in the substrate of the T328S, T318S, TP19, and P19 treatments first increased and then decreased. In the T treatment, the available iron content in the substrate continued to increase. Meanwhile, in the 318S and 328S treatments, the available iron content in the substrate remained relatively stable (Figure 4G).

4. Discussion

This study conducted a statistical analysis of data on seedling crown width, plant height, ground diameter, biomass, and mycorrhizal infection rate following fertilization, combined with changes in substrate nutrient content, to evaluate the effects of various treatments on seedling growth and mycorrhizal infection. The results indicate that pH has a significant impact on mycorrhizal formation, with the highest number of mycorrhizae observed at a pH of 6.53 to 6.86. This pH range is more conducive to the synthesis of *T. indicum* mycorrhizae. Research has shown that various truffle species also grow in slightly alkaline soils, and the pH of the soil in *T. indicum* truffle beds gradually shifts from slightly acidic to slightly alkaline during the maturation stage, indicating that it can regulate the pH of its growing environment [54–57]. Furthermore, *T. indicum* can form mycorrhizae under slightly acidic conditions, suggesting that the pH range for *P. armandii* inoculated with *T. indicum* to form mycorrhizae is relatively broad. However, excessively acidic or alkaline conditions can inhibit mycorrhizal formation, making neutral to slightly alkaline substrates the most suitable for growth. Within the tested pH range (6.18–7.63), mycorrhizal formation showed no significant correlation with pH variation across treatments ($p > 0.05$), suggesting that the observed differences in colonization rates were primarily driven by fertilizer types rather than pH effects. This aligns with field observations showing *T. indicum*'s broad ecological adaptability to soils ranging from pH 4.78 to 8.5 [58,59]. The homogeneity of pH across treatments suggests it was a permissive rather than limiting factor in this study.

Our results demonstrate that slow-release fertilizers (Aolu 318S and 328S) significantly enhanced seedling growth, with T328S increasing aboveground biomass by 30.7% (3.19 g vs. 2.44 g in the control) and mycorrhizal infection rates by 2.0-fold (50% vs. 24% in T-only) at 270 days (Table 2 and Figure 2). These improvements align with the sustained nutrient release of slow-release formulations, which matched the prolonged nutrient demands of both *P. armandii* and *T. indicum*. In contrast, quick-release Youguduo (19-19-19 NPK) led to transient nutrient availability, resulting in 12% lower biomass and 20% lower infection rates compared to slow-release treatments, likely due to leaching losses (Figure 4D,E) and microbial immobilization [25].

In this study's statistical analysis, it was found that fertilization provided abundant nutrients for the mycelium and mycorrhizae, and mycorrhizal treatment effectively increased the organic matter content of the substrate, promoting the absorption and utilization of nitrogen, phosphorus, and potassium by *P. armandii* seedlings. This is because *P. armandii–T. indicum* mycorrhizal seedlings have a strong ability to convert unavailable nitrogen into available nitrogen during growth, and they can quickly activate insoluble phosphorus in the substrate, thereby promoting the growth of *P. armandii* seedlings and *T. indicum–P. armandii* mycorrhizal seedlings, with the latter showing significant growth advantages [56,59,60]. Research has shown that soil nitrogen nutrition plays a crucial role in the growth and development of truffles during their maturation period, and phosphorus may regulate truffle growth by affecting the fungal and bacterial communities in the soil [61–63], although the exact mechanism requires further study. In fact, ectomycorrhizae and host plants have a mutually beneficial symbiotic relationship. Kang [64] indicates that truffle inoculation can increase the nitrate nitrogen content in the rhizosphere soil of *Quercus*, reduce the available phosphorus content, enhance the activity of *Quercus* roots, and promote the colonization of denitrifying bacteria in the rhizosphere soil, thus affecting the soil's physical and chemical properties.

The superior performance of T318S/T328S was correlated with elevated levels of substrate organic matter (46.34 g/kg versus 43.25 g/kg in the control) and available phosphorus (87.32 mg/kg versus 83.69 mg/kg in the control), which directly facilitated the expansion of fungal hyphae and the development of host roots (Figure 4B–D). The 11% higher phosphorus content in T328S likely stimulated mycelial growth, given that Tuber spp. upregulate phosphatase activity under phosphorus-limiting conditions [19]. Similarly, the increased concentration of exchangeable magnesium (1.57 mg/kg versus 1.17 mg/kg in the control) may have promoted enzymatic processes in both plant and fungal metabolism, considering magnesium's essential role in ATP synthesis [59]. Notably, the transient peak of iron in T328S (37.3 mg/kg at 90 days, Figure 4G) might enhance the stress responses of fungi, as iron functions as a cofactor for multiple antioxidant enzymes in fungal cells [65].

Nitrogen, phosphorus, and potassium are essential nutrients for plant growth and development, and within a certain range, they can promote the growth of both aboveground parts and root systems [66,67]. This experiment showed that moderate levels of available phosphorus, quick-acting potassium, exchangeable magnesium, and available iron promote the formation of *T. indicum* mycorrhizae. Chen's research on artificial truffle cultivation techniques pointed out that an excessive organic matter content in the substrate may inhibit mycorrhizal formation, while organic nitrogen sources such as aspartic acid and moderate levels of phosphorus significantly promote mycorrhizal synthesis [29]. The growth-promoting effects of different metal ions on truffle mycelium follow the order of $Ca^{2+} > Na^{+} > K^{+} > Mg^{2+}$ [68–71]. The experimental results indicate significant differences between fertilization treatments. The quick-release fertilizer Youguduo may lead to nutrient loss before the mycelium and mycorrhizae have fully absorbed them. The fertilizer efficiency differences produced by applying different fertilizers may be potentially affected by various biotic and abiotic factors. For quick-release fertilizers such as Youguduo (19-19-19) in this research, rapid nutrient dissolution likely increased initial leaching risks, particularly for mobile N and K, as evidenced by the sharper decline in available P and K in P19/TP19 treatments compared to slow-release formulations (Aolu 318S/328S). Similar patterns were reported in truffle agroforestry systems, where irrigation accelerated nutrient leaching in sandy substrates [18]. Microbial immobilization may have further reduced available N, as ectomycorrhizal fungi (EMF) and associated bacteria compete for inorganic N [7]. The higher organic matter in T318S/T328S treatments (Figure 4B) suggests fungal-mediated nutrient retention, aligning with findings that EMF hyphae reduce leaching by binding nutrients [70]. Considering that there is no significant difference in substrate pH values during different stages of Youguoduo treatment, the possibility of non-biological nutrient loss caused by this is relatively small. These also explain the superior performance of slow-release fertilizers in maintaining the long-term nutritional supply of *P. armandii* and *T. indicum* symbiosis.

The experimental results indicate significant differences between fertilization treatments. The quick-release fertilizer Youguduo may lead to nutrient loss before the mycelium and mycorrhizae have fully absorbed them. Therefore, it is recommended to use the slow-release fertilizers Aolu 318S and Aolu 328S, which are more suitable for mycelium germination and promote fruiting body growth.

Under the context of climate change, the distribution range of *P. armandii* is gradually shrinking. The most concerning issue is that the suitable habitats for *P. armandii* in the Wushan region, which connects the Daba Mountains and the Yunnan–Guizhou Plateau, will gradually disappear, leading to the disruption of the current ecological corridor [72]. Fertilization and mycorrhization of seedlings play crucial roles in promoting the growth of *P. armandii*, enhancing stress resistance, improving soil structure, increasing nutrient utilization, and maintaining the health and development of forest ecosystems [73–75]. Therefore, protecting and restoring these critical habitats is of utmost importance. To address the ecological challenges posed by the shrinking distribution of *P. armandii*, it is essential to research and promote biotechnological measures based on ectomycorrhizal fungi. Ectomycorrhizal fungi not only enhance the stress resistance of *P. armandii* but also improve the soil's physical and chemical properties, promote nutrient cycling, and enhance plants' adaptability to environmental changes [10,76,77]. Moreover, by adopting appropriate fertilization strategies, particularly the use of slow-release fertilizers, the long-term healthy growth of seedlings can be supported, nutrient loss can be reduced, and soil fertility can be increased. This approach will help maintain and restore *P. armandii* populations while ensuring the stability and biodiversity of the entire forest ecosystem. Future research should further explore the optimal combination of mycorrhization treatments and fertilization strategies under different environmental conditions to provide scientific support for the conservation and sustainable development of *P. armandii*.

5. Conclusions

Under conditions of inoculation with *Tuber indicum* ascocarp inoculum at a spore concentration of 5×10^6 spores/mL, the optimal fertilization regime for *Pinus armandii–T. indicum* mycorrhizal synthesis and seedling development was determined as follows: Aolu 318S and 328S exhibited optimal efficacy at 2 g per plant, applied three months post-inoculation by uniform incorporation into the substrate. For Youguduo, the optimal dosage was 0.66 g per plant, administered three months after inoculation via the root application of a 600-fold water dilution. Comparative analysis revealed Aolu 318S to be the most effective formulation for promoting mycorrhizal formation and seedling growth. These findings provide robust scientific evidence and practical guidelines for cultivating *P. armandii–T. indicum* mycorrhizal seedlings, demonstrating that optimized fertilizer type, dosage, and application method can significantly enhance mycorrhizal establishment and seedling quality, thereby maximizing yields in truffle artificial cultivation. The superior performance of slow-release Aolu 318S over quick-release Youguduo offers critical insights for fertilizer selection and fertilization strategy development, enabling the optimization of tree-fungal symbiosis under diverse cultivation scenarios.

Author Contributions: L.H. (Li Huang), visualization, methodology, software, and formal analysis; R.W., writing—review and editing and data curation; R.L., writing—review and editing; S.Y., methodology, conceptualization, and supervision; F.Y., resources, supervision, and project administration; C.H., data curation and visualization; L.H. (Lanlan Huang), methodology and resources; D.L., supervision, validation, and writing—review and editing; S.W., validation, funding acquisition, and writing—review and editing. All authors have read and agreed to the published version of the manuscript.

Funding: This work was financially supported by the Young Talents Special Project of the Xing Dian Talent Support Program of Yunnan Province (XDYC-QNRC-2023-0415), the National Natural Science Foundation of China (No. 32060008), and the Yunnan Technology Innovation Program (202205AD160036).

Data Availability Statement: Data are contained within the article.

Conflicts of Interest: The authors declare no conflicts of interest.

References

1. Bonito, G.; Smith, M.E.; Nowak, M.; Healy, R.A.; Guevara, G.; Cazares, E.; Kinoshita, A.; Nouhra, E.R.; Dominguez, L.S.; Tedersoo, L.; et al. Historical biogeography and diversification of truffles in the *Tuberaceae* and their newly identified southern hemisphere sister lineage. *PLoS ONE* **2013**, *8*, E52765. [CrossRef] [PubMed]
2. Zambonelli, A.; Iotti, M.; Murat, C. *True Truffle (Tuber spp.) in the World*; Springer International Publishing: Berlin/Heidelberg, Germany, 2016; pp. 3–18. [CrossRef]
3. Fan, L.; Li, T.; Xu, Y.Y.; Yan, X.Y. Species diversity, phylogeny, endemism and geography of the truffle genus *Tuber* in China based on morphological and molecular data. *Persoonia* **2022**, *48*, 175–202. [CrossRef] [PubMed]
4. Li, Y.; Fan, L.; Wang, K.; Song, M.S.; Wang, X.C.; Zhuang, W.Y.; Yao, Y.J. Status and Protection of Threatened Truffle (*Tuberaceae*) Resources in China. *Acta Edulis Fungi* **2020**, *27*, 109–119. [CrossRef]
5. Guerin-Laguette, A. Successes and challenges in the sustainable cultivation of edible mycorrhizal fungi–furthering the dream. *Mycoscience* **2021**, *62*, 10–28. [CrossRef] [PubMed]
6. Reddy, G.; Reddy Bokka, S.K.; Chellem, S.R.; Kalpana, K.; Indrani, K.; Lakshmi Hima Bindu, T.P.; Sri Navya, D. Mutualistic Relationships between Plants and Mycorrhizal Fungi Impacts on Ecosystem Functioning: A Review. *Uttar Pradesh J. Zool.* **2024**, *45*, 242–261. [CrossRef]
7. Li, Y.; Zhang, T.; Zhou, Y.; Zou, X.; You, Y.; Hui, L.; Liying, L.; Songzhu, Z. Ectomycorrhizal symbioses increase soil calcium availability and water use efficiency of Quercus acutissima seedlings under drought stress. *Eur. J. For. Res.* **2021**, *140*, 1039–1048. [CrossRef]
8. Pickles, B.J.; Simard, S.W. Mycorrhizal networks and forest resilience to drought. In *Mycorrhizal Mediation of Soil Fertility, Structure, and Carbon Storage*; Academic Press: Cambridge, MA, USA, 2017; pp. 319–339. [CrossRef]
9. Chot, E.; Reddy, M. Role of Ectomycorrhizal Symbiosis Behind the Host Plants Ameliorated Tolerance Against Heavy Metal Stress. *Front. Microbiol.* **2022**, *13*, 855473. [CrossRef] [PubMed]
10. Smith, S.E.; Read, D.J. *Mycorrhizal Symbiosis*, 3rd ed.; Academic Press: New York, NY, USA, 2008; p. 787. [CrossRef]
11. Zhang, H.Y.; Lü, X.T.; Hartmann, H.; Keller, A.; Han, X.G.; Trumbore, S.; Phillips, R.P. Foliar nutrient resorption differs between arbuscular mycorrhizal and ectomycorrhizal trees at local and global scales. *Glob. Ecol. Biogeogr.* **2018**, *27*, 875–885. [CrossRef]
12. Ning, C.; Mueller, G.M.; Egerton-Warburton, L.M.; Wilson, A.W.; Yan, W.; Xiang, W. Diversity and Enzyme Activity of Ectomycorrhizal Fungal Communities Following Nitrogen Fertilization in an Urban-Adjacent Pine Plantation. *Forests* **2018**, *9*, 99 [CrossRef]
13. Burke, D.; Carrino-Kyker, S.R.; Chervenak, C.F.; Hoke, A.J.; Hewins, C.R. The function of root mat fungal communities: Changes in response to pH and phosphorus addition. *Plants People Planet* **2021**, *3*, 653–666. [CrossRef]
14. Yang, N.; Hua, J.N.; Zhang, J.B.; Liu, D.; Bhople, P.; Li, X.X.; Zhang, Y.; Ruan, H.H.; Xing, W.; Mao, L.F. Soil nutrients and plant diversity affect ectomycorrhizal fungal community structure and functional traits across three subalpine coniferous forests. *Front Microbiol.* **2022**, *13*, 820144. [CrossRef] [PubMed]
15. Averill, C.; Hawkes, C. Ectomycorrhizal fungi slow soil carbon cycling. *Ecol. Lett.* **2016**, *19*, 937–947. [CrossRef] [PubMed]
16. Hawkins, H.; Cargill, R.; Nuland, M.; Hagen, S.; Field, K.; Sheldrake, M.; Soudzilovskaia, N.; Kiers, E. Mycorrhizal mycelium as a global carbon pool. *Curr. Biol.* **2023**, *33*, R560–R573. [CrossRef] [PubMed]
17. Stuart, E.; Singan, V.; Amirebrahimi, M.; Na, H.; Ng, V.; Grigoriev, I.; Martin, F.; Anderson, I.; Plett, J.; Plett, K. Acquisition of host-derived carbon in biomass of the ectomycorrhizal fungus *Pisolithus microcarpus* is correlated to fungal carbon demand and plant defences. *FEMS Microbiol. Ecol.* **2023**, *99*, fiad037. [CrossRef] [PubMed]
18. Leonardi, M.; Iotti, M.; Pacioni, G.; Hall, I.R.; Zambonelli, A. Truffles: Biodiversity, Ecological Significances, and Biotechnological Applications. In *Industrially Important Fungi for Sustainable Development. Volume 1: Biodiversity and Ecological Perspectives*; Abdel-Azeem, A.M., Yadav, A.N., Yadav, N., Usmani, Z., Eds.; Springer: Cham, Switzerland, 2021; pp. 107–146. [CrossRef]
19. Martin, F.M.; van der Heijden, M.G.A. The mycorrhizal symbiosis: Research frontiers in genomics, ecology, and agricultural application. *New Phytol.* **2024**, *242*, 1486–1506. [CrossRef] [PubMed]

20. Jaillard, B.; Barry-Etienne, D.; Colinas, C.; Miguel, A.M.D.; Genola, L.; Libre, A.; Neveu, P.; Oliach, D.; Saenz, W.; Saez, M.; et al. Alkalinity and structure of soils determine the truffle production in the Pyrenean Regions. *For. Syst.* **2014**, *23*, 364–377. [CrossRef]
21. Cahanovitc, R.; Livne-Luzon, S.; Angel, R.; Klein, T. Ectomycorrhizal fungi mediate belowground carbon transfer between pines and oaks. *ISME J.* **2022**, *16*, 1420–1429. [CrossRef] [PubMed]
22. van der Linde, S.; Suz, L.M.; Orme, C.D.L.; Cox, F.; Andreae, H.; Asi, E.; Atkinson, B.; Benham, S.; Carroll, C.; Cools, N.; et al. Environment and host as large-scale controls of ectomycorrhizal fungi. *Nature* **2018**, *558*, 243–248. [CrossRef] [PubMed]
23. Garcia-Barreda, S.; Sánchez, S.; Marco, P.; Serrano-Notivoli, R. Agro-climatic zoning of Spanish forests naturally producing black truffle. *Agric. For. Meteorol.* **2019**, *269–270*, 231–238. [CrossRef]
24. Castrignano, A.; Goovaerts, P.; Lulli, L.; Bragato, G. A geostatistical approach to estimate probability of occurrence of *Tuber melanosporum* in relation to some soil properties. *Geoderma* **2000**, *98*, 95–113. [CrossRef]
25. Barou, V.; Rincón, A.; Parladé, J. Modelling environmental drivers of *Tuber melanosporum* extraradical mycelium in productive holm oak plantations and forests. *For. Ecol. Manag.* **2024**, *563*, 121988. [CrossRef]
26. Guo, W.J.; Ding, J.X.; Wang, Q.T.; Yin, M.Z.; Zhu, X.Y.; Liu, Q.; Zhang, Z.L.; Yin, H.J. Soil fertility controls ectomycorrhizal mycelial traits in alpine forests receiving nitrogen deposition. *Soil. Biol. Biochem.* **2021**, *161*, 108386. [CrossRef]
27. Hendrick, R.L.; Pregitzer, K.S.; Fogel, R. Effects of nitrogen addition on ectomycorrhizal fungi in forest ecosystems. *Front. Plant Sci.* **2010**, *10*, 1658. [CrossRef]
28. Pang, L.; Zhang, Y.; Zhou, Z.C.; Feng, Z.P.; Chu, D.Y. Effects of simulated nitrogen deposition on root exudates and phosphorus efficiency in *Pinus massoniana* families under low phosphorus stress. *Chin. J. Plant Ecol.* **2014**, *38*, 27–35. [CrossRef]
29. Jones, M.D.; Phillips, L.A.; Treu, R.; Ward, V.; Berch, S.M. Functional responses of ectomycorrhizal fungal communities to long-term fertilization of lodgepole pine (*Pinus contorta* Dougl. ex Loud. var. *latifolia Engelm.*) stands in central British Columbia. *Appl. Soil. Ecol.* **2012**, *60*, 29–40. [CrossRef]
30. Song, P.; Zhang, Y.; Zhang, R.; Zhuo, Z.C.; Feng, Z.P. Responses of phosphorus efficiency to simulated nitrogen deposition under phosphorus deficiency in *Pinus massoniana* clones. *J. Plant Nutr.* **2017**, *23*, 502–511. [CrossRef]
31. Hao, L.F.; Hao, W.Y.; Liu, T.Y.; Zhang, M.; Xu, J.K. Responses of root morphology and nutrient content of *Pinus sylvestris* var. mongolica seedlings to nitrogen addition and inoculation treatments. *J. Beijing For. Univ.* **2021**, *43*, 1–7. [CrossRef]
32. Wang, J.Y.; Hao, L.F.; Bai, S.L.; Bao, H.G. Responses of *Pinus sylvestris* seedlings to inoculation with ectomycorrhizal fungi and exponential fertilization. *Soil. Fertil. Sci. China* **2022**, 193–200. [CrossRef]
33. Chen, Q.X. Study on Key Techniques df *Cyclobalanopsis glauca* Container Seedling. Ph.D. Thesis, Beijing Forestry University, Beijing, China, 2010.
34. Dupre, C.; Chevalier, G.; Morizet, J.; Leblevenec, L. Effect of nitrogen and phosphorus on mycorrhizal association between *Quercus pubescens* Willd. and *Tuber melanasporum* Vitt. Under controlled conditions. *Colloques-de-1'INRA* **1982**, *13*, 156–158.
35. Chen, Y.L.; Gong, M.Q.; Chen, Y.; Wang, F.Z. Effects of Inoculation with 11 Ectomycorrhizal Fungal Isolates on Growth and Photosynthesis of *Castanopsis hystrix* Saplings. *For. Res.* **2001**, *14*, 515–522. [CrossRef]
36. Liu, Q.B.; Li, X.M.; Bai, H.F.; Mu, L.Q.; Xiong, H.K. Establishment and management techniques of truffle ecological plantation in Chuxiong Prefecture. *Edible Fungi* **2022**, *44*, 42–45. [CrossRef]
37. Leff, J.W.; Bardgett, R.D.; Wilkinson, A.; Jackson, B.G.; Pritchard, W.J.; De Long, J.R.; Oakley, S.; Mason, K.E.; Ostle, N.J.; Johnson, D.; et al. Predicting the structure of soil communities from plant community taxonomy, phylogeny, and traits. *ISME J.* **2018**, *12*, 1794–1805. [CrossRef] [PubMed]
38. Porazinska, D.L.; Farrer, E.C.; Spasojevic, M.J.; Bueno de Mesquita, C.P.; Sartwell, S.A.; Smith, J.G.; White, C.T.; King, A.J.; Suding, K.N.; Schmidt, S.K. Plant diversity and density predict belowground diversity and function in an early successional alpine ecosystem. *Ecology* **2018**, *99*, 1942–1952. [CrossRef] [PubMed]
39. Sweeney, C.J.; Baer, S.G.; Blair, J.M. Long-term agricultural management alters soil fungal community composition and functional diversity. *Agric. Ecosyst. Environ.* **2021**, *316*, 107446. [CrossRef]
40. Geml, J.; Wagner, H. Changes in fungal diversity and composition along a chronosequence of *Eucalyptus grandis* plantations in Ethiopia. *Fungal Ecol.* **2018**, *39*, 328–335. [CrossRef]
41. Jin, W.H.; Tu, J.Y.; Sheng, W.X.; Xing, J.J.; Peng, L.Y.; Ma, X.M.; Chen, J.H.; Li, Y.F.; Qin, H. The conversion of mycorrhizal types closely associated with the changes in microbial keystone taxa and potential function in subtropical forests. *Plant Soil.* **2023**, *482*, 599–615. [CrossRef]
42. Alexis, S. Effects of Truffle Inoculation on Root Physiology and Mycorrhizosphere Microbial Communities of *Carya illinoinensis* Seedlings. *Forests* **2021**, *14*, 2078. [CrossRef]
43. Huang, L.L.; Wan, S.P.; Wang, Y.L.; Yu, Y.M.; Shi, X.F.; Yu, F.Q. Mycorrhiza Synthesis of *Pinus thunbergii* with Eight Ectomycorrhizal Fungi Species. *Acta Edulis Fungi* **2022**, *29*, 98–106. [CrossRef]
44. Bu, W.S.; Wang, F.C.; Zhang, C.C.; Bruelheide, H.; Fang, X.M.; Wang, H.M.; Chen, F.S. The contrasting effects of nitrogen and phosphorus fertilizations on the growth of Cunninghamia lanceolata depend on the season in subtropical China. *For. Ecol. Manag.* **2020**, *482*, 118874. [CrossRef]

45. Fox, T.R.; Allen, H.L.; Albaugh, T.J.; Rubilar, R.; Carlson, C.A. Tree nutrition and forest fertilization of pine plantations in the southern United States. *South. J. Appl. For.* **2007**, *31*, 5–11. [CrossRef]
46. Sanborn, P.; Ott, P.K. Tree growth, foliar nutrition, and soil properties in mixtures of lodgepole pine and Sitka alder in the central interior of British Columbia, Canada: 25-year results. *Front. For. Glob. Change* **2022**, *5*, 867247. [CrossRef]
47. Harrison, K.A.; Bardgett, R.D. Nutrient co-limitation of primary producer communities. *Ecol. Lett.* **2011**, *14*, 852–862. [CrossRef]
48. Feng, W.; Shan, C.D.; Liu, J.M.; Huang, X.S.; Fu, L.S.; Kong, H.P.; Yao, Z.Y. Effects of exponential fertilization on the biomass and nutrient accumulation and allocation of *Pinus armandii* container seedlings. *J. Cent. South. Univ. For. Technol.* **2023**, *43*, 62–69. [CrossRef]
49. Yu, J.X.; Zhang, H.; Pan, T.C.; Qiu, Z.B.; Gao, X.X.; Zhang, S.L. Study on the container seedling substrate ratio and fertilization of *Pinus armandii* on the north slope of the Qinling mountains based on regression rotation analysis. *J. Cent. South. Univ. For. Technol.* **2021**, *41*, 109–116. [CrossRef]
50. Pan, T.C.; Zhang, S.L.; Yu, J.X.; Qiu, Z.B.; Gao, X.X. Substrate Compositions and Fertilizer Formulations for Breeding Container Seedings of *Pinus armandii* Based on Quadratic Regression Universal Rrotation Combination Design. *J. Northwest For. Univ.* **2020**, *35*, 85–90.
51. Zambonelli, A.; Bonito, G.M. *Edible Ectomycorrhizal Mushrooms: Current Knowledge and Future Prospects*; Springer: Berlin/Heidelberg, Germany, 2013; pp. 227–240.
52. Allen, G.C.; Flores-Vergara, M.A.; Krasynanski, S.; Kumar, S.; Thompson, W.F. A modified protocol for rapid DNA isolation from plant tissues using cetyltrimethylammonium bromide. *Nat. Protoc.* **2006**, *1*, 2320–2325. [CrossRef] [PubMed]
53. Bao, S.D. *Soil and Agrochemical Analysis*, 3rd ed.; China Agriculture Press: Beijing, China, 2000.
54. Zen, W.J.; Wang, J.; Yang, Y.H.; He, Y.W.; Liu, Q.B.; Kang, C.; Yang, L. Soil Nutrition and Microbial Diversity Changes of Truffle Pond in the Mature Stage of *Tuber indicum*. *J. West. China For. Sci.* **2022**, *51*, 26–33. [CrossRef]
55. Li, Q.; Yan, L.; Ye, L.; Zhou, J.; Zhang, B.; Peng, W.; Zhang, X.; Li, X. Chinese black truffle (*Tuber indicum*) alters the ectomycorrhizosphere and endoectomycosphere microbiome and metabolic profiles of the host tree *Quercus aliena*. *Front. Microbiol.* **2018**, *9*, 2202. [CrossRef] [PubMed]
56. Geng, L.Y.; Wang, X.H.; Yu, F.Q.; Deng, X.J.; Tian, X.F.; Shi, X.F.; Xie, X.D.; Liu, P.G.; Shen, Y.Y. Mycorrhizal synthesis of *Tuber indicum* with two indigenous hosts, *Castanea mollissima* and *Pinus armandii*. *Mycorrhiza* **2009**, *19*, 461–467. [CrossRef] [PubMed]
57. Innangi, M.; Fioretto, A.; Fondon, C.; García-Montero, L.; Marzaioli, R.; Pinto, S.; Rutigliano, F.; Menta, C. *Tuber aestivum* is associated with changes in soil chemistry and reduced biological quality in a Quercus pubescens stand in Northern Italy. *Pedobiologia* **2020**, *80*, 150648. [CrossRef]
58. Ye, L.; Yang, X.Z.; Zhang, B.; Zhou, J.; Tian, H.; Zhang, X.P.; Li, X.L. Seasonal succession of fungal communities in native truffle (*Tuber indicum*) ecosystems. *Appl. Environ. Microbiol.* **2023**, *89*, 599–615. [CrossRef] [PubMed]
59. Liu, D.; Herrera, M.; Zhang, P.; He, X.H.; Pérez-Moreno, J.; Chater, C.; Yu, F.Q. Truffle species strongly shape their surrounding soil mycobiota in a *Pinus armandii* forest. *Arch. Microbiol.* **2021**, *203*, 6303–6314. [CrossRef] [PubMed]
60. Wang, Y.; Zhang, Q.; Li, H.; Chen, X. Fertilization combined with mycorrhization significantly improves organic matter content in substrate and enhances nutrient uptake in high-demand crops. *PLoS ONE* **2023**, *18*, e0256879. [CrossRef]
61. Antony-Babu, S.; Deveau, A.; Van Nostrand, J.D.; Zhou, J.; Le Tacon, F.; Robin, C.; Frey-Klett, P.; Uroz, S. Black truffle-associated bacterial communities during the development and maturation of *Tuber melanosporum* ascocarps and putative functional roles. *Env. Microbiol.* **2014**, *16*, 2831–2847. [CrossRef] [PubMed]
62. Li, Y.; Zhu, X.; Jia, Y.; Liu, J.; Zhang, Q.; Li, S. Differential mechanisms underlying responses of soil bacterial and fungal communities to nitrogen and phosphorus inputs in a subtropical forest. *PeerJ* **2018**, *6*, e5986. [CrossRef] [PubMed]
63. Siebyła, M.; Hilszczańska, D. Species diversity of bacteria associated with fungi of the genus *Tuber* (truffles). *Adv. Microbiol.* **2019**, *56*, 28–32. [CrossRef]
64. Kang, Z.; Li, X.; Li, Y.; Ye, L.; Zhang, B.; Zhang, X.; Penttinen, P.; Gu, Y. Black truffles affect Quercus aliena physiology and root-associated nirK- and nirS-type denitrifying bacterial communities in the initial stage of inoculation. *Front. Microbiol.* **2022**, *13*, 792568. [CrossRef] [PubMed]
65. Emri, T.; Antal, K.; Varga, K.; Gila, B.C.; Pócsi, I. The Oxidative Stress Response Highly Depends on Glucose and Iron Availability in Aspergillus fumigatus. *J. Fungi* **2024**, *10*, 221. [CrossRef] [PubMed]
66. Fang, X.Y.; Yang, Y.; Zhao, Z.G.; Zhou, Y.; Liao, Y.; Guan, Z.Y.; Chen, S.M.; Fang, W.M.; Chen, F.D.; Zhao, S.A. Optimum nitrogen, phosphorous, and potassium fertilizer application increased chrysanthemum growth and quality by reinforcing the soil microbial community and nutrient cycling function. *Plants* **2023**, *12*, 4062. [CrossRef] [PubMed]
67. Xu, X.; Du, X.; Wang, F.; Sha, J.; Chen, Q.; Tian, G.; Zhu, Z.; Ge, S.; Jiang, Y. Effects of potassium levels on plant growth accumulation and distribution of carbon, and nitrate metabolism in apple dwarf rootstock seedlings. *Front. Plant Sci.* **2020**, *11* 904. [CrossRef] [PubMed]
68. Joner, E.J.; Jakobsen, I. Growth and extracellular phosphatase activity of arbuscular mycorrhizal hyphae as influenced by soil organic matter. *Soil. Biol. Biochem.* **1995**, *27*, 1153–1159. [CrossRef]

69. Fu, Y. Effects of arbuscular mycorrhizal fungi on carbon assimilation and ecological stoichiometry of maize under combined abiotic stresses. *J. Plant Ecol.* **2023**, *16*, 234–246. [CrossRef]
70. Gryndler, M.; Larsen, J.; Hrselova, H.; Rezacova, V.; Gryndlerova, H.; Kubat, J. Organic and mineral fertilization, respectively, increase and decrease the development of external mycelium of arbuscular mycorrhizal fungi in a long-term field experiment. *Mycorrhiza* **2006**, *16*, 159–166. [CrossRef] [PubMed]
71. Chen, Y.L. Cultivation Techniques for European Truffles (*Tuber* spp.). *Edible Fungi China* **2002**, *21*, 7–9+11. [CrossRef]
72. Ning, H.; Ling, L.; Sun, X.; Kang, X.; Chen, H. Predicting the future redistribution of Chinese white pine *Pinus armandii* Franch. Under climate change scenarios in China using species distribution models. *Glob. Ecol. Conserv.* **2021**, *25*, e01420. [CrossRef]
73. Brundrett, M. Mycorrhizal associations and other means of nutrition of vascular plants: Understanding the global diversity of host plants by resolving conflicting information and developing reliable means of diagnosis. *Plant Soil.* **2009**, *320*, 37–77. [CrossRef]
74. Shi, L.; Guttenberger, M.; Kottke, I.; Hampp, R. The effect of drought on mycorrhizas of beech (*Fagus sylvatica* L.): Changes in community structure, and the content of carbohydrates and nitrogen storage bodies of the fungi. *Mycorrhiza* **2002**, *12*, 303–311. [CrossRef] [PubMed]
75. Rillig, M.C.; Mummey, D.L. Mycorrhizas and soil structure. *New Phytol.* **2006**, *171*, 41–53. [CrossRef] [PubMed]
76. Liu, Y.; Li, X.; Kou, Y. Ectomycorrhizal fungi: Participation in nutrient turnover and community assembly pattern in forest ecosystems. *Forests* **2020**, *11*, 453. [CrossRef]
77. Garcia, K. Potassium nutrition of ectomycorrhizal *Pinus pinaster*: Overexpression of the *Hebeloma cylindrosporum* HcTrk1 transporter affects the translocation of both K^+ and phosphorus in the host plant. *New Phytol.* **2014**, *201*, 951–960. [CrossRef] [PubMed]

agronomy

MDPI

Review

Effects of Microbial Agents on Soil Improvement—A Review and Bibliometric Analysis

Mengdi Tan [1,2], Tianjiao Feng [1,2,3,*], Cong Wang [4], Xiaozhen Hao [1,2] and Hang Yu [1,2]

[1] School of Soil and Water Conservation, Beijing Forestry University, Beijing 100083, China; tan0323@bjfu.edu.cn (M.T.)
[2] Jixian National Forest Ecosystem Observation and Research Station, Linfen 042200, China
[3] Engineering Technology Innovation Center for Ecological Protection and Restoration in the Middle Yellow River, Ministry of Natural Resources, Taiyuan 030006, China
[4] State Key Laboratory of Mycology, Institute of Microbiology, Chinese Academy of Sciences, Beijing 100101, China
* Correspondence: fengtianjiao1991@bjfu.edu.cn

Abstract: Microbial agents play a crucial role in improving soil quality, increasing soil fertility, enhancing crop yields, and reducing the incidence of diseases. The ecological benefits of these products contribute to environmental protection and to the promotion of sustainable agricultural development. Since the beginning of the 21st century, research in the academic community on the use of microbial agents for soil improvement has increased, yet a systematic summary of the progress in this field is lacking. In this paper, we review trends in microbial agent applications, focusing on their classification, mechanisms of action, and practical implementations. To achieve this, we conduct a bibliometric analysis based on the SCI-EXPANDED database of the Web of Science, using tools such as VOSviewer for visualization. We focus on microbial agents for soil improvement and analyze publication trends, research hotspots, and annual variations in relevant studies published between 2003 and 2024. The results show that (1) the number of publications on microbial soil improvement has steadily increased over the years, indicating that the academic community has maintained a high level of interest in this field. Keywords such as "soil", "diversity", "carbon", and "nitrogen" have been central research hotspots in the past 20 years. The research has been highly concentrated in a few countries, including China and the United States, as well as in key institutions such as the Chinese Academy of Sciences and the United States Department of Agriculture. (2) We further analyze the principles governing microbial agent efficacy, address limitations in their application, and propose strategies to advance research in this field. Finally, several suggestions are proposed to promote the further development of research on microbial agents for soil improvement.

Keywords: soil microorganisms; microbial agents; soil improvement; soil health; bibliometric analysis

1. Introduction

Soil, essential for human survival, is a critical component of Earth's environment, influencing plant, animal, and human health while serving as a key resource for food security [1,2]. Its condition is interconnected with air, water quality, and microbial activity. However, population growth, climate change, industrial expansion, and unsustainable land use have exacerbated soil pollution and degradation, threatening ecosystem stability and societal development [3]. Globally, 33% of soils face degradation from urbanization, erosion, nutrient depletion, salinization, desertification, and pollution. In China, degradation is concentrated in 13 provinces across the northwest, north, and northeast, positioning the country among those most severely impacted [4,5]. Addressing soil degradation through improved physicochemical properties and sustainable management remains a priority for soil science research [6].

Historically, agricultural reliance on chemical fertilizers boosted crop yields and organic matter content but led to soil acidification, organic loss, and environmental pollution [7]. This environmental toll has accelerated the search for eco-friendly alternatives, with microbial remediation emerging as a promising strategy due to its potential to enhance soil health through nutrient cycling, aggregate stabilization, and pollutant degradation [8,9]. To counter degradation, research focuses on understanding soil ecosystem dynamics and developing eco-friendly solutions. Current remediation strategies include biological, chemical, and physical methods. Biological approaches, leveraging plant, animal, and microbial activities to enhance soil health, have gained prominence due to their sustainability, cost-effectiveness, and environmental benefits.

Among biological technologies, phytoremediation is the most established, while soil amendments target desertification. Microbial remediation, a newer approach, employs industrially produced microbial inoculants to regulate nutrient cycles, stabilize soil aggregates, and degrade pollutants [8]. Soil microbes, integral to soil formation and nutrient cycling, act as nutrient transformers, pollutant purifiers, and ecosystem stabilizers [9,10]. Microbial agents are defined as formulated products containing live microorganisms (e.g., *Bacillus*, *Pseudomonas*, arbuscular mycorrhizal fungi) or their bioactive metabolites, specifically designed to enhance soil functions such as organic matter decomposition, nutrient cycling (C, N, P), and pollutant degradation [10,11]. These agents directly contribute to ecosystem services including soil carbon sequestration, crop yield stability, and heavy metal immobilization. Microbial agents improve fertility, reduce chemical inputs, suppress pathogens, and enhance crop yields. Their integration with phytoremediation and other techniques offers a pathway to sustainable agriculture [11,12].

Advances in microbial agent development stem from methodological innovations. Early isolation and cultivation techniques in the 19th century gave rise to microbial studies, while 1970s biomass quantification methods spurred progress in nutrient turnover analysis [13,14]. Late 20th-century molecular biology revolutionized soil microbiology, expanding microbial agent diversity, functionality, and global application [15,16]. Current research focuses on agent development, microbial interactions, biocrusts, plant–microbe synergy, and hybrid remediation technologies [17].

Despite these advancements, a critical gap persists—the existing literature either focuses on mechanistic reviews of microbial functions or bibliometric analyses of global trends, rarely integrating both to identify how research hotspots align with on-the-ground challenges in soil improvement [17–19]. Traditional literature analysis struggles to track trends in this vast field. Bibliometrics, combined with tools like VOSviewer, enables quantitative analysis and visualization of research progress [18]. This study uses Web of Science (WoS) and China National Knowledge Infrastructure (CNKI) data to conduct a bibliometric review of microbial soil improvement over the two decades, identifying hotspots, gaps, and future directions to guide researchers and deepen understanding of microbial agents' ecological significance.

2. Materials and Methods

2.1. Data Sources

In this paper, we establish a database using the Science Citation Index (WoS TM Core Collection) as the source of international bibliographic information. A search query of TS = ("microbial agent" OR "microbe" OR "bacterial agent") AND TS = ("improve" OR "restoration" OR "amendment") AND TS = ("soil" OR "soil fertility" OR "crop yield" OR "soil-borne pathogen") was applied, excluding review articles to prioritize original research contributions and minimize citation bias, with a time span from 1 January 2003 to 31 December 2024. Data cleaning involved removing duplicates using EndNote X20 (95% title–author similarity threshold) and manually excluding irrelevant studies (e.g., marine or medical applications). A total of 3964 papers related to microbial improvement of soil were included in the analysis (Figure 1). The Chinese literature data source was determined through keyword searches in the China National Knowledge Infrastructure (CNKI) database, restricted to journals indexed in the Chinese Science Citation Database (CSCD), Peking University Core, or Engineering Index (EI) to ensure quality. In the advanced search box, the query "microbial agents + soil improvement" was used with a time limit from 2003 to 2024, and non-English publications were cross-checked by bilingual researchers to align with the study scope. After rigorous screening, 118 relevant papers from Chinese sources were retained. All raw datasets are publicly accessible via the WoS and CNKI platforms, ensuring transparency and reproducibility. Data validation included duplicate removal and manual verification by independent researchers to confirm consistency with inclusion criteria.

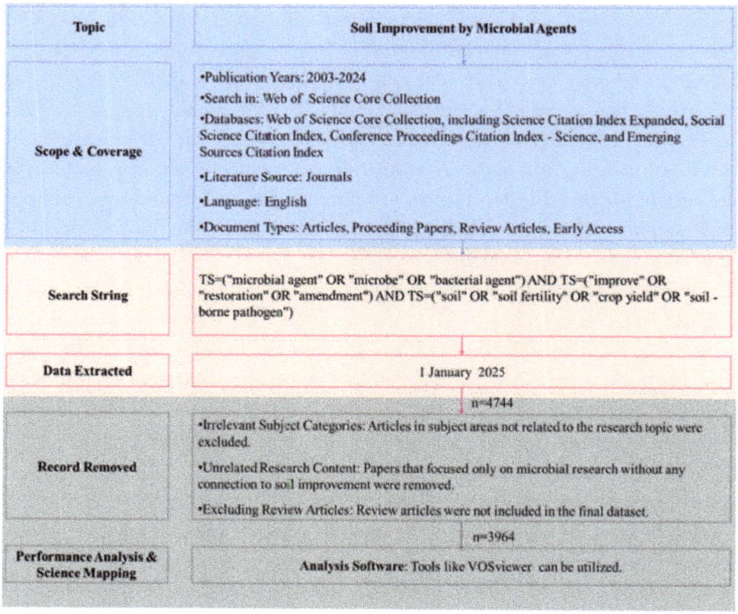

Figure 1. Research methodology steps.

2.2. Analysis Method

VOSviewer (version 1.6.20) software analyzed publication volume, annual keyword trends, and institutional collaborations in the Web of Science [19] (Figure 2). The literature from the CNKI database was not considered because the actual number after screening to highlight duplicates and the irrelevant literature was below 60, making it difficult to form a statistical pattern. Keyword co-occurrence/clustering networks were constructed to visualize research hotspots, with a minimum keyword occurrence threshold set to 12 to filter low-frequency noise and enhance thematic clarity. Temporal keyword evolution maps highlighted shifting priorities, while country–institution collaboration networks were also generated to assess global partnerships. VOSviewer settings utilized the Louvain clustering algorithm and association strength normalization to optimize network interpretability. Annual publication counts and top journal rankings since 2003 were plotted using Origin Pro (2021). These bibliometric approaches revealed the field's current landscape, research foci, and future directions [20].

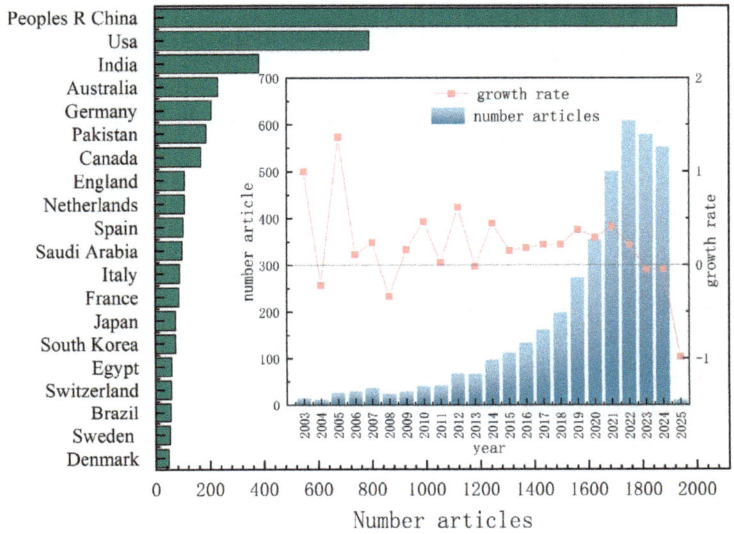

Figure 2. Statistics on publication counts and annual publication counts in 20 countries, 2003 to 2024. Shows the change in the number of articles published in the field of microbial agents for soil improvement in the last 20 years and the number of articles published in each country. This sentence is translated into English to ensure scientific accuracy. Data from the Web of Science TM Core Collection.

3. Bibliometric Analysis

3.1. Global Publication Trend Analysis

From 2003 to 2024, SCI-indexed publications on soil-improving microbial agents increased annually, with distinct growth phases (Figure 2). The period 2003–2014 showed high growth rates, while 2015–2024 exhibited slower relative growth but larger absolute increases due to an expanded publication base. Post-2015 trends followed an exponential trajectory (Figure 2), indicating accelerated research output alongside methodological diversification. While the overall publication trend shows an increase, the temporary decline in citations after 2022 may reflect research constraints due to the impact of the epidemic and delays in indexing articles in the Web of Science database.

Clarivate's 2024 Journal Citation Reports identified contributions from 104 countries, with significant output disparities (Figures 2 and 3). China led with 1923 publications (continuous 20-year growth), followed by the United States (789 publications, 19.8% of global output), both maintaining upward trends.

At the institutional level, the Chinese Academy of Sciences (CAS) is the most prolific institution in the world, with 477 papers. The U.S. Department of Agriculture's Agricultural Research Service (USDA ARS) and University of Florida published 45 and 43 papers, respectively (Figure S1 and Table S2).

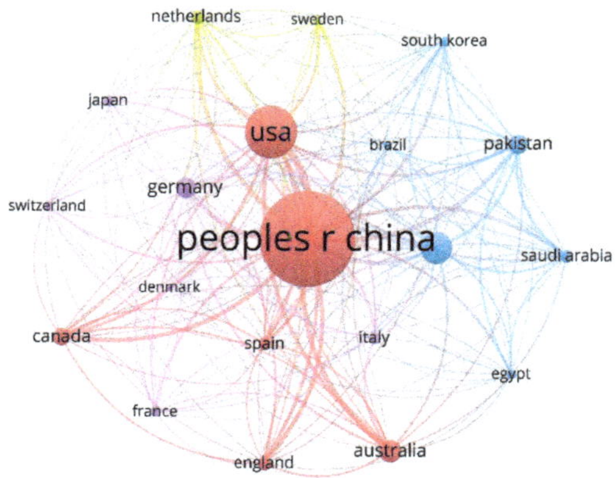

Figure 3. Analysis of cooperation networks in the top 20 countries (2003–2024). Collaboration network of the top 20 most productive countries. The node size increases with the number of publications of each country. The distance between two nodes decreases with the increase in their association strength. The thickness of the linking lines increases with the number of co-authored documents published by two countries. The different colors of the nodes represent the different clusters of countries.

3.2. Visualization Analysis of High-Frequency Keyword Co-Occurrence

Between 2003 and 2024, 9821 keywords were identified in microbial agent research for soil improvement, with 169 high-frequency keywords (occurrence > 12, defined as hotspots) highlighting research hotspots after manual deletion of duplicate keywords (Figure 4, Table 1 and Table S4). During 2013–2023, terms including "soil", "diversity", "nitrogen", "carbon", "rhizosphere", "plant-growth", and "bacteria" emerged as central nodes in the keyword network.

Table 1. The top 10 keywords for research on soil improvement by microbial agents.

Rank	Keyword	Frequency	Total Link Strength
1	rhizosphere	156	254
2	biochar	159	202
3	microbial community	159	184
4	soil microbes	120	118
5	phytoremediation	103	133
6	bacteria	105	197
7	sustainable agriculture	68	101
8	nitrogen	55	102
9	heavy metals	57	82
10	soil microbial community	50	64

Note: the top 10 most frequently occurring keywords extracted by the bibliometric analysis software and their total link strength.

The co-occurrence network clustered into three modules, representing core research themes. The red module ("biochar", "microbial community", "phytoremediation", "heavy metals", "bioremediation") focused on biochar-induced microbial community dynamics and their roles in nutrient cycling, particularly phytoremediation mechanisms for heavy metal contamination. The green module ("microbiome", "rhizosphere", "sustainable agriculture", "biocontrol", "metagenomics", "plant-microbe interactions", "arbuscular mycorrhizal fungi") emphasized rhizosphere microbiome functions, exploring stress resistance enhancement through microbial symbiosis and arbuscular mycorrhizal fungi to advance sustainable agriculture. The blue module ("soil microbes", "soil organic carbon", "bacteria", "fungi", "microbial biomass", "phosphorus", "nitrogen") underscored the microbial regulation of soil organic carbon cycling and nutrient (N/P) transformation mechanisms.

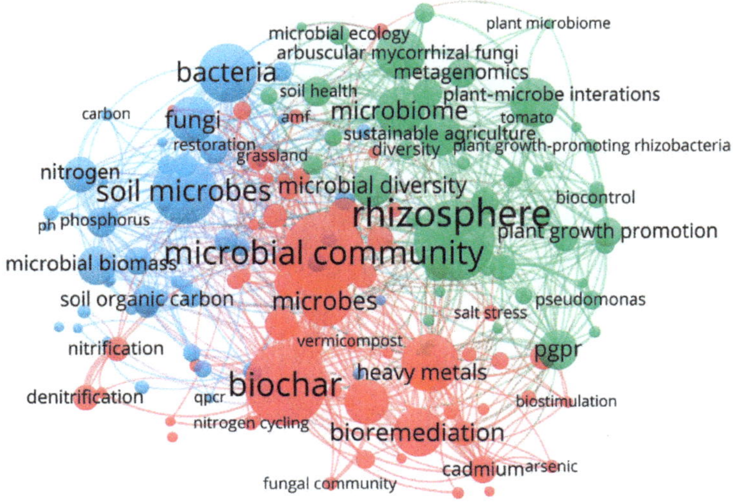

Figure 4. Co-occurrence network analysis of keywords in soil microbial research. Keywords are represented by tags and circles. The sizes of the tags and circles indicate the importance of the keywords. The greater the weight of keywords is, the larger the tags and circles of the project are. The color of the item is determined by the cluster to which the item belongs. The lines between items represent links. The closer the distance between keywords and the denser the lines are, the stronger the correlation.

3.3. *Analysis of Research Hotspots and Emerging Trends*

Since keywords effectively summarize the main research themes, we further employed keyword clustering networks to provide an overview of changes in research hotspots within the field of microbial agents for soil improvement over the past 20 years. Figure 5a illustrates the annual variations in the keyword co-occurrence network, showing the keywords that were predominant between 2010 and 2023, which indicates a significant increase in research hotspots during this period (Table S3). This trend aligns with the rapid growth in publication volume observed from 2015 to 2023, as shown in Figure 2, Figure 5b shows the annual changes in keyword hotspots from 2018 to 2024 (Table S3), which transition from purple to green to yellow. The purple keywords include "microbial biomass", "nitrogen", "nutrient cycling", "rhizobacteria", "heavy metals", and "remediation"; the green keywords are "microbial diversity", "soil microbes", "rhizosphere", "biochar", and "sustainable agriculture"; and the yellow keywords include "bacterial community", "soil health", "ecological restoration", "microbial agent" and "metagenomics". The results demonstrate a shift in research trends within this field: around 2018, the focus was primarily on the role of microorganisms in soil remediation and nutrient cycling. After 2021, the attention shifted toward the application of molecular biological techniques, as well as the diversity and community composition of microorganisms. More recently, emphasis has been placed on the role of microorganisms in plant disease resistance and growth promotion.

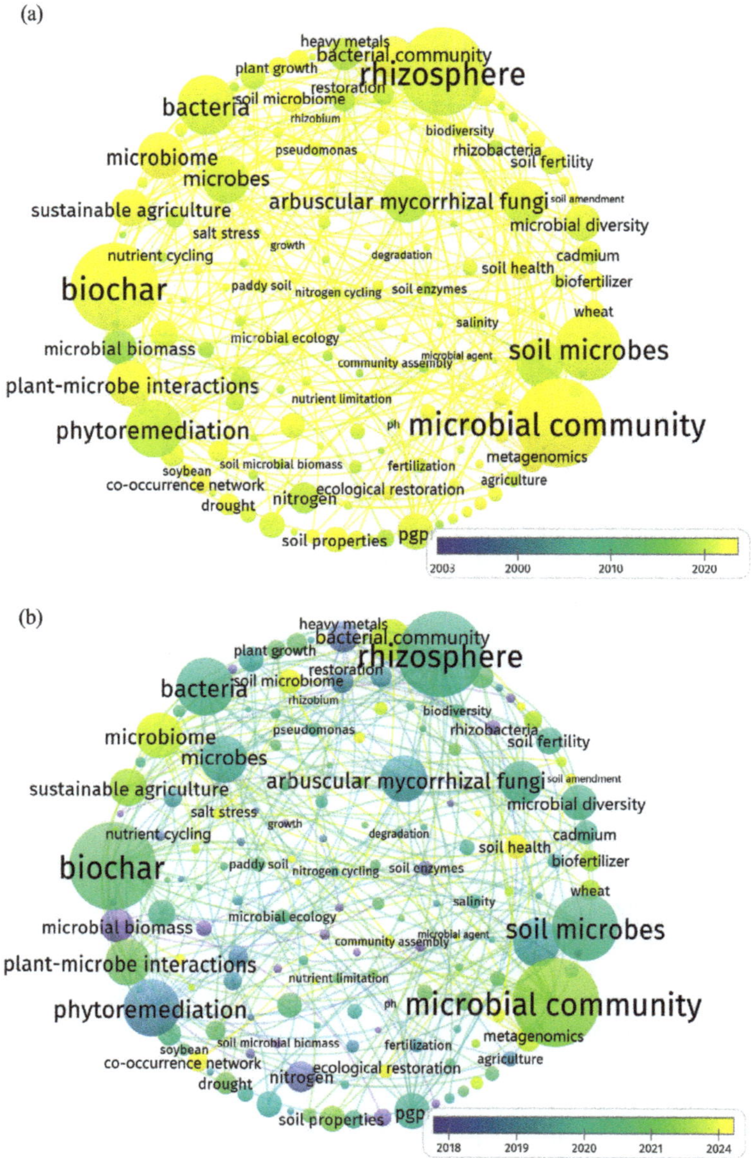

Figure 5. Interannual evolution of hotspots in the soil microbial research keyword co-occurrence network from (**a**) 2003–2024 and (**b**) 2018–2024. The image shows the year-to-year change in keywords (research hotspots); the color shift from dark blue to yellow indicates the transition of research hotspots appearing from 2003 to 2024 (for example, (**a**) dark blue a = 2003, yellow = 2024; (**b**) dark blue a = 2018, yellow = 2024), and the size of the circle indicates the weight of the keyword—the bigger the circle, the higher the weight.

4. Classification of Microbial Agents

Microbial agents and fertilizers are widely utilized for soil improvement globally, demonstrating benefits in agricultural productivity and environmental sustainability [21]. To elucidate their mechanisms, challenges, and applications, microbial agents are classified through multiple criteria based on global research, industry standards, and bibliometric findings (Table 2). These distinctions are classified by form as liquid or solid formulations, composition as single-strain or composite agents, functional roles encompassing organic composting, soil remediation, disease resistance, and growth promotion, and microbial types including bacterial, fungal, or actinomycete agents [22]. Proper application of these agents enhances crop yields, soil quality, and ecological protection, offering a theoretical foundation for optimizing agricultural practices [23].

Table 2. Classification of functional characteristics of microbial agents.

Classification	Categorization	Functional Characteristic	Application	References
Dosage form	Liquid agents	(1) Advantages such as ease of preparation, ease of application, short activation time, and rapid effectiveness. (2) The ability to prevent soil-borne diseases, maintain the balance of rhizosphere microbial communities, and degrade toxic substances.	Applicable for seed/root treatment, drip irrigation/fertigation, and water purification/pollutant degradation.	[24]
	Solid agents	(1) Promoting plant growth and increasing yield. (2) Enhancing the activity of urease and transaminase in soil, increasing the content of available nitrogen, and promoting nitrogen absorption.	Enhances long-term soil enhancement with latent-phase efficacy while functioning as basal fertilizer.	[24,25]
Composite method	Single agents	(1) Single function, with limited adaptability. (2) Targeted antagonism against pathogens, pests, and soil-borne diseases.	Resolving single soil environment issues.	[25]
	Composite agents	(1) Compared to single microbial agents, composite agents are more efficient and stable in soil, promoting crop growth. (2) They can significantly increase the nitrogen, phosphorus, and soluble leaf protein content in plants, thereby promoting plant growth.	Improve soil health and crop growth; inoculate leguminous crops to suppress wilt.	[26]
Functional characteristic	Organic material composting agents	(1) Induces rapid thermophilic phase, enhancing organic matter conversion. (2) Eradicate pathogens, nematode eggs, and compost contaminants. (3) Effectively degrade odor-producing organic macromolecules and suppress putrefactive microorganisms.	Decompose diverse organic materials (e.g., straw, manure, distillers' grains, plant litter, sewage sludge).	[27]
	Soil remediation agents	(1) Promote the formation of soil aggregates, increase soil porosity, improve soil structure, and alleviate compaction. (2) Regulate soil pH, degrade pesticide residues, and immobilize heavy metals, mitigating salinization and alkalization. (3) Enhance soil water and nutrient retention capacity, promoting stable and high crop yields.	Remediate compacted/acidic soils and agrochemical/heavy metal co-contaminated sites.	[28]

Table 2. *Cont.*

Classification	Categorization	Functional Characteristic	Application	References
Functional characteristic	Disease-resistance agents	(1) Inhibit the proliferation of soil pathogens, effectively preventing root rot, damping-off, root-knot nematodes, and other replant diseases. (2) Enhance crop stress resistance, improving tolerance to early frost, cold, and lodging.	Applicable to field crops such as wheat, corn, and soybeans, as well as economic crops like vegetables, fruit trees, and medicinal plants.	[29,30]
	Growth-promoting agents	(1) Provide or activate nutrients, stimulate root growth, and increase yield. (2) Improve the quality of agricultural products by participating in the synthesis of plant cell materials, enhancing the content of beneficial components such as proteins, sugars, vitamins, and amino acids.	Applicable to field crops such as wheat, corn, and soybeans, as well as economic crops like vegetables, fruit trees, and medicinal plants.	[31,32]
Microbial species	Bacterial agents	(1) Diverse in species, with strong reproductive ability and complex metabolism. (2) Enhance soil fertility, improve soil structure, increase soil bacterial population, and combat soil pests and diseases.	Enhances phosphorus/potassium solubilization and soil transformation while suppressing pathogenic microbes.	[33]
	Fungal agents	(1) Form a network-like mycelial structure, increasing contact opportunities with contaminants on particulate surfaces. (2) Degrade diverse organic pollutants, including polycyclic aromatic hydrocarbons, petroleum hydrocarbons, and halogenated hydrocarbons. (3) Enrich indigenous microbiota, synergizing with native microbes-plant roots to boost remediation.	It is suitable for the remediation of sites with organic and composite contamination, as well as for land with poor soil structure.	[34]
	Actinomycete agents	(1) Assist plants in nutrient absorption. (2) Control the spread of plant pathogens, exhibiting strong antagonistic and parasitic effects on pathogenic organisms. (3) Actinomycetes secondary metabolites exhibit antimicrobial activity, enhance crop growth/yield, ameliorate soil ecology, and demonstrate eco-friendliness.	It can produce antibiotics and is used for wastewater treatment, as well as for the chemical control of root-knot nematodes.	[35]

Note: The table summarizes four microbial agent classification methods, their features, and applications. Microbial agents may belong to multiple categories (e.g., a liquid formulation of *Bacillus* spp. can function as both a growth-promoting and disease-resistance agent).

4.1. Microbial Agent Formulation Types

Microbial agents can be categorized by their formulation type into liquid and solid microbial agents. Liquid microbial agents are typically prepared by suspending or concentrating microbial strains obtained from fermentation, which are then mixed with other components, such as nutrients and trace elements. This type of microbial agent has advantages such as ease of preparation process and short activation and initiation times, allowing for rapid activation of the reactor and improved processing efficiency. Therefore, it holds significant potential for development and application in practice [26]. For example, Previous studies isolated the FE-1 fungal strain from farmland and formulated it into a liquid microbial agent [36]. This agent effectively degrades fomesafen in the soil and promotes root growth in plants, thereby reducing the impact of pesticide residues on crops. Compared with liquid microbial agents, solid microbial agents have a longer shelf life and lower strain loss rate, making them easier to store and transport. This is significant for reducing the transportation and usage costs of the agents. Solid formulations have a shelf life of up to 12 months, but activation takes 24–48 h, which may delay farm time; liquid formulations have an activation time of <6 h but require cold-chain transportation. Recent research has shown that, even after different periods of storage, the composting effects of solid microbial agents remain similar [37]. These agents can effectively improve compost maturity and shorten the composting cycle, further demonstrating the advantages of solid microbial agents in terms of shelf-life and strain stability. Furthermore, research has demonstrated that the application of both liquid and solid microbial agents can significantly increase soil urease activity, thereby increasing the total nitrogen content in plants, expanding the leaf area, and substantially increasing the chlorophyll content (as measured by a soil–plant analysis development (SPAD) meter). This leads to an increase in the net photosynthetic rate of leaves, with yield improvements of 26.9% and 34.4%, respectively, compared with those of treatments without microbial agents [38].

4.2. Classification of Microbial Species and Functional Characteristics of Microbial Agents

Microbial agents are categorized by microbial type into bacterial, fungal, and actinomycete agents. Bacterial agents, including *Bacillus* spp., lactic acid bacteria, rhizobia, and *Pseudomonas* spp., enhance soil fertility and structure through metabolic diversity and rapid reproduction [39]. For example, *Bacillus siamensis* promotes garlic growth and alters soil enzyme activity via protease activity, siderophore production, and phosphorus solubilization [40]. Fungal agents such as arbuscular mycorrhizal (AM) fungi improve crop nitrogen uptake while reducing soil N_2O emissions through regulation of nitrification/denitrification genes [41]. Actinomycete agents, notably *Streptomyces* spp., suppress plant pathogens via antagonism and parasitism, with Streptomyces albus demonstrating superior biocontrol efficacy against cucumber root-knot nematodes [35].

Functionally, microbial agents are classified as growth-promoting, composting, disease-resistance, or bioremediation agents. Growth-promoting agents enhance nutrient uptake and stress tolerance through hormone secretion and nutrient solubilization. Their combined application with bioorganic fertilizers elevates soil organic matter, phosphorus, potassium, and microbial biomass in tea chrysanthemum systems [42]. Composting agents, including *Bacillus* subtilis–Aspergillus niger consortia, accelerate organic matter degradation and compost maturation [43]. Disease-resistance agents utilize actinomycetes—responsible for 67% of antibiotic production—to combat pathogens through antibiotic synthesis and nutrient competition [44]. Bioremediation agents such as *Bacillus* licheniformis mitigate heavy metal toxicity and pollutant residues, restoring soil microecosystem functionality [45].

4.3. Microbial Agents Can Be Classified on the Basis of Their Composition

Microbial agents can be classified into single microbial agents and composite microbial agents on the basis of their composition. Single microbial agents (e.g., preparations containing only *Bacillus* subtilis) have limited functionality and environmental adaptability but are highly targeted and suitable for specific disease control; compound microbial agents (e.g., EM microbial agents) can improve adaptability through multibacterial synergy, but at an increased cost. Many studies have demonstrated that the degradation or transformation of many organic substances requires the synergistic action of multiple microorganisms [39]. Composite microbial agents, which are prepared by mixing two or more strains with different functions at a specific ratio, offer stronger adaptability and stability than single microbial agents do, fully leveraging the synergistic advantages of the microbial community for optimal application results [46]. Since the 1970s, in countries such as the United States, Japan, and those in Europe, various composite microbial agents have been successfully developed, with the most successful being the effective microorganism (EM) agent developed by Professor Teruo Higa of the University of Ryukyu in the early 1980s, which has been widely used in more than 90 countries for agriculture, animal husbandry, and environmental remediation, yielding significant economic, ecological, and social benefits [46]. The research on microbial agents in China started later but has developed rapidly, and several microbial agents have now been introduced, such as the successfully developed "Shenwei" microbial ecological formulation, which, after extensive experimental application, has begun to be widely used in agriculture, animal husbandry, and environmental applications, achieving good societal, economic, and ecological benefits [47]. The use of this microbial agent not only reduces the amount of chemical fertilizers needed but also degrades harmful substances in the soil, enhances soil fertility, and improves soil resistance and resilience, thereby improving the soil environment.

5. Soil Improvement Effects of Microbial Agents and Influencing Factors

5.1. Role of Microorganisms in the Regulation of Soil Physicochemical Properties

Microorganisms play a crucial role in regulating nutrient cycling and transformation in soil and plants, serving as indispensable and highly active biological agents in soil. To varying degrees, the abundance and diversity of microorganisms in the soil control the direction of soil nutrient transformation, the types of compounds produced, and the exchange flux in the soil. Soil microorganisms play a key role in nutrient cycling, particularly for elements such as C, N, and P, by participating in processes such as autotrophic carbon fixation, by driving the soil carbon cycle, and by regulating soil phosphorus availability, making them important contributors to soil health regulation [48] (Figure 6 and Table 3).

Table 3. Microbial species involved in the regulation of soil physicochemical properties and their functions.

Microbial Function	Mechanism	Genus/Species	References	First Author	Year
Carbon Cycle	Microbial decomposition of organic matter, CO_2 fixation via photosynthesis.	*Cyanobacteria, Streptomyces*	[49]	Falkowski, P.G.	2008
Nitrogen Cycle	Nitrogen fixation, nitrification, denitrification, and ammonification.	*Rhizobium, Nitrosomonas*	[50]	Kuypers, M.M.M.	2018
Phosphorus Cycle	Solubilization, mineralization, and uptake via phosphatases.	*Pseudomonas,Bacillus*	[51]	Richardson, A.E.	2009
Soil Microbes Resist Pests/Diseases	Antibiotic production, induced systemic resistance (ISR).	*Pseudomonas Fluorescens, Bacillus Subtilis*	[52]	Pieterse, C.M.J.	2014

Table 3. *Cont.*

Microbial Function	Mechanism	Genus/Species	References	First Author	Year
Microbial Degradation of Petroleum	Hydrocarbon degradation via oxygenases and biosurfactants.	*Pseudomonas Putida, Alcanivorax Borkumensis*	[53]	Denaro, R.	2021
Microbial Degradation of Pesticides	Enzymatic breakdown (e.g., hydrolases, oxidoreductases).	*Arthrobacter, Pseudomonas*	[54]	Singh, B.K.	2009
Microbial Heavy Metal Tolerance	Biosorption, bioaccumulation, and detoxification via metallothioneins.	*Geobacter, Cupriavidus Metallidurans*	[55]	Nies, D.	1999
Plastic Degradation	Enzymatic hydrolysis	*Ideonella Sakaiensis, Aspergillus Niger*	[56]	Yoshida, S.	2016

Note: The content of the table explains the various functions and processes in which microorganisms are involved in regulating the physical and chemical properties of soil shown in Figures 6 and 7. These include the carbon cycle, nitrogen cycle, phosphorus cycle, the resistance of microorganisms to soil pests and diseases, the degradation of pesticide pollution, petroleum degradation, enhanced tolerance to heavy metals, plastic degradation, as well as regulatory mechanisms and references.

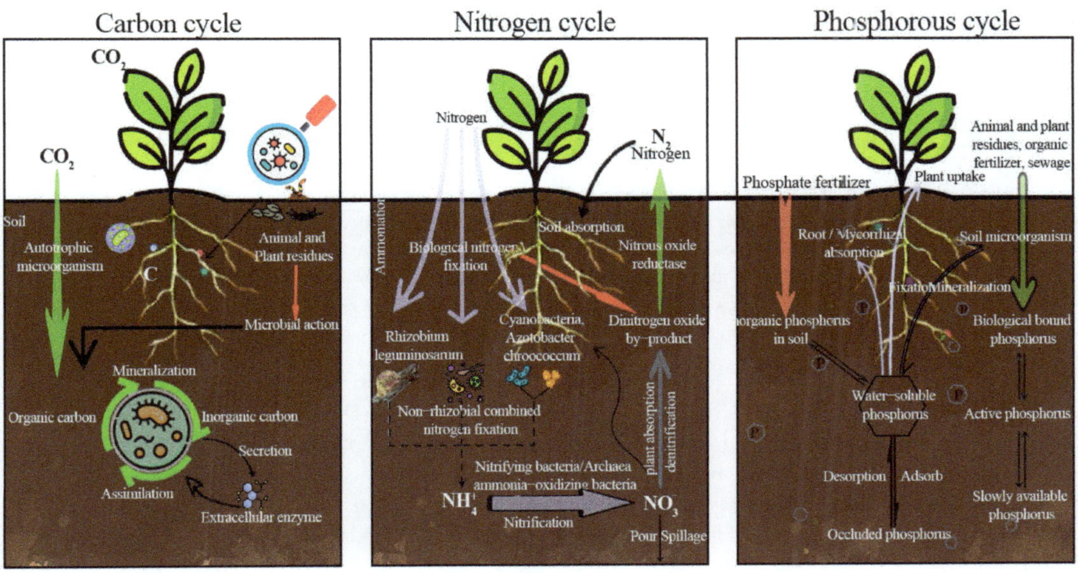

Figure 6. The main process of microbially mediated regulation of soil physical and chemical properties. The figure depicts the essential roles of soil microorganisms in mediating biogeochemical cycles (carbon, nitrogen, and phosphorus cycles), highlighting their indispensable contributions to soil formation, development, and amelioration processes.

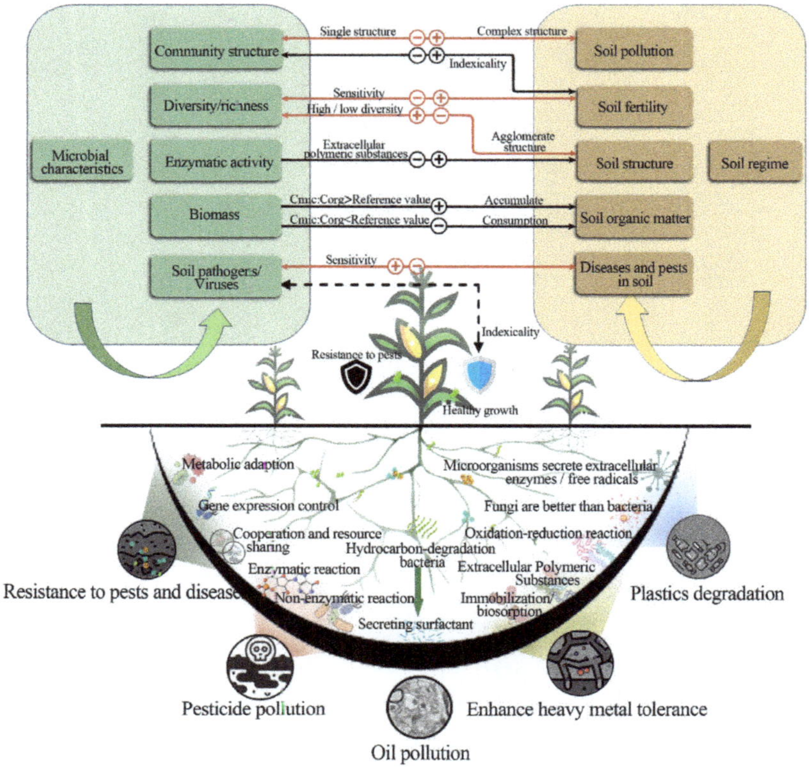

Figure 7. The main functional framework showing the mechanisms by which microorganisms affect soil health. Soil microorganisms are sensitive to changes in soil conditions, the relationship between soil microorganisms and soil health, and the mechanisms by which microorganisms purify, repair, and improve soil physicochemical properties by participating in a variety of biochemical processes and thereby maintaining soil health (for example, microbial communities dominated by fungi degraded microplastics significantly more efficiently than bacteria, a finding that provides direct evidence for targeted regulation of functional microbes for remediation of contaminated soils.).

5.1.1. Microorganisms Are Involved in the Assimilation and Mineralization of Organic Carbon

Microorganisms regulate the soil carbon balance by participating in the assimilation and mineralization of organic carbon. The extracellular and intracellular enzymes secreted by microorganisms are key factors contributing to soil enzymatic activity, with the extracellular enzymes that catalyze the transformation of organic and inorganic substances in soil being the primary agents driving soil organic carbon mineralization [57] (Figure 6 and Table 3). Recent studies on the soil carbon pump theory suggest that, after the exterior modification of recalcitrant organic substrates into small, easily absorbable molecules through extracellular enzymes, microorganisms carry out internal turnover, leading to the iterative accumulation of microbial residues and promoting the formation of organic matter. This process ultimately results in the stabilization of such compounds in the soil via the microbial "entombing effect", which significantly contributes to the accumulation of soil organic carbon (SOC) [58]. The physiological characteristics of individual microorganisms and the ecological traits of microbial communities regulate the diversity of soil ecological functions, thereby influencing soil mineralization. The production of microbial residues is controlled by three physiological characteristics: the microbial growth rate (MGR), the carbon use efficiency (CUE), and the microbial biomass turnover rate (MTR) [59]. Studies have shown a significant correlation between soil CUE and soil SOC, and since the composition and diversity of soil microbial communities are key factors influencing soil CUE, the contribution of soil microorganisms to CUE and SOC is highly important. However, the mechanisms by which individual microorganisms and microbial communities drive soil carbon accumulation through assimilation still require further investigation [60,61].

5.1.2. Regulatory Role of Microorganisms in Soil Biological Nitrogen Fixation

Biological nitrogen fixation (BNF) in soil mainly occurs through the following pathways: (1) symbiosis with leguminous plants to form root nodules; (2) associative nitrogen fixation without root nodule formation; and (3) nitrogen fixation by bacteria such as cyanobacteria, Azotobacter, and Clostridia [62,63] (Figure 6 and Table 3). In addition, the efficiency of biological nitrogen fixation can be enhanced through the exogenous application of organic materials, microbial agents, and other measures. Inoculation with nitrogen-fixing bacteria can also promote biological nitrogen fixation; however, challenges include competition for nutrients with indigenous microbial communities, unstable host-symbiont compatibility, and difficulty in ensuring nitrogen fixation capacity throughout the entire lifecycle [64]. Biological nitrogen fixation involves the conversion of atmospheric nitrogen (N_2) into ammonium (NH_4^+), which is then oxidized in the soil by ammonia-oxidizing archaea and bacteria to form nitrate (NO_3^-) or further converted into nitrate (NO_3^-) through nitrite oxidation by nitrifying bacteria. However, biological nitrogen fixation produces the byproduct N_2O, a potent greenhouse gas that exacerbates the atmospheric greenhouse effect. Soil microorganisms carrying genes for N_2O reductase (encoded by *nosZ*) can reduce N_2O to N_2 [63,65] (Figure 6 and Table 3). Thus, it is evident that microorganisms play crucial roles in driving the soil nitrogen cycle and regulating soil nitrogen availability.

5.1.3. Microbial Regulation of Soil Phosphorus Availability

Plants acquire phosphorus primarily from soil, where it exists as organic (bound in recalcitrant compounds) and inorganic forms. While organic phosphorus requires microbial mineralization to release plant-available inorganic phosphate (Pi), inorganic phosphorus dominates (60–80% of total soil phosphorus) yet is largely immobilized through adsorption or mineral fixation [66]. Only 1% of inorganic phosphorus exists as soluble forms (e.g., water-soluble Pi bound to low-molecular-weight organics), directly accessible to plants. Plant roots enhance Pi mobilization via phosphatase secretion, aided by microbial activity. Soil microorganisms sustain phosphorus availability through three key functions: (1) solubilizing mineral phosphorus, (2) mineralizing organic phosphorus, and (3) facilitating plant uptake and translocation [67] (Figure 7 and Table 3). By balancing organic-inorganic phosphorus conversion and enriching soil organic phosphorus pools, microbes ensure sustained nutrient supply for soil–plant ecosystem health.

5.1.4. Microbial Improvement of Soil Physical Structure

Soil aggregates, critical for nutrient storage and ecosystem function, maintain water stability, root development, and erosion resistance (Figure 7 and Table 3). Their formation requires cementing agents (e.g., humus, polysaccharides, microbial secretions) and aggregation forces (e.g., wet–dry cycles, biological activity). Humus, the primary stabilizing agent, originates from microbial transformation of organic matter. Biological forces—particularly microbial activity, root growth, and burrowing fauna—drive aggregate development [68].

Microorganisms enhance aggregation through three mechanisms: (1) secreting extracellular polymeric substances (EPS) to bind particles, (2) utilizing cell surface charges for electrostatic adhesion, and (3) entangling particles via fungal/actinobacterial hyphae [69] (Figure 7 and Table 3). Microbial spatial heterogeneity further facilitates aggregate formation, with diverse communities synergistically improving soil structure. Thus, microorganisms are pivotal in establishing stable aggregate networks [70].

5.2. Microbial Improvement of the Soil Biological Environment

The ecological effects of microorganisms on improving the soil environment are reflected primarily in an increase in the population of beneficial microorganisms, an inhibition of pathogen growth, an increase in soil resistance, the degradation of harmful substances in the soil, and restoration of the soil biological environment.

5.2.1. Microbial Resistance to Stress and Control of Soil Pests and Diseases

Soil microorganisms play vital roles in countering biotic and abiotic stresses, directly influencing soil health and plant resilience [69]. Beneficial microbes enhance soil disease resistance through competition, predation, parasitism, and secretion of antagonistic compounds (Figure 7 and Table 3). Their adaptive strategies to environmental stress include: (1) metabolic adaptation via altered pathways to produce stress-specific metabolites; (2) dynamic regulation of gene expression timing and intensity; and (3) cooperative resource-sharing networks that suppress pathogen colonization [70]. For instance, bacterial-fungal synergies enable resource exchange while maintaining ecological equilibrium [71]. Such interconnected microbial networks enhance soil–plant health, underscoring microbial biotherapy as a promising strategy to enrich beneficial taxa, reduce pathogens, and improve soil resilience.

5.2.2. Microbial Control of Pollution and Degradation of Harmful Substances

Soil microorganisms critically transform and detoxify pollutants such as petroleum hydrocarbons, microplastics, and heavy metals, which disrupt microbial communities and soil ecosystem functions [72,73]. Petroleum remediation relies on bioaugmentation with microbial agents, as biostimulation alone often fails to degrade complex hydrocarbons. Enzymatic processes—including oxidation, reduction, and decarboxylation—break hydrocarbons into CO_2, H_2O, and benign byproducts [74]. For microplastics, extracellular enzymes and radicals depolymerize plastics into oligomers, which are mineralized into CO_2, H_2O, and N_2 [75] (Figure 7 and Table 3). Fungal-dominated communities exhibit superior degradation efficiency compared to bacteria, highlighting the importance of enhancing fungal diversity [76,77].

Heavy metal remediation leverages microbial bioadsorption, immobilization, and redox reactions. Microorganisms chelate metal ions via extracellular polymeric substances (EPS) or intracellular sequestration, offering an eco-friendly alternative to physicochemical methods that risk secondary pollution [78]. These mechanisms—combined with metal ion-EPS complexation—enable efficient, low-cost immobilization of contaminants [79,80].

5.3. Microorganisms as Predictors of Soil Health

Soil microorganisms naturally respond strongly to various physicochemical changes in soil, showing direct or indirect reactions to changes in the soil environment, making them sensitive indicators of soil health [81]. Following the inoculation of microbial agents into the soil, these agents interact with the soil ecosystem through various mechanisms, including antagonism, competition, metabolic inhibition, and symbiosis, and their proper application can positively influence microbial activity, diversity, and community structure within soil [82]. The diversity, biomass, abundance, health and community structure of microorganisms in soil, as determined by metagenomics techniques, are closely related to soil environmental conditions, such as soil-borne diseases, soil pollution, nutrient content, soil texture and soil structure. For example, studies have shown that soil pathogens and viruses can serve as indicators for diagnosing soil health and quality, while the activity of soil enzymes can reflect the physical and chemical properties, microbial activity, and biodiversity of soil, all of which are closely linked to the functional diversity of soil [83]. Additionally, research has shown that the ratio of microbial biomass carbon to soil organic carbon (Cmic:Corg) serves as an indicator of soil organic carbon status, with values exceeding the reference ratio found in mature forests, indicating an accumulation of organic carbon, whereas values below this threshold suggest a depletion [84] (Figure 7). Microorganisms are sensitive indicators of soil health, and specific microbial groups are strongly correlated with soil variables, playing a crucial role in guiding the application of targeted microbial inoculants or the use of microbiome engineering techniques to modulate soil microbial communities and improve soil health.

Determination of soil microbiota through microbial red macrogenomic engineering techniques and prediction of soil health based on soil microbiota has become a prominent research focus. In studies on the comprehensive assessment and prediction of soil health, microbial communities have demonstrated predictive accuracy rates ranging from 50% to 95% for soil quality indices and 80% for overall soil health indices [85]. Recent studies have combined macrogenomics (e.g., functional gene *phoD* abundance to predict phosphorus effectiveness) and stable isotope probes (SIPs, which track ^{13}C-labeled microbial metabolic activity) to develop predictive models of soil health. For example, generalized linear modeling (GLM) allows quantitative assessment of soil denitrification potential through the negative correlation between *nosZ* gene abundance and N_2O emissions [64]. In addition, ^{15}N isotope tracing revealed the contribution of AM fungi to plant N uptake [65], providing new indicators of microbial regulation of soil health. These findings indicate the feasibility of using the microbiome to predict soil health status.

5.4. Limitations of Microbial Agent Application

Current research identifies six key limitations in microbial agent use for soil improvement: (1) competition with indigenous microbiota, poor environmental adaptability, and limited proliferation post-inoculation [86]; (2) variable efficacy under field conditions; (3) methodological constraints in microbial research requiring advanced technological integration; (4) high cultivation costs and complexity; (5) insufficient understanding of multi-strain synergies and long-term safety; (6) limited functionality under extreme environments, with rare efficient strain availability [87].

Globally, microbial agent technologies are well-established, whereas China's later-starting research has focused on strain selection but lags in field application and commercialization. Unregulated usage, including misuse and overapplication, further complicates implementation. Despite these challenges, expanding microbial agent applications in pollution control remains a critical priority amid escalating environmental crises.

6. Conclusions

Over the past three decades, global research on microbial inoculants for soil improvement has expanded rapidly, driven by technological advancements and societal demands. China, facing urgent ecological and economic challenges, emerges as the largest contributor to Web of Science publications in this field, followed by the United States, with both nations leading disciplinary frontiers (Figure S1). Bibliometric analysis (2003–2024) highlights the Chinese Academy of Sciences as the top institution in publication output, citations, and international collaboration (Figure S1). Key journals, including *Frontiers in Microbiology*, *Science of the Total Environment*, *Soil Biology & Biochemistry*, and *Applied Soil Ecology*, underpin scholarly advancements (Figure S2 and Table S1). While China demonstrates strengths in theoretical research, harmonizing ecological restoration with sustainable development remains a critical priority, it needs to establish a 'microbe-plant–soil' interdisciplinary research platform, prioritize funding for long-term field trials (e.g., the Saline Fungus Combined Restoration Project), and incorporate the ecological benefits of microbial agents into its agricultural subsidy policy in order to promote the deeper integration of ecological restoration and sustainable agricultural development(Figure S3).

International research has shown quantifiable progress with microbial inoculants, with an approximate 38-fold increase in relevant SCI publications from 14 in 2003 to 552 in 2024. Studies have demonstrated significant progress in microbial inoculants for soil improvement, focusing on biodiversity, soil carbon/nitrogen cycling, microbial community assembly, and pollution bioremediation [88,89]. Macrogenomics has revealed the central role of functional genes in petroleum hydrocarbon degradation, while heterologous expression of laccase genes (laccases) has led to an increase in the efficiency of polyethylene degradation by fungi. In addition, synthetic biology tools are being used to construct highly efficient composite fungicides, providing a new direction for bioremediation technology [88,89]. China, despite its extensive theoretical research on soil microbial zonation characteristics, rhizosphere microbial interactions, and functional diversity [90], started relatively late in applied studies of microbial inoculants. The CNKI database reveals only over a hundred related publications, indicating insufficient empirical research and an urgent need to refine practical application frameworks.

Bibliometric analysis highlights three core research priorities: (1) enhancing functional strain screening via high-throughput sequencing to isolate key species adaptable to soil environments, capable of improving fertility and ecosystem health [88]; (2) deciphering microbial interspecies interactions and soil-microbe signaling pathways [89]; (3) identifying environmental drivers of inconsistent field efficacy through inoculation trials [91]. Future research should focus on three major directions: (1) cross-domain bacterial agent development: constructing bacteria–fungus–archaea consortium (e.g., *Bacillus-Aspergillus*–Methanotrophs synergistic system), which has been shown to have a higher yield-increasing effect than a single bacterial agent in field trials; (2) intelligent optimization technology: combined with machine learning models (e.g., Random Forest algorithm to predict fungicide adaptability), it realizes dynamic regulation of inoculation strategy; (3) eco-certification system: establish soil health grading standards based on microbial biomarkers (e.g., Cmic:Corg ratio, *nosZ* gene abundance) to promote both policy- and market-driven incentives. Current limitations stem from incomplete mechanistic understanding, delayed effects of inoculants, and field performance variability. Integrating laboratory and field experiments to develop synergistic "microbial–physicochemical remediation" systems will advance cost-effective, eco-friendly solutions. This approach not only addresses the shortcomings of conventional techniques but also drives innovation in soil restoration engineering, supporting China's ecological agriculture and sustainable development goals.

Supplementary Materials: The following supporting information can be downloaded at: https://www.mdpi.com/article/10.3390/agronomy15051223/s1, Figure S1: Collaboration network of the top 39 most productive institutes; Figure S2: Collaborative networks of the top 20 journals with the highest number of publications; Figure S3: Collaborative network of the top 75 authors with the highest number of publications;Table S1: Top 10 most productive journals; TableS2: Top 15 most productive journals; Table S3: Average year of the 20 most frequent keywords; Table S4: Keyword Frequency Table.

Author Contributions: Conceptualization, M.T. and T.F.; methodology, M.T.; software, M.T.; validation, M.T., T.F., H.Y., and X.H.; formal analysis, M.T., X.H., and H.Y.; investigation, M.T., X.H., and H.Y.; resources, T.F.; data curation, M.T.; writing—original draft preparation, M.T.; writing—review and editing, M.T., T.F., C.W., X.H., and H.Y.; visualization, M.T.; supervision, T.F.; project administration, T.F.; funding acquisition, T.F. All authors have read and agreed to the published version of the manuscript.

Funding: This study was supported by the National Key Research and Development Program of China (No. 2022YFB4202103), the National Natural Science Foundation of China (No. 42371114), the Open Fund Project of the Engineering Technology Innovation Center for Ecological Protection and Restoration in the Middle Yellow River, Ministry of Natural Resources (No. 2025034). The funding agencies had no role in the study design, data interpretation, or decision to submit the manuscript for publication.

Data Availability Statement: The original contributions presented in this study are included in the article. Further inquiries can be directed to the corresponding author.

Conflicts of Interest: The authors declare no conflicts of interest. The funders had no role in the design of the study; in the collection, analyses, or interpretation of data; in the writing of the manuscript; or in the decision to publish the results.

References

1. Fierer, N.; Wood, S.A.; Bueno de Mesquita, C.P. How microbes can, and cannot, be used to assess soil health. *Soil Biol. Biochem.* **2021**, *153*, 108111. [CrossRef]
2. Liu, Z.; Yin, Y.; Zhang, Y.; Shi, S. Legal System of Soil Pollution Remediation in China and Its Regulation and Guidance to Soil Pollution Remediation. *Sustainability* **2023**, *15*, 11504. [CrossRef]
3. Sheer, A.; Sardar, M.F.; Younas, F.; Zhu, P.; Noreen, S.; Mehmood, T.; Farooqi, Z.U.R.; Fatima, S.; Guo, W.H. Trends and social aspects in the management and conversion of agricultural residues into valuable resources: A comprehensive approach to counter environmental degradation, food security, and climate change. *Bioresour. Technol.* **2024**, *394*, 130258. [CrossRef] [PubMed]
4. Teng, Y.; Wu, J.; Lu, S.; Wang, Y.; Jiao, X.; Song, L. Soil and soil environmental quality monitoring in China: A review. *Environ. Int.* **2014**, *69*, 177–199. [CrossRef]
5. Bastida, F.; Zsolnay, A.; Hernández, T.; García, C. Past, present and future of soil quality indices: A biological perspective. *Geoderma* **2008**, *147*, 159–171. [CrossRef]
6. Wang, G.; Wang, X.; Wu, B.; Qi, L. Desertification and its mitigation strategy in China. *J. Resour. Ecol.* **2012**, *3*, 97–104. [CrossRef]
7. Rahman, K.M.A.; Zhang, D. Effects of Fertilizer Broadcasting on the Excessive Use of Inorganic Fertilizers and Environmental Sustainability. *Sustainability* **2018**, *10*, 759. [CrossRef]
8. Lwin, C.S.; Seo, B.H.; Kim, H.U.; Owens, G.; Kim, K.R. Application of soil amendments to contaminated soils for heavy metal immobilization and improved soil quality—A critical review. *Soil Sci. Plant Nutr.* **2018**, *64*, 156–167. [CrossRef]
9. Grandy, A.S.; Daly, A.B.; Bécu, T.; Cardinael, R.; Fontaine, S.; Jilling, A.; MacLaren, C.; Phillips, R.P. A microbial framework for nitrogen cycling solutions in agroecosystems. *One Earth* **2024**, *7*, 2103–2107. [CrossRef]
10. Six, J.; Frey, S.D.; Thiet, R.K.; Batten, K.M. Bacterial contribution to carbon sequestration in soils. *J. Plant Nutr. Soil Sci.* **2006**, *169*, 350–353. [CrossRef]
11. Sun, X.; Wang, W.; Yi, S.; Zheng, F.; Zhang, Z.; Alharbi, S.A.; Filimonenko, E.; Wang, Z.; Kuzyakov, Y. Microbial composition in saline and alkaline soils regulates plant growth with P-solubilizing bacteria. *Appl. Soil Ecol.* **2024**, *203*, 105653. [CrossRef]
12. Liu, W.; He, C.; Han, S.; Lin, B.; Liu, W.; Dang, Y.P.; Zhao, X.; Zhang, H. Enhancing soil ecosystem multifunctionality through combined conservation tillage and legume-based crop rotation in the North China Plain. *Agric. Ecosyst. Environ.* **2025**, *379*, 109355. [CrossRef]
13. Yaashikaa, P.R.; Devi, M.K.; Kumar, P.S. Engineering microbes for enhancing the degradation of environmental pollutants: A detailed review on synthetic biology. *Environ. Res.* **2022**, *214*, 113868. [CrossRef] [PubMed]
14. Mokrani, S.; Houali, K.; Yadav, K.K.; Arabi, A.I.A.; Eltayeb, L.B.; AwjanAlreshidi, M.; Benguerba, Y.; Cabral-Pinto, M.M.; Nabti, E.H. Bioremediation techniques for soil organic pollution: Mechanisms, microorganisms, and technologies—A comprehensive review. *Ecol. Eng.* **2024**, *207*, 107338. [CrossRef]
15. Semenov, M.; Blagodatskaya, E.; Stepanov, A.; Kuzyakov, Y. DNA-based determination of soil microbial biomass in alkaline and carbonaceous soils of semi-arid climate. *J. Arid Environ.* **2018**, *150*, 54–61. [CrossRef]
16. Eren, A.M.; Banfield, J.F. Modern microbiology: Embracing complexity through integration across scales. *Cell* **2024**, *187*, 5151–5170. [CrossRef]
17. He, Y.; Lan, Y.; Zhang, H.; Ye, S. Research characteristics and hotspots of the relationship between soil microorganisms and vegetation: A bibliometric analysis. *Ecol. Indic.* **2022**, *141*, 109145. [CrossRef]
18. Zhu, J.; Hua, W. Visualizing the knowledge domain of sustainable development research between 1987 and 2015: A bibliometric analysis. *Scientometrics* **2017**, *110*, 893–914. [CrossRef]

19. Nakagawa, S.; Samarasinghe, G.; Haddaway, N.R.; Westgate, M.J.; O'Dea, R.E.; Noble, D.W.; Lagisz, M. Research weaving: Visualizing the future of research synthesis. *Trends Ecol. Evol.* **2019**, *34*, 224–238. [CrossRef]
20. Sun, S.; Xue, R.; Liu, M.; Wang, L.; Zhang, W. Research progress and hotspot analysis of rhizosphere microorganisms based on bibliometrics from 2012 to 2021. *Front. Microbiol.* **2023**, *14*, 1085387. [CrossRef]
21. Nkongolo, K.K.; Narendrula-Kotha, R. Advances in monitoring soil microbial community dynamic and function. *J. Appl. Genet.* **2020**, *61*, 249–263. [CrossRef] [PubMed]
22. Singh, J.S.; Pandey, V.C.; Singh, D.P. Efficient soil microorganisms: A new dimension for sustainable agriculture and environmental development. *Agric. Ecosyst. Environ.* **2011**, *140*, 339–353. [CrossRef]
23. Banerjee, S.; Van Der Heijden, M.G. Soil microbiomes and one health. *Nat. Rev. Microbiol.* **2023**, *21*, 6–20. [CrossRef] [PubMed]
24. Sivaram, A.K.; Abinandan, S.; Chen, C.; Venkateswartlu, K.; Megharaj, M. Microbial inoculant carriers: Soil health improvement and moisture retention in sustainable agriculture. *Adv. Agron.* **2023**, *180*, 35–91. [CrossRef]
25. Kumar, A.; Das, A.; Singh, D.; Das, M.K.; Srivastava, G.P.; Singh, J.P.; Thapa, S.; Das, S.; Chakdar, H. Soil health restoration in degraded lands: A microbiological perspective. *Land Degrad. Dev.* **2023**, *34*, 5155–5170. [CrossRef]
26. Deng, L.; Wang, T.; Luo, W.; He, L.; Liang, Z. Effects of a compound microbial agent and plants on soil properties, enzyme activities, and bacterial composition of Pisha sandstone. *Environ. Sci. Pollut. Res.* **2021**, *28*, 53353–53364. [CrossRef]
27. Yang, Q.; Zhang, S.; Li, X.; Rong, K.; Li, J.; Jiang, L. Effects of microbial inoculant and additives on pile composting of cow manure. *Front. Microbiol.* **2023**, *13*, 1084171. [CrossRef]
28. Pan, L.; Mao, L.; Zhang, H.; Wang, P.; Wu, C.; Xie, J.; Yu, B.; Sial, M.U.; Zhang, L.; Zhang, Y.; et al. Modified biochar as a more promising amendment agent for remediation of pesticide-contaminated soils: Modification methods, mechanisms, applications, and future perspectives. *Appl. Sci.* **2022**, *12*, 11544. [CrossRef]
29. Veliz, E.A.; Martínez-Hidalgo, P.; Hirsch, A.M. Chitinase-producing bacteria and their role in biocontrol. *AIMS Microbiol.* **2017**, *3*, 689. [CrossRef]
30. Funahashi, F.; Myrold, D.D.; Parke, J.L. The effects of soil solarization and application of a Trichoderma biocontrol agent on soil fungal and prokaryotic communities. *Soil Sci. Soc. Am. J.* **2022**, *86*, 369–383. [CrossRef]
31. Compant, S.; Duffy, B.; Nowak, J.; Clément, C.; Barka, E.A. Use of plant growth-promoting bacteria for biocontrol of plant diseases: Principles, mechanisms of action, and future prospects. *Appl. Environ. Microbiol.* **2005**, *71*, 4951–4959. [CrossRef] [PubMed]
32. Morales-Cedeño, L.R.; del Carmen Orozco-Mosqueda, M.; Loeza-Lara, P.D.; Parra-Cota, F.I.; de Los Santos-Villalobos, S.; Santoyo, G. Plant growth-promoting bacterial endophytes as biocontrol agents of pre-and post-harvest diseases: Fundamentals, methods of application and future perspectives. *Microbiol. Res.* **2021**, *242*, 126612. [CrossRef] [PubMed]
33. Yang, L.; Jie, G.; She-Qi, Z.; Long-Xiang, S.; Wei, S.; Xun, Q.; Man-Li, D.; Ya-Nan, Y.; Xiao-Juan, W. Effects of adding compound microbial inoculum on microbial community diversity and enzymatic activity during co-composting. *Environ. Eng. Sci.* **2018**, *35*, 270–278. [CrossRef]
34. Hart, M.M.; Antunes, P.M.; Chaudhary, V.B.; Abbott, L.K. Fungal inoculants in the field. *Funct. Ecol.* **2018**, *32*, 126–135. [CrossRef]
35. Li, X.; Yang, X.; Yuan, X.; Chen, S. Effect of three actinomycetes on cucumber root-knot nematode disease. *J. Northwest A F Univ.-Nat. Sci. Ed.* **2022**, *50*, 1–8. [CrossRef]
36. Cui, N.; Wang, S.; Khorram, M.S.; Fang, H.; Yu, Y. Microbial degradation of fomesafen and detoxification of fomesafen-contaminated soil by the newly isolated strain *Bacillus* sp. FE-1 via a proposed biochemical degradation pathway. *Sci. Total Environ.* **2018**, *616*, 1612–1619. [CrossRef]
37. Gu, J.; Qi, X.; Li, X.; Ren, Y.; Wang, X. Preparation of solid microbial inoculants and its application in aerobic composting. *Chin. J. Environ. Eng.* **2020**, *14*, 253–261. [CrossRef]
38. Wen, D.; Wang, X.; Sun, K.N.; Wang, K.A.; Gao, J.W.; Zhang, W.; Yang, N. Effects of different forms of microbial agents on the growth and quality of *Brassica rapa* L. ssp. chinensis Makino (non-heading Chinese cabbage). *Chin. J. Appl. Ecol.* **2021**, *32*, 1777–1782. [CrossRef]
39. Afanador-Barajas, L.N.; Navarro-Noya, Y.E.; Luna-Guido, M.L.; Dendooven, L. Impact of a bacterial consortium on the soil bacterial community structure and maize (*Zea mays* L.) cultivation. . *Sci. Rep.* **2021**, *11*, 13092. [CrossRef]
40. Zhang, Y.; Wang, K.; Niu, B.; Wang, L.; Zhang, M.; Chen, Y.; Wang, J.; Gao, M. Functional characteristics of *Bacillus siamensis* 37402-1 and its growth-promoting effect on garlic. *Soil Fertil. Sci. China* **2024**, *2*, 185–192. [CrossRef]
41. Wang, Y.; Liu, J.; Li, Z.; Liu, L. Effects of arbuscular mycorrhizal fungi on nitrogen uptake of maize and soil N_2O emissions in cinnamon soil. *Acta Ecol. Sin.* **2024**, *44*, 1972–1984. [CrossRef]
42. Wang, Y.; Li, P.; Wu, W.; Jin, Q.; Wang, R.; Zhang, L.; Gao, F.; Zhao, Y.; Wang, W. Effects of bio-organic fertilizer and microbial agent on the growth of tea chrysanthemum and soil fertility under continuous cropping cultivation system in the mountainous area of Beijing. *Soil Fertil. Sci. China* **2023**, *12*, 107–113. [CrossRef]
43. Dong, W.; Zhou, R.; Li, X.; Yan, H.; Zheng, J.; Peng, N.; Zhao, S. Effect of simplified inoculum agent on performance and microbiome during cow manure-composting at industrial-scale. *Bioresour. Technol.* **2024**, *393*, 130097. [CrossRef] [PubMed]

44. Beneduzi, A.; Ambrosini, A.; Passaglia, L.M. Plant growth-promoting rhizobacteria (PGPR): Their potential as antagonists and biocontrol agents. *Genet. Mol. Biol.* **2012**, *35*, 1044–1051. [CrossRef]
45. Song, J.; Brookes, P.C.; Shan, S.; Xu, J.; Liu, X. Effects of remediation agents on microbial community structure and function in soil aggregates contaminated with heavy metals. *Geoderma* **2022**, *425*, 116030. [CrossRef]
46. Cheng, Z.; Lu, L.; Kennes, C.; Ye, J.; Yu, J.; Chen, D.; Chen, J. A composite microbial agent containing bacterial and fungal species: Optimization of the preparation process, analysis of characteristics, and use in the purification for volatile organic compounds. *Bioresour. Technol.* **2016**, *218*, 751–760. [CrossRef]
47. Singh, V.; Kumar, B. A review of agricultural microbial inoculants and their carriers in bioformulation. *Rhizosphere* **2024**, *29*, 100843. [CrossRef]
48. Bhattacharyya, S.S.; Furtak, K. Soil–Plant–Microbe interactions determine soil biological fertility by altering rhizospheric nutrient cycling and biocrust formation. *Sustainability* **2022**, *15*, 625. [CrossRef]
49. Falkowski, P.G.; Fenchel, T.; Delong, E.F. The microbial engines that drive Earth's biogeochemical cycles. *Science* **2008**, *320*, 1034–1039. [CrossRef]
50. Kuypers, M.M.; Marchant, H.K.; Kartal, B. The microbial nitrogen-cycling network. *Nat. Rev. Microbiol.* **2018**, *16*, 263–276. [CrossRef]
51. Richardson, A.E.; Hocking, P.J.; Simpson, R.J.; George, T.S. Plant mechanisms to optimise access to soil phosphorus. *Crop Pasture Sci.* **2009**, *60*, 124–143. [CrossRef]
52. Pieterse, C.M.; Zamioudis, C.; Berendsen, R.L.; Weller, D.M.; Van Wees, S.C.; Bakker, P.A. Induced systemic resistance by beneficial microbes. *Annu. Rev. Phytopathol.* **2014**, *52*, 347–375. [CrossRef] [PubMed]
53. Denaro, R.; Di Pippo, F.; Crisafi, F.; Rossetti, S. Biodegradation of Hydrocarbons in Marine Environment. In *Water Pollution and Remediation: Organic Pollutants*; Inamuddin, Ahamed, M.I., Lichtfouse, E., Eds.; Environmental Chemistry for a Sustainable World; Springer: Cham, Switzerland, 2021; Volume 54. [CrossRef]
54. Singh, B. Organophosphorus-degrading bacteria: Ecology and industrial applications. *Nat. Rev. Microbiol.* **2009**, *7*, 156–164. [CrossRef]
55. Nies, D. Microbial heavy-metal resistance. *Appl. Microbiol. Biotechnol.* **1999**, *51*, 730–750. [CrossRef]
56. Yoshida, S.; Hiraga, K.; Takehana, T.; Taniguchi, I.; Yamaji, H.; Maeda, Y.; Toyohara, K.; Miyamoto, K.; Kimura, Y.; Oda, K. A bacterium that degrades and assimilates poly(ethylene terephthalate). *Science* **2016**, *351*, 1196–1199. [CrossRef]
57. Vázquez, M.M.; César, S.; Azcón, R.; Barea, J.M. Interactions between arbuscular mycorrhizal fungi and other microbial inoculants (Azospirillum, Pseudomonas, Trichoderma) and their effects on microbial population and enzyme activities in the rhizosphere of maize plants. *Appl. Soil Ecol.* **2000**, *15*, 261–272. [CrossRef]
58. Liang, C.; Zhu, X. The soil microbial carbon pump as a new concept for terrestrial carbon sequestration. *Sci. China Earth Sci.* **2021**, *64*, 545–558. [CrossRef]
59. Tao, F.; Huang, Y.; Hungate, B.A.; Manzoni, S.; Frey, S.D.; Schmidt, M.W.; Reichstein, M.; Carvalhais, N.; Ciais, P.; Jiang, L.; et al. Microbial carbon use efficiency promotes global soil carbon storage. *Nature* **2023**, *618*, 981–985. [CrossRef]
60. Bhattacharyya, S.S.; Ros, G.H.; Furtak, K.; Iqbal, H.M.; Parra-Saldívar, R. Soil carbon sequestration–An interplay between soil microbial community and soil organic matter dynamics. *Sci. Total Environ.* **2022**, *815*, 152928. [CrossRef]
61. Wang, M.; Guo, X.; Zhang, S.; Xiao, L.; Mishra, U.; Yang, Y.; Zhu, B.; Wang, G.; Mao, X.; Qian, T.; et al. Global soil profiles indicate depth-dependent soil carbon losses under a warmer climate. *Nat. Commun.* **2022**, *13*, 5514. [CrossRef]
62. Jiang, S.; Jardinaud, M.F.; Gao, J.; Pecrix, Y.; Wen, J.; Mysore, K.; Xu, P.; Sanchez-Canizares, C.; Ruan, Y.; Li, Q.; et al. NIN-like protein transcription factors regulate leghemoglobin genes in legume nodules. *Science* **2021**, *374*, 625–628. [CrossRef] [PubMed]
63. Wang, B.; An, S.; Liang, C.; Liu, Y.; Kuzyakov, Y. Microbial necromass as the source of soil organic carbon in global ecosystems. *Soil Biol. Biochem.* **2021**, *162*, 108422. [CrossRef]
64. Sepp, S.K.; Vasar, M.; Davison, J.; Oja, J.; Anslan, S.; Al-Quraishy, S.; Al-Quraishy, S.; Bahram, M.; Bueno, C.G.; Cantero, J.J.; et al. Global diversity and distribution of nitrogen-fixing bacteria in the soil. *Front. Plant Sci.* **2023**, *14*, 1100235. [CrossRef] [PubMed]
65. Ren, R.; Ma, K.; Jiang, Y.; Chen, J.; Kou, Y.; Ge, Z.; Chen, Z.; Wei, X.; Yu, L. Endothelial miR-196b-5p regulates angiogenesis via the hypoxia/miR-196b-5p/HMGA2/HIF1α loop. *Am. J. Physiol.-Cell Physiol.* **2023**, *324*, C407–C419. [CrossRef]
66. Zhu, J.; Li, M.; Whelan, M. Phosphorus activators contribute to legacy phosphorus availability in agricultural soils: A review. *Sci. Total Environ.* **2018**, *612*, 522–537. [CrossRef]
67. Pang, F.; Li, Q.; Solanki, M.K.; Wang, Z.; Xing, Y.X.; Dong, D.F. Soil phosphorus transformation and plant uptake driven by phosphate-solubilizing microorganisms. *Front. Microbiol.* **2024**, *15*, 1383813. [CrossRef]
68. Guhra, T.; Stolze, K.; Totsche, K.U. Pathways of biogenically excreted organic matter into soil aggregates. *Soil Biol. Biochem.* **2022**, *164*, 108483. [CrossRef]
69. Wang, F.; Zhang, L.; Zhou, J.; Rengel, Z.; George, T.S.; Feng, G. Exploring the secrets of hyphosphere of arbuscular mycorrhizal fungi: Processes and ecological functions. *Plant Soil* **2022**, *481*, 1–22. [CrossRef]

70. Wu, D.; Wang, W.; Yao, Y.; Li, H.; Wang, Q.; Niu, B. Microbial interactions within beneficial consortia promote soil health. *Sci. Total Environ.* **2023**, *870*, 165801. [CrossRef]
71. Deveau, A.; Bonito, G.; Uehling, J.; Paoletti, M.; Becker, M.; Bindschedler, S.; Hacquard, S.; Hervé, V.; Labbé, J.; Lastovetsky, O.A.; et al. Bacterial–fungal interactions: Ecology, mechanisms and challenges. *FEMS Microbiol. Rev.* **2018**, *42*, 335–352. [CrossRef]
72. Sun, Y.; Duan, C.; Cao, N.; Ding, C.; Huang, Y.; Wang, J. Biodegradable and conventional microplastics exhibit distinct microbiome, functionality, and metabolome changes in soil. *J. Hazard. Mater.* **2022**, *424*, 127282. [CrossRef] [PubMed]
73. Mrozik, A.; Piotrowska-Seget, Z. Bioaugmentation as a strategy for cleaning up of soils contaminated with aromatic compounds. *Microbiol. Res.* **2010**, *165*, 363–375. [CrossRef] [PubMed]
74. Wilhelm, R.C.; van Es, H.M.; Buckley, D.H. Predicting measures of soil health using the microbiome and supervised machine learning. *Soil Biol. Biochem.* **2022**, *164*, 108472. [CrossRef]
75. Muhonja, C.N.; Makonde, H.; Magoma, G.; Imbuga, M. Biodegradability of polyethylene by bacteria and fungi from Dandora dumpsite Nairobi-Kenya. *PLoS ONE* **2018**, *13*, e0198446. [CrossRef]
76. Solanki, S.; Sinha, S.; Singh, R. Myco-degradation of microplastics: An account of identified pathways and analytical methods for their determination. *Biodegradation* **2022**, *33*, 529–556. [CrossRef]
77. Yuan, J.; Ma, J.; Sun, Y.; Zhou, T.; Zhao, Y.; Yu, F. Microbial degradation and other environmental aspects of microplastics/plastics. *Sci. Total Environ.* **2020**, *715*, 136968. [CrossRef]
78. Priya, A.K.; Gnanasekaran, L.; Dutta, K.; Rajendran, S.; Balakrishnan, D.; Soto-Moscoso, M. Biosorption of heavy metals by microorganisms: Evaluation of different underlying mechanisms. *Chemosphere* **2022**, *307*, 135957. [CrossRef]
79. Li, W.W.; Yu, H.Q. Insight into the roles of microbial extracellular polymer substances in metal biosorption. *Bioresour. Technol.* **2014**, *160*, 15–23. [CrossRef]
80. Mahto, K.U.; Priyadarshanee, M.; Samantaray, D.P.; Das, S. Bacterial biofilm and extracellular polymeric substances in the treatment of environmental pollutants: Beyond the protective role in survivability. *J. Clean. Prod.* **2022**, *379*, 134759. [CrossRef]
81. Hu, M.; Li, Y.; Ge, C.; Zhang, Y.; Yao, H. Research status and application prospects of combined nitrogen fixation in gramineous plants. *Chin. J. Eco-Agric.* **2021**, *29*, 1815–1826. [CrossRef]
82. Philippot, L.; Chenu, C.; Kappler, A.; Rillig, M.C.; Fierer, N. The interplay between microbial communities and soil properties. *Nat. Rev. Microbiol.* **2024**, *22*, 226–239. [CrossRef] [PubMed]
83. Dubey, A.; Malla, M.A.; Khan, F.; Chowdhary, K.; Yadav, S.; Kumar, A.; Sharma, S.; Khare, P.K.; Khan, M.L. Soil microbiome: A key player for conservation of soil health under changing climate. *Biodivers. Conserv.* **2019**, *28*, 2405–2429. [CrossRef]
84. Hermans, S.M.; Buckley, H.L.; Case, B.S.; Curran-Cournane, F.; Taylor, M.; Lear, G. Using soil bacterial communities to predict physico-chemical variables and soil quality. *Microbiome* **2020**, *8*, 79. [CrossRef] [PubMed]
85. Zhang, L.; Guan, Y. Microbial investigations of new hydrogel-biochar composites as soil amendments for simultaneous nitrogen-use improvement and heavy metal immobilization. *J. Hazard. Mater.* **2022**, *424*, 127154. [CrossRef]
86. Toju, H.; Peay, K.G.; Yamamichi, M.; Narisawa, K.; Hiruma, K.; Naito, K.; Fukuda, S.; Ushio, M.; Nakaoka, S.; Onoda, Y.; et al. Core microbiomes for sustainable agroecosystems. *Nat. Plants* **2018**, *4*, 733. [CrossRef]
87. Pirttilä, A.M.; Mohammad Parast Tabas, H.; Baruah, N.; Koskimäki, J.J. Biofertilizers and biocontrol agents for agriculture: How to identify and develop new potent microbial strains and traits. *Microorganisms* **2021**, *9*, 817. [CrossRef]
88. Manfredini, A.; Malusà, E.; Costa, C.; Pallottino, F.; Mocali, S.; Pinzari, F.; Canfora, L. Current methods, common practices, and perspectives in tracking and monitoring bioinoculants in soil. *Front. Microbiol.* **2021**, *12*, 698491. [CrossRef]
89. O'Callaghan, M.; Ballard, R.A.; Wright, D. Soil microbial inoculants for sustainable agriculture: Limitations and opportunities. *Soil Use Manag.* **2022**, *38*, 1340–1369. [CrossRef]
90. Thompson, I.P.; Van Der Gast, C.J.; Ciric, L.; Singer, A.C. Bioaugmentation for bioremediation: The challenge of strain selection. *Environ. Microbiol.* **2005**, *7*, 909–915. [CrossRef]
91. Kaminsky, L.M.; Trexler, R.V.; Malik, R.J.; Hockett, K.L.; Bell, T.H. The inherent conflicts in developing soil microbial inoculants. *Trends Biotechnol.* **2019**, *37*, 140–151. [CrossRef]

agronomy

MDPI

Article

Bacterial Community Dynamics in Oil-Contaminated Soils in the Hyper-Arid Arava Valley

Varsik Martirosyan [1], Ilan Stavi [2,3,*], Tirza Doniger [4], Itaii Applebaum [4], Chen Sherman [4], May Levi [4] and Yosef Steinberger [4,*]

1 C-Lab LLC Testing Laboratory, Artashisyan 105, Yerevan 0038, Armenia; varsikmartirosyan@gmail.com
2 Dead Sea and Arava Science Center, Yotvata 88820, Israel
3 Eilat Campus, Ben-Gurion University of the Negev, Eilat 88100, Israel
4 The Mina and Everard Goodman Faculty of Life Sciences, Bar-Ilan University, Ramat-Gan 5290002, Israel; tirza.doniger@biu.ac.il (T.D.); itaiiapplebaum@gmail.com (I.A.)
* Correspondence: istavi@adssc.org (I.S.); yosef.steinberger@biu.ac.il (Y.S.)

Abstract: Petroleum pollution has become a substantial challenge in soil ecology. The soil bacterial consortia play a major role in the biodegradation of petroleum hydrocarbons. The main objective of this study was to assess changes in bacterial composition and diversity in oil-contaminated dryland soils. The Illumina MiSeq high-throughput sequencing technique was used to study the bacterial diversity and structural change in hyper-arid oil-contaminated soil in the Arava Valley of Israel. The diversity and abundance of soil bacteria declined significantly following oil pollution. The dominant phyla in the petroleum-contaminated soils were *Proteobacteria* (~33% higher vs. control soil) and *Patescibacteria* (~2.5% higher vs. control soil), which are oil-associated and hydrocarbon-degrading bacteria. An opposite trend was found for the *Actinobacteria* (~8%), *Chloroflexi* (12%), *Gemmatimonadetes* (3%), and *Planctomycetes* (2%) phyla, with the lower abundances in contaminated soil vs. control soil. Investigation of long-term contaminated sites revealed significant genus-level taxonomic restructuring in soil bacterial communities. The most evident changes were observed in *Mycobacterium*, *Alkanindiges*, and *uncultured bacterium-145*, which showed marked abundance shifts between spill and control soils across decades. Particularly, hydrocarbon-degrading genera such as *Pseudoxanthomonas* demonstrated persistent dominance in contaminated sites. While some genera (e.g., *Frigoribacterium*, *Leifsonia*) declined over time, others—particularly *Nocardioides* and *Streptomyces*—exhibited substantial increases by 2014, suggesting potential ecological succession or adaptive selection. Minor but consistent changes were also detected in stress-tolerant genera like *Blastococcus* and *Quadrisphaera*. The effect of oil contamination on species diversity was greater at the 1975 site compared to the 2014 site. These patterns highlight the dynamic response of bacterial communities to chronic contamination, with implications for bioremediation and ecosystem recovery. The study results provide new insights into oil contamination-induced changes in soil bacterial community and may assist in designing appropriate biodegradation strategies to alleviate the impacts of oil contamination in drylands.

Keywords: bacterial community; crude oil; environmental contamination; hyper-arid ecosystem; natural bioremediation; total petroleum hydrocarbons (TPHs)

1. Introduction

Contamination of soil and water by crude oil, a worldwide environmental challenge [1], causes severe damage to ecosystems [2,3].

Such pollution degrades or destroys soils, leads to secondary pollution of air and groundwater, decreases soil–water permeability, and reduces or inhibits the growth of soil organisms [4,5]. Further, crude oil contains many mutagenic and carcinogenic compounds that may cause fatal effects at the genetic level, due to their toxicity [6]. The main components of crude oil are aromatic and saturated hydrocarbons, asphaltene, and colloids [7], which decompose slowly after entering the soil [8].

The reactive components of oil can interact with inorganic phosphorus and nitrogen, inhibiting both dephosphorization and nitrification processes, and reducing the availability of these essential nutrients in the soil [9]. Microorganisms require nutrients like nitrogen and phosphorus for their growth and metabolism activities during hydrocarbon degradation. Consequently, this nutrient limitation can reduce the absorption of water and essential nutrients by plant roots, ultimately inhibiting primary productivity [10]. Continued existence of oil also leads to the accumulation of heavy metals and inorganic salts, increasing the risk of groundwater contamination [11,12].

Crude oil has low water solubility, limiting its availability to microorganisms for degradation [13]. This challenge is further intensified in xeric environments, where water scarcity hampers microbial activity and contaminant transport. Additionally, xeric environments often experience extreme temperatures and high salinity due to evaporation, further stressing indigenous microorganisms, inhibiting enzyme activity, and reducing overall microbial diversity and functionality [14]. As a result, xeric ecosystems are more fragile than mesic ones, taking longer to recover from disturbances. Oil pollution in dryland environments can therefore cause severe ecological damage and significantly alter soil properties [15]. In these systems, where biotic activity is already low due to sharp fluctuations in water availability, radiation, and temperature, the biodegradation of hydrocarbon contaminants by native soil microorganisms is a slow and complex process [16].

Since microorganisms are effective in degrading petroleum products, it is important to understand changes that take place in the microbial composition following oil contamination [17]. The main microbial hydrocarbon decomposition pathways take place under aerobic conditions. In aerobic conditions, these microorganisms convert the dangerous organic pollutants from crude oil into compounds such as CO_2 and H_2O [18]. Nonetheless, bacteria can also utilize hydrocarbons anaerobically [19]; they convert the dangerous organic pollutants from crude oil into CH_4. Interestingly, it has been demonstrated that bacteria can decompose aliphatic and aromatic compounds, whereas fungi are also able to degrade polycyclic aromatics. Moreover, bacteria utilize specific metabolic pathways to degrade hydrocarbons, including those involving dioxygenase and alkane monooxygenase. Additionally, some hydrocarbon-degrading microorganisms adapt by producing and secreting surfactant compounds that emulsify hydrocarbon and form micelles, which are subsequently taken up later through various mechanisms [7]. Such bacteria become more abundant in oil-polluted soils. Salinity plays a significant role in the distribution of microorganisms, particularly with petroleum pollution. It tends to reduce the abundance of non-salt-tolerant microorganisms while fostering an increase in salt-tolerant species. This shift reflects the adaptability of microbial communities to changing environmental conditions [20]. Yet, changes in the microbial composition can have serious deleterious effects on soil functions. Therefore, the modified abundance of microorganisms comprises an important indicator for restoring soil vitality [21,22].

Many studies have discovered that there are numerous hydrocarbon-degrading bacteria in oil-rich environments [14,23]. Their community composition was found to be related to the petroleum hydrocarbon types and the environmental conditions [24,25]. More than 79 genera of bacteria capable of degrading petroleum hydrocarbons were identified. These include *Mycobacterium*, *Achromobacter*, *Marinobacter*, *Acinetobacter*, *Arthrobacter*, *Alkanindiges*,

Alteromonas, Burkholderia, Streptococcus, Enterobacter, Pseudomonas, Staphylococcus, Streptobacillus and *Rhodococcus* [18,26–28]. Interestingly, "conditionally rare taxa" in soil, such as *Alkanindiges* spp., were reported to undergo rare-to-dominant shifts which are greatly affected by environmental constraints, including diesel pollution [24]. In addition, some Obligate Hydrocarbon Clastic Bacteria, including *Thallassolituus, Cycloclasticus, Alcanivorax, Marinobacter, Oleispira,* and others, were scarce or undetectable before petroleum pollution and dominant after it [29].

Significant knowledge gaps remain in understanding the long-term ecological impacts of oil pollution and bioremediation in xeric ecosystems. The objective of the present study was to assess the effects of crude oil contamination on the soil microbial community composition and diversity. The study was conducted in the Evrona Nature Reserve, which is found in the hyper-arid (sandy loam soil) Arava Valley of southern Israel, where two petroleum hydrocarbon spills occurred. The first pollution event occurred 47 years ago (1975) and was never treated. This presents a unique time-point perspective in relation to the second, comparatively recent (2014) contamination event. Thus, soil samples were collected 47 years and 7 years after the spills, from the two contamination sites. The comparison of the two pollution sites enabled an investigation into the extent and progression of natural bioremediation over time. Molecular tools were used to analyze microbial diversity as well as to determine the effect of oil contamination on the soil bacterial community after 47 years [30]. We hypothesized that the bacterial community undergoes changes following oil spill contamination, and only bacterial species that adapt to the new, harsh, contaminated environment survive.

2. Materials and Methods

2.1. Study Site

The Evrona Nature Reserve (29.670349 N; 25.003633 E; 30 m.a.s.l.) is located in the southern Arava Valley of Israel. The valley's lithology is highly varied, and mainly comprises marine calcareous rocks, such as chalk, limestone, and marl, alongside granite and magmatic rocks. The region's soils are composed of coarse desert alluvium and sandy alluvium, with sandy loam texture being predominant. The climate is hyper-arid, with mild winters (mean daily temperature of 15 °C in January) and very hot summers (mean daily temperature of 32 °C in July). Annual precipitations are rather erratic, and average annual rainfall is 30 mm. Mean annual potential evapotranspiration is 2600 mm.

2.2. Soil Sampling

Topsoil samples (0–10 cm) were collected in October 2021. The samples were obtained from the 1975 and the 2014 contamination sites, 47 and 7 years after the spills, respectively. For each of the sites, soil samples were obtained from three oil-contaminated ephemeral stream channels and three nearby non-contaminated (control) ones. In each contaminated and control area, a 50 m^2 plot was delineated. Soil was collected from five randomly selected locations across each plot and pooled into one sample (~1 kg) which represents the plot. In total, 12 samples representing the 12 plots were taken. The samples were placed in separate plastic bags and kept in a cooler box until they arrived at the laboratory. The samples were sieved with a 2 mm mesh and divided into two parts. One part was kept at −20 °C for genetic analysis, and the second was kept at 4 °C for abiotic analysis.

2.3. Soil Moisture (SM) Content and Electrical Conductivity (EC)

SM was determined gravimetrically by drying soil samples for 24 h at 105 °C (expressed as percentage of dry weight) [31].

EC was determined by an auto-ranging EC/temp meter (TH2400, El-Hamma, Algeria) in the filtered supernatant (1:10 soil–double-distilled water ratio). This was followed by shaking for 30 min (160 rpm) and overnight incubation [32].

2.4. Organic Matter (OM) Content

Soil samples were placed in a muffle furnace at 400 °C for 6 h, to determine soil organic matter content.

2.5. Soil pH

Soil pH was determined using a pH electrode in a filtered supernatant of a mixture containing 20 g soil and 40 mL distilled water (1:2 soil–water ratio). This was followed by shaking for 10 min (160 rpm) and overnight incubation at room temperature.

2.6. Microbial Community Diversity Determination

DNA extraction: DNA was extracted from 0.5 g of soil, using an Exgene soil DNA minikit (GeneAll, Seoul, Republic of Korea); 550 µL buffer SL (extraction buffer), 900 µL buffer TB (tissue binding), 500 µL buffer NW (wash buffer N), 50 µL buffer RH, 300 µL buffer PD, and 50 µL elution buffer were used. Samples were centrifuged between each extraction step in a 5810R Eppendorf centrifuge.

Polymerase chain reaction (PCR): Partial sequences of the bacterial 16S ribosomal RNA (rRNA) gene region from the extracted DNA were amplified by PCR, using 1.0 µL extracted DNA, 1 µL CS1-515F (ACACTGACGACATGGTTCTACAGTGCCAGCMGCCGCGGT), and 1 µL CS2-806R (TACGGTAGCAGAGACTTGGTCTGGACTACHVGGGTWTCT) [33]. PCR was carried out in an Applied BiosystemsVeriti^{TM} 96-Well Thermal Cycler (Thermo Fisher Scientific, Waltham, MA, USA). The 25 µL reaction mix used for the PCR runs consisted of 1.0 µL extracted DNA, 1 µL forward and reverse primers, 12.5 µL HS Taq Mix Red (PCR Biosystems, London, UK), and 9.5 µL ultrapure water. The PCR conditions used were 3 m at 98 °C, 20 cycles of 30 s at 98 °C, 30 s at 55 °C, and 60 s at 72 °C, followed by 5 m at 72 °C. The amplified DNA was stored at −20 °C until sequencing.

The final PCR products were detected by Hylabs Inc. (Rehovot, Israel), using the Fluidigm Access Array primers for Illumina to generate libraries compatible for sequencing on the Miseq. The samples were assessed for concentration by Qubit and for size by Tape Station. They were then sequenced on the Illumina Miseq using a Miseq V2 sequencing kit (500 cycles) to generate 2 × 250 paired end reads. Sequencing data have been deposited to NCBI under accession number PRJNA841836.

2.7. Data Analysis

The data were de-multiplexed using the Illumina Base Space cloud to generate two FASTQ files for each sample. The FASTQ files were imported into CLC-bio and analyzed: reads were trimmed for quality and adaptor sequences, merged, and subjected to OTU picking in order to generate abundance tables. Reads were processed in QIIME, version 1.7.0 [34]. The reads were first clustered into operational taxonomic units (OTUs) at the >97% similarity level using UCLUST, with the open reference protocol. Taxonomy was assigned to each OTU using the Greengenes [35] database. Downstream diversity analyses (including alpha and beta analyses) were run using Qiime. A 480 bp fragment was amplified and sequenced. Bacterial diversity was grouped at the class level. All bacterial classes with an abundance lower than 1% were eliminated.

Data were subjected to statistical analysis of variance using the SAS model (ANOVA, Duncan's multiple range test, Pearson correlation coefficients) to assess differences in the studied properties. ANOVA was followed by Tukey's HSD test for assessing the significance of differences, using the statistical package Statistica 4.3. Differences were considered significant at $p < 0.05$.

3. Results

3.1. Soil Physicochemical Properties

Results of abiotic analysis of soils are presented in Table 1. The mean soil moisture content (SM) was significantly higher ($p < 0.05$), and mean organic matter content (OM) was not significantly higher ($p < 0.5$) in the spill areas (1975 and 2014) than in the control soils (Table 1, Figure 1).

Table 1. Mean (±SD; n = 3) values of physicochemical properties of soil samples obtained from the 1975 and 2014 sites.

	Soil Moisture (%)	Organic Matter (%)	pH	Electrical Conductivity ($\mu S\ cm^{-1}$)
1975 Control	0.28 ± 0.04	0.13 ± 0.01	8.12 ± 0.11	111.60 ± 16.57
1975 Spill	0.41 ± 0.06	0.28 ± 0.04	7.95 ± 0.05	87.80 ± 23 90
2014 Control	0.49 ± 0.03	0.13 ± 0.03	7.91 ± 0.05	272.67 ± 67.20
2014 Spill	0.75 ± 0.21	0.59 ± 0.09	7.93 ± 0.14	348.67 ± 158.60

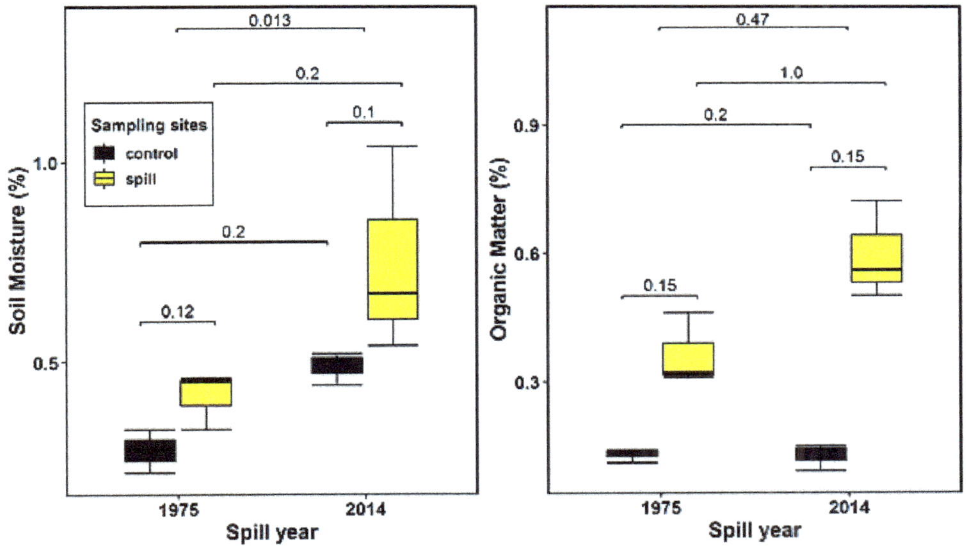

Figure 1. Soil moisture (SM) and organic matter (OM) content of soils from spill and control samples.

The mean SM value was lower in the 1975 control ($p < 0.2$) and spill soils ($p < 0.2$) than in the 2014 control and spill soils, respectively. Similar mean OM values were recorded for the control soils of the 1975 and 2014 spills, whereas OM for the 2014 spill samples was higher than that in the 1975 spill soil (Table 1, Figure 1). No significant differences between the control and contaminated soils were found for pH. At the same time, mean pH values of the 1975 control soil (8.12 ± 0.11) were higher than in the 1975 spill soil (7.95 ± 0.05). An opposite trend was recorded for the 2014 soil, in which the mean pH of control soil was lower (7.91 ± 0.05) than that in the spill soil (7.93 ± 0.14) (Table 1, Figure 2). Mean electric conductivity (EC) was significantly higher ($p < 0.05$) in the control and spill soils from 2014 than in the 1975 control and spill soils. In the 1975 site, the mean EC values of the control soil were higher than in the spill soil. An opposite trend was recorded for the 2014 spill soil, in which mean EC values were lower in the control soils than those in the spill soils (Table 1, Figure 2).

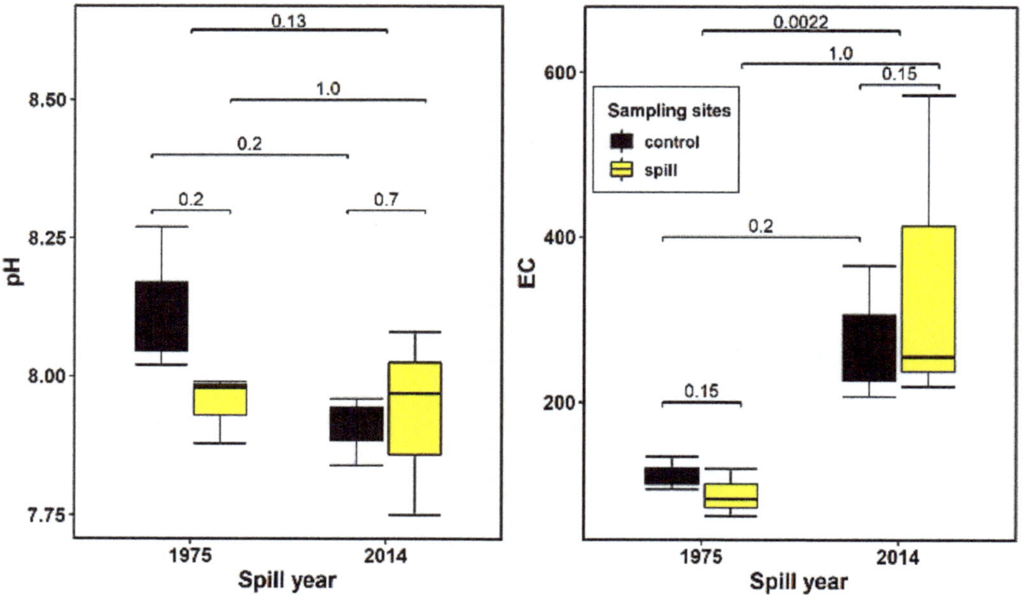

Figure 2. Measured pH and electric conductivity (EC) parameters of soils from spill and control samples.

3.2. Miseq Sequencing Results and Bacterial Community Structures

Amplification of the 16S rRNA gene of bacteria enabled application of the Illumina high-throughput sequencing-by-synthesis approach [36]. This approach provides thorough identification of bacterial community structures, including those that cannot be cultured or detected using traditional approaches [37]. In total, 107,318 of the 16S rRNA gene operational taxonomic units (OTUs) were observed. They had an average length of 250 bp. The RDP classifier was used in hierarchical clustering analysis at a similarity threshold of 97%. The sequence information and bacterial diversity indices are presented in Table 2.

Table 2. Mean (±SD; n = 3) sequence information and bacterial diversity indices of soil samples obtained from the 1975 and 2014 sites.

	Mean No. of OTUs ± SD	Evenness	Observed Species	Chao1	Shannon–Weaver
1975 control	7373.3 ± 1090.5	0.6 ± 0.03	931.7 ± 75.1	1285.4 ± 55.7	5.8 ± 0.06
1975 spill	11,234.7 ± 2585.2	0.8 ± 0.01	338.7 ± 119.6	472.2 ± 171.8	3.5 ± 0.31
2014 control	8334.0 ± 524.7	0.6 ± 0.05	806.0 ± 28.3	1010.72 ± 59.1	5.2 ± 0.13
2014 spill	11,608.7 ± 981.6	0.8 ± 0.02	360.7 ± 57.6	477.4 ± 47.9	3.6 ± 0.39

The number of observed OTUs was higher in the contaminated soils compared to both control soils (Table 2). Soil contaminated in 1975 hosted fewer species than soil contaminated in 2014, whereas an opposite trend was documented for the control samples; the 1975 control soil harbored fewer observed bacterial species than the 2014 control soil (Table 2).

The taxonomic diversity of the 16S rRNA bacterial genus detected in the soil samples, whose relative abundance was greater than 1%, is presented in Figure 3 (genus abundance greater than 1% in one soil sample but less in another was considered as greater than 1%). The dominating genus detected in the 1975 spill soil was *Alkanindiges* (14.45%, belonging to *Proteobacteria* phylum). This genus was absent in the 1975 control soil, very rare in the 2014 control soil (≤0.07%; Figure 3), and accounted for 4.61% in the 2014 spill soil (Figure 3). The second dominating genus for the 1975 spill soil was *Mycobacterium* (9.78%, belonging to *Actinobacteria* phylum), which was almost absent in the control soil (≤0.3%; Figure 3), but in 2014 spill soil, its abundance was 4.10%. The third dominating genus for the 1975 spill soil was *Acinetobacter* (9.23%; also belonging to *Proteobacteria* phylum); its value was double that of the 1975 control soil (4.53%). We also found *Acinetobacter* in the 2014 spill soil, but its trend was the opposite: abundance in the spill soil was lower than in the 2014 control soil (2.21% and 7.28%, respectively; Figure 3). The fourth dominating genus for the 1975 spill was *Pseudoxanthomonas* (8.43%; also belonging to *Proteobacteria* phylum), which was almost absent in the 1975 control and 2014 control soils (≤0.01%; Figure 3) and accounted for 6.57% in the 2014 spill soil (Figure 3). The fifth dominating genus for the 1975 spill was uncultured bacterium-145 (9.76%; also belonging to *Proteobacteria* phylum), which was almost absent in the 1975 control and 2014 control soils (≤0.04%; Figure 3), and was 31.92% higher in the 2014 spill soil (Figure 3).

Ambiguous taxa-025 genus (belonging to *Actinobacteria* phylum) was also found in 1975 spill soil (2.71%), 2014 spill soil (2.91%), and to a lesser extent in both control soils (≤0.3%; Figure 3). *Frigoribacterium, Leifsonia, Microbacterium* (belonging to *Actinobacteria* phylum), and *uncultured bacterium-092* (belonging to *Patescibacteria* phylum) existed only in spill soils though in lesser percentages than the abovementioned genera, and were almost absent in the control soil. *Blastococcus* genus (belonging to *Actinobacteria* phylum) existed in all samples in similar percentages (1975 spill—4.42%; 1975 control—3.08%; 2014 spill—5.69%; 2014 control—4.30%) (Figure 3). Minor changes also were observed in the following bacterial genera: *Quadrisphaera* (1975 spill soil 1.93% and in control soil 0.36%; 2014 spill soil 1.08% and in control soil 0.50%), *Nocardioides* (1975 spill soil 1.11% and in control soil 0.81%; 2014 spill soil 5.56% and in control soil 0.96%), and *Streptomyces* (1975 spill soil 0.06% and in control soil 0.28%; 2014 spill soil 3.56% and in control soil 0.11%). Interestingly, the two most recent genera (*Nocardioides, Streptomyces*) showed a primary increase in the 2014 spill site.

The analysis of genomic DNA isolated from the 1975 and 2014 spill soils grouped the obtained sequences into 23 different phyla. These included two phyla that dominated the sum proportion of total phyla, accounting for ~92% in the spill soils and ~67% in the control soils. For the 1975 and 2014 control groups, dominating phyla included *Proteobacteria* and *Actinobacteria* (~67%), where the *Proteobacteria* were less abundant (~25%) than the *Actinobacteria* (~42%; Figure 4). For the 1975 and 2014 spill groups, dominant phyla also included *Proteobacteria* and *Actinobacteria* (~92%), but the *Proteobacteria* was found in higher percentages (~58%) than the *Actinobacteria* (~34%).

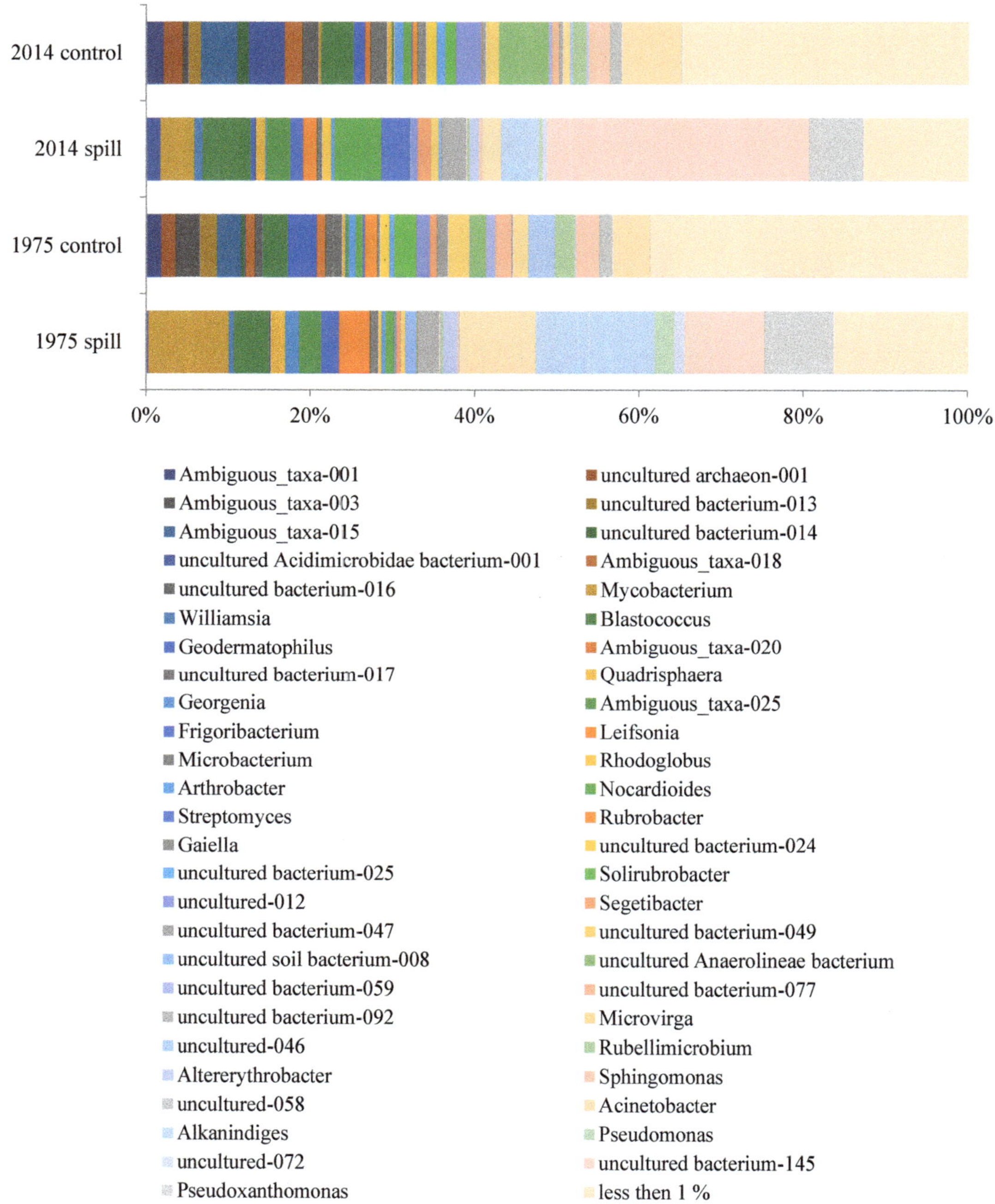

Figure 3. The relative abundance of the 16S rRNA genus detected in soils from spill and control samples.

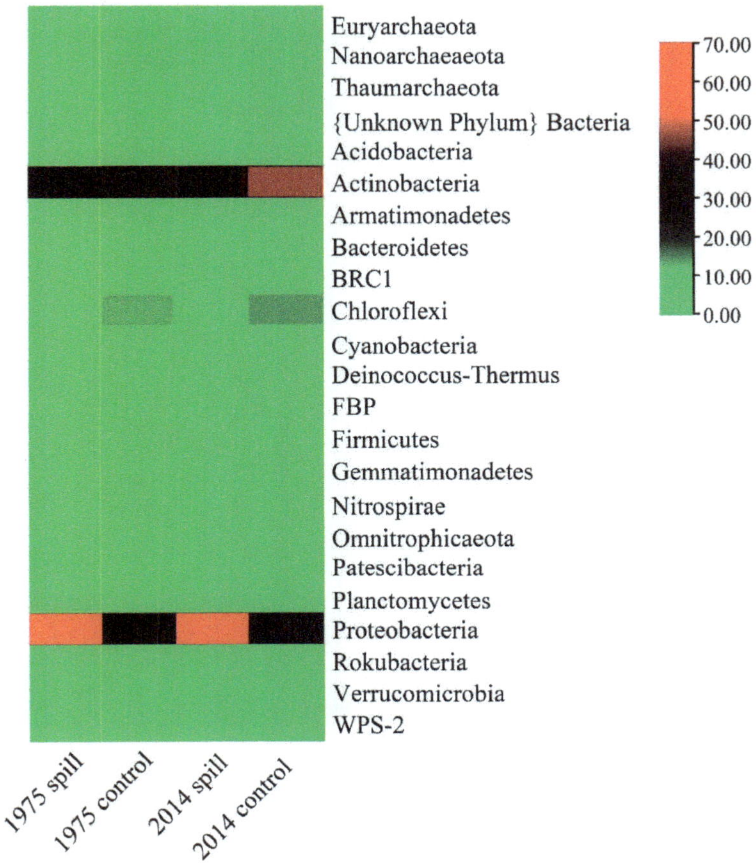

Figure 4. Heatmap of the 16S rRNA gene phyla detected in soil from spill and control samples, by percentage.

3.3. Alpha Diversity Within Samples and Rarefaction Curves

OTU-based alpha diversity or Shannon index was calculated using QIIME software (Table 2) in order to better present the mean species diversity in the two sites. The results confirmed that the sequencing depth [38] enabled us to accurately describe bacterial diversities in our database. The Shannon index and the information on the entropy of the observed OTU abundances accounting for evenness indicated that the bacterial community was more diverse in the control soils than in the spill soils (Table 2). The species diversity was as follows: 1975 control > 2014 control > 2014 spill > 1975 spill. Note that two metrics were used to calculate α-diversity in QIIME: (a) Chao1 metric estimated species richness, and (b) observed species metric (Figure 5).

3.4. Beta Diversity (Pairwise Sample Dissimilarity) Among Samples, and Principal Component Analysis

Analysis of β-diversity was performed to assess OTU distribution. QIIME was used to calculate β-diversity for each of the four bacterial communities. Principal coordinate analysis (PCoA) plots showed similarity between the spill soils (blue circles) and control soils (purple circles; Figure 6). For each site (i.e., 1975 and 2014), the distance between the spill and control clusters was relatively small (differences along the second principal component axis: PC2—11.42%). Greater differences between spill and control clusters for each site were observed along the first principal component axis (PC1—55.91%).

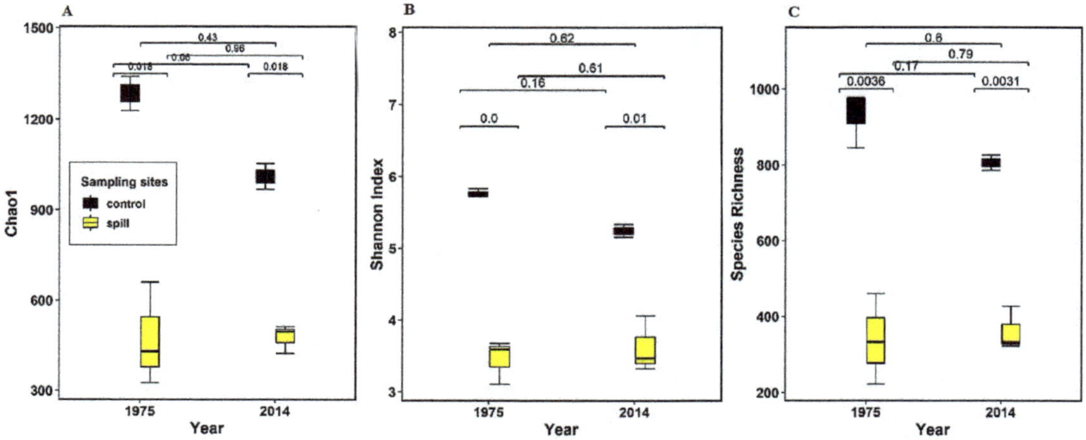

Figure 5. Soil bacterial α-diversity parameters: (**A**)—diversity as a function of sampling—Chao1 matrix; (**B**)—the Shannon–Wiener index; (**C**)—species richness.

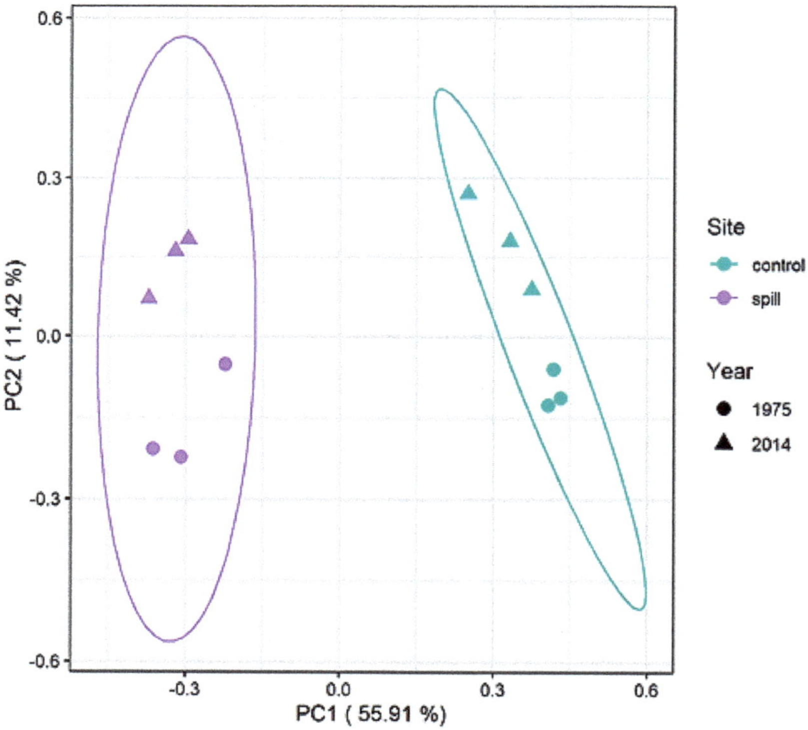

Figure 6. Principal components analysis (PCA) of 16S rRNA gene sequencing of soil microbes obtained from spill and control samples. PC1 and PC2 account for 55.91% and 11.42% of the variation, respectively.

4. Discussion

This study demonstrates that oil pollution significantly reduced both the diversity and abundance of soil bacteria, while substantially altering the relative abundance of major bacterial populations. These findings are consistent with existing literature, which shows that long-term crude oil contamination leads to a decline in native microbial communities while promoting the survival or proliferation of those that have developed adaptive mechanisms [39–41].

The significant decrease in dominance in the oil-contaminated soils demonstrates the substantial impact of oil spills on the soil microbial diversity. The oil contamination's impact on species diversity was more pronounced for the 1975 site than that for the 2014 site. This confirmed the gradual long-term deleterious effect of oil on soil microbial populations. The potential of natural bioremediation of soil contaminated by petroleum hydrocarbons is affected not only by biotic factors, but also by abiotic factors such as moisture, salinity, soil chemistry, and nutrient availability. These abiotic factors are influenced by hydrocarbon contamination and substantially affect microbial pollution catabolism.

The effects of oil on the physical and chemical properties of soil decline continuously. However, the reasons for this trend vary. Hydrocarbons, alkanes, aromatic, etc., comprise carbon sources for soil microorganisms, thus promoting their growth and metabolism. This results in the production of organic acids and CO_2, and a decrease in the soil pH [42]. Yet, the buffering capacity of soil resists changes in pH. Therefore, the soil first experiences a change in its properties and later faces changes in its microbial community. Increased electrical conductivity in the contaminated soil indicates that the viscous hydrocarbons and soluble minerals found in oil not only increase soil salinity but also exacerbate soil slab formation and reduce water permeability [5,42].

The lack of water has a negative impact on soil microbial communities. It hampers their interactions with each other and with the environment, which may lead to a decrease in abundance and diversity [21]. Our results are in accordance with Truskewycz et al. [21], who reported that soil moisture content and organic matter parameters were lower in non-polluted samples than in spill samples, and the number of OTUs was also lower in non-polluted samples (Tables 1 and 2). A direct correlation between the abiotic and biotic parameters was found, where lower soil moisture and organic matter values corresponded with fewer OTUs in the samples.

The soil pH is another crucial factor for the biodegradation of petroleum hydrocarbons and the biotransformation of toxic metals. Changes in pH can affect metal ions' behavior and alter fungal and bacterial enzyme activities and community structure [43]. A strong relationship between soil pH and microbial communities is expected, where higher pH correlates with wider microbial diversity, especially in dryland environments [44]. Our data are in accordance with these previous studies, where high pH value led to an increased Shannon index for bacterial diversity (in the 1975 control sample). According to Samira et al. [45], after 18 years of oil contamination in Kuwait, the chemical composition of soil exhibited an increase in resins, a higher concentration of polycyclic aromatic hydrocarbons (PAHs), and a decrease in aromatic compounds. *Actinobacteria*, *Proteobacteria*, and *Firmicutes* are known oil-degrading microorganisms. These bacteria can use crude oil as their sole source of carbon and energy [46]. Huang et al. [22] identified eight core genera in an oil-contaminated soil. This is in partial accordance with our findings as presented in Figure 3.

Moreover, the dominance of *Proteobacteria* and the decrease in the diversity of microbial communities in pervaded arid soils indicate a specialized community adapted to hydrocarbon stress. While this shift increases degradation efficiency, it could also compromise wider ecosystem functions, such as those determining nutrient cycling and bioremediation capacity. The loss of *Acidobacteria* and *Bacteroidetes*—big players in carbon and nitrogen cycling—could shake up soil fertility. In Proteobacteria-based systems, hydrocarbon degradation may be favored over organic matter mineralization, thereby limiting nutrient availability to plants [47]. While hydrocarbon-degrading taxa (e.g., *Pseudomonas*, *Rhodococcus*) can flourish, their dominance may decrease functional redundancy and render the ecosystem susceptible to consequential additional perturbations [47].

Unlike non-arid ecosystems, arid ecosystems exhibit limited functional flexibility in response to oil pollution. After contamination, arid systems demonstrate greater loss of diversity than temperate soils [15], likely resulting from an inherently lower baseline resilience. In humid areas, *Actinobacteria* and fungi (e.g., *Phanerochaete*) are the primary degraders [15], while *Proteobacteria* and halotolerant genera (*Marinobacter*) are dominant in arid environments. Non-arid ecosystems are still able to retain higher functional diversity despite pollution [48], which in turn provides buffering for nutrient cycling [48].

Overall, insights of this study suggest that oil pollution changes the soil microbial community dramatically and poses a threat to soil health. Identifying petroleum hydrocarbon-degrading bacteria in contaminated soils can comprise an effective strategy for alleviating oil pollution. These bacterial strains can comprise the basic microbial agent for natural bioremediation. Also, future studies might investigate changes in the composition and diversity of Archaea, which are of similar appearance and numerous similarities in oil-contaminated dryland soils, as Archaea play a crucial role in the transformation of petroleum compounds.

5. Conclusions

This study examined the long-term changes in the soil microbial populations following oil spill contamination in the Evrona Nature Reserve in the Arava Valley, Israel. Over a span of 47 years, microbial biodiversity substantially decreased in the oil-contaminated soils. Most of the original bacterial species disappeared, replaced by species capable of utilizing oil as their sole source of carbon and energy. This shift occurred gradually, as evidenced by comparisons between soils contaminated by oil in 1975 and 2014. Focusing future research on identifying microbial genera well-suited for petroleum degradation may improve the effective degradation of oil in contaminated environments.

Author Contributions: Conceptualization, Y.S.; Methodology, V.M., I.S., T.D., I.A., C.S. and M.L.; Software, T.D.; Validation, V.M., T.D., I.A., C.S. and Y.S.; Investigation, V.M., I.S., I.A., C.S. and M.L.; Resources, I.S.; Data curation, Y.S.; Writing—original draft, V.M., T.D., I.A., M.L. and Y.S.; Writing—review & editing, I.S. and Y.S.; Project administration, Y.S. All authors have read and agreed to the published version of the manuscript.

Funding: This research received no external funding.

Informed Consent Statement: Not applicable.

Data Availability Statement: All data generated or analyzed during this study are included in this article. Raw data may be available upon proper request to the corresponding author.

Acknowledgments: The authors gratefully acknowledge Michelle Finzi for proofreading the manuscript.

Conflicts of Interest: The authors declare that they have no competing interests.

References

1. Kuppusamy, S.; Maddela, N.R.; Megharaj, M.; Venkateswarlu, K. Impact of Total Petroleum Hydrocarbons on Human Health. In *Total Petroleum Hydrocarbons*; Springer: Cham, Switzerland, 2020; pp. 139–165. [CrossRef]
2. Tedesco, P.; Balzano, S.; Coppola, D.; Esposito, F.P.; de Pascale, D.; Denaro, R. Bioremediation for the recovery of oil polluted marine environment, opportunities and challenges approaching the Blue Growth. *Mar. Pollut. Bull.* **2024**, *200*, 116157. [CrossRef] [PubMed]
3. Li, Y.; Li, C.; Xin, Y.; Huang, T.; Liu, J. Petroleum pollution affects soil chemistry and reshapes the diversity and networks of microbial communities. *Ecotoxicol. Environ. Saf.* **2022**, *246*, 114129. [CrossRef]
4. Sui, X.; Wang, X.; Li, Y.; Ji, H. Remediation of Petroleum-Contaminated Soils with Microbial and Microbial Combined Methods: Advances, Mechanisms, and Challenges. *Sustainability* **2021**, *13*, 9267. [CrossRef]
5. Oruabena, B.; Elechi, O. Clean-up of crude oil-contaminated soils: Bioremediation option. *Int. J. Environ. Sci. Technol.* **2019**, *17*, 1185–1198.
6. Maciejczyk, M.; Tyrpień-Golder, K.; Janoszka, B.; Gierat, B.; Muzyka, R. Mutagenic and carcinogenic polycyclic aromatic hydrocarbons (PAHs) in food—Occurrence, human health effects, and assessment methods of exposure. *Environ. Med.* **2023**, *26*, 8–15. [CrossRef]
7. Elgazali, A.; Althalb, H.; Elmusrati, I.; Ahmed, H.M.; Banat, I.M. Remediation Approaches to Reduce Hydrocarbon Contamination in Petroleum-Polluted Soil. *Microorganisms* **2023**, *11*, 2577. [CrossRef]
8. Adedeji, J.A.; Tetteh, E.K.; Amankwa, M.O.; Asante-Sackey, D.; Ofori-Frimpong, S.; Armah, E.K.; Rathilal, S.; Mohammadi, A.H.; Chetty, M. Microbial Bioremediation and Biodegradation of Petroleum Products—A Mini Review. *Appl. Sci.* **2022**, *12*, 12212. [CrossRef]
9. Cai, P.; Ning, Z.; Liu, Y.; He, Z.; Shi, J.; Niu, M. Diagnosing bioremediation of crude oil-contaminated soil and related geochemical processes at the field scale through microbial community and functional genes. *Ann. Microbiol.* **2020**, *70*, 36. [CrossRef]
10. Chen, G.; Yuan, M.; Ma, B.; Ren, Y. Responses of Petroleum Contamination at Different Sites to Soil Physicochemical Properties and Indigenous Microbial Communities. *Water Air Soil Pollut.* **2023**, *234*, 494. [CrossRef]
11. Borah, G.; Deka, H. Crude oil associated heavy metals (HMs) contamination in agricultural land: Understanding risk factors and changes in soil biological properties. *Chemosphere* **2022**, *310*, 136890. [CrossRef]
12. Ullah, Z.; Rashid, A.; Ghani, J.; Nawab, J.; Zeng, X.-C.; Shah, M.; Alrefaei, A.F.; Kamel, M.; Aleya, L.; Abdel-Daim, M.M.; et al. Groundwater contamination through potentially harmful metals and its implications in groundwater management. *Front. Environ. Sci.* **2022**, *10*, 1021596. [CrossRef]
13. McKew, B.A.; Coulon, F.; Yakimov, M.M.; Denaro, R.; Genovese, M.; Smith, C.J.; Osborn, A.M.; Timmis, K.N.; McGenity, T.J. Efficacy of intervention strategies for bioremediation of crude oil in marine systems and effects on indigenous hydrocarbonoclastic bacteria. *Environ. Microbiol.* **2007**, *9*, 1562–1571. [CrossRef] [PubMed]
14. Kebede, G.; Tafese, T.; Abda, E.M.; Kamaraj, M.; Assefa, F. Factors influencing the bacterial bioremediation of hydrocarbon contaminants in the soil: Mechanisms and impacts. *J. Chem.* **2021**, *2021*, 9823362. [CrossRef]
15. Shi, L.; Liu, Z.; Yang, L.; Fan, W. Effects of oil pollution on soil microbial diversity in the Loess hilly areas, China. *Ann. Microbiol.* **2022**, *72*, 26. [CrossRef]
16. Maestre, F.T.; Eldridge, D.J.; Soliveres, S.; Kéfi, S.; Delgado-Baquerizo, M.; Bowker, M.A.; García-Palacios, P.; Gaitán, J.; Gallardo, A.; Lázaro, R.; et al. Structure and functioning of dryland ecosystems in a changing world. *Annu. Rev. Ecol. Evol. Syst.* **2016**, *47*, 215–237. [CrossRef]
17. Abbasian, F.; Lockington, R.; Megharaj, M.; Naidu, R. The Biodiversity Changes in the Microbial Population of Soils Contaminated with Crude Oil. *Curr. Microbiol.* **2016**, *72*, 663–670. [CrossRef]
18. Paul, S.; Ingti, B.; Das, D.; Joshi, S.R. Microbial Responses to Hydrocarbon Contamination and Its Bioremediation. In *Mineral Transformation and Bioremediation by Geo-Microbes*; Joshi, S.R., Verma, P., Banerjee, S., Eds.; Interdisciplinary Biotechnological Advances; Springer: Singapore, 2025. [CrossRef]
19. Widdel, F.; Knittel, K.; Galushko, A. Anaerobic Hydrocarbon-Degrading Microorganisms: An Overview. In *Handbook of Hydrocarbon and Lipid Microbiology*; Timmis, K.N., Ed.; Springer: Berlin/Heidelberg, Germany, 2010. [CrossRef]
20. Xu, X.; Liu, W.; Wang, W.; Tian, S.; Jiang, P.; Qi, Q.; Li, F.; Li, H.; Wang, Q.; Li, H.; et al. Potential biodegradation of phenanthrene by isolated halotolerant bacterial strains from petroleum oil polluted soil in Yellow River Delta. *Sci. Total. Environ.* **2019**, *664*, 1030–1038. [CrossRef]
21. Truskewycz, A.; Gundry, T.D.; Khudur, L.S.; Kolobaric, A.; Taha, M.; Aburto-Medina, A.; Ball, A.S.; Shahsavari, E. petroleum hydrocarbon contamination in terrestrial ecosystems—Fate and microbial responses. *Molecules* **2019**, *24*, 3400. [CrossRef] [PubMed]
22. Huang, L.; Ye, J.; Jiang, K.; Wang, Y.; Li, Y. Oil contamination drives the transformation of soil microbial communities: Co-occurrence pattern, metabolic enzymes and culturable hydrocarbon-degrading bacteria. *Ecotoxicol. Environ. Saf.* **2021**, *225*, 112740. [CrossRef]

23. Yerulker, G.; Patel, P.; Chafale, A.; Rathod, V.; Das, S.; Pandey, P.; Khan, N.A.; Devi, A.; Munshi, N.S.; Dhodapkar, R.; et al. Comparative assessment of soil microbial community in crude oil contaminated sites. *Environ. Pollut.* **2023**, *328*, 121578. [CrossRef]
24. Fuentes, S.; Barra, B.; Caporaso, J.G.; Seeger, M. From rare to dominant: A fine-tuned soil bacterial bloom during petroleum hydrocarbon bioremediation. *Appl. Environ. Microbiol.* **2016**, *82*, 888–896. [CrossRef] [PubMed]
25. Varjani, S.J.; Gnansounou, E. Microbial dynamics in petroleum oilfields and their relationship with physiological properties of petroleum oil reservoirs. *Bioresour. Technol.* **2017**, *245*, 1258–1265. [CrossRef]
26. Varjani, S.; Upasani, V.N.; Pandey, A. Bioremediation of oily sludge polluted soil employing a novel strain of Pseudomonas aeruginosa and phytotoxicity of petroleum hydrocarbons for seed germination. *Sci. Total. Environ.* **2020**, *737*, 139766. [CrossRef]
27. Tremblay, J.; Yergeau, E.; Fortin, N.; Cobanli, S.; Elias, M.; King, T.L.; Lee, K.; Greer, C.W. Chemical dispersants enhance the activity of oil- and gas condensate-degrading marine bacteria. *ISME J.* **2017**, *11*, 2793–2808. [CrossRef] [PubMed]
28. Xu, X.; Zhai, Z.; Li, H.; Wang, Q.; Han, X.; Yu, H. Synergetic effect of bio-photocatalytic hybrid system: G-C3N4 and Acinetobacter sp. JLS1 for enhanced degradation of C16 alkane. *Chem. Eng. J.* **2017**, *323*, 520–529. [CrossRef]
29. Yakimov, M.M.; Timmis, K.N.; Golyshin, P.N. Obligate oil-degrading marine bacteria. *Curr. Opin. Biotechnol.* **2007**, *18*, 257–266. [CrossRef] [PubMed]
30. Dane, J.H.; Topp, G.C. Methods of Soil Analysis. In *Part 4: Physical Methods*; Soil Science Society of America, Inc.: Madison, WI, USA, 2002.
31. Rhoades, J.D. Salinity: Electrical Conductivity and Total Dissolved Solids. In *Methods for Soil Analysis, Part 3: Chemical Methods*; Sparks, R.L., Ed.; Soil Science Society of America: Madison, WI, USA, 1996; pp. 417–435.
32. Girsowicz, R.; Koryachenko, O.; Sherman, C.; Mayzlish-Gati, E.; Doniger, T.; Steinberger, Y. Impact of oil-spill contamination on a soil bacterial community: A 40-year history of rehabilitation in the arava valley. *Soil Sediment Contam. Int. J.* **2018**, *27*, 175–185. [CrossRef]
33. Lane, D.J. 16S/23S rRNA sequencing. In *Nucleic Acid Techniques in Bacterial Sys-Tematics*; Stackebrandt, E., Goodfellow, M., Eds.; Wiley: New York, NY, USA, 1991; pp. 115–147.
34. Caporaso, J.G.; Kuczynski, J.; Stombaugh, J.; Bittinger, K.; Bushman, F.D.; Costello, E.K.; Fierer, N.; Gonzalez Peña, A.; Goodrich, J.K.; Gordon, J.I.; et al. QIIME allows analysis of high-throughput community sequencing data. *Nat. Methods* **2010**, *7*, 335–336. [CrossRef] [PubMed]
35. DeSantis, T.Z.; Hugenholtz, P.; Larsen, N.; Rojas, M.; Brodie, E.L.; Keller, K.; Huber, T.; Dalevi, D.; Hu, P.; Andersen, G.L. Greengenes, a chimera-checked 16S rRNA gene database and workbench compatible with ARB. *Appl. Environ. Microbiol.* **2006**, *72*, 5069–5072. [CrossRef]
36. Quail, M.A.; Smith, M.; Coupland, P.; Otto, T.D.; Harris, S.R.; Connor, T.R.; Bertoni, A.; Swerdlow, H.P.; Gu, Y. A tale of three next generation sequencing platforms: Comparison of Ion Torrent, Pacific Biosciences and Illumina MiSeq Sequencers. *BMC Genom.* **2012**, *13*, 341. [CrossRef]
37. Lou, J.; Liu, Y.; Li, Y. Review of high-throughput sequencing techniques in studies of soil microbial diversity. *Chin. Agri-Cult. Sci. Bull.* **2014**, *30*, 256–260.
38. Smith, D.P.; Peay, K.G. Sequence depth, not pcr replication, improves ecological inference from next generation DNA sequencing. *PLoS ONE* **2014**, *9*, e90234. [CrossRef] [PubMed]
39. Balachandran, C.; Duraipandiyan, V.; Balakrishna, K.; Ignacimuthu, S. Petroleum and polycyclic aromatic hydrocarbons (PAHs) degradation and naphthalene metabolism in Streptomyces sp. (ERI-CPDA-1) isolated from oil contaminated soil. *Bioresour. Technol.* **2012**, *112*, 83–90. [CrossRef]
40. Zhang, S.; Wang, Q.; Xie, S. Bacterial and archaeal community structures in phenanthrene amended aquifer sediment microcosms under oxic and anoxic conditions. *Int. J. Environ. Res.* **2012**, *6*, 1077–1088.
41. Grządziel, J.; Gałązka, A. Fungal biodiversity of the most common types of polish soil in a long-term microplot experiment. *Front. Microbiol.* **2019**, *10*, 6. [CrossRef]
42. Wang, X.; Li, Y.; Zhang, Y.; Pan, Y.-R.; Li, L.; Liu, J.; Butler, D. Stepwise pH control to promote synergy of chemical and biological processes for augmenting short-chain fatty acid production from anaerobic sludge fermentation. *Water Res.* **2019**, *155*, 193–203 [CrossRef]
43. Kocaoba, S. Adsorption of Fe(II) and Fe(III) from aqueous solution by using sepiolite: Speciation studies with MINEQL+ computer program. *Sep. Sci. Technol.* **2019**, *55*, 896–906. [CrossRef]
44. Rousk, J.; Bååth, E.; Brookes, P.C.; Lauber, C.L.; Lozupone, C.; Caporaso, J.G.; Knight, R.; Fierer, N. Soil bacterial and fungal communities across a pH gradient in an arable soil. *ISME J.* **2010**, *4*, 1340–1351. [CrossRef]
45. Samira, A.; Omar, S.; Bhat, N.R.; Asem, A. Critical assessment of the environmental consequences of the invasion of Kuwait the Gulf War, and the aftermath. In *Environmental Consequences of War and Aftermath—The Handbook of Environmental Chemistry*; Kassim, T.A., Barcelo, D., Eds.; Springer: Berlin/Heidelberg, Germany, 2009; pp. 141–170.

46. Kostka, J.E.; Prakash, O.; Overholt, W.A.; Green, S.J.; Freyer, G.; Canion, A.; Delgardio, J.; Norton, N.; Hazen, T.C.; Huettel, M. Hydrocarbon-degrading bacteria and the bacterial community response in gulf of mexico beach sands impacted by the deepwater horizon oil spill. *Appl. Environ. Microbiol.* **2011**, *77*, 7962–7974. [CrossRef]
47. Qiu, L.; Zhang, Q.; Zhu, H.; Reich, P.B.; Banerjee, S.; A van der Heijden, M.G.; Sadowsky, M.J.; Ishii, S.; Jia, X.; Shao, M.; et al. Erosion reduces soil microbial diversity, network complexity and multifunctionality. *ISME J.* **2021**, *15*, 2474–2489. [CrossRef]
48. Liang, Y.; Van Nostrand, J.D.; Deng, Y.; He, Z.; Wu, L.; Zhang, X.; Li, G.; Zhou, J. Functional gene diversity of soil microbial communities from five oil-contaminated fields in China. *ISME J.* **2010**, *5*, 403–413. [CrossRef] [PubMed]

agronomy

MDPI

Review

Crop–Mushroom Rotation: A Comprehensive Review of Its Multifaceted Impacts on Soil Quality, Agricultural Sustainability, and Ecosystem Health

Tingting Dou [1,2], Kaixuan Zhang [2], Xiaofei Shi [2], Wei Liu [2], Fuqiang Yu [2,*] and Dong Liu [2,*]

[1] School of Ecology and Environmental Science, Yunnan University, Kunming 650500, China; doutingting@mail.kib.ac.cn

[2] The Germplasm Bank of Wild Species & Yunnan Key Laboratory for Fungal Diversity and Green Development, Kunming Institute of Botany, Chinese Academy of Sciences, Kunming 650201, China; zhangkaixuan@mail.kib.ac.cn (K.Z.); xiaofeishi@mail.kib.ac.cn (X.S.); weiliu@mail.kib.ac.cn (W.L.)

* Correspondence: fqyu@mail.kib.ac.cn (F.Y.); liudongc@mail.kib.ac.cn (D.L.)

Abstract: Cross-kingdom rotation offers several agronomic and ecological benefits, including enhanced soil nutrient availability, reduced pest and disease prevalence, improved soil structure, and minimized chemical inputs, which contribute to a dynamic and resilient soil ecosystem, thereby fostering biodiversity and ecological balance. Additionally, crop diversity encourages plant root exudates that feed a wider range of beneficial soil microbes, ultimately leading to a balanced soil food web. Integrating rice cultivation with the edible mushroom *Stropharia rugosoannulata* further improves soil fertility and enhances organic carbon sequestration. This rotation introduces organic matter into the soil, affecting microbial community structure and supporting the decomposition of complex organic materials via lignocellulose-decomposing fungi. These processes contribute to soil organic carbon accumulation, nutrient cycling, and long-term soil health. The study emphasizes the importance of microbial communities (including live biomass and necromass) in maintaining ecosystem stability and highlights the potential of the rice–*S. rugosoannulata* rotation model as a sustainable agricultural practice. Further research is needed to clarify how fungal necromass contributes to soil carbon accumulation and to optimize agricultural practices for improving soil health and carbon sequestration in response to climate change. These findings provide valuable insights for developing sustainable agricultural strategies that balance productivity with environmental conservation.

Keywords: crop rotation; soil organic carbon; mushroom cultivation; fertility; fungal necromass

1. Introduction

Terrestrial ecosystems represent the most extensive carbon reservoirs, serving as critical components in the global carbon cycle. Within these ecosystems, soil carbon stocks surpass the quantities of carbon sequestered in aboveground biomass and atmospheric carbon, positioning soil as the largest and most stable repository of carbon. This vast accumulation of soil carbon underscores its essential role in mitigating climate change by acting as a long-term carbon sink. The stability and magnitude of soil carbon reserves are vital for maintaining ecological balance and regulating the global carbon budget, thereby highlighting the importance of soil conservation and sustainable land management practices in preserving this crucial carbon store [1,2]. Soil organic carbon (SOC) represents a significant and dynamic component of this soil carbon pool, accounting for about 60% to 74% [3]. Minor fluctuations in SOC levels can lead to significant changes in atmospheric

https://doi.org/10.3390/agronomy1503056

greenhouse gases; therefore, understanding and enhancing the stability and accumulation of SOC are critical in mitigating the increase in atmospheric CO_2 levels [4]. Furthermore, soil organic carbon plays a key role in regulating soil properties, directly impacting soil quality [5], supporting nutrient provision for crops, and fostering consistent, high yields. As crop growth is intimately connected to the condition of farm soils, maintaining optimal soil health is essential for sustainable agricultural productivity. Rice, widely recognized as one of the world's most essential staple foods, serves as the primary sustenance for nearly half of the global population [6], thus forming a cornerstone of food security and nutrition for billions. The cultivation and production of rice are not only critical for meeting the dietary needs of large populations but also carry significant economic and cultural weight across various regions [7]. Among these, the southwestern region of China emerges as a particularly important area for rice cultivation, playing a substantial role in the nation's overall rice output [8]. The region's advantageous climatic conditions, rich agricultural traditions, and adoption of advanced farming practices have established it as a key contributor to the growth and development of rice farming. The role of the southwestern region, especially in Hani Rice Terraces, in rice production highlights its vital importance within the broader framework of agricultural development and sustainability, both within China and on a global scale. This comprehensive assessment is strategically designed to scrutinize the diverse and far-reaching implications of this agricultural practice, with a focused and distinct emphasis on its profound impacts on soil quality, agricultural productivity, and ecological equilibrium.

The crop–mushroom rotation has emerged as a highly viable and sustainable agricultural paradigm. Comprehensively understanding the numerous and complex aspects of this practice is crucial for facilitating its broader acceptance and optimizing its integration into agricultural systems. By meticulously collating and critically evaluating current research findings, the potential benefits and challenges associated with this innovative practice can be elucidated. An in-depth exploration has been conducted on the mechanisms through which crop–mushroom rotation influences the soil carbon reservoir. This necessitates a detailed examination of how mushroom cultivation, in harmonious synergy with crop interactions, modulates the cycling and sequestration of carbon within the complex soil matrix. The effects on various organic carbon components have been precisely dissected, revealing the processes that govern carbon dynamics. Furthermore, the significant role of this rotation modality in reshaping and restructuring soil microbial communities has been thoroughly investigated. Given the fundamental functions of soil microbes in nutrient cycling, organic matter decomposition, and soil health maintenance, understanding their responses to crop–mushroom rotation is of critical importance. This review provides a comprehensive assessment of the implications of crop–mushroom rotation for soil fertility and agricultural productivity. The practice has been shown to intricately modulate nutrient availability, soil structure, and crop growth and yield parameters. Furthermore, the potential of this rotation strategy in fostering ecological balance has been thoroughly investigated. This includes its contributions to biodiversity conservation, mitigation of environmental pollution, and enhanced resilience to climate change challenges. It provides practical, evidence-based recommendations for implementing crop–mushroom rotation, identifies critical knowledge gaps that require further research, and ultimately contributes to the development of sustainable agricultural strategies that are environmentally friendly and economically viable in the long term.

1.1. Challenges of Intensive Agriculture and the Importance of Sustainable Soil Management Practices

In traditional intensive agricultural systems, the extensive use of chemical fertilizers and pesticides has led to soil functional degradation, a loss of fertility, and escalating soil pollution issues [9,10]. This problem is particularly pronounced in countries and regions with limited agricultural resources and dense populations. The long-term accumulation of residual pesticides and other organic pollutants in farmland can inhibit soil microbial activity, disrupt the soil ecological balance, and adversely affect the growth and development of crops [11]. Moreover, excessive application of fertilizers during the planting process can lead to nutrient overaccumulation in the soil, resulting in soil compaction and reduced fertility. This not only increases production costs but also pollutes groundwater and surface water bodies, exacerbating ecological degradation [12]. Enhancing soil carbon sequestration capacity and reducing the use of pesticides and fertilizers are crucial issues for the sustainable development of agriculture, requiring the implementation of reasonable measures. Proper fertilizer management and cultivation practices are important technical measures in agricultural production and are among the effective strategies to ensure the long-term sustainability of agriculture [13].

Soil cultivation practices are significant technical measures in agricultural production, playing a crucial role in improving the quality of the tillage layer and enhancing soil fertility [14]. Crop rotation refers to the sequential planting of crops with varying nutrient requirements on the same plot of land across seasons and years, combining the use of land with its restoration in a biological strategy [15]. This practice disrupts the reproductive cycles of insects and pathogens, thereby controlling their life cycles and achieving pest and disease prevention [16]. In Western Canada, field rotations of peas, barley, and canola compared to continuous canola cultivation have shown that the inclusion of wheat, barley, and peas in rotation systems not only increases canola yield but also significantly reduces pests and diseases such as root maggots [17].

1.2. Benefits and Environmental Implications of Paddy–Upland Crop Rotation in China

Paddy–upland crop rotation involves the orderly alternation of planting rice and dryland crops on the same field and is one of the major crop production systems in China [18]. Rice, as a staple crop in many regions, plays a central role in this system. This rotation method not only enhances ecological conditions for crop growth but also improves soil's physical and chemical properties. It increases soil organic carbon content and its components [19], enhances soil permeability and water retention, and ultimately elevates soil fertility levels, promoting sustainable crop development. However, the rotation process generates a significant amount of agricultural waste, such as crop residues, which are traditionally disposed of by on-site burning [20]. This practice leads to the loss of valuable resources and contributes to carbon dioxide emissions [21], exacerbating global warming and threatening environmental sustainability. An alternative approach is to utilize organic waste as a substrate for cultivating saprotrophic edible fungi, which promotes resource recycling and sustainable agricultural practices. For instance, agricultural straw, which is often burned, can be used as a substrate for mushroom cultivation. Instead of releasing carbon dioxide through burning, the straw is decomposed and converted by mushrooms through a natural biological process. This not only reduces straw burning but also sequesters carbon in the form of mushroom biomass and residual substrate. In the context of crop–mushroom rotation, rice residues can be effectively combined with mushroom cultivation. After the rice harvest, the residues can serve as substrates for mushroom growth. Once the mushrooms are harvested, the spent substrate can be returned to the paddy fields, enriching the soil with nutrients and beneficial microorganisms. This cycle improves soil fertility and promotes the growth of subsequent rice crops, creating a sustainable agricultural system.

1.3. Mushroom–Rice Rotation: Definition and Overview

The "mushroom-rice" rotation is an agricultural practice that involves alternating the cultivation of mushrooms with the growth of traditional crops. This approach aims to improve soil health, increase soil organic carbon, and enhance agricultural productivity [22]. The concept of mushroom–crop rotation has a relatively recent history. As concerns about soil degradation, pesticide use, and sustainable farming methods have grown, innovative agricultural practices like this rotation system have emerged. While it may not have a long historical record in the traditional sense, its development is a response to modern agricultural challenges. When implementing mushroom–crop rotation, farmers typically start by preparing the land for mushroom cultivation. This involves creating a suitable substrate, often a mixture of compost, straw, and other organic materials [23,24]. Once the mushrooms are harvested, the spent substrate can be incorporated into the soil. This enriched substrate provides a boost of nutrients and beneficial microorganisms to the soil, preparing it for subsequent crop cultivation. For the crop phase, a variety of plants can be grown depending on the region, climate, and market demand. The choice of crops should be carefully considered to complement the benefits provided by the mushroom phase. For instance, some crops may be particularly well-suited to the improved soil structure and nutrient availability resulting from mushroom cultivation.

The selection and preparation of the substrate for mushrooms is crucial. It should provide the right balance of nutrients, moisture, and air porosity to support mushroom growth. Composting and sterilization processes may be employed to ensure a clean and fertile substrate [25]. Different mushroom species require specific growing conditions. Farmers need to be familiar with the appropriate temperature, humidity, and lighting requirements for the mushrooms they are cultivating. This may involve the use of specialized growing facilities or modifications to existing agricultural structures [26,27]. After the mushroom phase, proper soil management is essential to make the most of the enriched substrate. This may include soil testing to determine nutrient levels, adjusting pH if necessary, and incorporating organic matter to maintain soil fertility. Careful selection of crops for the rotation is important to maximize the benefits of the system. Consideration should be given to crops that have different nutrient requirements and pest susceptibilities to reduce the risk of pest and disease buildup.

1.4. Regarding the Origin of Mushrooms and Their Impacts on Soil Quality

Mushrooms are fascinating organisms with a mysterious origin. They are believed to have evolved over millions of years from simple single-celled organisms. There are countless different types of mushrooms, each with its own unique characteristics. Some mushrooms are edible and highly prized for their taste and nutritional value, while others are poisonous and can be deadly [28]. The regions of China are rich in fungal diversity, with different types of mushrooms thriving in different habitats. In certain areas, mushrooms grow in forests, while in others they can be found in grasslands or even on decaying logs [29]. These mushrooms may have adapted to the local climate and soil conditions and could play an important role in the ecosystem. Understanding the origin and types of mushrooms in this region can help us better appreciate the natural world and potentially discover new sources of food and medicine.

In China, several types of mushrooms are prevalent. *Agaricus bisporus*, also known as the common button mushroom, is widely cultivated due to its popularity in the culinary world. *Pleurotus ostreatus*, or the oyster mushroom, is another common species. It is highly adaptable and can grow on a variety of substrates, making it a favorite among farmers. Mushrooms play a significant role in improving soil quality. Different types of mushrooms exert diverse impacts on the soil ecosystem. These different types of mushrooms do indeed affect soil fertility in distinct ways. Mycorrhizal mushrooms, such as *Cantharellus* spp., *Boletus edulis*, and *Tricholoma matsutake*, form symbiotic associations with plant roots. This symbiosis enhances nutrient uptake, particularly phosphorus, for plants. As a result, it indirectly improves soil structure by promoting better root growth and stability. The increased root exudates from healthier plants also attract beneficial soil microorganisms, further enhancing soil fertility [30].

Saprophytic mushrooms such as *Agaricus bisporus*, *Pleurotus ostreatus*, and *Coprinus comatus*, can break down organic matter, accelerating the decomposition process and releasing nutrients back into the soil. They contribute to the recycling of nutrients and the maintenance of soil fertility [31]. Additionally, mushrooms can secrete enzymes and organic acids that help solubilize minerals and make them more available to plants.

Certain mushrooms, such as *S. rugosoannulata*, can decompose lignocellulosic materials in agricultural residues, enhancing nutrient availability in the soil for subsequent crops. This species has gained significant attention in the context of crop–mushroom rotation due to its potential to improve soil quality and agricultural productivity. By directly promoting crop growth, this process can contribute to higher yields. However, if not properly managed, some mushrooms may compete with crops for resources. For example, fast-growing mushrooms could consume significant amounts of soil nutrients and water, potentially negatively affecting crop growth. When integrated into a well-planned crop–mushroom rotation system, the overall benefits of mushroom cultivation typically outweigh these potential drawbacks. The presence of mushrooms in the soil can increase microbial activity, fostering a more diverse and active soil microbial community. These complex interactions between mushrooms and soil have significant implications for sustainable agriculture and ecosystem health.

1.5. Multifunctional Benefits and Cultivation Practices of Saprotrophic Edible Mushroom S. rugosoannulata in Sustainable Agriculture

Saprotrophic edible mushrooms play crucial ecological roles. These mushrooms help break down dead organic matter in the environment, recycling nutrients and contributing to the health of ecosystems [32]. By decomposing plant and animal remains, they speed up the natural process of decay and release essential elements back into the soil, which benefits other plants and organisms. In terms of human consumption, saprotrophic edible mushrooms are a great source of nutrition. They are rich in proteins, vitamins, and minerals [33,34]. Some varieties are known for their high levels of antioxidants, which can help protect the body against damage from free radicals and reduce the risk of chronic diseases [35]. *S. rugosoannulata* is a saprotrophic edible mushroom that has garnered significant attention due to its diverse array of beneficial properties, including its nutritional, medicinal, bioremediation, and biocontrol potential. This species is characterized by its ability to thrive in environments rich in organic matter, where it plays a crucial role in the decomposition of complex lignocellulosic materials [36]. The *Stropharia* is recognized not only for its high nutritional value, which includes essential vitamins, minerals, and proteins, but also for its medicinal properties, which have been linked to various health benefits [37] (Figure 1).

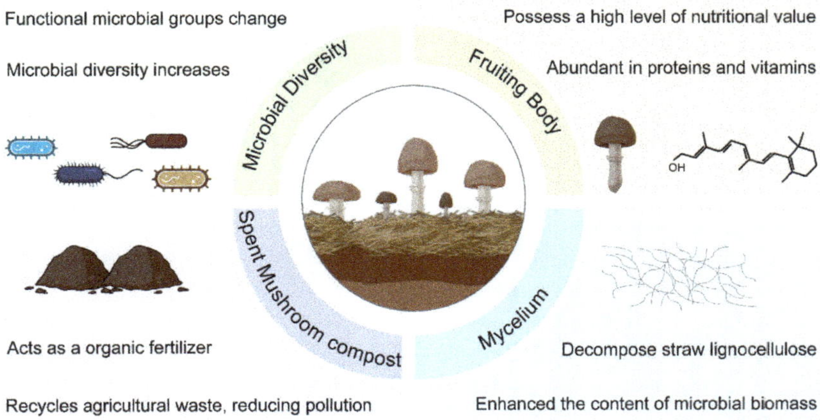

Figure 1. Benefits of mushroom *S. rugosoannulata*.

Stropharia demonstrates considerable efficacy in bioremediation, effectively degrading environmental pollutants and enhancing soil health [38]. Its robust lignocellulosic degradation capabilities make it ideal for breaking down agricultural residues and organic waste [39], contributing to sustainable agriculture. Additionally, its increasing cultivation highlights its significant role in agriculture and environmental management as a multifunctional organism. The cultivation model for saprotrophic edible mushrooms differs from typical edible mushroom cultivation. These mushrooms grow in a cultivation substrate that has undergone microbial fermentation, primarily composed of agricultural waste such as crop residues, livestock manure, and forestry waste [40]. After harvesting the mushrooms, the spent mushroom substrate can be returned to the fields as organic fertilizer [41]. This practice leverages the ecological function of saprotrophs as decomposers of carbon sources in nature, significantly enhancing the decomposition rate of rice straw and the sequestration efficiency of soil organic carbon [42]. This promotes more stable carbon components in rice paddy soils, providing a scientific basis for the management of the soil organic carbon reservoir.

1.6. The Novelty of This Study

This research uniquely integrates crop science and mycology, exploring the intricate interactions within the crop–mushroom rotation system. This interdisciplinary approach offers a fresh perspective, as most previous studies have focused on either crop–crop rotations or single-species mushroom cultivation in isolation. By examining the combined effects of crop and mushroom growth, new insights are revealed into how these two kingdoms interact in an agricultural context. The study emphasizes the role of saprotrophic mushrooms, particularly *S. rugosoannulata*, in soil ecosystems. While the general concept of mushrooms improving soil quality is well-documented, this research specifically investigates the multifaceted impacts of *S. rugosoannulata*. Its ability to decompose lignocellulosic materials, enhance soil carbon sequestration, and influence microbial communities within the context of crop–mushroom rotation has not been comprehensively explored before.

An in-depth analysis of the impact of this rotation on soil microbial communities is conducted, with a special focus on fungal necromass. Understanding how dead fungal biomass contributes to soil organic carbon formation and nutrient cycling represents a relatively new area of research. This review synthesizes existing knowledge and highlights the importance of considering fungal necromass in sustainable agriculture, addressing a significant gap in the literature. Finally, a comprehensive review of the economic, environmental, and ecological aspects of crop–mushroom rotation is provided. While previous reviews may have focused on one or two of these aspects, this study offers a holistic perspective. By simultaneously considering factors such as cost-effectiveness, environmental pollution reduction, and biodiversity conservation, a more complete understanding of the potential of this rotation system is achieved, contributing valuable insights to the field.

Overall, the mushroom–rice rotation offers a promising approach to sustainable agriculture by combining the benefits of mushroom cultivation with traditional crop production. With proper implementation and management, it can contribute to improved soil health, increased productivity, and a more resilient agricultural system.

This review synthesizes and conducts an in-depth exploration of existing knowledge on crop–mushroom rotation, aiming to provide valuable insights for farmers, researchers, and policymakers. By focusing on the complex aspects of this practice, it seeks to enhance understanding and support the development of sustainable agricultural strategies.

2. The Impact of Crop–Mushroom Rotation on the Soil Carbon Reservoir

2.1. The Effects of Crop–Mushroom Rotation on Soil Organic Carbon Components

The soil carbon reservoir comprises inorganic and organic carbon [43], with organic carbon being particularly sensitive to climate change and changes in land use [44]. Even minor variations can cause significant fluctuations in atmospheric greenhouse gases [45]; inorganic carbon, due to its turnover generally on a geological scale, is widely considered a relatively stable and inert carbon store [46]. Soil organic carbon is composed of organic matter with molecular structures ranging from simple to complex, typically formed through the long-term accumulation of plant and animal residues by natural oxidation and decomposer activity [47]. Organic carbon components can be further divided into active and inactive organic carbon, with active carbon sub-divided into dissolved organic carbon, particulate organic carbon, easily oxidized organic carbon, and microbial biomass carbon [48,49]. These components are more bioavailable to soil microorganisms compared to other forms of organic carbon. Dissolved organic carbon is readily soluble in water, exhibiting significant transitivity in the soil, and is easily oxidized by air, making it a major form of carbon migration and leaching in soils [50]. Easily oxidized organic carbon represents the first fraction of active organic carbon to be oxidized in the soil, serving as a key indicator of soil organic carbon turnover and nutrient change [51]. Microbial biomass carbon is a crucial component of the soil carbon pool, reflecting the state of soil microbiota by synthesizing microbial biomass from plant-derived small molecules through processes of growth, proliferation, and death, thereby forming microbial residue carbon. It directly drives the decomposition of humus, mineralization of soil nutrients, and their migration [52–54].

The effects of mushrooms on soil organic carbon components can be divided into the following aspects:

(i) For dissolved organic carbon. Mushroom cultivation enhances the breakdown of organic matter, increasing soluble carbon levels. This process is driven by the decomposition of easily degradable plant residues and fungal exudates, which contribute to the rise in dissolved organic carbon. Additionally, the stabilization of insoluble carbon may be linked to fungal interactions that promote soil aggregation and carbon sequestration [22].

(ii) For particulate and easily oxidized organic carbon. The mycelium of mushrooms binds soil particles together, retaining particulate organic matter and preventing its loss through erosion or other processes. This contributes to the long-term storage of carbon in the soil. Additionally, mushroom cultivation increases the amount of easily oxidized organic carbon, which is more readily available for microbial activity. This enhances soil structure, boosts soil biological activity, and promotes nutrient release and carbon cycling within the soil ecosystem.

(iii) For microbial biomass carbon. Mushrooms create a favorable environment for the growth of various soil microbes. As these microorganisms multiply, the microbial biomass carbon increases. This is important as microbial biomass plays a crucial role in nutrient cycling, decomposition of organic matter, and soil health. In addition, the interaction between mushrooms and soil microbes can lead to synergistic effects that further enhance soil carbon sequestration and overall soil quality.

Studies have shown that crop–mushroom rotation enhances the activities of carbon-cycling enzymes, such as cellulase and lignin-degrading enzymes, which promote the breakdown of complex organic compounds in the soil and the release of more readily available carbon forms [55]. This process is further supported by an increase in microbial biomass carbon content after the rotation [32]. Additionally, the fungal mycelium from mushrooms plays a crucial role in forming soil aggregates, which protect the organic carbon within them. As a result, the soil organic carbon pool becomes more stable, and the overall quality of soil carbon components improves, creating a more favorable environment for subsequent crop growth.

2.2. Effects of Crop–Mushroom Rotation on Soil Carbon Reservoir and Soil Aggregate Structure

Crop rotation can profoundly impact the soil carbon reservoir, significantly enhancing organic carbon content. Numerous studies indicate that crop rotation can increase soil organic carbon content [56–58]. Specifically, in North China, the peanut and winter wheat–summer maize rotation system has been shown to enhance SOC stability through the accumulation of recalcitrant carbon [56]. Furthermore, a modeling study in Southwest Germany demonstrates that rotations with cover crops can improve soil organic carbon (SOC) change rates and crop yields while reducing nitrogen leaching through site-specific management recommendations [57]. A study shows that in the past 40 years, the rate of increase in soil organic carbon (SOC) in continuous crop rotation systems in river delta croplands was 57% of that observed in the North China Plain [58]. Haruna et al. [59] found that rotations including leguminous plants resulted in higher soil organic matter and nitrogen levels compared to monoculture cultivation, significantly enhancing the soil carbon-to-nitrogen ratio. However, different crop rotations, due to their unique root system development characteristics, induce changes in soil physical properties [60]. Long-term monoculture tobacco fields exhibit a significant reduction in soil bulk density and an increase in total porosity after diverse crop rotations [61]. Diverse crop rotations can effectively reduce the humification of soluble organic matter, increase the sequestration and accumulation of soil organic carbon, and improve soil fertility. Furthermore, Yang and others [62] observed that after six years of multi-crop rotations, soil organic carbon content significantly increased, emissions of greenhouse gases such as CO_2 and N_2O were notably reduced compared to the control group, and the soil health index significantly improved. However, previous research has primarily focused on plant-to-plant rotations, with few studies exploring rotations involving crops and saprotrophic macrofungi.

Crop–mushroom rotation has notable impacts on the soil carbon reservoir and organic carbon content:

(i) Organic matter input augmentation. In crop–mushroom rotation systems, the use of mushroom substrates plays a critical role in enhancing soil organic matter. Straw-based mixtures, commonly used in mushroom cultivation, are particularly significant. When these substrates are incorporated into the soil during rotation, they directly contribute to the soil carbon reservoir. For example, a study conducted at an agricultural experimental station demonstrated that after multiple cycles of crop–mushroom rotation with straw-based substrates, soil organic carbon content increased by approximately 15% compared to continuous crop cultivation without mushroom integration [32]. This highlights the substantial impact of mushroom substrates on enriching the soil carbon pool.

Straw-based substrates provide a rich source of carbon and introduce diverse organic compounds into the soil. These compounds fuel the metabolic activities of soil microorganisms, stimulating the decomposition and transformation of organic matter. This creates a positive feedback loop: increased microbial activity further enhances carbon cycling and sequestration. Additionally, the fibrous nature of straw improves soil structure and porosity by binding soil particles into aggregates, which promote soil aeration and water infiltration [63,64]. These changes create a more favorable environment for root growth and nutrient uptake, ultimately boosting agroecosystem health and productivity.

(ii) Enhancement of microbial activity and soil aggregate structure.

The growth of mushrooms in the crop–mushroom rotation system fosters a symbiotic association with soil microorganisms, with the mushroom mycelium playing a central role. During its growth phase, the mycelium releases a diverse array of extracellular enzymes into the soil environment [65,66]. These enzymes are catalytically active agents that target and break down complex organic substances present in the soil matrix.

The enzymatic breakdown of complex organic matter serves a dual purpose: it liberates essential nutrients for mushroom growth and development while also stimulating the growth and metabolic activity of other soil microbes. A comprehensive field study demonstrated that the population density of beneficial bacteria and fungi was significantly higher in soils under crop–mushroom rotation compared to control plots without rotation [32]. These microorganisms play a critical role as nature's recyclers, decomposing organic matter into simpler forms and facilitating nutrient cycling within the soil ecosystem. Increased microbial activity also improves soil structure and fertility. The metabolic byproducts of these microorganisms act as binding agents, promoting the formation of soil aggregates [22]. These aggregates improve soil porosity, water-holding capacity, and aeration, creating a more favorable environment for root growth and nutrient uptake by plants. In essence, promoting microbial activity in crop–mushroom rotation systems is a cornerstone of sustainable agriculture. It enriches the soil with organic matter, stabilizes carbon, and enhances the overall health and productivity of the agroecosystem.

2.3. *Effects of Crop–Mushroom Rotation on Soil Fertility and Organic Carbon Sequestration*

Crop–mushroom rotation positively impacts soil fertility and organic carbon sequestration. During mushroom cultivation, the decomposition of organic substrates enriches the soil with residual organic matter, nutrients, and beneficial microorganisms. For example, spent mushroom compost contributes to soil health by providing these essential components, which enhance soil structure and promote sustainable agricultural practices [67]. When incorporated into the soil, it improves soil structure and water-holding capacity, enhancing fertility. In a study of fields with crop–mushroom rotation, the soil's porosity increased, facilitating better root penetration and nutrient uptake by subsequent crops [22].

Regarding organic carbon sequestration, mushroom mycelium plays a key role. It forms a network within the soil that helps in aggregating soil particles. This process protects organic carbon from decomposition. In addition, the mycelium secretes enzymes that break down complex organic compounds in the soil, converting them into more stable forms of carbon [68]. Field experiments have demonstrated that crop–mushroom rotation systems can increase the soil's organic carbon content over time compared to traditional cropping systems, thereby contributing to carbon sequestration [32]. After harvesting saprotrophic edible mushrooms, returning the spent mushroom substrate as organic fertilizer to the soil can significantly enhance soil fertility. During the cultivation of saprotrophic edible fungi, returning the mycelial residues (known as fungal necromass) to the cultivation substrate increases the content of microbial biomass carbon in the soil, which directly drives the decomposition of humus, mineralization of soil nutrients, and sequestration of organic carbon. Wang et al. [2], using nuclear magnetic resonance spectroscopy, found that soils receiving long-term organic fertilizer applications exhibited changes in the chemical composition of soil organic carbon, with reduced resistance indicators (aromaticity and aliphaticity), slower decomposition rates of soil organic carbon, and significantly reduced greenhouse gas emissions. Kang et al. [69], using Fourier transform infrared spectroscopy, discovered that straw as an organic material input into the soil accelerated the turnover of soil organic carbon, promoted the formation of large aggregates enriched in active carbon fractions such as aliphatic carbon, improved soil physical structure, increased soil porosity, and enhanced soil water retention capacity.

In the context of field cultivation of saprotrophic mushrooms, a critical question arises: does the introduction of exogenous organic matter significantly alter the composition and dynamics of microbial communities? Furthermore, this practice may create a synergistic effect that enhances the accumulation and stabilization of soil organic carbon. These potential impacts warrant comprehensive investigation to elucidate the underlying mechanisms and their broader ecological implications.

3. The Impact of Crop–Mushroom Rotation on Soil Microbial Communities

3.1. The Essential Role of Soil Microbial Communities in Ecosystem Functionality

Soil microbial communities play an indispensable and multifaceted role in maintaining ecosystems. They drive essential processes such as nutrient cycling, transforming and mobilizing key elements like nitrogen, phosphorus, and potassium, making them available for plant uptake and growth. Additionally, these microorganisms decompose organic matter, breaking down complex compounds into simpler substances that enhance soil fertility and structure. They also support energy flow by driving biochemical reactions that sustain life within the soil ecosystem. Several studies have revealed that soils under diverse rotation systems including rice with legumes, maize or rapeseed, etc., produced higher microbial richness [70,71]. Microbial diversity promotes ecosystem functioning in terms of nutrient cycling and resilience to stress [72]. Remarkably, soil microbial communities regulate approximately 80% to 90% of these essential soil functions, highlighting their profound impact on ecosystem productivity and resilience [73].

Crop rotation significantly influences soil microbial communities, which are essential for ecosystem functionality. Different crops alter the input of organic matter, root exudates, and soil structure. For instance, leguminous plants increase populations of nitrogen-fixing bacteria. Soil microbes play a critical role in nutrient cycling by decomposing organic matter and releasing nutrients such as nitrogen, phosphorus, and carbon. In crop rotation systems, diverse plant residues provide varied substrates, further enhancing microbial activity. Additionally, microbes contribute to soil aggregation, improving soil structure and water infiltration, while their symbiotic relationships with plants support plant health and productivity. Understanding how these microbial communities respond to crop rotation is crucial for sustainable agriculture and maintaining ecosystem balance [74,75].

3.2. The Impact of Crop Rotation on Soil Microbial Communities: Microbial Community Structure as an Indicator of Soil Health

Crop rotation significantly influences soil microbial community structure, which in turn reflects soil health. Different crops in rotation introduce diverse root exudates, residues, and management practices. For instance, legume–cereal rotations are common. Legumes can enrich the soil with nitrogen, altering the nutrient availability for microbes. The diversity within the microbial community structure is a key aspect. A healthy soil typically has a high diversity of bacteria, fungi, and other microorganisms. In crop rotation systems, this diversity can be enhanced as compared to monoculture. For example, some bacteria are specialized in decomposing specific organic compounds from certain crop residues. Rotation provides a variety of these substrates, promoting the growth of different bacterial groups [76]. Fungal-to-bacterial ratios also change with crop rotation. Fungi are often more dominant in soils with more complex organic matter inputs, such as those from crop rotations with high-residue crops. These fungi play crucial roles in soil aggregation and nutrient cycling [77]. Moreover, the presence of beneficial microbial groups like mycorrhizal fungi can be promoted by appropriate crop rotations. Their symbiotic relationships with plants improve plant nutrient uptake and contribute to soil stability.

Overall, monitoring the microbial community structure under crop rotation helps in assessing and maintaining soil health for sustainable agriculture. Therefore, assessments of microbial community dynamics can provide information for effective soil management and conservation strategies that aim to maintain or improve ecosystem services and functions in the long term [78,79].

3.3. The Impact of Crop Rotation on Soil Microbial Communities: The Ecological Roles of Bacteria and Fungi in Soil Carbon Cycling

Crop rotation has a profound impact on soil microbial communities, especially regarding the roles of bacteria and fungi in soil carbon cycling. Bacteria play diverse roles in carbon cycling within the soil. For instance, many heterotrophic bacteria are responsible for decomposing simple carbon compounds derived from crop residues. In a crop rotation system, different types of crop residues provide a range of carbon-based substrates for these bacteria. Some bacteria are also involved in the mineralization process, converting organic carbon into inorganic forms that can be reused in the ecosystem [80,81]. Fungi, on the other hand, have significant functions in soil carbon cycling as well. Saprotrophic fungi are key decomposers of complex organic matter like lignin and cellulose.

In crop rotation, the variety of crop residues with different lignocellulosic contents can stimulate the growth and activity of these fungi. Mycorrhizal fungi form symbiotic associations with plants in rotation. They help in the uptake of nutrients by plants and in return receive carbon from the host plants [82]. This carbon is then stored in the soil through the extensive mycelial network of the fungi, contributing to the long-term storage of soil carbon. Overall, crop rotation can alter the composition and activity of these bacteria and fungi, thereby influencing the efficiency and balance of soil carbon cycling processes.

In contrast, fungi assume a more critical ecological role as the pool of readily available nutrients becomes exhausted. Unlike their bacterial counterparts, fungi are slow-growing but highly efficient in their utilization of nutrients. Their metabolic capabilities enable them to degrade more complex and stable carbon compounds that are resistant to bacterial decomposition. This efficiency in breaking down recalcitrant organic matter positions fungi as key players in the process of soil organic carbon accumulation. By degrading these stable carbon components, fungi contribute to the long-term sequestration of carbon within the soil, thereby enhancing soil carbon stocks and influencing the broader carbon cycle. The interplay between bacteria and fungi in soil ecosystems is thus characterized by a dynamic balance, where the rapid exploitation of easily accessible carbon by bacteria is complemented by the slower, but more sustained, carbon degradation activities of fungi. This balance is crucial for maintaining soil health and fertility, as it supports the continuous turnover of organic matter and the stabilization of soil organic carbon, which are essential processes for sustaining productive and resilient ecosystems. The differential strategies employed by these microbial groups underscore the complexity of microbial interactions in soil environments and highlight the importance of both bacterial and fungal communities in the regulation of soil biogeochemical processes [83].

3.4. Impact of Rotation and Crop Residue Management on Soil Microbes and Organic Matter

In crop rotation systems, the specialized metabolic products of different plants selectively recruit specific microbes and suppress pathogenic ones, thereby fostering microbial communities conducive to plant growth. Concurrently, the Shannon diversity index of soil microbes significantly increases, enhancing soil nutrient content [62,84]. Compared to continuous monoculture, crop rotation significantly increases soil fungal biomass and the bacterial Shannon index, resulting in a 13.4% increase in microbial biomass carbon and a 16.6% increase in microbial biomass nitrogen [85].

Furthermore, the practice of returning crop residues to the soil induces substantial alterations in both soil organic carbon levels and the dynamics of soil microbial communities. This agronomic strategy not only enhances the SOC content but also significantly influences the composition, activity, and functional diversity of soil microorganisms. The decomposition of these crop residues provides a continuous supply of organic substrates, which in turn fuels microbial activity and drives shifts in microbial community structure. These changes in the microbial community can lead to varied outcomes, including the stimulation of microbial processes related to carbon cycling and nutrient availability. A study by Hong Qi et al. [86] has demonstrated that returning the residue of *S. rugosoannulata* mushrooms to the field can effectively neutralize soil pH and significantly increase the SOC content in fields. After the mycelium enters the soil, it suppresses the growth of other fungi, reducing soil fungal diversity. Mushroom cultivation can decrease the number of plant pathogens and enhance the soil disease suppression capability.

Soil fungi, as a critical part of global biodiversity, play a pivotal role in agricultural ecosystems [87]. When straw is incorporated into the soil and crop root residues begin to undergo decomposition, fungi that have evolved to specialize in the breakdown of fibrous plant materials are among the first organisms to flourish in this specific ecological niche [88]. These fungi, due to their unique physiological adaptations and specialized metabolic mechanisms, are particularly well-suited to thrive under such conditions. Their capacity to efficiently colonize and thrive in this environment is primarily due to their specialized enzyme systems [89], which are precisely adapted to decompose the complex and resilient fibrous components of dead plants. This enzymatic specialization underscores their critical role in the decomposition process within ecosystems. Through the action of their unique enzymatic machinery, they can systematically convert these tough, fibrous structures into simpler polysaccharides. This process not only facilitates the mineralization of organic matter but also plays a crucial role in altering the physical and chemical composition of the soil [90]. Moreover, the breakdown of these complex lignocelluloses into simpler carbohydrates creates a more favorable environment for the subsequent colonization and growth of other fungal species. These secondary fungi, which primarily utilize sugars as their carbon source, depend on the initial breakdown of fibrous materials by these pioneer decomposers. The breakdown of complex plant residues into accessible polysaccharides creates the biochemical conditions needed for the growth of diverse fungal species. Therefore, adding saprotrophic edible fungi to crop rotation systems enhances the decomposition and cycling of soil organic matter (Figure 2).

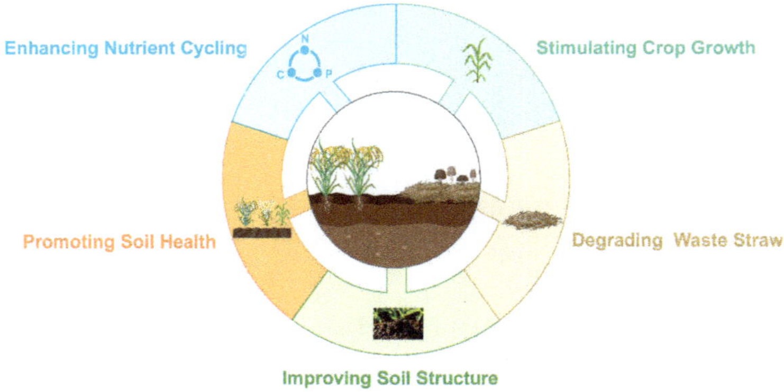

Figure 2. The benefits of soil microorganisms in crop–mushroom rotation process crop–mushroom rotation.

4. The Role of Saprotrophic Mushrooms in Soil Ecosystems

4.1. The Critical Role of Saprotrophic Mushrooms in Fungal Community Dynamics and Decomposition

The early-stage decomposition activities of saprotrophic mushrooms (known as lignocellulose-decomposing fungi) play a pivotal role in the initial breakdown of plant residues. These processes not only initiate the transformation of organic matter but also create the foundation for the subsequent development and interactions of the broader fungal community. These processes underscore the interconnectedness of fungal communities and the sequential nature of decomposition, where the metabolic activities of one group of organisms pave the way for the subsequent activity of others, ultimately leading to the efficient recycling of organic matter and the enhancement of soil fertility. This process promotes nutrient transformation and material cycling within the soil ecosystem. The chain reaction initiated by lignocellulose-decomposing fungi has profound and significant effects on nutrient conversion and material cycling in soil ecosystems. Specifically, the polysaccharides produced by these fungi provide a readily available carbon source and energy for other fungi, spurring their rapid growth and reproduction. During their metabolic processes, these fungi decompose organic matter into simpler compounds through a series of complex biochemical reactions, thereby releasing previously bound nutrients in the soil, such as nitrogen, phosphorus, and potassium. These nutrients become accessible for plant root absorption, facilitating efficient nutrient conversion.

4.2. Impact of Lignocellulose-Decomposing Mushroom on Nutrient Transformation and Material Cycling in Soil Ecosystems

In the context of material cycling, saprotrophic mushrooms play a highly dynamic and essential role in accelerating the decomposition and mineralization of organic matter present in the soil. This process is of fundamental importance as it enables the cycling of crucial elements, most notably carbon and nitrogen, within the complex soil–crop–atmosphere continuum [91,92].

A prime example of this intricate process is the generation of polysaccharides as a byproduct of the decomposition of crop residues by lignocellulose-decomposing saprotrophic mushrooms. These polysaccharides serve as a vital substrate for other fungal species within the soil ecosystem. The subsequent saprotrophic fungal species such as *S. rugosoannulata* then metabolize these polysaccharides, initiating a series of biochemical reactions that convert them into simpler compounds, with carbon dioxide and water being the end products. The carbon dioxide released during this metabolic process is emitted into the atmosphere, where it becomes an integral and influential component of the global carbon cycle. This emission has far-reaching implications as it directly impacts atmospheric carbon levels and, consequently, plays a significant role in global climate regulation. By actively participating in the carbon cycle in this manner, saprotrophic fungi contribute to the balance of carbon dioxide concentrations in the atmosphere, which is crucial for maintaining a stable climate [93,94]. Simultaneously, the breakdown of polysaccharides through the process of mineralization leads to the release of inorganic nutrients into the soil. These nutrients, which include essential elements such as nitrogen, phosphorus, and potassium, are the lifeblood of soil fertility. Their presence in the soil is vital for sustaining the growth and development of future plant communities. The replenishment of these inorganic nutrients provides a solid material foundation for the next cycle of crop growth. This, in turn, enables the continuation of primary productivity, which is the cornerstone of the functioning of the broader ecosystem. It is through this cyclical process that the interconnectedness of soil microbial communities, plant growth, and ecosystem dynamics becomes vividly evident. The decomposition of organic matter by lignocellulose-decomposing saprotrophic fungi is not only a key contributor to the global carbon cycle but also serves as a key in sustaining the nutrient cycles that underpin terrestrial ecosystems. This dual role highlights the ecological significance of saprotrophic fungi in maintaining the health and functionality of ecosystems [95,96]. Their activities ensure the efficient recycling of nutrients and carbon, promoting the stability and productivity of the entire ecosystem. Without the active participation of fungi in these processes, the soil would gradually become depleted of essential nutrients, leading to a decline in plant growth and a disruption of the delicate ecological balance.

4.3. The Role of Mushroom Necromass in SOC Accumulation

Through the conversion of complex organic materials into simpler forms and releasing essential nutrients back into the soil, these fungi facilitate the ongoing renewal of soil resources, ensuring the resilience and productivity of ecosystems over time. This complex and multifaceted interaction between decomposition processes, nutrient cycling, and overall ecosystem function underscores the critical importance of microbial activity in maintaining the health, stability, and resilience of environmental systems. Microorganisms play a central role in the decomposition of organic matter, driving the breakdown and conversion of complex compounds into simpler forms that are more easily assimilated by plants and other organisms. This microbial-mediated nutrient cycling is crucial for sustaining ecosystem productivity and long-term viability, as it facilitates the ongoing replenishment of essential nutrients within the soil.

Apart from the benefits of live lignocellulose-decomposing fungi, the role of dead fungal necromass in SOC accumulation has garnered increasing attention [97]. The increased input of straw carbon in rice paddies can accelerate fungal community turnover and succession, enhance microbial growth efficiency, and thereby increase the contribution of fungal necromass carbon to SOC formation and accumulation [98]. However, due to the varying responses of soil fungal community succession and nutritional strategies to the priming effect of straw carbon, the quantity and accumulation characteristics of fungal necromass generated from assimilated exogenous substrates can differ. These changes directly impact the amount and contribution ratio of fungal necromass within the organic carbon pool, consequently affecting variations in SOC accumulation and its stability. Further research is needed on the mechanisms and processes by which fungal necromass contributes to organic carbon formation following the input of exogenous straw carbon in typical soils of China.

In the context of crop–mushroom rotation, the comprehensive fertility of the soil is of utmost importance. It plays a multiple and critical role in determining the quantity and quality of agricultural production, especially in rice cultivation. The accumulation and dynamic changes in microbial residues within the soil have a substantial influence on the soil organic carbon pool [99]. In this rotation system, returning straw after cultivating *S. rugosoannulata* to the field significantly alters the soil microbial community structure and the soil carbon pool. These changes enhance soil fertility and nutrient availability, underscoring the need to manage soil health and microbial activity. Such management is critical for sustainably implementing crop–mushroom rotation systems and improving long-term agricultural productivity.

5. Potential Benefits and Challenges from Cross-Kingdom (Crop–Mushroom) Rotation

5.1. Improving Soil Quality

(i) Balancing soil nutrients: Different types of crops have different nutrient requirements in the soil. Through cross-kingdom rotation, continuous planting of a single crop and excessive consumption of certain specific nutrients in the soil can be avoided. For example, leguminous crops can fix nitrogen in the air and increase the nitrogen content in the soil, providing sufficient nitrogen sources for non-leguminous crops planted subsequently [100].

(ii) Reducing soil pests and diseases: Long-term planting of a single crop easily leads to the accumulation and reproduction of specific pests and diseases in the soil. Cross-kingdom rotation breaks the life cycle of pests and diseases, making it difficult for them to adapt to the growth environment of different crops, thereby reducing the occurrence probability of pests and diseases [101].

(iii) Improving soil structure: The root growth patterns of different crops vary. Some have deep and developed roots, while others have shallow and extensive roots. In the context of crop–mushroom rotation, the promotion of microbial diversity indeed leads to enhanced organic matter decomposition and the production of binding agents like glomalin and polysaccharides, which in turn facilitate the formation of stable soil aggregates [102]. In the context of crop–mushroom rotation, the promotion of microbial diversity indeed leads to enhanced organic matter decomposition and the production of binding agents like glomalin and polysaccharides, which in turn facilitate the formation of stable soil aggregates [103]. These aggregates play a fundamental role in determining soil structure. Stable aggregates improve soil porosity, allowing for better air and water flow within the soil matrix. This not only benefits plant root growth and nutrient uptake but also enhances the soil's ability to retain water, reducing the risk of erosion and improving overall soil fertility [104,105].

5.2. Improving Agricultural Production Efficiency

(i) Increasing crop yields: By improving soil quality and reducing pests and diseases, cross-kingdom rotation can increase the yield and quality of crops. The complementary roles of different crops in rotation enable more sufficient utilization of nutrients in the soil, thereby promoting the growth and development of crops [106].

(ii) Reducing production costs: Reducing the occurrence of pests and diseases means reducing the use of pesticides and lowering agricultural production costs. At the same time, reasonable rotation can also reduce the input of chemical fertilizers because the nutrient requirements of different crops can be partially met by the other crops in rotation [107].

(iii) Improving land utilization rate: Cross-kingdom rotation can make full use of land resources and avoid idle and wasteland. Planting different crops in different seasons can realize the full-year utilization of land and improve the economic benefits of agricultural production [108].

5.3. Promoting Ecological Balance

(i) Protecting biodiversity: Cross-kingdom rotation enriches the farmland ecosystem and provides different living environments for various organisms. This helps increase biodiversity in farmland, including insects, birds, microorganisms, etc. The increase in biodiversity can promote the stability and balance of the ecosystem and improve the ecological service function of farmland.
(ii) Reducing environmental pollution: Reasonable cross-kingdom rotation can reduce the pollution of chemical fertilizers and pesticides to the environment. By reducing the use of chemical fertilizers and pesticides, the risk of agricultural non-point source pollution is reduced, and the quality of soil, water sources, and air is protected.
(iii) Adapting to climate change: Different crops have different adaptability to climate change. Cross-kingdom rotation can make agricultural production more adaptable to the impacts of climate change and reduce the losses of agricultural disasters caused by climate.

Through increased soil carbon sequestration, more carbon is stored in the soil, reducing the amount of carbon dioxide released into the atmosphere. The reduced use of chemical fertilizers and pesticides also decreases the energy-intensive production and application processes, which are significant sources of greenhouse gas emissions. Additionally, improved soil structure and fertility can enhance crop resilience to climate change, reducing the need for energy-consuming agricultural practices to address crop failures (Figure 3).

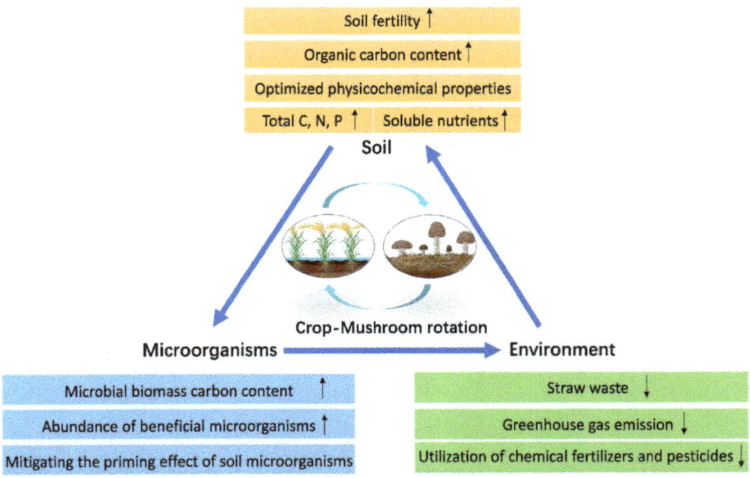

Figure 3. Effects of crop–mushroom rotation on soil, microorganisms, and environment.

5.4. Potential Challenges of Large-Scale Implementation

However, several challenges may hinder the large-scale implementation of this rotation system. Logistically, establishing and managing infrastructure for mushroom cultivation, such as climate-controlled growing facilities, can be complex and resource-intensive. Economically, the initial investment for mushroom-specific equipment, substrates, and fungal strains is substantial, and market fluctuations may pose additional risks to farmers. Environmentally, specific conditions, such as temperature and humidity requirements for mushroom growth, could limit the widespread adoption of this system in certain regions.

5.5. Comparison of Crop–Mushroom Rotation with Other Sustainable Agricultural Strategies

A comparison between crop–mushroom rotation and other sustainable agricultural strategies is presented (Table 1), focusing on their ability to balance productivity with environmental conservation. This analysis highlights the unique advantages of crop–mushroom rotation, such as its potential to enhance soil health, promote resource recycling, and reduce environmental impacts. By contrasting these strategies, the table provides a comprehensive overview of their respective contributions to sustainable agriculture.

Table 1. Comparison of crop–mushroom rotation with other sustainable agricultural strategies.

Sustainable Agricultural Strategies	Comparison with Crop-Mushroom Rotation	Advantages	Limitations
Traditional Crop Rotation (e.g., Wheat—Maize Rotation)	- **Soil Carbon Storage:** It can increase the soil organic carbon content. However, in crop-mushroom rotation, the addition of mushroom substrates can introduce more organic carbon sources, which is more conducive to carbon sequestration. - **Microbial Community**: It changes the structure of the microbial community. Nevertheless, the growth of mushrooms in crop-mushroom rotation can promote the growth of specific microorganisms, resulting in a more complex impact on the community structure. - **Pest and Disease Control**: It reduces pests and diseases to a certain extent. Crop-mushroom rotation not only breaks the life cycle of pests and diseases but also mushrooms may secrete substances that inhibit pests and diseases.	- The planting technology is relatively mature and easy for farmers to master. - It can adapt well to different regional climates and soil conditions.	- The effect of enhancing soil fertility is relatively limited. - Long-term monoculture crop rotation may still cause some pest and disease problems.
Organic Agriculture (without the use of chemical fertilizers and pesticides, relying on organic fertilizers and natural pest and disease control methods)	- **Soil Quality**: It improves soil structure and fertility. Based on this, crop-mushroom rotation further optimizes the physical and chemical properties of the soil through the decomposition of organic matter by mushrooms. - **Ecological Balance**: Both focus on ecological balance. In crop-mushroom rotation, mushrooms can also utilize agricultural waste, improving resource utilization efficiency. - **Agricultural Product Quality**: Both can enhance the quality of agricultural products. The mushrooms produced in crop-mushroom rotation also have economic and nutritional value.	- The agricultural products produced have no pesticide residues and high market value. - It is conducive to protecting the ecological environment and biodiversity.	- The production cost is relatively high, and the acquisition and use of organic fertilizers are costly. - The yield is relatively low, making it difficult to meet large-scale market demands.
Precision Agriculture (using modern information technology for precise management of agricultural production)	- **Resource Utilization Efficiency**: Precision agriculture precisely invests resources, while crop-mushroom rotation improves efficiency through resource recycling, with different focuses. - **Soil Management**: Precision agriculture precisely monitors the soil, and crop-mushroom rotation can improve the overall soil ecology, and the two can complement each other. - **Economic Benefits**: Precision agriculture improves economic benefits, and crop-mushroom rotation reduces production costs, jointly enhancing profitability.	- It can accurately control resource input and reduce waste. - It can enhance production efficiency and product quality with the help of technology.	- The upfront investment in technical equipment is large, and professional technicians are required for operation. - The application of technology is limited by regions and the quality of farmers.

Future research could focus on optimizing the crop–mushroom rotation system for different regions and climates. This includes identifying the best crop–mushroom combinations, determining the optimal rotation timing, and developing more efficient integration methods. Additionally, further studies are needed to understand the long-term effects of this rotation system on soil health and ecosystem stability.

6. Conclusions

The present study underscores the multifaceted benefits of crop–mushroom rotation as a sustainable agricultural practice, offering a comprehensive approach to enhancing soil quality, agricultural productivity, and ecological balance. Key findings from the research highlight the significant role of this cross-kingdom rotation in balancing soil nutrients, reducing pest and disease prevalence, and improving soil structure, which are crucial for maintaining long-term soil health and fertility. One of the most notable outcomes of integrating crop–mushroom rotation is the enhancement of soil organic carbon sequestration, a critical factor in mitigating climate change and ensuring the resilience of agricultural systems. The introduction of organic matter through mushroom cultivation not only affects microbial community structure but also supports the decomposition of complex organic materials, contributing to soil organic carbon accumulation and nutrient cycling. The study also emphasizes the importance of microbial communities, including live biomass and necromass, in maintaining ecosystem stability and the potential of the rice–*S. rugosoannulata* rotation model as a sustainable agricultural practice. The multifunctional benefits of saprotrophic edible mushrooms like *S. rugosoannulata* extend beyond soil improvement to include nutritional and medicinal values, further solidifying their role in sustainable agriculture. However, the research also identifies the need for further exploration into the complex interactions among microbial communities, plant residues, and soil management practices within the context of crop–mushroom rotation. Understanding these dynamics is essential for optimizing agricultural practices aimed at improving soil health and carbon sequestration in response to climate change. Taken together, the crop–mushroom rotation model emerges as a promising strategy for sustainable agriculture, providing a balance between productivity and environmental conservation. By harnessing the synergistic effects of crop diversity and mushroom cultivation, this practice offers a pathway to enhance soil fertility, increase agricultural yields, and promote ecological balance. Future research should focus on unraveling the intricate relationships between soil microorganisms, plant residues, and soil management to fully capitalize on the potential of crop–mushroom rotation in supporting sustainable agricultural systems and climate change mitigation efforts.

Author Contributions: T.D.: writing—original draft, investigation; K.Z.: writing—review and editing; X.S.: methodology, resources, W.L.: resources; F.Y.: Resources, supervision, validation, funding acquisition; D.L.: funding acquisition, supervision, writing—review and editing. All authors have read and agreed to the published version of the manuscript.

Funding: This work received funding from the "Strategic Priority Research Program" of the Chinese Academy of Sciences (XDA26050302) to Fuqiang Yu, the Yunnan Revitalization Talent Support Program "Young Talent" Project (YNQRQNRC-2019–025) to Dong Liu, the Yunnan Technology Innovation Program (202205AD160036) to Fuqiang Yu, the Clean Conversion of Agricultural Waste and Efficient Recycling through Edible Mushroom Cultivation (2023520000240179) and the Yunnan Revitalization Talent Support Program to Jesús Perez-Moreno and Xinhua He.

Data Availability Statement: No new data were created or analyzed in this study.

Conflicts of Interest: The authors declare no conflicts of interest.

References

1. Liu, T.Q.; Guo, L.J.; Cao, C.G.; Tan, W.; Li, C. Long-term rice-oilseed rape rotation increases soil organic carbon by improving functional groups of soil organic matter. *Agric. Ecosyst. Environ.* **2021**, *319*, 107548. [CrossRef]
2. Wang, E.; Yu, B.; Zhang, J.; Gu, S.; Yang, Y.; Deng, Y.; Guo, X.; Wei, B.; Bi, J.; Sun, M.; et al. Low carbon loss from long-term manure-applied soil during abrupt warming is realized through soil and microbiome interplay. *Environ. Sci. Technol.* **2024**, *58*, 9658–9668. [CrossRef] [PubMed]
3. Smith, P. Soils as carbon sinks: The global context. *Soil. Use Manag.* **2004**, *20*, 212–218. [CrossRef]
4. Wang, Y.; Li, C.; Sun, Q.; Chang, Y. Research on SOC estimation of lithium-ion batteries based on robust full order proportional integral observer. *Int. J. Electrochem. Sc.* **2024**, *19*, 100645. [CrossRef]
5. Gadermaier, F.; Berner, A.; Fliessbach, A.; Friedel, J.K.; Mäder, P. Impact of reduced tillage on soil organic carbon and nutrient budgets under organic farming. *Renew. Agr. Food Syst.* **2012**, *27*, 68–80. [CrossRef]
6. Yuan, S.; Linquist, B.A.; Wilson, L.T.; Cassman, K.G.; Stuart, A.M.; Pede, V.; Miro, B.; Saito, K.; Agustiani, N.; Aristya, V.E.; et al. Sustainable intensification for a larger global rice bowl. *Nat. Commun.* **2021**, *12*, 7163. [CrossRef]
7. Cui, Y.; Zhao, K.; He, J.; Qu, M. Effect of capital endowment on farmers' decision-making in protecting cultivated land in a rice-growing area: An empirical study based on a double-hurdle model. *Chin. J. Eco-Agric.* **2019**, *27*, 959–970.
8. Lei, Y. Analysis on the control of regional rice frequent pests and diseases. *Seed Sci. Technol.* **2017**, *35*, 108–109.
9. Chen, S.; Meng, W.; Li, S.; Zheng, H.; Lei, X.; Sun, X.; Wang, L. Current situation and problems of heavy metal pollution prevention and control in farmland soil in China. *Geosci. Front.* **2019**, *26*, 35–41. [CrossRef]
10. Garnett, T.; Appleby, M.C.; Balmford, A.; Bateman, I.J.; Benton, T.G.; Bloomer, P.; Burlingame, B.; Dawkins, M.; Dolan, L.; Fraser, D.; et al. Sustainable intensification in agriculture: Premises and policies. *Science* **2013**, *341*, 33–34. [CrossRef]
11. Li, C.; Zhu, H.; Li, C.; Qian, H.; Yao, W.; Guo, Y. The present situation of pesticide residues in China and their removal and transformation during food processing. *Food Chem.* **2021**, *354*, 129552. [CrossRef] [PubMed]
12. Yang, L.; Feng, Y.; Shi, W.; Xue, L.H.; Wang, S.Q.; Song, X.F.; Chang, Z.Z. Review of the advances and development trends in agricultural non-point source pollution control in China. *Chin. J. Eco-Agric.* **2013**, *1*, 96–101. [CrossRef]
13. Chen, Z.; Li, X.; Liu, T.; Fu, H.; Yuan, X.; Cheng, Q.; Liao, Q.; Zhang, Y.; Li, W.; Sun, Y.; et al. Strategies for fertilizer management to achieve higher yields and environmental and fertilizer benefits of rice production in China. *Sci. Total Environ.* **2023**, *904*, 166325. [CrossRef] [PubMed]
14. Liu, Z.; Cao, S.; Sun, Z.; Wang, H.; Qu, S.; Lei, N.; He, J.; Dong, Q. Tillage effects on soil properties and crop yield after land reclamation. *Sci. Rep.* **2021**, *11*, 4611. [CrossRef]
15. Zou, Y.; Liu, Z.; Chen, Y.; Wang, Y.; Feng, S. Crop Rotation and Diversification in China: Enhancing Sustainable Agriculture and Resilience. *Agriculture* **2024**, *14*, 1465. [CrossRef]
16. Shah, K.K.; Modi, B.; Pandey, H.P.; Subedi, A.; Aryal, G.; Pandey, M.; Shrestha, J. Diversified crop rotation: An approach for sustainable agriculture production. *Adv. Agric.* **2021**, *2021*, 8924087. [CrossRef]
17. Harker, K.; O Donovan, J.; Turkington, T.; Blackshaw, R.E.; Lupwayi, N.Z.; Smith, E.G.; Johnson, E.N.; Gan, Y.; Kutcher, H.R.; Dosdall, L.M.; et al. Canola rotation frequency impacts canola yield and associated pest species. *Can. J. Plant Sci.* **2015**, *95*, 9–20. [CrossRef]
18. Fan, M.; Jiang, R.; Zhang, F.; Lü, S.H.; Liu, X.J. Nutrient management strategy of paddy rice-upland crop rotation system. *Chin. J. Appl. Ecol.* **2008**, *19*, 424–432.
19. Yang, B.; Sun, D.; Zhang, Y.; Zhong, C.; Huang, G.Q. Effects of different paddy-upland multiple cropping rotation systems on soil organic carbon and its fractions in paddy field. *Chin. J. Appl. Ecol.* **2019**, *2*, 456–462. [CrossRef]
20. Fang, Y.; Xu, K.; Guo, X.; Hong, Y. Identifying determinants of straw open field burning in northeast China: Toward greening agriculture base in newly industrializing countries. *J. Rural. Stud.* **2020**, *74*, 111–123. [CrossRef]
21. He, G.; Liu, T.; Zhou, M. Straw burning, PM2.5, and death: Evidence from China. *J. Dev. Econ.* **2020**, *145*, 102468. [CrossRef]
22. Hu, Q.; Li, C.; Cao, C.; Huang, J.; Ding, H.; Yu, M.; Liu, J.; Liu, T. The rice–edible mushroom pattern promotes the transformation of composted straw-C to soil organic carbon. *Agr. Ecosyst. Environ.* **2023**, *353*, 108560. [CrossRef]
23. Solovyev, N.; Prakash, N.T.; Bhatia, P.; Prakash, R.; Drobyshev, E.; Michalke, B. Selenium-rich mushrooms cultivation on a wheat straw substrate from seleniferous area in Punjab, India. *J. Trace Elem. Med. Biol.* **2018**, *50*, 362–366. [CrossRef] [PubMed]
24. Wang, F.; Fang, Y.; Wang, L.; Xiang, H.; Chen, G.; Chang, X.; Liu, D.; He, X.; Zhong, R. Effects of residual monensin in livestock manure on nitrogen transformation and microbial community during "crop straw feeding-substrate fermentation-mushroom cultivation" recycling system. *Waste Manag.* **2022**, *149*, 333–344. [CrossRef]
25. Aditya; Neeraj; Jarial, R.S.; Jarial, K.; Bhatia, J. Comprehensive review on oyster mushroom species (Agaricomycetes): Morphology nutrition, cultivation and future aspects. *Heliyon* **2024**, *10*, e26539. [CrossRef]
26. Moonmoon, M.; Uddin, M.N.; Ahmed, S.; Shelly, N.J.; Khan, A. Cultivation of different strains of king oyster mushroom (*Pleurotus eryngii*) on saw dust and rice straw in Bangladesh. *Saudi J. Biol. Sci.* **2010**, *17*, 341–345. [CrossRef]

27. Öztürk, C.; Atila, F. Changes in lignocellulosic fractions of growing substrates during the cultivation of *Hypsizygus ulmarius* mushroom and its effects on mushroom productivity. *Sci. Hortic.* **2021**, *288*, 110403. [CrossRef]
28. El Sheikha, A.F.; Hu, D. How to trace the geographic origin of mushrooms? *Trends Food Sci. Tech.* **2018**, *78*, 292–303. [CrossRef]
29. Wu, F.; Zhou, L.; Yang, Z.; Bau, T.; Li, T.-H.; Dai, Y.-C. Resource diversity of Chinese macrofungi: Edible, medicinal and poisonous species. *Fungal Divers.* **2019**, *98*, 1–76. [CrossRef]
30. Jayaraman, S.; Yadav, B.; Dalal, R.C.; Naorem, A.; Sinha, N.K.; Rao, C.S.; Dang, Y.; Patra, A.; Datta, S.; Rao, A.S. Mushroom farming: A review Focusing on soil health, nutritional security and environmental sustainability. *Farming Syst.* **2024**, *2*, 100098. [CrossRef]
31. Tang, Q.C.; Liu, W.J.; Huang, H.; Peng, Z.; Deng, L. Responses of Crop Yield, Soil Fertility, and Heavy Metals to Spent Mushroom Residues Application. *Plants* **2024**, *13*, 663. [CrossRef] [PubMed]
32. Hao, H.; Yue, Y.; Chen, Q.; Yang, Y.; Kuai, B.; Wang, Q.; Xiao, T.; Chen, H.; Zhang, J. Effects of an efficient straw decomposition system mediated by *Stropharia rugosoannulata* on soil properties and microbial communities in forestland. *Sci. Total Environ.* **2024**, *916*, 170226. [CrossRef] [PubMed]
33. Akyüz, M.; Kirbag, S. Nutritive value of wild edible and cultured mushrooms. *Turk. J. Biol.* **2010**, *34*, 97–102. [CrossRef]
34. Thakur, A.; Pandey, K.K.; Kharka, K.; Borker, S.S.; Kumar, B.; Bhatt, A.; Kumar, R. Comparative analysis of substrate components, nutritive value, ergosterol distribution, and vitamin D_2 enrichment in shiitake mushrooms cultivated on wood substrates. *J. Food Compos. Anal.* **2024**, *134*, 106508. [CrossRef]
35. Ghobad-Nejhad, M.; Antonin, V.; Moghaddam, M.; Langer, E. Resources of Iranian agarics (Basidiomycota) with an outlook on their antioxidant potential. *Front. Microbiol.* **2022**, *13*, 1015440. [CrossRef]
36. Guo, M.; Ma, X.; Zhou, Y.; Bian, Y.; Liu, G.; Cai, Y.; Huang, T.; Dong, H.; Cai, D.; Wan, X.; et al. Genome sequencing highlights the plant cell wall degrading capacity of edible mushroom *Stropharia rugosoannulata*. *J. Microbiol.* **2023**, *61*, 83–93. [CrossRef]
37. Lai, M.D.; Ong, K.C.; Arumugam, B.; Kuppusamy, U.R. Nutritional composition, efficacy and mechanisms of oyster mushrooms (*Pleurotus* spp.) in preventing metabolic syndrome: Insights into perspectives and challenges. *Food Biosci.* **2024**, *61*, 104768. [CrossRef]
38. Sahithya, K.; Mouli, T.; Biswas, A. Remediation potential of mushrooms and their spent substrate against environmental contaminants: An overview. *Biocatal. Agr. Biotech.* **2022**, *42*, 102323. [CrossRef]
39. Dos Santos Bazanella, G.C.; de Souza, D.F.; Castoldi, R.; Oliveira, R.F.; Bracht, A.; Peralta, R.M. Production of laccase and manganese peroxidase by *Pleurotus pulmonarius* in solid-state cultures and application in dye decolorization. *Folia Microbiol.* **2013**, *58*, 641–647. [CrossRef]
40. Kumla, J.; Suwannarach, N.; Sujarit, K.; Penkhrue, W.; Kakumyan, P.; Jatuwong, K.; Vadthanarat, S.; Lumyong, S. Cultivation of Mushrooms and Their Lignocellulolytic Enzyme Production Through the Utilization of Agro-Industrial Waste. *Molecules* **2020**, *25*, 2811. [CrossRef]
41. Carpio, M.J.; Andrades, M.S.; Herrero-Hernández, E.; Marín-Benito, J.M.; Sánchez-Martín, M.J.; Rodríguez-Cruz, M.S. Changes in vineyard soil parameters after repeated application of organic-inorganic amendments based on spent mushroom substrate. *Environ. Res.* **2023**, *221*, 115339. [CrossRef] [PubMed]
42. Owaid, M.N.; Abed, I.A.; Al-Saeedi, S.S.S. Applicable properties of the bio-fertilizer spent mushroom substrate in organic systems as a byproduct from the cultivation of *Pleurotus* spp. *Inf. Process. Agric.* **2017**, *4*, 78–82. [CrossRef]
43. Zamanian, K.; Zhou, J.; Kuzyakov, Y. Soil carbonates: The unaccounted, irrecoverable carbon source. *Geoderma* **2021**, *384*, 114817. [CrossRef]
44. Luo, Z.; Feng, W.; Luo, Y.; Baldock, J.; Wang, E. Soil organic carbon dynamics jointly controlled by climate, carbon inputs, soil properties and soil carbon fractions. *Glob. Change Biol.* **2017**, *23*, 4430–4439. [CrossRef]
45. Ruddiman, W.F. The anthropogenic greenhouse era began thousands of years ago. *Clim. Change* **2003**, *61*, 261–293. [CrossRef]
46. Zamanian, K.; Pustovoytov, K.; Kuzyakov, Y. Pedogenic carbonates: Forms and formation processes. *Earth-Sci. Rev.* **2016**, *157*, 1–17. [CrossRef]
47. Keskin, H.; Grunwald, S.; Harris, W.G. Digital mapping of soil carbon fractions with machine learning. *Geoderma* **2019**, *339*, 40–58. [CrossRef]
48. Chen, L.; Zhou, S.; Zhang, Q.; Zou, M.; Yin, Q.; Qiu, Y.; Qin, W. Effect of organic material addition on active soil organic carbon and microbial diversity: A meta-analysis. *Soil. Tillage Res.* **2024**, *241*, 106128. [CrossRef]
49. Lin, H.; Zhou, M.; Zeng, F.; Xu, P.; Ma, S.; Zhang, B.; Li, Z.; Wang, Y.; Zhu, B. How do soil organic carbon pool, stock and their stability respond to crop residue incorporation in subtropical calcareous agricultural soils? *Agr. Ecosyst. Environ.* **2022**, *332*, 107927. [CrossRef]
50. Ren, T.; Ukalska-Jaruga, A.; Smreczak, B.; Cai, A. Dissolved organic carbon in cropland soils: A global meta-analysis of management effects. *Agr. Ecosyst. Environ.* **2024**, *371*, 109080. [CrossRef]

51. Wang, Z.; Wang, S.; Xiao, B.; Xie, L.; Guo, Z.; Zhang, K.; Zhang, L.; Fan, Y.; Guo, X.; Luo, S.; et al. Response of soil organic carbon accumulation and allocation dynamics to ant nesting activities in xishuangbanna tropical forests. *Ecol. Environ. Sci.* **2024**, *33*, 35–44.
52. Jiang, H.; Yuan, C.; Wu, Q.; Heděnec, P.; Zhao, Z.; Yue, K.; Ni, X.; Wu, F.; Peng, Y. Effects of transforming multiple ecosystem types to tree plantations on soil microbial biomass carbon, nitrogen, phosphorus and their ratios in China. *Appl. Soil. Ecol.* **2024**, *193*, 105145. [CrossRef]
53. Liang, C.; Schimel, J.P.; Jastrow, J.D. The importance of anabolism in microbial control over soil carbon storage. *Nat. Microbiol.* **2017**, *2*, 17105. [CrossRef] [PubMed]
54. Tuo, Y.; Tan, H.; Liang, J.; Li, J.; Xiang, P.; Yang, Q.; He, X. Optimization of water and fertilizer management of Panax pseudoginseng based on changes in soil microbial biomass carbon and nitrogen and enzyme activities. *Appl. Soil. Ecol.* **2024**, *196*, 105282. [CrossRef]
55. Meng, L.; Li, W.; Zhang, S.; Wu, C.; Lv, L. Feasibility of co-composting of sewage sludge, spent mushroom substrate and wheat straw. *Bioresour. Technol.* **2017**, *226*, 39–45. [CrossRef]
56. Sun, Q.; Zheng, Y.; Li, S.; Yang, J.; Zhao, X.; Du, L.; He, K.; Liu, J. Diversified crop rotation: Synergistically enhancing peanut yield and soil organic carbon stability. *Agric. Ecosyst. Environ.* **2025**, *382*, 109497. [CrossRef]
57. Liu, D.; Gong, H.; Li, J.; Liu, Z.; Wang, L.; Ouyang, Z.; Xu, L.; Wang, T. Continuous crop rotation increases soil organic carbon stocks in river deltas: A 40-year field evidence. *Sci. Total Environ.* **2024**, *906*, 167749. [CrossRef]
58. Attia, A.; Marohn, C.; Shawon, A.R.; de Kock, A.; Strassemeyer, J.; Feike, T. Do rotations with cover crops increase yield and soil organic carbon?—A modeling study in southwest Germany. *Agric. Ecosyst. Environ.* **2024**, *375*, 109167. [CrossRef]
59. Haruna, S.I.; Nkongolo, N.V. Tillage, cover crop and crop rotation effects on selected soil chemical properties. *Sustainability* **2019**, *11*, 2770. [CrossRef]
60. Zhou, X.; Han, T.; Shen, Y.; Shi, Z.; He, B. Response characteristics of soil microecology in long-term continuous cropping tobacco field under 4 rotation patterns. *J. Agric. Sci. Tech.* **2024**, *26*, 174–187.
61. Chen, X.; Liu, Y.; Chen, M.; Wang, W. Impact of crop diversity (rotation) on quantity and spectral characteristics of latosolic dissolved organic matter. *J. Agric. Resour. Environ.* **2024**, *42*, 149–158.
62. Yang, X.; Xiong, J.; Du, T.; Ju, X.; Gan, Y.; Li, S.; Xia, L.; Shen, Y.; Pacenka, S.; Steenhuis, T.S.; et al. Diversifying crop rotation increases food production, reduces net greenhouse gas emissions and improves soil health. *Nat. Commun.* **2024**, *15*, 198. [CrossRef]
63. Wang, C.; Ma, J.; Wang, Y.; Li, Z.; Ma, B. The influence of wheat straw mulching and straw length on infiltration, runoff and soil loss. *Hydrol. Process* **2022**, *36*, e14561. [CrossRef]
64. Zhang, P.; Wei, T.; Jia, Z.; Han, Q.; Ren, X. Soil aggregate and crop yield changes with different rates of straw incorporation in semiarid areas of northwest China. *Geoderma* **2014**, *230–231*, 41–49. [CrossRef]
65. Carrasco, J.; Preston, G.M. Growing edible mushrooms: A conversation between bacteria and fungi. *Environ. Microbiol.* **2020**, *22*, 858–872. [CrossRef] [PubMed]
66. Wang, X.; Li, Z.; Jin, H.; Huang, X.; Xiao, X.; Zi, S.; Liu, T. Effects of mushroom-tobacco rotation on microbial community structure in continuous cropping tobacco soil. *J. Appl. Microbiol.* **2023**, lxad088. [CrossRef]
67. Leong, Y.K.; Ma, T.; Chang, J.; Yang, F. Recent advances and future directions on the valorization of spent mushroom substrate (SMS): A review. *Bioresour. Technol.* **2022**, *344*, 126157. [CrossRef]
68. Andrew, C.; Halvorsen, R.; Heegaard, E.; Kuyper, T.W.; Heilmann-Clausen, J.; Krisai-Greilhuber, I.; Bässler, C.; Egli, S.; Gange, A.C.; Høiland, K.; et al. Continental-scale macrofungal assemblage patterns correlate with climate, soil carbon and nitrogen deposition. *J. Biogeogr.* **2018**, *45*, 1942–1953. [CrossRef]
69. Kang, B.; Zhang, J.; Tang, P.; Zhang, F. Effect of long-term cotton straw returning on the chemical structure of soil organic carbon *North. Hortic.* **2024**, *10*, 78–84. [CrossRef]
70. Sun, Y.; Yang, X.; Elsgaard, L.; Du, T.; Siddique, K.H.; Kang, S.; Butterbach-Bahl, K. Diversified crop rotations improve soil microbial communities and functions in a six-year field experiment. *J. Environ. Manag.* **2024**, *370*, 122604. [CrossRef]
71. Venter, Z.S.; Jacobs, K.; Hawkins, H. The impact of crop rotation on soil microbial diversity: A meta-analysis. *Pedobiologia* **2016**, *59*, 215–223. [CrossRef]
72. Xiong, C.; Lu, Y. Microbiomes in agroecosystem: Diversity, function and assembly mechanisms. *Environ. Microbiol. Rep* **2022**, *14*, 833–849. [CrossRef] [PubMed]
73. Bastida, F.; Eldridge, D.J.; García, C.; Png, G.K.; Bardgett, R.D.; Delgado-Baquerizo, M. Soil microbial diversity–biomass relationships are driven by soil carbon content across global biomes. *ISME J.* **2021**, *15*, 2081–2091. [CrossRef] [PubMed]
74. Dai, H.; Zhang, H.; Xue, Y.; Gao, Y.; Qian, X. Response of fungal community and function to different tillage and straw returning methods. *Sci. Agric. Sin.* **2019**, *13*, 2280–2294.
75. Wang, R.; He, X.; Zhang, Q.; Li, B.; Shu, Z.; Chu, C. Inconsistent elevational patterns of soil microbial biomass, diversity, and community structure on four elevational transects from subtropical forests. *Appl. Soil. Ecol.* **2024**, *201*, 105462. [CrossRef]

76. Jin, X.; Yang, X.; Peng, S.; Ma, E.; Zhang, H.; Lin, X.; Wang, Y.; Li, J. Cropping rotation improved the bacterial diversity and N-cycling genes in tobacco fields through a 19-year long-term experiment. *Appl. Soil. Ecol.* **2024**, *193*, 105165. [CrossRef]
77. Marro, N.; Cofré, N.; Grilli, G.; Alvarez, C.; Labuckas, D.; Maestri, D.; Urcelay, C. Soybean yield, protein content and oil quality in response to interaction of arbuscular mycorrhizal fungi and native microbial populations from mono- and rotation-cropped soils. *Appl. Soil. Ecol.* **2020**, *152*, 103575. [CrossRef]
78. Gao, C.; Bezemer, T.M.; de Vries, F.T.; van Bodegom, P.M. Trade-offs in soil microbial functions and soil health in agroecosystems. *Trends Ecol. Evol.* **2024**, *39*, 895–903. [CrossRef]
79. Zhao, L.; Xiao, S.; Liu, C.; Liu, Y.; Xue, C. Effect of different tillage practices on fungal community and diversity in maize-growing soil. *China Plant Prot.* **2017**, *4*, 5–14.
80. Séneca, J.; Söllinger, A.; Herbold, C.W.; Pjevac, P.; Prommer, J.; Verbruggen, E.; Sigurdsson, B.D.; Peñuelas, J.; A Janssens, I.; Urich, T.; et al. Increased microbial expression of organic nitrogen cycling genes in long-term warmed grassland soils. *ISME Commun.* **2021**, *1*, 69. [CrossRef]
81. Xue, S.; Li, X.; Fu, Y.; Zhu, P.; Liu, J.; Kou, W.; Huang, D.; Gao, Y.; Wang, X. New insights into organic carbon mineralization: Combining soil organic carbon fractions, soil bacterial composition, microbial metabolic potential, and soil metabolites. *Soil. Till Res.* **2024**, *244*, 106243. [CrossRef]
82. Frew, A.; Powell, J.R.; Glauser, G.; Bennett, A.E.; Johnson, S.N. Mycorrhizal fungi enhance nutrient uptake but disarm defences in plant roots, promoting plant-parasitic nematode populations. *Soil. Biol. Biochem.* **2018**, *126*, 123–132. [CrossRef]
83. Song, X.; Liu, X.; Liang, G.; Li, S.; Li, J.; Zhang, M.; Zheng, F.; Ding, W.; Wu, X.; Wu, H. Positive priming effect explained by microbial nitrogen mining and stoichiometric decomposition at different stages. *Soil. Biol. Biochem.* **2022**, *175*, 108852. [CrossRef]
84. Guo, C.; Yang, C.; Fu, J.; Song, Y.; Chen, S.; Li, H.; Ma, C. Effects of crop rotation on sugar beet growth through improving soil physicochemical properties and microbiome. *Ind. Crop Prod.* **2024**, *212*, 118331. [CrossRef]
85. Liu, Q.; Zhao, Y.X.; Li, T.; Chen, L.; Chen, Y.; Sui, P. Changes in soil microbial biomass, diversity, and activity with crop rotation in cropping systems: A global synthesis. *Appl. Soil. Ecol.* **2023**, *186*, 104815. [CrossRef]
86. Hong, Q.; Zhang, J.; Hao, H.; Zhao, J.; Chen, H. Effects of in-situ returning of *Stropharia rugosoannulata* residue on soil physicochemical properties and fungal diversity. *Shanghai J. Agric. Sci.* **2023**, *39*, 1–7. [CrossRef]
87. Banerjee, S.; Zhao, C.; Garland, G.; Edlinger, A.; García-Palacios, P.; Romdhane, S.; Degrune, F.; Pescador, D.S.; Herzog, C.; Camuy-Velez, L.A.; et al. Biotic homogenization, lower soil fungal diversity and fewer rare taxa in arable soils across Europe. *Nat. Commun.* **2024**, *15*, 327. [CrossRef]
88. Kang, P.; Pan, Y.; Ran, Y.; Li, W.; Shao, M.; Zhang, Y.; Ji, Q.; Ding, X. Soil saprophytic fungi could be used as an important ecological indicator for land management in desert steppe. *Ecol. Indic.* **2023**, *150*, 110224. [CrossRef]
89. Wang, L.; Wang, J.; Tang, Z.; Wang, J.; Zhang, Y. Long-term organic fertilization reshapes the communities of bacteria and fungi and enhances the activities of C- and P-cycling enzymes in calcareous alluvial soil. *Appl. Soil. Ecol.* **2024**, *194*, 105204. [CrossRef]
90. Nie, S.; Lei, X.; Zhao, L.; Brookes, P.C.; Wang, F.; Chen, C.; Yang, W.; Xing, S. Fungal communities and functions response to long-term fertilization in paddy soils. *Appl. Soil. Ecol.* **2018**, *130*, 251–258. [CrossRef]
91. Crowther, T.W.; Boddy, L.; Hefin Jones, T. Functional and ecological consequences of saprotrophic fungus–grazer interactions. *ISME J.* **2012**, *6*, 1992–2001. [CrossRef] [PubMed]
92. Marañón-Jiménez, S.; Radujković, D.; Verbruggen, E.; Grau, O.; Cuntz, M.; Peñuelas, J.; Richter, A.; Schrumpf, M.; Rebmann, C. Shifts in the abundances of saprotrophic and ectomycorrhizal fungi with altered leaf litter inputs. *Front. Plant Sci.* **2021**, *12*, 682142. [CrossRef] [PubMed]
93. Juan-Ovejero, R.; Briones, M.J.I.; Öpik, M. Fungal diversity in peatlands and its contribution to carbon cycling. *Appl. Soil. Ecol.* **2020**, *146*, 103393. [CrossRef]
94. Schreiner, K.M.; Blair, N.E.; Levinson, W.; Egerton-Warburton, L.M. Contribution of Fungal Macromolecules to Soil Carbon Sequestration. In *Soil Carbon*; Hartemink, A.E., McSweeney, K., Eds.; Springer International Publishing: Cham, Switzerland, 2014; pp. 155–161, ISBN 978-3-319-04084-4.
95. Miyauchi, S.; Navarro, D.; Grisel, S.; Chevret, D.; Berrin, J.-G.; Rosso, M.-N. The integrative omics of white-rot fungus Pycnoporus coccineus reveals co-regulated CAZymes for orchestrated lignocellulose breakdown. *PLoS ONE* **2017**, *12*, e0175528. [CrossRef]
96. van der Wal, A.; Geydan, T.D.; Kuyper, T.W.; de Boer, W. A thready affair: Linking fungal diversity and community dynamics to terrestrial decomposition processes. *FEMS Microbiol. Rev.* **2013**, *37*, 477–494. [CrossRef]
97. Liu, D.; Zhou, Z.; Iqbal, S.; Dou, T.T.; Bonito, G.; Liu, W.; An, S.; Chater, C.C.; Perez-Moreno, J.; Che, R.; et al. Fungal necromass contribution to carbon sequestration in global croplands: A meta-analysis of driving factors and conservation practices. *Sci. Total Environ.* **2024**, *949*, 174954. [CrossRef]
98. Tian, J.; Dungait, J.A.J.; Hou, R.; Deng, Y.; Hartley, I.P.; Yang, Y.; Kuzyakov, Y.; Zhang, F.; Cotrufo, M.F.; Zhou, J. Microbially mediated mechanisms underlie soil carbon accrual by conservation agriculture under decade-long warming. *Nat. Commun.* **2024**, *15*, 377. [CrossRef]

99. Li, N.; Wang, Y.; Wei, L.; Wang, X.; Zhang, Q.; Guo, T.; Xu, X.; Zhao, N.; Xu, S. Variations in microbial residue carbon and its contribution to soil organic carbon after vegetation restoration on farmland: The case of Guinan County. *Org. Geochem.* **2024**, *189*, 104753. [CrossRef]
100. Zhao, J.; Chen, J.; Beillouin, D.; Lambers, H.; Yang, Y.; Smith, P.; Zeng, Z.; Olesen, J.E.; Zang, H. Global systematic review with meta-analysis reveals yield advantage of legume-based rotations and its drivers. *Nat. Commun.* **2022**, *13*, 4926. [CrossRef]
101. He, H.; Liu, L.; Munir, S.; Bashir, N.H.; Wang, Y.; Yang, J.; Li, C.-Y. Crop diversity and pest management in sustainable agriculture. *J. Integr. Agr.* **2019**, *18*, 1945–1952. [CrossRef]
102. Ouda, S.; Zohry, A.; Noreldin, T. (Eds.) Crop Rotation Maintains Soil Sustainability. In *Crop Rotation: An Approach to Secure Future Food*; Springer International Publishing: Cham, Switzerland, 2018; pp. 55–76.
103. See, C.R.; Keller, A.B.; Hobbie, S.E.; Kennedy, P.G.; Weber, P.K.; Pett-Ridge, J. Hyphae move matter and microbes to mineral microsites: Integrating the hyphosphere into conceptual models of soil organic matter stabilization. *Glob. Change Biol.* **2022**, *28*, 2527–2540. [CrossRef] [PubMed]
104. Okebalama, C.B.; Marschner, B. Reapplication of biochar, sewage waste water, and NPK fertilizers affects soil fertility, aggregate stability, and carbon and nitrogen in dry-stable aggregates of semi-arid soil. *Sci. Total Environ.* **2023**, *866*, 161203. [CrossRef] [PubMed]
105. Wang, Y.; Liang, A.; Xu, Y.; Zhang, W.; Li, M.; Wang, J. Maize residue types and soil fertility levels influence sequestration of newly associated carbon in aggregates with in situ experiments. *Soil. Till Res.* **2023**, *233*, 105783. [CrossRef]
106. Zhang, L.; Yuan, J.; Zhang, M.; Zhang, Y; Wang, L.; Li, J. Long term effects of crop rotation and fertilization on crop yield stability in southeast China. *Sci. Rep.* **2022**, *12*, 14234. [CrossRef]
107. Yang, T.; Siddique, K.H.M.; Liu, K. Cropping systems in agriculture and their impact on soil health—A review. *Glob. Ecol. Conserv.* **2020**, *23*, e01118. [CrossRef]
108. Tsai, H.; Lee, Y. Effects of land use change and crop rotation practices on farmland ecosystem service valuation. *Ecol. Indic.* **2023**, *155*, 110998. [CrossRef]

agronomy

MDPI

Article

Impact of Organic Fertilizer Substitution and Chemical Nitrogen Fertilizer Reduction on Soil Enzyme Activity and Microbial Communities in an Apple Orchard

Yuyang Yan [1,2,3,4,†], Xinran Zhang [1,2,†], Yuan Liu [5], Lei Hou [6], Zengchao Geng [1,2], Feinan Hu [3,4] and Chenyang Xu [1,2,*]

1 College of Natural Resources and Environment, Northwest A&F University, Yangling 712100, China; yanyuyang@nwafu.edu.cn (Y.Y.); zhangxinran@nwafu.edu.cn (X.Z.); gzch@nwafu.edu.cn (Z.G.)
2 Key Laboratory of Plant Nutrition and Agri-Environment in Northwest China, Ministry of Agriculture and Rural Affairs, Northwest A&F University, Yangling 712100, China
3 State Key Laboratory of Soil Erosion and Dryland Farming on the Loess Plateau, Northwest A&F University, Yangling 712100, China; hufn@nwafu.edu.cn
4 Institute of Soil and Water Conservation, Chinese Academy of Sciences, Ministry of Water Resources, Yangling 712100, China
5 Farmland Irrigation Research Institute, Chinese Academy of Agricultural Sciences, Xinxiang 453000, China; liuyuan01@caas.cn
6 Tibet Agriculture and Animal Husbandry University, Nyingchi 860000, China; houlei@nefu.edu.cn
* Correspondence: xuchenyang@nwafu.edu.cn
† These authors contributed equally to this work.

Abstract: To mitigate the issues of soil quality degradation and environmental pollution caused by excessive fertilizer use in apple orchards, the present study investigated the effects of organic fertilizer substitution combined with chemical nitrogen (N) fertilizer reduction on soil nutrient status, enzyme activity, and microbial communities (bacteria, fungi and archaea) over one year in an apple orchard. Five fertilization treatments were implemented, including 100% chemical fertilizer (CK), 80% chemical fertilizer + 20% liquid humic fertilizer (S1), 60% chemical fertilizer + 40% liquid humic fertilizer (S2), 60% chemical fertilizer + 20% liquid humic fertilizer (S3), and 40% chemical fertilizer + 40% liquid humic fertilizer (S4). Substituting chemical fertilizers with liquid humic fertilizers effectively enhanced the soil organic matter (SOM) content in the topsoil (0–20 cm) for all treatments. Compared to CK, the amounts of available N (NO_3^--N and NH_4^+-N) were decreased in the topsoil and the amounts of total N, total phosphorous and available phosphorous were increased in the subsoil (20–40 cm) for all treatments. The β-diversity of bacterial communities exhibited the highest sensitivity to soil environmental changes, followed by that of archaea, whereas fungi demonstrated the least susceptibility. The higher soil carbon/nitrogen ratio and SOM content in S2 altered the abundance of microorganisms (*Proteobacteria*, *Ascomycota*, and *Crenarchaeota*) that were closely related to the decomposition and mineralization of SOM and N, enhancing the efficiency of SOM decomposition. The activities of sucrase (SUC), urease (UE), and phosphatase were increased, also promoting the conversion efficiency of SOM and improving N fixation and soil fertility. In the organic fertilizer substitution treatments (S1 and S2), the abundance of dominant *Actinobacteriota*, *Ascomycota* and *Crenarchaeota* phyla were increased, as well as the activities of SUC and UE, accelerating the decomposition and mineralization of SOM and improving soil fertility. In the top, organic fertilizer substitution combined with reduced chemical N fertilizer (S3 and S4) treatments increased the abundance of bacteria and fungi. In addition, RDA showed that total potassium content could significantly affect changes in the bacterial and fungal community structure in subsoil. Overall, organic fertilizer substitution enhanced the content of soil available nutrients and improved soil nutrient retention. It is recommended to promote organic fertilizer substitution + chemical N fertilizer reduction (S4) with the supplementation of potassium fertilizer in the subsoil. The findings provide a theoretical basis and practical guidance for improving orchard soil management and achieving sustainable development in the apple industry.

Keywords: organic fertilizer substitution; N reduction; apple orchard soil; microorganisms

1. Introduction

China is the largest apple producer in the world, with its apple production accounting for nearly half of the global total [1]. However, China's current apple production is confronted with the significant challenge of high input and low output [2], which hampers the sustainable, high-quality development of the apple industry. Studies have shown that the average nitrogen (N) application rate in apple orchards on the Loess Plateau region has reached as high as 1220 kg/hm^2, far exceeding the recommended application rate of 150–250 kg/hm^2, and is 5 to 12 times higher than that in developed apple-producing countries (100–220 kg/hm^2) [3]. The overuse of chemical N fertilizers in apple orchards not only leads to resource inefficiency but also contributes to the degradation of soil structure and the contamination of downstream aquatic ecosystems [4]. Reducing chemical fertilizer use through combining organic and inorganic fertilizers is a viable approach to promoting green and sustainable development in the apple industry.

Organic fertilizer substitution can increase soil organic matter (SOM) content, enhance soil biological activity, and maintain soil productivity [5]. In a crop rotation system of winter wheat (*Triticum aestivum* L.) and summer maize (*Zea mays* L.), a 20% organic fertilizer substitution increased the content of SOM, total nitrogen (TN), and total phosphorus (TP), and increased the abundance of bacterial flora in typical calcareous alluvial soil [6]. Organic fertilizers can provide more carbon and energy material for microorganisms. The application of biogas slurry combined with chemical fertilizer in a 50:50 ratio could modify the apple orchard soil bacterial community structure and increase the relative abundance of beneficial flora (e.g., *Acidobacteria* and *Parasegetibacter*). Meanwhile, it also enhanced functional groups involved in carbon and N cycling, such as *chemoheterotrophy*, *cellulolysis*, and N fixation [7]. A 50% substitution of pig manure for chemical fertilizer increased the abundance of *Clostridiales* and *Bacillales* in peanut soils, and the co-occurrence network structure of soil bacteria became more complex, with a more stable community structure [8]. In a winter wheat–summer maize double-cropping system, long-term 30% pig manure substitution could improve soil fertility, leading to stable crop yield and soil enzyme activity [9].

Chemical N fertilizer reduction could affect the pH and SOM content of the rhizosphere soil in tobacco, affecting the activities of sucrase (SUC), catalase (CAT), urease (UE), and phosphatase (PHO), and altering soil bacterial community dynamics [10]. The substitution of pigeon manure organic fertilizer combined with a 20% reduction in N application significantly increased the SOM content and the activities of CAT and UE in vegetable fields [11]. Compared to bacteria, fungi were more sensitive to changes in soil physicochemical properties caused by N reduction practices, whereas chicken manure substitution of chemical fertilizers could increase the stability of fungal community structure in maize fields [12–14]. Substituting chemical fertilizers with commercial organic fertilizers over a one-year period in an apple orchard had shown that a 40% substitution with organic fertilizer improved apple fruit yield and quality [15]. A 25% reduction in N fertilizer combined with bio-organic fertilizer also decreased soil pH, enhanced soil nutrient levels, and increased red raspberry yield [16]. Overall, organic fertilizer substitution and chemical N fertilizer reduction can increase SOM and total N content, improve soil fertility [17], and enhance the abundance of soil bacteria and fungi, stabilizing the soil microenvironment under extreme conditions such as water and nutrient stress [18]. However, there is limited research on the effects of organic fertilizer substitution and N fertilizer reduction on soil microbial community structure with respect to bacteria, fungi and archaea in orchards.

This study investigated the impact of organic fertilizer substitution and chemical N fertilizer reduction on soil physicochemical properties, nutrient content, extracellular enzyme activities, and the microbial communities of bacteria, fungi, and archaea. The scientific hypothesis is that organic fertilizer substitution and chemical N fertilizer reduction not only enhance the activities of SUC, UE, and PHO, but also promote the transformation of SOM and improve N fixation capacity. Additionally, these changes influence the diversity of soil microorganisms, alter their community structure, and increase the abundance of microorganisms related to the decomposition of SOM and N. The findings offer theoretical insights and practical guidance for optimizing integrated soil management practices in apple orchards, thereby supporting green and sustainable development in the apple industry.

2. Materials and Methods

2.1. Study Area

This study was conducted at the Baishui Apple Experimental Station in Weinan city, Shaanxi Province, China (109°16′–109°45′ E, 35°4′–35°27′ N) (Figure S1), which is located in the transition zone between the Guanzhong Plain and the Northern Shaanxi Plateau, characterized by the typical loess hills and gullies of the Weibei Loess Plateau. This region has a warm temperate continental monsoon climate, with an average annual rainfall of 577 mm and elevation of 1080 m above sea level, an average annual temperature of 9.2 °C, an annual total of 2552 h of sunshine, and a frost-free period of 180 days. The soil type is black loessial soil, based on the Chinese Soil Genetic Classification System, which is developed from loess. The soil texture is classified as clay loam based on the soil texture classification system of United States Department of Agriculture. According to the World Reference Base for Soil Resource [19], black loessial soil is classified as Cambisols.

The Baishui Apple Experimental Station encompasses 10.67 ha of land. Specifically, an area of 420 m^2 had been designated for conducting fertilizer research, with each plot covering 24 m^2. Six-year-old dwarf Ruiyang apple trees were selected as the research material. Healthy, high-quality, and mature trees with consistent growth were chosen. The planting spacing was 1.5 × 4 m. Four trees were used for each treatment, with three replicates per treatment; thus, 60 trees were adopted. The replicate groups were arranged randomly, and buffer rows were set between every two treatments (Figure S2).

2.2. Organic Fertilizer Substitution and N Reduction Treatments

The field fertilization control experiment was initiated from August of 2020, prior to which 59,970 kg/ha of well-decomposed sheep manure had been applied to enhance soil quality. Fertilizer was applied using a trenching method, where trenches were dug on both sides of the apple trees. The trenches were located 40 cm from the tree trunks within the row spacing, with a trench width of 40 cm and a depth of 30 cm. Other agronomic practices followed the standard practices of the experimental station, including weeding, summer pruning, etc. Basal fertilizer was applied in September, bud-promoting fertilizer in April of the following year, and fruit-promoting fertilizer in June, accounting for 30%, 20%, and 50% of the total chemical fertilizer application, respectively.

Five fertilization treatments were implemented, including 100% chemical fertilizer (CK), 80% chemical fertilizer + 20% liquid humic fertilizer (S1), 60% chemical fertilizer + 40% liquid humic fertilizer (S2), 60% chemical fertilizer + 20% liquid humic fertilizer (S3), and 40% chemical fertilizer + 40% liquid humic fertilizer (S4) (Table 1). In the control (CK) treatment, the chemical fertilizer used was a balanced fertilizer containing 14% N, 15% P_2O_5, and 16% K_2O. For S2–S4 treatments, a commercial humic acid fertilizer made from weathered coal was applied, with a pH of 6.5 and N, P_2O_5, and K_2O contents of 14.10%, 1.43%, and 4.68%, respectively. In CK, chemical N fertilizer was applied in the form of balanced fertilizer, so reducing chemical N fertilizer in other treatments also led to a reduction in phosphorus (P) and potassium (K) application, which were supplemented with calcium superphosphate (16% P_2O_5) and potassium sulfate (50% K_2O). Therefore, compared to CK, the total N application amounts in S1 and S2 treatments remained the same, but 20% and 40% of the chemical N fertilizers were replaced with organic fertilizers, respectively. In S3 and S4 treatments, based on the organic fertilizer substitution in S1 and

S2, the total N application amounts were reduced to 80% of the total N application. The application amounts of P and K were the same for all treatments.

Table 1. Different fertilization modes for soils in the apple orchard.

Treatment	Fertilization Mode	Basal Fertilizer kg/ha	Ratio of Chemical and Organic Fertilizer	Application Amounts of Chemical Fertilizer (kg/tree)			Application Amount of Liquid Humic Fertilizer (kg/tree)
		Well-Decomposed Sheep Manure		Balanced Fertilizer	Superphosphate	Potassium Sulfate	
CK	Control	59,970	100% chemical fertilizer	2.5	0	0	0
S1	Organic fertilizer substitution mode	59,970	80% chemical fertilizer + 20% liquid humic fertilizer	2	0.424	0.114	0.496
S2		59,970	60% chemical fertilizer + 40% liquid humic fertilizer	1.5	0.849	0.227	0.994
S3	Organic fertilizer substitution + N reduction mode	59,970	60% chemical fertilizer + 20% liquid humic fertilizer	1.5	0.894	0.274	0.496
S4		59,970	40% chemical fertilizer + 40% liquid humic fertilizer	1	1.318	0.387	0.994

2.3. Soil Sampling

In October 2021, after one year of fertilization, soil samples were collected from the apple orchard. For each treatment group, soil samples were collected at a distance of 1.2 m from the tree trunk along the direction of the fertilization trench. Topsoil (0–20 cm) and subsoil (20–40 cm) were sampled using a stainless steel soil auger. Each treatment included four apple trees, from which eight soil core samples were collected. These samples were thoroughly mixed to form one bulk soil sample. A total of 30 bulk samples, comprising 15 topsoil and 15 subsoil samples, were collected. A portion of the soil was placed in sterile bags, transported in insulated boxes, and stored in an ultra-low temperature freezer at −80 °C for microbial analysis. Another portion was stored in a refrigerator at 4 °C for the determination of soil enzyme activity. The remaining soil was air-dried, sieved, and used for the determination of soil physicochemical properties.

The basic physical and chemical properties of soils were analyzed with standard soil analysis methods [20]. Soil pH was measured at a 1:2.5 soil-to-water ratio using a pH meter (FE28, Mettler Toledo, Zurich, Switzerland). Soil electrical conductivity (EC) was measured at a 1:5 soil-to-water ratio using electrical conductivity (Mettler Toledo, Zurich, Switzerland). The SOM content was quantified through the $K_2Cr_2O_7$ oxidation method. The cation exchange capacity (CEC) of the soil was measured with the exchange method. TN content in the soil was determined using the Kjeldahl method. Both NH_4^+-N and NO_3^--N were extracted with KCl solution, with the resulting extracts analyzed with a continuous flow analyzer (FIA Compact, Berlin, Germany). TP and available phosphorous (AP) levels were measured using the molybdenum blue method. Total potassium (TK) and available potassium (AK) contents were analyzed using a flame photometer (PinAAciie 900F, Waltham, MA, USA). The basic physical and chemical properties reported in the study were the average result of nine measurements. The apple fruit yield under each fertilization treatment was also measured.

2.4. Determination of Soil Enzyme Activity

The activities of four soil extracellular enzymes, CAT, SUC, UE, and PHO, were measured. CAT activity was determined using the K permanganate titration method; SUC activity was measured using the 3,5-dinitrosalicylic acid method; UE activity was determined using the phenol–sodium hypochlorite colorimetric method, and PHO activity was measured using the disodium phenyl phosphate colorimetric method [21].

2.5. Analysis of Soil Microbial Community

High-throughput sequencing was used to analyze the types and abundance of bacteria, fungi, and archaea in the soil [22]. The primers used for bacteria, fungi, and archaea were 341F (CCTAYGGGRBGCASCAG) and 806R (GGACTACNNGGGTATCTAAT), ITS5-1737F (GGAAGTAAAAGTCGTAACAAGG) and ITS2-2043R (GCTGCGTTCTTCATCGATGC), and U519F (CAGYMGCCRCGGKAAHACC) and 806R (GGACTACNSGGGTMTCTAAT), respectively.

Soil samples were processed using a magnetic bead soil genomic DNA extraction kit (Tiangen Biotech Co., Ltd., Beijing, China) to extract genomic DNA. PCR products were detected using 2% agarose gel electrophoresis. Qualified PCR products were purified using magnetic beads and quantified using enzyme labeling. Equal amounts of PCR products were pooled based on their concentrations, thoroughly mixed, and checked using 2% agarose gel electrophoresis. The target bands were recovered using a universal DNA purification recovery kit (Tiangen Biotech Co., Ltd., Beijing, China). Library construction was performed using a NEB Next Ultra II DNA Library Prep Kit (New England Biolabs, Inc., Ipswich, MA, USA), and the constructed library was quantified using Qubit and Q-PCR. Once the library met the quality standards, sequencing was performed on the NovaSeq6000 platform.

The raw data were obtained by matching the barcode sequences and PCR amplification primer sequences from the sequenced data. After removing the barcode and primer sequences, the data were compared with the database to detect and eliminate chimeras, yielding effective data (Effective Tags). The Effective Tags were denoised using the DADA2 module in QIIME2, with sequences with an abundance of less than five being filtered out. The ASV (Amplicon Sequence Variant) sequences were obtained and subjected to species annotation.

2.6. Statistical Analysis

Data analysis was performed using SPSS 22.0 (SPSS Inc., Chicago, IL, USA) and Origin 2021 (OriginLab, Northampton, MA, USA). One-way analysis of variance (ANOVA) was used to compare the significant differences ($p < 0.05$) in soil physicochemical properties, enzyme activities and α-diversity (Pielou, Shannon, and Simpson) between treatments at the same depth. The microbial community β-diversity was compared using the R based *vegan* package. Significant differences ($p < 0.05$) between different treatments were compared using Duncan's test. Pielou index measures the uniformity of species distribution within a community. The Shannon index assigns greater weight to rare species, whereas the Simpson index is more influenced by dominant species [23]. Redundancy analysis (RDA) was employed to explore the correlations between soil microbial communities and soil physicochemical properties and nutrient contents.

3. Results

3.1. Soil Physicochemical Properties and Nutrient Content Under Different Fertilization Treatments

The physicochemical properties and nutrient contents of orchard soil under different fertilization treatments are shown in Figure 1. In the topsoil, the pH value was highest in the CK treatment, while the pH in the S3 treatment was significantly lower than in other treatments. In the subsoil, the pH in all fertilization treatments was significantly lower than in CK (Figure 1a). Compared to CK, the EC (Figure 1b) was increased in the subsoil and the CEC (Figure 1c) in the topsoil was reduced for all treatments. As soil depth increased, SOM content decreased. The SOM content in the topsoil was higher in all treatments compared to CK, with the S1 and S4 treatments having SOM contents of 25.56 g/kg and 25.27 g/kg, respectively, representing increases of 13.6% and 12.3%. In the 20–40 cm subsoil, the SOM content in the CK treatment was lower than those in the S1, S2, and S3 treatments but significantly higher than that in S4 (Figure 1d) The TN, TP, TK, NO_3^--N, AP, and AK contents decreased with increasing soil depth (Figure 1f–k). In the S4 treatment, the NH_4^+-N content in the topsoil was higher than in other treatments (Figure 1e). The S2 treatment significantly increased NO_3^--N and NH_4^+-N contents in the subsoil compared to other treatments (Figure 1e,f). Compared to CK, all treatments except S4 reduced the total nutrient content in the topsoil and increased the TN and TP contents in the subsoil, but TK content was reduced in the subsoil (Figure 1g–k), even though this was not statistically significant. Meanwhile, the apple yield of CK was measured to be 30,071.51 kg/ha, while there was an insignificant decrease of 1.27% for S1, and the yields were increased by 11.84% 1.11% and 7.58% in the S2, S3 and S4 treatments, respectively. Thus, the fertilization treatments applied in this study resulted in negligible reductions in yield.

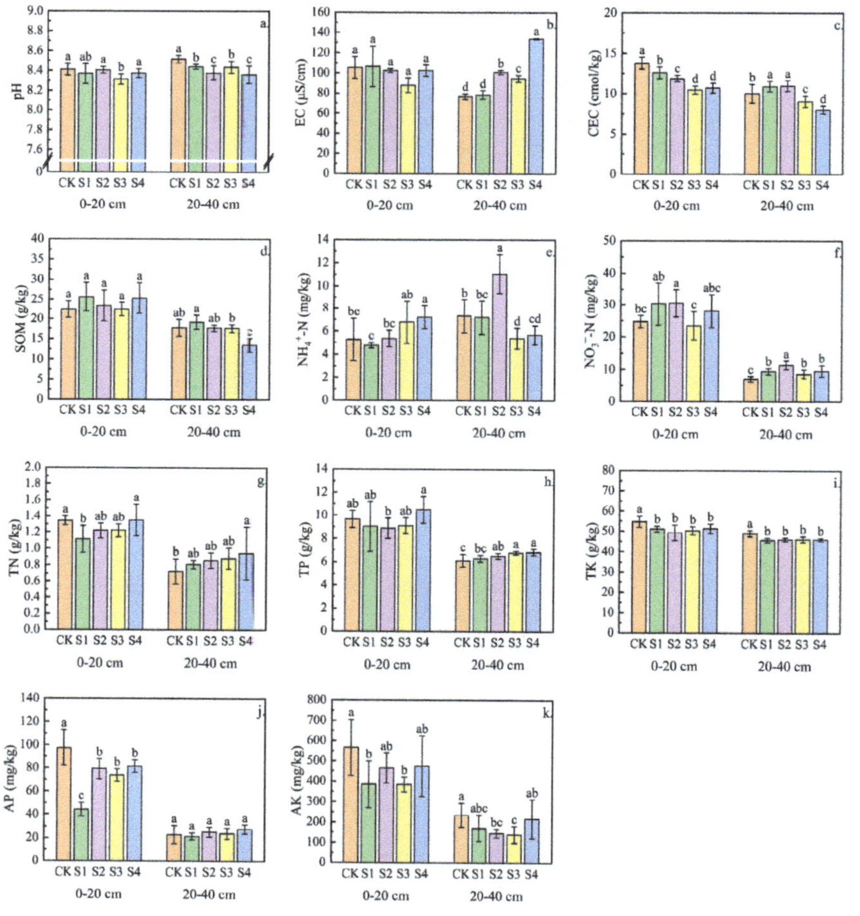

Figure 1. Soil physiochemical properties and nutrient contents under different fertilization treatments. (**a**) pH; (**b**) EC; (**c**) CEC; (**d**) SOM; (**e**) NH_4^+-N; (**f**) NO_3^--N; (**g**) TN; (**h**) TP; (**i**) TK; (**j**) AP; (**k**) AK. Different lowercase letters indicate significant differences among treatments at $p < 0.05$.

3.2. Soil Enzyme Activity Under Different Fertilization Treatments

The effects of different fertilization treatments on soil enzyme activity are shown in Figure 2. As soil depth increased, CAT activity showed an increasing trend, while SUC, UE, and PHO activities showed a decreasing trend. In both the topsoil and subsoil, CAT activities under all fertilization treatments were lower than those in CK. However, SUC activity was higher, especially in the topsoil in the S2 treatment, where it was 78.45% greater than in CK. In the topsoil, UE activity was significantly higher in all fertilization treatments compared to CK. However, no significant differences were detected among the treatments in the subsoil. PHO activity in the topsoil was higher in the S1 treatment compared to CK, while it was significantly lower in the S4 treatment. In the subsoil, PHO activity in the CK treatment was lower than in all other fertilization treatments.

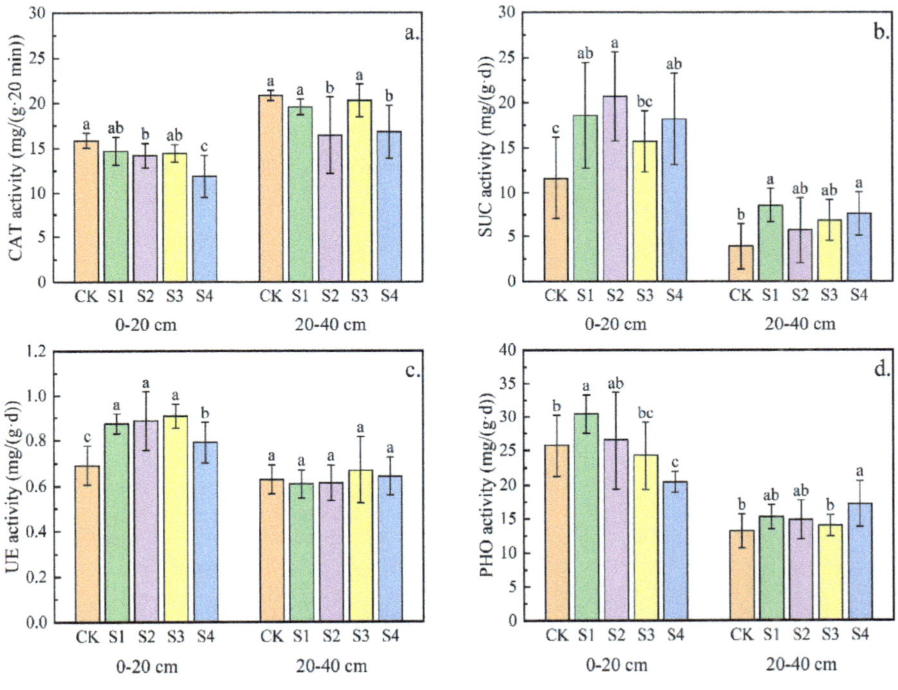

Figure 2. Soil enzymes activity under different fertilization treatments. (**a**) CAT activity; (**b**) SUC activity; (**c**) UE activity; (**d**) PHO activity. Different lowercase letters indicate significant differences among treatments at $p < 0.05$.

The correlation analysis of enzyme activity with soil physicochemical properties and nutrient contents, as shown in Figure 3, indicated that soil CAT activity was significantly negatively correlated with SOM (−0.76), TN (−0.86), NO_3^--N (−0.86), TP (−0.87), AP (−0.75), and AK (−0.73), while it was positively correlated with soil pH. This suggests that higher pH promoted CAT activity. Conversely, SUC, UE, and PHO activities were positively correlated with EC, SOM, TN, NO_3^--N, TP, AP, TK, and AK, but negatively correlated with pH and NH_4^+-N content. Additionally, CAT activity showed negative correlations with SUC (−0.83), UE (−0.71), and PHO (−0.71) activities.

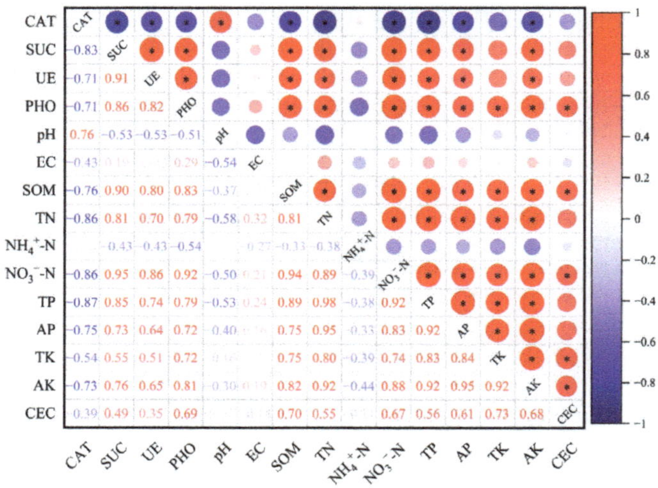

Figure 3. Heatmap of correlations between enzyme activities and soil physicochemical properties and nutrient contents. * indicates significant correlations at $p < 0.05$.

3.3. Soil Microbial Community Structure Under Different Fertilization Treatments

3.3.1. Changes in Soil Bacterial Community Structure

Table S1 presents the effects of different fertilization treatments on the α-diversity of soil microorganisms. α-diversity reflects the diversity of the soil microbial community within a specific habitat. In the topsoil, the Pielou, Shannon, and Simpson indices were higher in all fertilization treatments compared to CK, indicating that organic fertilizer substitution and N reduction increased the evenness and diversity of the soil bacterial community in the topsoil. In the subsoil, the Pielou indices were decreased in all fertilization treatments compared to CK, while the Shannon indices increased in the S1 and S3 treatments, and the Simpson indices showed no significant changes. These differences were not statistically significant, suggesting that organic fertilizer substitution and N reduction had a minimal impact on the bacterial community structure in the subsoil.

The β-diversity analysis of the soil bacterial community structure is shown in Figure 4a. β-diversity effectively measures the variation in species composition within a habitat. The first principal component (PC1) and the second principal component (PC2) explained 6.96% and 4.30% of the total variability, respectively. In the topsoil, there were differences between the S1 and S2 treatments, revealing that a high proportion of organic fertilizer substitution significantly influenced the species composition of the soil bacterial community. There were also significant differences between S2 and S4, indicating that N reduction significantly affected the species composition of the soil bacterial community. In the subsoil, CK formed a distinct group compared to the other treatments, demonstrating that all treatments significantly altered the species composition of the bacterial community in the subsoil.

Figure 5a shows the relative abundance of bacteria under different fertilization treatments. In the topsoil, *Proteobacteria* was the most abundant bacterial phylum, with an average proportion of 33.38%. In addition, *Actinobacteriota*, *Acidobacteriota*, and *Bacteroidota* also had relatively high abundance. Compared to CK, a higher proportion of organic substitution (S2) increased the abundance of *Proteobacteria* and *Actinobacteriota*, while a lower proportion of organic substitution (S1) reduced the abundance of *Proteobacteria*, *Acidobacteriota*, and *Bacteroidota*. In the topsoil, all treatments increased the abundance of *Actinobacteriota* while decreasing the abundance of *Proteobacteria* (except S2), *Acidobacteriota*, and *Bacteroidota*. In the subsoil, the most abundant bacterial phyla were *Proteobacteria*, *Acidobacteriota*, and *Actinobacteriota*. Compared to CK, all treatments reduced the abundance of *Actinobacteriota* in the subsoil, while increasing the relative abundance of *Proteobacteria*, *Acidobacteriota*, and *Gemmatimonadota*.

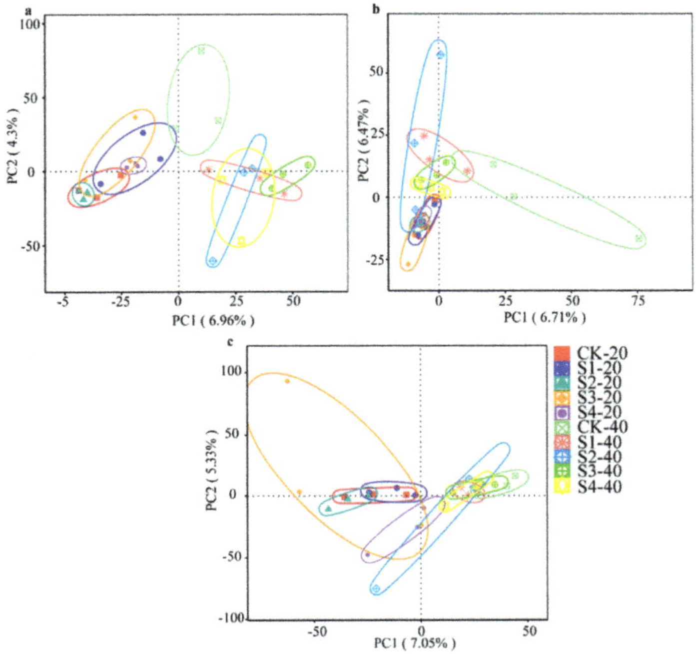

Figure 4. Principal component analysis plot of microbial community composition. (**a**) bacteria; (**b**) fungi; (**c**) archaea. The numbers 20 and 40 represent 0–20 and 20–40 cm, respectively.

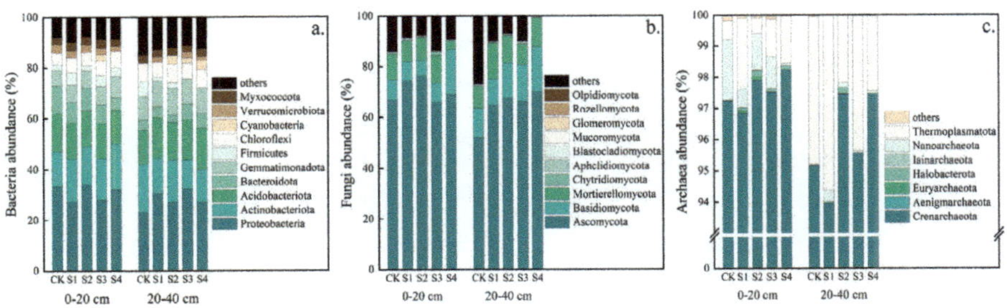

Figure 5. Relative abundance of soil microorganisms under different fertilization treatments. (**a**) bacteria; (**b**) fungi; (**c**) archaea.

Linear Discriminant Analysis Effect Size (LEfSe) analysis is a method for discovering and interpreting high-dimensional biomarkers, aiding in the identification of microbial taxa with notable differences across various samples or treatments [24]. Due to significant disparities in β-diversity between the S2 and S4 treatments, LEfSe analysis identified that the bacterial taxa differed notably among treatments (Figure S3). Significant discrepancies in microbial phyla were observed in the CK treatment, where *Bacteroidota* was predominant in the topsoil, while *Actinobacteriota* was more abundant in the subsoil. In the S2 treatment, significant differences were noted with *Proteobacteria* in the topsoil and *Cyanobacteria* in the subsoil. Figure 6a shows the redundancy analysis (RDA) of soil physicochemical properties and nutrient contents within the bacterial community structure. In the topsoil, the main factors influencing the bacterial community structure were SOM, TP, and pH. In the subsoil, the primary factors driving changes in the bacterial community structure were soil pH, TN, TP, and TK.

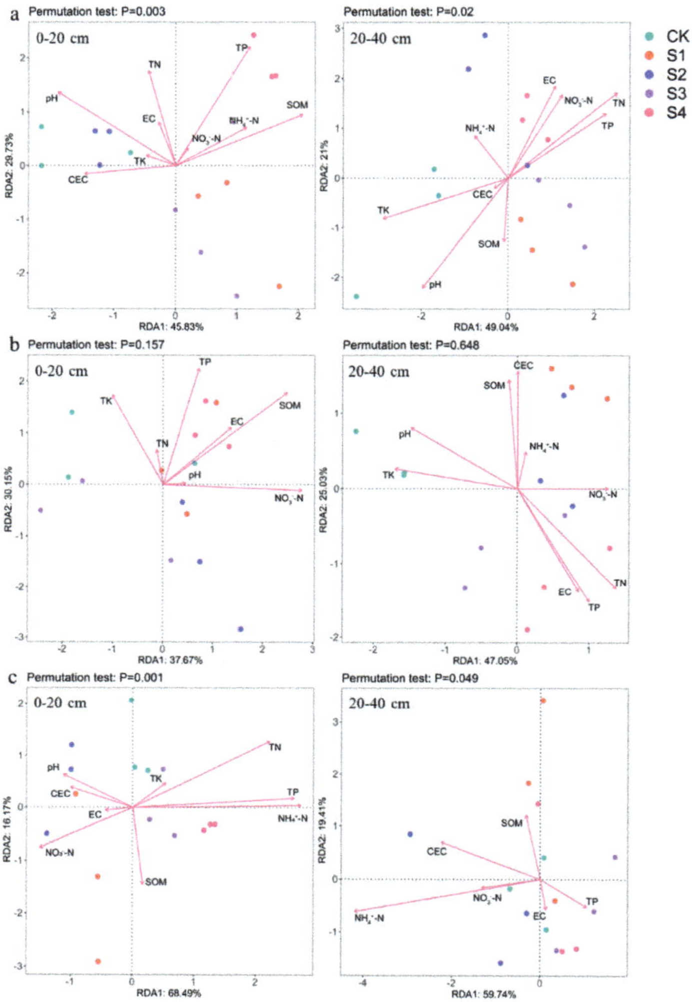

Figure 6. Redundancy analysis of microbial communities in relation to soil physicochemical properties and nutrient contents. (**a**) bacteria; (**b**) fungi; (**c**) archaea.

3.3.2. Changes in Soil Fungal Community Structure

In the topsoil, compared to CK, the Pielou, Shannon, and Simpson indices were decreased in the S2 and S4 treatments, indicating that a high proportion of organic fertilizer substitution reduced fungal community diversity. In contrast, the S1 and S3 treatments had little effect on the Pielou and Simpson indices but increased the Shannon index, suggesting that a low proportion of organic fertilizer substitution enhanced fungal community diversity (Table S1). In the subsoil, the Pielou, Shannon, and Simpson indices in the S3 and S4 treatments were lower than in CK, indicating that organic fertilizer substitution and N reduction reduced fungal community diversity in the subsoil (Table S1).

The β-diversity analysis of the fungal community structure is shown in Figure 4b. PC1 explained 6.71% of the total variation, and PC2 explained 6.47% of the total variation.

Significant differences in fungal communities were observed between the topsoil and subsoil, except for treatment S2, while the differences between fungal communities among treatments in the topsoil and subsoil were relatively small, suggesting that organic fertilizer substitution and N reduction had a minimal impact on changes in fungal composition.

Figure 5b shows the relative abundance of fungi in different fertilization treatments. Among the fungal phyla, the relative abundance of *Ascomycota* reached 50% in all treatments, followed by *Basidiomycota* and *Mortierellomycota*. In the topsoil, organic fertilizer substitution treatments (S1 and S2) increased the relative abundance of *Ascomycota* compared to CK. In the subsoil, the relative abundance of *Ascomycota* in all treatments was higher than in CK. RDA (Figure 6b) revealed that SOM and TP were the main factors influencing fungal community structure in the topsoil, whereas soil pH, TN, and TK were the primary factors affecting fungal community structure in the subsoil.

3.3.3. Changes in Soil Archaeal Community Structure

Compared to CK, the S3 and S4 treatments significantly increased the Pielou, Shannon, and Simpson indices in the topsoil, indicating that organic fertilizer substitution and N reduction increased the evenness and diversity of the archaeal community in the topsoil. In contrast, the indices decreased in the subsoil across all treatments (Table S1). The β-diversity analysis of the archaeal community structure is shown in Figure 4c. PC1 explained 7.05% of the total variation, and PC2 explained 5.33%. The results revealed significant differences in archaeal communities between the topsoil and subsoil (except for the S4 treatment). In the topsoil, the S4 treatment formed a distinct group compared to the S1 and S2 treatments, indicating that a high proportion of organic substitution and N reduction significantly affected the β-diversity of the archaeal community structure. In the subsoil, differences in archaeal community structure among treatments were smaller, suggesting that organic fertilizer substitution and N reduction had a minimal impact on changes in archaeal composition in the subsoil.

Figure 5c shows the relative abundance of archaea under different fertilization treatments. A total of eight archaeal phyla were detected, with *Crenarchaeota* being the dominant phylum, accounting for over 90% of the relative abundance in all treatments, followed by *Thermoplasmatota*. In the topsoil, *Crenarchaeota* abundance was higher in the S2 and S4 treatments compared to CK. The relative abundance of *Nanoarchaeota* was higher in the CK and S2 treatments. In the subsoil, *Crenarchaeota* abundance was higher in the S2 and S4 treatments compared to CK, and organic fertilizer substitution treatments (S1 and S2) increased the abundance of *Nanoarchaeota*. RDA analysis (Figure 6c) indicated that soil TP, TN, and $NH_4{}^+$-N were the main factors influencing changes in the archaeal community structure in the topsoil, while $NH_4{}^+$-N was the main factor driving changes in the subsoil archaeal community structure.

4. Discussion

4.1. Effects of Different Fertilization Treatments on Soil Physicochemical Properties and Enzyme Activities

The reduction in soil pH could primarily be attributed to the application of humic fertilizer, which had a lower pH (6.5) than both the topsoil (pH 8.4) and the subsoil (pH 8.5) [25]. Additionally, the nitrification process, known to produce H^+, further exacerbated this reduction in pH [26]. The SOM content in the topsoil was higher in all treatments compared to CK, indicating that the partial substitution of chemical fertilizer with humic acid fertilizer increased SOM content, consistent with previous research [27,28]. The decreased SOM content in the subsoil in the S2, S3, and S4 treatments could be attributed to the increased C/N ratio resulting from both high organic fertilizer substitution and N reduction, which enhanced microbial mineralization and decomposition of SOM [29]. Organic fertilizer substitution (S1 and S2) improved CEC in the subsoil, enhancing its buffering capacity and nutrient retention ability. The S4 treatment was most effective in increasing subsoil TN and TP contents. The reductions in TK and AK contents were caused by enhanced microbial activity in this treatment, which led to the utilization of some K in microbial metabolism [30]. The reduction in AK content may also be attributed to the application of organic fertilizer converting some free K^+ into adsorbed K^+ [31]. Figure 1 suggests that liquid humic fertilizer may primarily infiltrate into the subsoil, where apple roots are most densely distributed [32], benefiting apple production.

The CAT activity was inhibited in different fertilization treatments, likely because the humic acid derived from weathered coal could reduce CAT activity. On the other hand, these treatments increased SUC and UE activities, which is consistent with previous research [33]. This can be explained by the fact that organic fertilizer increased SOM content, root exudates, and microbial activity, thereby enhancing soil enzyme activity. Additionally, under N reduction conditions, the reduced $NH_4{}^+$-N and $NO_3{}^-$-N contents in the soil accelerated microbial decomposition, further increasing UE activity. The application of organic fertilizer also enhanced the PHO activity of the subsoil, primarily alkaline PHO, and its activity was negatively correlated with pH, in line with previous research findings [34].

4.2. Effects of Different Fertilization Treatments on Changes in the Soil Microbial Community Structure

Organic fertilizer substitution and N reduction treatments increased the richness and diversity of the soil bacterial community, particularly in the subsoil, where organic fertilizer substitution significantly altered bacterial community structure. In the topsoil, *Actinobacteriota*, *Acidobacteriota*, and *Cyanobacteria* were the key contributors to bacterial community differences, while in the subsoil, *Proteobacteria* played a dominant role. *Cyanobacteria* participated in N fixation, whereas *Actinobacteriota*, *Acidobacteriota*, and *Proteobacteria* were involved in SOM decomposition in the soil [35]. *Actinobacteriota*, *Proteobacteria*, and *Bacteroidota* are copiotrophic *r*-strategists that thrive in high-fertility soils, mainly consuming unstable soil carbon pools; in contrast, *Acidobacteriota* are oligotrophic *k*-strategists [36–38]. *Proteobacteria*, *Acidobacteriota*, and *Bacteroidota* were highly sensitive to changes in soil chemical properties [39]. The S2 treatment increased the abundance of *Proteobacteria* in the topsoil, indicating that the application of humic acid fertilizer could enhance SOM content. The S3 treatment showed higher bacterial diversity in the subsoil, suggesting that the combined application of humic acid fertilizer and reduced N improves the stability of the subsoil bacterial community structure.

Fungal community diversity was increased in the topsoil, while significant differences in fungal community structure were observed in the subsoil under high organic fertilizer substitution treatments (S2 and S4), suggesting that fungal community structure was influenced by SOM and N content. P is a crucial nutrient for both bacterial and fungal growth, directly affecting the composition and function of the soil microbial community. Across all treatments, the dominant fungal phyla were *Ascomycota*, *Basidiomycota*, and *Mortierellomycota*, consistent with findings in other alkaline soils [40]. *Ascomycota* were organic decomposers and typically increased with the application of organic fertilizers.

The organic fertilizer substitution combined with N reduction treatment in the topsoil increased the richness, evenness, and diversity of the archaeal community, while archaeal diversity was decreased in the subsoil across all treatments. Fertilization treatments had a significant impact on the diversity of archaea, with high community similarity likely caused by the limited diversity of archaea species and the dominance of *Crenarchaeota*. In both the topsoil and subsoil, the S2 treatment increased the abundance of *Crenarchaeota* and *Nanoarchaeota*. *Crenarchaeota* participated in the decomposition of SOM, breaking down proteins and glucose, while *Nanoarchaeota* decomposed glycogen into glucose [41]. Current research on the ecological roles of archaea is still limited, and further studies are needed to understand their function in complex soil ecosystems.

5. Conclusions

Organic fertilizer substitution effectively increased SOM content in the topsoil and nutrient levels in the subsoil. The β-diversity of bacterial communities showed the greatest sensitivity to soil environmental changes, followed by archaea, while fungi were the least susceptible. All treatments significantly affected the α-diversity of soil bacteria or fungi. S2 increased the soil C/N ratio and SOM content, altering the abundance of microorganisms closely related to the decomposition and mineralization of SOM and N (such as *Proteobacteria, Ascomycota,* and *Crenarchaeota*), and improving the efficiency of SOM decomposition. Soil enzyme activity data supported these findings, showing that the subsoil enzyme activities of SUC, UE, and PHO increased, promoting the conversion of SOM, enhancing N fixation capacity, and improving soil fertility. In the subsoil, the combination of organic fertilizer substitution and N reduction (S3 and S4) increased the abundance of bacteria and fungi. Additionally, RDA showed that microbial abundance was highly correlated with TK content. Therefore, when promoting organic fertilizer substitution combined with N reduction, K supplementation in the subsoil should be considered. Future research should focus on the long-term effects of organic fertilizer substitution and N reduction on apple orchard soil quality and its application to diverse soil types, sustaining green development in the global apple industry.

Supplementary Materials: The following supporting information can be downloaded at: https://www.mdpi.com/article/10.3390/agronomy14122917/s1, Table S1. The α-diversity of soil microorganisms in the apple orchard. Figure S1. Location of the study area. Figure S2. Design of the study plots. Figure S3. Microbiological LEfSe analysis (a. 0–20 cm, b. 20–40 cm).

Author Contributions: Conceptualization, Y.Y., X.Z., C.X., Y.L., L.H., Z.G. and F.H.; methodology, C.X.; software, Y.Y.; formal analysis, X.Z.; investigation, X.Z.; resources, Y.Y.; writing—original draft, Y.Y.; writing—review and editing, C.X. and F.H.; visualization, C.X. and Y.Y.; funding acquisition, C.X., Z.G. and F.H. All authors have read and agreed to the published version of the manuscript.

Funding: This work was supported by Natural Science Foundation of Shaanxi Province (2023-JC-YB-263), the National Natural Science Foundation of China (42277311), the Shaanxi Provincial Agricultural Science and Technology Innovation Program Project (NYKJ-2022-YL(XN)49), and the Tibet Agriculture and Animal Husbandry University–Northwest A&F University Joint Project (XNLH2022-04).

Data Availability Statement: The authors state that all data generated or analyzed during this study are included in this article.

Conflicts of Interest: The authors declare that they have no known competing financial interests or personal relationships that could have appeared to influence the work reported in this paper.

References

1. Jiang, S.; Yang, C.; Guo, Y.; Jiao, X.Q. Integrated horticultural practices for improving apple supply chain sustainability: A case study in the North China Plain. *Agronomy* **2021**, *11*, 1975. [CrossRef]
2. Wang, N.; Wolf, J.; Zhang, F.S. Towards sustainable intensification of apple production in China—Yield gaps and nutrient use efficiency in apple farming systems. *J. Integr. Agric.* **2016**, *15*, 716–725. [CrossRef]
3. Glover, J.D.; Reganold, J.P.; Andrews, P.K. Systematic method for rating soil quality of conventional, organic, and integrated apple orchards in Washington State. *Agr. Ecosyst. Environ.* **2000**, *80*, 29–45. [CrossRef]
4. Wan, L.; Zhao, Y.M.; Xia, L.D.; Hu, J.; Xue, T.X.; Lv, H.F.; Yao, Z.S.; Meng, F.Q.; Li, G.Y.; Butterbach-Bahl, K. Assessing the environmental sustainability of different soil disinfestation methods used in solar greenhouse vegetable production systems. *Sci. Total Environ.* **2023**, *885*, 163962. [CrossRef]
5. Tian, S.Y.; Zhu, B.J.; Yin, R.; Wang, M.W.; Jiang, Y.J.; Zhao, C.Z.; Li, D.M.; Chen, X.Y.; Kardol, P.; Liu, M.Q. Organic fertilization promotes crop productivity through changes in soil aggregation. *Soil Biol. Biochem.* **2022**, *165*, 108533. [CrossRef]
6. Wang, Y.J.; Xu, Y.; Jiang, L.H.; Yang, Y.; Guan, X.L.; Sun, T.; Zhao, H.Y.; Wang, Y.F.; Liu, Y.M. Effect of Mild Organic Substitution on Soil Quality and Microbial Community. *Agronomy* **2021**, *14*, 888. [CrossRef]
7. Zhang, H.; Ma, Y.; Shao, J.Z.; Di, R.; Zhu, F.; Yang, Z.C.; Sun, J.S.; Zhang, X.Y.; Zheng, C.Y. Changes in soil bacterial community and functions by substituting chemical fertilizer with biogas slurry in an apple orchard. *Front. Plant Sci.* **2022**, *13*, 1013184. [CrossRef]
8. Ren, T.B.; Feng, H.L.; Xu, C.S.; Xu, Q.; Fu, B.; Azwar, E.; Wei, Y.W.; Lam, S.S.; Liu, G.S. Exogenous application and interaction of biochar with environmental factors for improving functional diversity of rhizosphere's microbial community and health. *Chemosphere* **2022**, *294*, 133710. [CrossRef]

9. Ma, L.; Li, Z.S.; Li, Y.; Wei, J.L.; Zhang, L.F.; Zheng, F.L.; Liu, Z.H.; Tan, D.S. Variations in crop yield caused by different ratios of organic substitution are closely related to microbial ecological clusters in a fluvo-aquic soil. *Field Crops Res.* **2024**, *306*, 109239. [CrossRef]
10. Shen, M.C.; Zhang, Y.Z.; Bo, G.D.; Yang, B.B.; Wang, P.; Ding, Z.G.; Wang, Z.B.; Yang, J.M.; Zhang, P.; Yuan, X.L. Microbial responses to the reduction of chemical fertilizers in the rhizosphere soil of flue-cured tobacco. *Front. Bioeng. Biotechnol.* **2022**, *9*, 812316. [CrossRef]
11. Ning, C.C.; Gao, P.; Wang, B.Q.; Lin, W.P.; Jiang, N.H.; Cai, K. Impacts of chemical fertilizer reduction and organic amendments supplementation on soil nutrient, enzyme activity and heavy metal content. *J. Integr. Agric.* **2017**, *16*, 1819–1831. [CrossRef]
12. Liu, Z.P.; Zhou, H.P.; Xie, W.Y.; Yang, Z.X.; Zhang, P.F. Effects of returning different organic materials in combination with inorganic fertilizers on the diversity of eukaryotic microorganisms in semi-arid Northern China. *Agronomy* **2022**, *12*, 3116. [CrossRef]
13. Jin, N.; Jin, L.; Wang, S.Y.; Li, J.W.; Liu, F.H.; Liu, Z.C.; Luo, S.L.; Wu, Y.; Lyu, J.; Yu, J.H. Reduced chemical fertilizer combined with bio-organic fertilizer affects the soil microbial community and yield and quality of lettuce. *Front. Microbiol.* **2022**, *13*, 863325. [CrossRef] [PubMed]
14. Ren, J.H.; Liu, X.L.; Yang, W.P.; Yang, X.X.; Li, W.G.; Xia, Q.; Li, J.H.; Gao, Z.Q.; Yang, Z.P. Rhizosphere soil properties, microbial community, and enzyme activities: Short-term responses to partial substitution of chemical fertilizer with organic manure. *J. Environ. Manage.* **2021**, *299*, 113650. [CrossRef] [PubMed]
15. Li, Q.; Chen, Y.A.; Zhu, J.D.; Liu, L.Z.; Liu, J.; Cheng, C.Z.; Li, L. Effects of organic substitution on the yield and quality of apples and residual nitrate-N leaching in soil. *Agronomy* **2024**, *14*, 415. [CrossRef]
16. Yuan, X.; Zhang, J.A.; Chang, F.Y.; Wang, X.Y.; Zhang, X.M.; Luan, H.A.; Qi, G.H.; Guo, S.P. Effects of nitrogen reduction combined with bio-organic fertilizer on soil bacterial community diversity of red raspberry orchard. *PLoS ONE* **2023**, *7*, e0283718. [CrossRef] [PubMed]
17. Zhu, Z.L.; Bai, Y.; Lv, M.L.; Tian, G.; Zhang, X.; Li, l.; Jiang, Y.M.; Ge, S.F. Soil Fertility, microbial biomass, and microbial functional diversity responses to four years fertilization in an apple orchard in North China. *Hortic. Plant J.* **2020**, *4*, 223–230. [CrossRef]
18. Ma, Y.J.; Shen, S.Z.; Wan, C.; Wang, S.Q.; Yang, F.X.; Zhang, K.Q.; Gao, W.X. Organic fertilizer substitution over six years improves the productivity of garlic, bacterial diversity, and microbial communities network complexity. *Appl. Soil Ecol.* **2023**, *182*, 104718. [CrossRef]
19. IUSS Working Group WRB. *World Reference Base for Soil Resources*; International Soil Classification System for Naming Soils and Creating Legends for Soil Maps; International Union of Soil Sciences (IUSS): Vienna, Austria, 2022.
20. Pansu, M.; Gautheyrou, J. *Handbook of Soil Analysis: Mineralogical, Organic and Inorganic Methods*; Springer: Berlin/Heidelberg, Germany, 2006.
21. Xiao, Q.Q.; He, B.P.; Wang, S. Effect of the different fertilization treatments application on paddy soil enzyme activities and bacterial community composition. *Agronomy* **2023**, *13*, 712. [CrossRef]
22. Lin, Y.X.; Ye, G.P.; Kuzyakov, Y.; Liu, D.Y.; Fan, J.B.; Ding, W.X. Long-term manure application increases soil organic matter and aggregation, and alters microbial community structure and keystone taxa. *Soil Biol. Biochem.* **2019**, *134*, 187–196. [CrossRef]
23. Wang, C.; Liu, D.W.; Bai, E. Decreasing soil microbial diversity is associated with decreasing microbial biomass under nitrogen addition. *Soil Biol. Biochem.* **2018**, *120*, 126–133. [CrossRef]
24. Lin, H.; Peddada, S.D. Analysis of microbial compositions: A review of normalization and differential abundance analysis. *Npj Biofilms Microbiomes* **2020**, *6*, 60. [CrossRef] [PubMed]
25. Kai, T.; Adhikari, D. Effect of organic and chemical fertilizer application on apple nutrient content and orchard soil condition. *Agriculture* **2021**, *11*, 340. [CrossRef]
26. Schroder, J.L.; Zhang, H.L.; Girma, K.; Raun, W.R.; Penn, C.J. Soil acidification from long-term use of nitrogen fertilizers on winter wheat. *Soil Sci. Soc. Am. J.* **2011**, *75*, 957–964. [CrossRef]
27. Shang, L.R.; Wan, L.Q.; Zhou, X.X.; Li, S.; Li, X.L. Effects of organic fertilizer on soil nutrient status, enzyme activity, and bacterial community diversity in Leymus chinensis steppe in Inner Mongolia, China. *PLoS ONE* **2020**, *15*, e0240559. [CrossRef]
28. Chen, Y.; Zhang, X.D.; He, H.B.; Xie, H.T.; Yan, Y.; Zhu, P.; Ren, J.; Wang, L.C. Carbon and nitrogen pools in different aggregates of a Chinese Mollisol as influenced by long-term fertilization. *J. Soils Sediments* **2010**, *10*, 1018–1026. [CrossRef]
29. Pinggera, J.; Geisseler, D.; Piepho, H.P.; Joergensen, R.G.; Ludwig, B. Effect of substrate quality on the N uptake routes of soil microorganisms in different soil depths. *Pedobiologia* **2015**, *58*, 211–218. [CrossRef]
30. Philippot, L.; Chenu, C.; Kappler, A.; Rillig, M.C.; Fierer, N. The interplay between microbial communities and soil properties *Nat. Rev. Microbiol.* **2024**, *22*, 226–239. [CrossRef]
31. Soumare, A.; Sarr, D.; Diedhiou, A.G. Potassium sources, microorganisms and plant nutrition: Challenges and future research directions. *Pedosphere* **2023**, *33*, 105–115. [CrossRef]
32. Zhao, Y.G.; Ma, J.J.; Sun, X.H.; Guo, X.H. Spatial distribution of soil moisture and fine roots of apple trees under water storage pit irrigation. *J. Irrig. Drain. Eng.* **2014**, *140*, 06013001. [CrossRef]
33. Yang, Y.H.; Li, M.J.; Wu, J.C.; Pan, X.Y.; Gao, C.M.; Tang, D.W.S. Impact of combining long-term subsoiling and organic fertilizer on soil microbial biomass carbon and N, soil enzyme activity, and water use of winter wheat. *Front. Plant Sci.* **2022**, *12*, 788651. [CrossRef] [PubMed]

34. Zhang, G.N.; Chen, Z.H.; Zhang, A.M.; Chen, L.J.; Wu, Z.J.; Ma, X.Z. Phosphorus composition and phosphatase activities in soils affected by long-term application of pig manure and inorganic fertilizers. *Commun. Soil Sci. Plant Anal.* **2014**, *45*, 1866–1876. [CrossRef]
35. Li, Y.L.; Tremblay, J.; Bainard, L.D.; Cade-Menun, B.; Hamel, C. Long-term effects of nitrogen and phosphorus fertilization on soil microbial community structure and function under continuous wheat production. *Environ. Microbiol.* **2020**, *22*, 1066–1088. [CrossRef] [PubMed]
36. Luo, J.P.; Liao, G.C.; Banerjee, S.; Gu, S.H.; Liang, J.B.; Guo, X.Y.; Zhao, H.P.; Liang, Y.C.; Li, T.Q. Long-term organic fertilization promotes the resilience of soil multifunctionality driven by bacterial communities. *Soil Biol. Biochem.* **2023**, *177*, 108922. [CrossRef]
37. Gao, R.P.; Duan, Y.; Zhang, J.; Ren, Y.F.; Li, H.C.; Liu, X.Y.; Zhao, P.Y.; Jing, Y.P. Effects of long-term application of organic manure and chemical fertilizer on soil properties and microbial communities in the agro-pastoral ecotone of North China. *Front. Environ. Sci.* **2022**, *10*, 993973. [CrossRef]
38. Ma, T.F.; He, X.H.; Chen, S.G.; Li, Y.J.; Huang, Q.W.; Xue, C.; Shen, Q.R. Long-term organic-inorganic fertilization regimes alter bacterial and fungal communities and rice yields in paddy soil. *Front. Microbiol.* **2022**, *13*, 890712. [CrossRef]
39. Shen, J.P.; Zhang, L.M.; Guo, J.F.; Ray, J.L.; He, J.Z. Impact of long-term fertilization practices on the abundance and composition of soil bacterial communities in Northeast China. *Appl. Soil Ecol.* **2010**, *46*, 119–124. [CrossRef]
40. Hua, L.Q.; Yang, S.Q.; Xia, Z.F.; Zeng, H. Application of sophora alopecuroides organic fertilizer changes the rhizosphere microbial community structure of melon plants and increases the fruit sugar content. *J. Sci. Food Agric.* **2023**, *103*, 164–175. [CrossRef]
41. Lv, X.F.; Yu, J.B.; Fu, Y.Q.; Ma, B.; Qu, F.Z.; Ning, K.; Wu, H.F. A meta-analysis of the bacterial and archaeal diversity observed in wetland soils. *Sci. World J.* **2014**, *2014*, 437684. [CrossRef]

agronomy

MDPI

Article

Leguminous Plants and Microbial Inoculation: An Approach for Biocatalytic Phytoremediation of Tebuthiuron in Agricultural Soil

Victor Hugo Cruz [1], Bruno Rafael de Almeida Moreira [2], Thalia Silva Valério [1], Yanca Araujo Frias [1], Vinícius Luiz da Silva [1], Eduardo Beraldo de Morais [3], Leonardo Gomes de Vasconcelos [3], Leandro Tropaldi [1], Evandro Pereira Prado [1], Renato Nallin Montagnolli [4] and Paulo Renato Matos Lopes [1,*]

[1] Department of Plant Production, College of Agricultural and Technological Sciences, São Paulo State University (Unesp), Dracena 17900-000, São Paulo, Brazil; hugo.cruz@unesp.br (V.H.C.); thalia.valerio@unesp.br (T.S.V.); yanca.frias@unesp.br (Y.A.F.); viniciusluizdasilva@gmail.com (V.L.d.S.); l.tropaldi@unesp.br (L.T.); evandro.prado@unesp.br (E.P.P.)

[2] Center of Crop Science, Queensland Alliance of Agriculture and Food Innovation (QAAFI), The University of Queensland, Brisbane QLD 4102, Australia; b.moreira@uq.edu.au

[3] Department of Chemistry, Institute of Exact and Earth Sciences, Federal University of Mato Grosso, Cuiabá 78060-900, Mato Grosso, Brazil; ebmorais@ufmt.br (E.B.d.M.); leonardo.vasconcelos@ufmt.br (L.G.d.V.)

[4] Department of Natural Sciences, Mathematics and Education, Center for Agricultural Sciences, Federal University of São Carlos, Araras 13600-000, São Paulo, Brazil; renatonm@ufscar.br

* Correspondence: prm.lopes@unesp.br

Abstract: Herbicides are important for weed control but can severely impact ecosystems, causing soil and water contamination, biodiversity loss, and harm to non-target organisms. Tebuthiuron, widely used in sugarcane cultivation, is highly soluble and persistent, posing significant environmental risks. Microbial inoculation has emerged as a sustainable strategy to mitigate such damage. This study investigated the phytoremediation potential of *Mucuna pruriens* and *Canavalia ensiformis* in tebuthiuron-contaminated soils, enhanced by fungal and bacterial inoculants. *Crotalaria juncea* served as a bioindicator plant, and *Lactuca sativa* was used in ecotoxicological bioassays. During a 140-day greenhouse experiment from September 2021 to March 2022, *M. pruriens* showed faster growth than *C. ensiformis* in uncontaminated soils but was more affected by tebuthiuron. Bacterial inoculants improved *M. pruriens* growth under stress, while fungal inoculants mitigated tebuthiuron's effects on *C. ensiformis*. *C. juncea* exhibited high sensitivity to tebuthiuron but grew beyond 100 cm with bacterial inoculants. Ecotoxicological assays showed that bacterial bioaugmentation significantly reduced soil toxicity. Natural attenuation further decreased tebuthiuron toxicity, and prior cultivation of *M. pruriens* enhanced soil detoxification. This integrated approach combining phytoremediation and bioaugmentation offers a sustainable method to degrade tebuthiuron, foster safer agriculture, and reduce environmental and health risks.

Keywords: bioaugmentation; bioremediation; *Canavalia ensiformis*; *Crotalaria juncea*; ecotoxicity; herbicide degradation; *Lactuca sativa*; *Mucuna pruriens*; sustainable agriculture

1. Introduction

The herbicide tebuthiuron raises significant environmental concerns due to its high water solubility (2.5 g L^{-1} at 25 °C) [1], prolonged soil persistence (log K_ow = 1.8) [2], and potential to contaminate terrestrial and aquatic ecosystems [3]. Given the increasing evidence of tebuthiuron's detrimental environmental impacts, there is an urgent need for innovative and sustainable remediation strategies. Phytoremediation has emerged as a promising approach, utilizing plants to detoxify environments contaminated by various pollutants, including heavy metals, hydrocarbons, dyes, and pesticides [4].

Our study investigates the tolerance and phytoremediation potential of two leguminous species, *Mucuna pruriens* (L.) DC. var. *pruriens* and *Canavalia ensiformis* L., in combination with microbial inoculants. These annual plants not only have the potential to degrade tebuthiuron but also fix atmospheric nitrogen, enriching the soil and improving its physical, chemical, and biological properties [5]. Previous studies have indicated that these species exhibit resilience to tebuthiuron, making them promising candidates for remediating contaminated environments.

Several studies have evaluated the potential of leguminous species for phytoremediation of tebuthiuron-contaminated soils. For instance, Mendes et al. [6] assessed the use of *Crotalaria spectabilis, C. ensiformis, Stizolobium aterrimum*, and *Lupinus albus* in soils treated with quinclorac and tebuthiuron. They found that all four species absorbed more tebuthiuron than quinclorac, with *C. ensiformis* identified as the most efficient species for remediating tebuthiuron-contaminated soils. Ferreira et al. [7] further advanced the field by identifying suitable phytoremediator organisms for tebuthiuron in agricultural soils. Their experiments involved *Cajanus cajan, C. ensiformis, M. pruriens*, and *Pennisetum glaucum*, which successfully removed tebuthiuron applied at 2 L·ha^{-1}, enabling subsequent growth of *Crotalaria juncea* and *Lactuca sativa* in the presence of residual herbicide. They also highlighted the potential of repurposing vinasse as a source of organic carbon to enhance plant development and improve the ecological viability of phytoremediation.

Contrastingly, Frias et al. [8] investigated the efficacy of *M. pruriens* as a phytoremediator in soil supplemented with vinasse and found it ineffective at removing tebuthiuron. *M. pruriens* was exposed to tebuthiuron at 0.5, 1, 1.5, and 2 L·ha^{-1} and vinasse at 75, 150, and 300 m^3·ha^{-1}. The herbicide caused phytotoxicity, severely inhibiting germination and growth. The addition of vinasse exacerbated damage to both photosynthetic and non-photosynthetic structures, reducing biomass production. Consequently, neither *C. juncea* nor *L. sativa* could grow in the presence of residual pesticide.

Despite its promise, phytoremediation faces challenges, as its success depends on factors such as soil characteristics, climate, and co-contaminants [9]. To optimize phytoremediation effectiveness in addressing tebuthiuron contamination, a comprehensive understanding of the interactions between selected plant species and the herbicide is essential [7,8]. To enhance remediation efficiency, we incorporate bioaugmentation as a pivotal strategy. Bioaugmentation involves introducing microorganisms into a contaminated environment to accelerate pollutant degradation [10]. By introducing selected microbial strains with high pesticide degradation capabilities, resilience, and adaptability [11], we establish a symbiotic alliance that expedites pesticide degradation in soil. The integration of phytoremediation and bioaugmentation offers a powerful synergy [9], significantly improving tebuthiuron degradation efficiency and overall remediation outcomes.

Our approach emphasizes safety and efficacy by evaluating the environmental toxicity levels of tebuthiuron during the remediation process. We designed tests to realistically predict the behavior of substances in the environment. Specifically, we employed *C. juncea* as a bioindicator species sensitive to tebuthiuron after prior cultivation of *M. pruriens* and *C. ensiformis*. Additionally, *L. sativa* was used in ecotoxicological assays to assess residual toxicity. These complementary experiments aimed to verify the presence of tebuthiuron in the soil [7–16].

Therefore, this research aims to investigate the tolerance and phytoremediation potential of *M. pruriens* and *C. ensiformis*, in conjunction with microbial inoculants, for the remediation of tebuthiuron-contaminated agricultural soil. Our findings will advance our understanding of the viability and effectiveness of these techniques in addressing tebuthiuron contamination and the associated ecological concerns. Innovative and sustainable approaches are crucial for the successful remediation of tebuthiuron-contaminated soils.

2. Materials and Methods

2.1. Soil, Tebuthiuron, and Microbial Inoculant

The soil used in this study was classified as a Dystrophic Red-Yellow Oxisol. This soil was sourced from an agricultural facility in the Dracena region with no recent history of phytosanitary treatments, ensuring minimal prior contamination. Upon acquisition, the soil was transported to a greenhouse, air-dried, sieved through a 2.0 mm mesh, and stored in hermetically sealed plastic containers for chemical characterization. Soil samples were collected both before and after the experiment to determine chemical composition (Table 1). For each analysis period, data from all treatments were averaged to present the overall soil properties. Additional soil samples were collected at the end of the experimental period for further chemical analysis.

Table 1. Soil chemical analysis at 0 and 70 DAS of *C. ensiformis* and *M. pruriens*.

Attributes	Unit	0 DAS	70 DAS	Indication
pH (H_2O)	-	4.0	7.5	Increased
Organic matter	g dm^{-3}	4.0	10	Increased
Potassium	mmol dm^{-3}	0.3	1.6	Increased
Calcium	mmol dm^{-3}	6	51	Increased
Magnesium	mmol dm^{-3}	2	23	Increased
Hydrogen + Aluminum	mmol dm^{-3}	33	8	Decreased
Aluminum^{3+}	mmol dm^{-3}	13	0	Decreased
Phosphor	mg dm^{-3}	1	6	Increased
Sulfur	mg dm^{-3}	7	-	Not detected
Boron	mg dm^{-3}	0.10	0.02	Decreased
Copper	mg dm^{-3}	0.1	0.2	Increased
Iron	mg dm^{-3}	4	2	Decreased
Manganese	mg dm^{-3}	1.8	1.2	Increased
Zinc	mg dm^{-3}	0.1	0.3	Increased
Sum of bases	mg dm^{-3}	8	75.6	Increased
Cation exchange capacity	mg dm^{-3}	41	83.6	Increased
Base saturation	%	20	90	Increased
Aluminum saturation	%	61	0	Decreased

Increased—the value in the soil was higher in 70 DAS (day after the sowing) than the initial time (0 DAS); Decreased—the value in the soil was lower in 70 DAS than the initial time (0 DAS); Not detected—there was no comparison between 0 and 70 DAS due to the minimum concentration of the parameter not detected by the analysis.

The herbicide used was Combine® 500SC (Batch: 041-14-2000) from Dow AgroSciences Industrial Ltda. (São Paulo, Brazil), a commercially available formulation of tebuthiuron (TBT).

To augment the soil microbial community and enhance phytoremediation, microbial inoculants were obtained from Microgreen® Ltda. (http://microgreen.agr.br/, Piracicaba, Brazil), a company specializing in soil microbial reclamation. Two types of inoculants were utilized: a bacterial inoculant (BACT) rich in actinomycetes, *Bacillus* spp., and lactic acid bacteria and a fungal inoculant (FUNG) containing *Trichoderma* spp., *Purpureocillium* spp., and *Beauveria* spp., whose application is for the restoration of microbiota in agricultural soils.

2.2. Plant Species: Phytoremediator Species, Indicator Plant, and Test Organism

Mucuna pruriens (MP) and *Canavalia ensiformis* (CE) were selected for their well-documented phytoremediation capabilities, especially in soils contaminated with tebuthiuron [6–8]. These leguminous species form symbiotic relationships with nitrogen-fixing bacteria, enhancing nutrient availability and improving soil fertility.

Sunn hemp (*Crotalaria juncea*) was chosen as a bioindicator plant due to its known sensitivity to tebuthiuron [15]. Acting as a sentinel species, *C. juncea* aids in assessing soil contamination levels and the efficacy of phytoremediation efforts. Seeds of *C. juncea*, *C.*

ensiformis, and *M. pruriens* were obtained from BR SEEDS® (Araçatuba, Brazil), ensuring uniformity and reliability of the experimental material.

For ecotoxicological bioassays, commercially available seeds of *Lactuca sativa* L. (variety Butterhead) were procured from Feltrin Sementes® (Caxias do Sul, Brazil).

2.3. Experimental Setup

The experiment was conducted using a completely randomized design with a 2 × 3 × 3 factorial scheme, comprising seven replicates and five analysis times, totaling 630 pots. The factors studied were tebuthiuron concentration (presence or absence), microbial inoculant type (bacterial, fungal, or none), and plant species (*C. ensiformis*, *M. pruriens*, or none). This comprehensive design allowed for a thorough investigation of the independent and combined effects of these variables on the study parameters (Figure 1). Randomization minimized bias and ensured equal representation of treatment groups across experimental units.

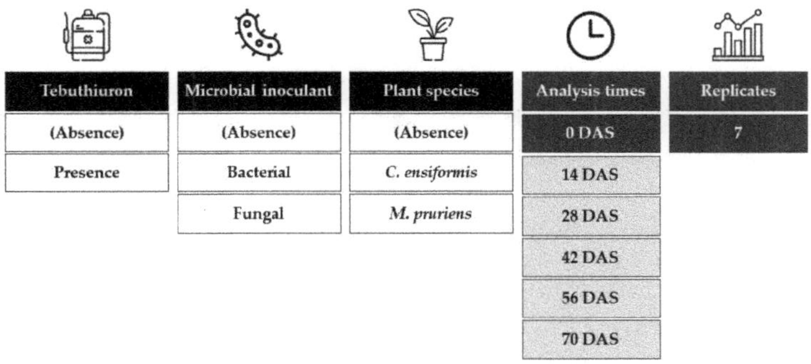

Figure 1. Summary of the experimental design with the three factors (tebuthiuron, microbial inoculant, and plant species), the five analysis times (0, 14, 28, 42, 56, and 70 DAS—days after sowing), and the number of replicates.

Prior to the experiment, the soil underwent a preparatory phase to adjust its acidity and fertility, following the procedures of Ferreira et al. [7] and Frias et al. [8]. For every 504 kg of soil, amendments were meticulously applied as follows: 454 g (1.8 t ha^{-1}) of limestone to regulate pH, 10 g (40 kg ha^{-1}) of urea as a nitrogen source, 56 g (222 kg ha^{-1}) of single superphosphate for phosphorus supplementation, and 13 g (52 kg ha^{-1}) of potassium chloride to ensure adequate potassium levels for optimal plant growth. After thorough mixing through uniform distribution, the soil was used to fill pots with a capacity of approximately 4.0 L each (19 cm × 15 cm × 19 cm).

The microbial inoculants were incorporated into the soil according to their respective treatment groups. The fungal inoculant (FUNG) was applied at a rate of 0.36 g (180 t ha^{-1}) per pot, while the bacterial inoculant (BACT) was added at a volume of 50 mL (25 m^3 ha^{-1}) per pot. Three days after inoculant incorporation, the herbicide Combine® was applied at the recommended dosage for sandy soils of 2.0 L ha^{-1} (1000 g active ingredient ha^{-1}). The application was performed using a laboratory sprayer equipped with four XR 11002 flat-fan nozzles (Jacto®, Pompéia, Brazil), operating at a pressure of 2 bar and a flow rate of 0.65 L· min^{-1}. Following the manufacturer's guidelines, the sprayer was calibrated to a speed of 5 km·h^{-1}, with the spray boom positioned 0.75 m above the pots, delivering an application volume of 156 L ha^{-1}. Environmental conditions during spraying were monitored, with a temperature of 27.2 °C and relative humidity of 63%. The control treatments without tebuthiuron received an equivalent volume of water to maintain consistency.

Seven days following herbicide application, three seeds of either *C. ensiformis* or *M. pruriens* were sown in each pot, according to the treatment design. Although three seeds were initially sown per pot, thinning was performed to retain only one plant per pot to ensure uniform growth conditions. These plants were cultivated for 70 days after sowing (DAS). Three days after harvesting the leguminous plants, three seeds of sunn hemp (*C. juncea*) were sown in each pot and cultivated for an additional 70 days, following the same thinning procedure.

All plants were cultivated in a greenhouse equipped with an automated irrigation system to maintain optimal growth conditions. Irrigation was performed daily to maintain the soil at 60% of its field capacity. The system was programmed via a digital timer to execute up to four irrigation cycles per day, each lasting approximately 40 ± 10 min. A micro-sprinkler located at the top of the greenhouse provided irrigation at a flow rate of approximately 80 L h^{-1} under a pressure of 2 bar, as indicated by the irrigation manometer. This controlled environment facilitated the investigation of the tolerance and phytoremediation capabilities of *C. ensiformis* and *M. pruriens*, as well as the influence of microbial inoculants in tebuthiuron-contaminated soil.

2.4. Plant Growth and Development Evaluation

Plant growth was monitored weekly by recording plant height in centimeters. At 70 DAS, the plants were harvested, the roots were gently cleaned to remove soil, and the samples were dried in a forced-air oven at 65 °C for 72 h. Dry biomass was then weighed and recorded in grams.

2.5. Ecotoxicological Bioassays

The ecotoxicological potential of the treatments was assessed at specific time points: 0, 20, 40, 60, 70 DAS (end of *M. pruriens* and *C. ensiformis* cultivation), and 140 DAS (end of *C. juncea* cultivation). Bioassays were conducted following the methodologies described in NBR 10006 [16] and Sobrero and Ronco [17].

For each treatment, five replicates of aqueous soil extracts were prepared. Superficial soil samples (approximately 2 cm deep) were collected from the edges of different pots within each treatment group. A 25 g sample of soil was mixed with 100 mL of deionized water in a 250 mL Erlenmeyer flask. The flasks were sealed with PVC plastic film, shaken at 120 rpm for five minutes, and then incubated in a Biochemical Oxygen Demand (BOD) chamber without lighting at 20 ± 2 °C for 7 days.

After incubation, 2.0 mL of the solubilized soil extract was applied to Petri dishes lined with filter paper and containing 10 seeds of *L. sativa*. The dishes were sealed with PVC plastic film to prevent moisture loss and incubated in a BOD chamber at 20 ± 2 °C with a 12 h photoperiod for 5 days. The positive control (CP) and negative control (CN) treatments were prepared using 0.05 M zinc sulfate solution and deionized water, respectively, to test seed sensitivity.

During the bioassays, seed germination, hypocotyl elongation, and root elongation were measured. These parameters were used to calculate the germination index (GI) of *L. sativa* seeds, as described in Equation (1) [18]:

$$GI = \frac{(G\% \times R\%)}{100} \tag{1}$$

This equation includes the following:
GI represents the germination index;
G% denotes the percentage of seed germination;
R% indicates the percentage of root elongation.

2.6. Statistical Data Analysis

Prior to conducting statistical analyses, the data were evaluated to ensure compliance with the assumptions of homogeneity of variances and normality. Homoscedasticity was assessed using Bartlett's test, while the Shapiro–Wilk test was employed to verify the normal distribution of residuals. Once these assumptions were confirmed, an Analysis of Variance (ANOVA) was performed at a 5% significance level ($p < 0.05$) to identify significant differences among treatment groups. When the ANOVA indicated significant effects, Tukey's Honest Significant Difference (HSD) post hoc test was utilized for pairwise comparisons to determine which specific means differed.

To model the growth dynamics of plant height and the germination index over time, we applied the Gompertz sigmoid function (Equation (2)). This nonlinear model is particularly suitable for describing sigmoidal growth patterns and has been widely used in ecological and biological studies due to its flexibility and interpretability [12,19]. The Gompertz function allowed us to estimate key growth parameters, such as the maximum attainable value (asymptote), growth rate constant, and inflection point, providing insights into the developmental processes under different treatment conditions. By fitting the Gompertz model to our data, we could predict growth trends beyond the observed time frames of 70 and 140 days after sowing (DAS), offering a more comprehensive understanding of the long-term effects of the treatments.

$$f_x = \alpha e^{-\beta e^{-kx}} \tag{2}$$

This equation includes the following:

fx represents the plant height in centimeters or the germination index;

x denotes the time in days after sowing (DAS);

α represents the upper asymptote or the maximum height development and germination index that the plants can reach;

β indicates the inflection point, which corresponds to the time when the growth rate starts to decrease;

k represents the exponential decay of the specific growth rate, indicating how quickly the growth rate decreases over time;

e represents Euler's constant, a mathematical constant approximately equal to 2.71828.

$$\gamma = \alpha + \beta + \sum((t_i \times f_{i1}) + (t_i \times f_{i2})) + \varepsilon \tag{3}$$

This equation includes the following:

γ represents the dependent variable (e.g., plant height or germination index);

α denotes the global intercept of the model;

β represents the random effect (tebuthiuron);

ti indicates the time in days after sowing (DAS);

f_{i1} represents the fixed effect 1 (green manure);

f_{i2} represents the fixed effect 2 (inoculant);

ε represents the residual term.

In cases where time was a significant variable, we extended the model to include time as a continuous covariate or as repeated measures, depending on the data structure. This allowed us to capture temporal trends and assess how the treatments influenced growth trajectories over time.

All statistical analyses were conducted using R software 4.2.2. for its advanced statistical capabilities and flexibility. The nlme package was utilized for fitting mixed-effects models (Equation (3)), while the nls function facilitated nonlinear regression for the Gompertz function. GraphPad Prism 9 and Microsoft Excel® 2019 were employed for data visualization, preliminary analyses, and to generate graphical representations of the results. These tools collectively enabled efficient data management, rigorous statistical testing, and clear presentation of findings.

2.6.1. Model Validation and Goodness-of-Fit and Consideration of Multiple Comparisons

Model adequacy was evaluated through diagnostic plots and goodness-of-fit statistics. For the Gompertz function, the coefficient of determination (R^2) and residual analysis were used to assess how well the model described the observed data. The Akaike Information Criterion (AIC) and Bayesian Information Criterion (BIC) were also considered for model selection and comparison. In the mixed-effects models, the significance of fixed effects was tested using likelihood ratio tests, and random effects were assessed through variance component analysis. Residuals were examined for homoscedasticity and normality to validate model assumptions. Given the multiple treatment groups and comparisons, we employed the Tukey–Kramer method in the post hoc analysis to control for Type I error rates associated with multiple testing. This approach ensures that the overall family wise error rate remains at the desired significance level, enhancing the reliability of the statistical conclusions.

2.6.2. Statistical Significance and Reporting

All statistical findings were interpreted in the context of the biological and ecological implications for phytoremediation practices; therefore, the results were considered statistically significant at $p < 0.05$.

3. Results and Discussion

3.1. *Soil's Properties: Enhancing Fertility*

As discussed in the preceding section, the soil used in this study was classified as a Dystrophic Red-Yellow Oxisol, a common soil type in tropical regions characterized by low natural fertility, acidic pH, and high levels of iron and aluminum oxides, which can influence nutrient and contaminant retention [20]. These properties make it an ideal substrate for simulating real-world scenarios of pesticide-contaminated agricultural soils.

Therefore, according to Table 1, significant improvements in the soil's chemical attributes were observed following amendments with lime and fertilizers, as well as the cultivation of *M. pruriens* and *C. ensiformis*. Various nutritional parameters increased numerically, indicating enhanced soil fertility. Notably, the concentration of exchangeable aluminum (Al^{3+}) did not increase, which is advantageous since high levels of aluminum can inhibit root growth and impair plant development [21]. The application of lime and fertilizers contributed to balancing soil pH and improving nutrient availability, thereby enhancing overall soil conditions.

Implementing appropriate agricultural practices, such as soil chemical correction and the cultivation of leguminous plants, can have multiple positive effects on soil health. Additionally, the presence of plants and soil microorganisms, combined with soil amendments, can lead to the release of organic acids. These organic acids aid in the solubilization of essential nutrients like phosphorus and potassium, making them more available for plant uptake. Furthermore, they contribute to an increase in the soil's cation exchange capacity, which helps reduce toxic levels of aluminum, ultimately benefiting plant growth and development [22].

Enhancing soil fertility through these practices creates a more conducive environment for plant growth, which is essential for successful phytoremediation efforts. Improved soil conditions can enhance plant vigor, allowing for more effective uptake and degradation of contaminants such as tebuthiuron.

3.2. *Production of Leguminous Plants with Microbial Inoculation: Unveiling the Biocatalytic Phytoremediation*

3.2.1. Growth Dynamics

The growth dynamics of *C. ensiformis* and *M. pruriens* exhibited distinct patterns (Figure 2). *M. pruriens* demonstrated a faster growth rate, indicated by a steeper growth curve compared to *C. ensiformis*. Table 2 presents the parameters of the growth curves for each treatment, including the maximum height (α), the inflection point (β), and the specific growth rate (k). A higher k value signifies a faster growth rate; both *M. pruriens* (k = 0.113) and *C. ensiformis* (k = 0.075) showed considerable growth rates.

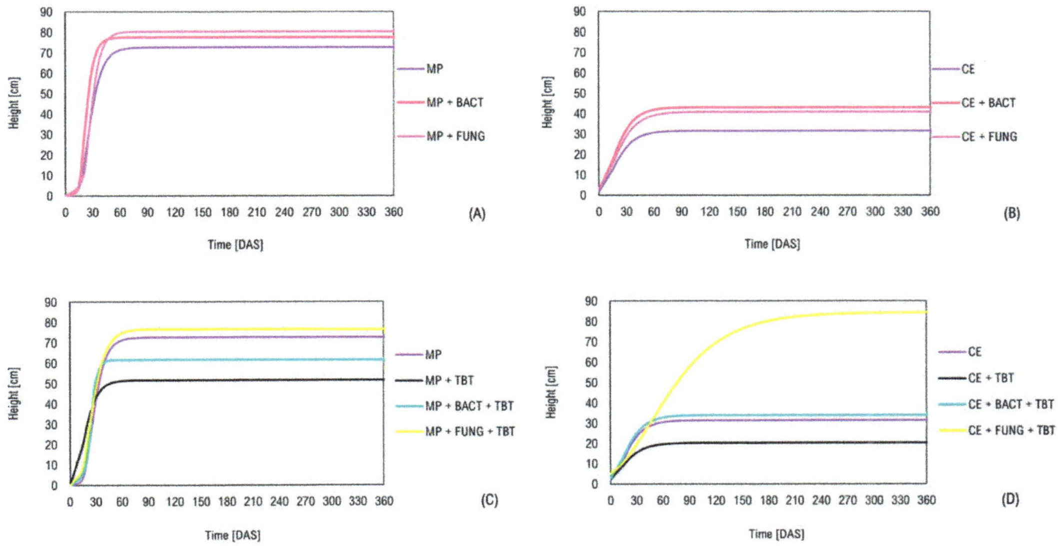

Figure 2. Kinetic growth of *M. pruriens* (MP) and *C. ensiformis* (CE) as potential phytoremediators of the herbicide tebuthiuron (TBT) in soil with fungal (FUNG) or bacterial (BACT) inoculants from the Gompertz model. (**A**,**B**) Treatments without TBT. (**C**,**D**) Treatments with TBT.

Table 2. Parameters of Gompertz kinetic models for the height of *M. pruriens* (MP) and *C. ensiformis* (CE) as potential phytoremediators of tebuthiuron (TBT) with fungal (FUNG) or bacterial (BACT) inoculants.

Treatments	Complexity			Adequacy		
	α	β	k	R^2	AIC	BIC
MP	72.72	15.49	0.1127	0.72	63.96	65.17
MP + TBT	51.69	4.76	0.1087	0.94	63.96	65.17
MP + BACT	77.66	32.71	0.1695	0.97	69.10	70.31
MP + FUNG	80.31	38.14	0.1425	0.99	56.33	57.54
MP + BACT + TBT	61.58	144.19	0.2272	0.89	78.80	80.01
MP + FUNG + TBT	76.63	9.15	0.0997	0.93	75.82	77.04
CE	31.64	2.72	0.0748	0.97	46.29	47.50
CE + TBT	20.27	2.13	0.0686	0.93	44.71	45.92
CE + BACT	43.10	2.73	0.0751	0.97	53.67	54.88
CE + FUNG	40.90	2.62	0.0679	0.97	50.41	51.62
CE + BACT + TBT	34.26	2.51	0.0707	0.96	49.03	50.24
CE + FUNG + TBT	84.52	2.81	0.0224	0.94	60.08	61.29

Parameters α, β, and k denote the superior asymmetry, the inflection point, and the exponential decay of the specific growth rate, respectively; $\beta = 1$ keeps the relative decrease with time constant; $\beta > 1$ accelerates the relative decrease with time; $\beta < 1$ slows the relative decrease with time; R^2: coefficient of determination; AIC: Akaike Information Criterion; BIC: Bayesian Information Criterion.

The application of microbial inoculants influenced the growth dynamics of both species. Within the *M. pruriens* group, the fungal inoculant treatment (MP + FUNG) resulted in faster growth compared to the bacterial inoculant (MP + BACT) and the control (MP alone). Similarly, for *C. ensiformis*, the bacterial inoculant (CE + BACT) promoted faster growth than the fungal inoculant (CE + FUNG) and the control (CE alone). These results suggest that the type of microbial inoculant can differentially affect plant growth, potentially due to specific interactions between the inoculant microorganisms and the plant species.

The introduction of tebuthiuron had a noticeable impact on plant growth, reducing plant height compared to the control treatments without the herbicide. As shown in Table 2, the CE + TBT treatment exhibited a greater reduction in height-related parameters, including maximum height (α) and specific growth rate (k), compared to MP + TBT. This indicates that *C. ensiformis* is more sensitive to tebuthiuron than *M. pruriens*. Previous studies by Ferreira et al. [7] and Belo et al. [23] also reported the sensitivity of *C. ensiformis* to tebuthiuron, suggesting that this species may not be ideal for the phytoremediation of soils contaminated with this herbicide.

Conversely, Mendes et al. [6] demonstrated that *C. ensiformis* can tolerate and even degrade tebuthiuron, potentially due to specific microorganisms present in its rhizosphere. This highlights the importance of considering plant–microbe interactions when selecting species for phytoremediation. For *M. pruriens*, Ferreira et al. [7] observed a 15% decrease in plant height when exposed to tebuthiuron. However, they found that the phytotoxicity of tebuthiuron was mitigated by the addition of vinasse (an industrial by-product) at 150 m^3 ha^{-1}, allowing the plant to develop even at high herbicide concentrations (2 L ha^{-1}).

Discrepancies between our results and previous studies may be attributed to factors such as differences in experimental conditions, soil types, moisture levels, environmental temperatures, and light exposure. These variables can directly influence phytoremediation efficiency and should be carefully considered when selecting plants for remediating tebuthiuron-contaminated soils [24].

An important finding in our study was the mitigating effect of microbial inoculants on plant sensitivity to tebuthiuron in contaminated soil. The inoculants appeared to enhance plant growth and increase height, particularly in *C. ensiformis*. For instance, the CE + BACT + TBT treatment showed a higher specific growth rate (k = 0.071) compared to CE + TBT without inoculants, although only the CE + FUNG + TBT treatment achieved the maximum height (α = 84.52). This suggests that fungal inoculants may be more effective in promoting growth under herbicide stress in *C. ensiformis*.

In *M. pruriens*, the MP + FUNG + TBT treatment displayed a steeper growth curve than MP + BACT + TBT, reaching a greater maximum height (α = 76.63). Although the bacterial inoculant (MP + BACT + TBT) exhibited a higher specific growth rate (k = 0.227 vs. k = 0.100), the overall growth performance was better with the fungal inoculant. These results could be explained by several factors: (a) Fungal inoculants may enhance plant resilience to herbicide-induced stress by improving nutrient uptake or producing growth-promoting substances, allowing plants to maintain or increase growth under adverse conditions. (b) The fungal inoculant may have the ability to degrade or metabolize tebuthiuron, reducing its toxicity in the soil environment. (c) The inoculants may stimulate beneficial soil microorganisms, creating a more favorable rhizosphere environment for plant growth.

Previous studies have demonstrated the effectiveness of such microbial combinations. Zhang et al. [25] reported successful remediation of soils contaminated with pentachloronitrobenzene using a fungal–bacterial inoculum in association with *Panax notoginseng*. Similarly, Madariaga-Navarrete et al. [26] observed substantial atrazine removal from soil within 40 days using *Trichoderma* sp. combined with *Phaseolus vulgaris*. These findings support the potential of microbial inoculants in enhancing phytoremediation efficiency.

3.2.2. Phytomass Accumulation

The application of microbial inoculants in the absence of tebuthiuron led to a significant increase in biomass production in both *M. pruriens* and *C. ensiformis* compared to the control treatments (Figure 3). This positive impact underscores the efficacy of microbial inoculants in promoting plant growth and enhancing phytomass accumulation, which is pivotal for the plants' capacity to tolerate and effectively remediate contaminated soils.

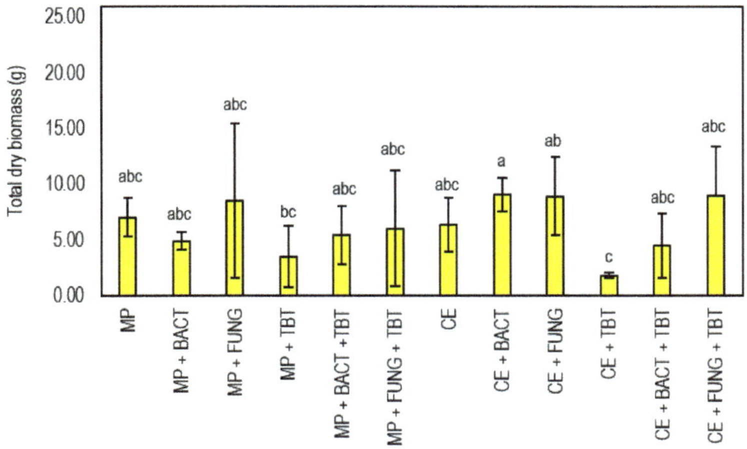

Figure 3. Production of total dry biomass of *M. pruriens* (MP) and *C. ensiformis* (CE) in soil associated or not with tebuthiuron (TBT) and/or fungal (FUNG) or bacterial (BACT) inoculants after 70 DAS. Different lowercase letters indicate statistical differences by the Tukey test ($p < 0.05$).

In contrast, the introduction of tebuthiuron had an adverse effect on biomass production in both species. The most pronounced reduction was observed in the MP + TBT treatment, indicating that *M. pruriens* was particularly affected by the herbicide. Tebuthiuron's negative influence on photosynthesis during plant development can impair biomass production and compromise the overall efficiency of phytoremediation processes [27], even though *M. pruriens* and *C. ensiformis* are not target plants for this herbicide in agroecosystems. Efficient biomass production is crucial for facilitating the transformation of pollutants into less toxic substances, a process optimized when plants grow without intense stress [28].

Further examination revealed that, in the presence of tebuthiuron, the bacterial inoculant did not significantly improve biomass production in *M. pruriens*. For *C. ensiformis*, microbial inoculation did not result in a substantial increase in phytomass when tebuthiuron was present, despite contributing to increased plant height. These observations emphasize the limitations of relying solely on variables like height and biomass accumulation to evaluate phytoremediation efficiency.

Moreover, the rhizospheric interactions between plants and microorganisms play a key role in the degradation of soil contaminants. Root exudates from leguminous plants like *M. pruriens* and *C. ensiformis* can enhance microbial activity by providing essential nutrients and signaling molecules that stimulate the growth and metabolic functions of degradative microbes [29]. These exudates may increase the bioavailability of tebuthiuron by altering soil pH and releasing chelating agents, thereby facilitating its uptake and degradation. Understanding the synergistic relationships within the rhizosphere is crucial, as it can lead to optimized phytoremediation strategies that harness both plant and microbial capabilities for more efficient contaminant removal.

Multiple uncontrolled factors can influence the bioavailability and environmental behavior of the herbicide, including soil properties, microbial community dynamics, and environmental conditions [30]. Therefore, relying exclusively on plant growth parameters may not provide a comprehensive assessment of phytoremediation effectiveness. Complementary approaches, such as cultivating bioindicator plants and implementing ecotoxicological bioassays, are indispensable for thoroughly evaluating environmental reclamation efforts [31].

In summary, our findings highlight the critical aspect of biomass accumulation in the context of phytoremediation. While microbial inoculants positively influenced biomass production in the absence of tebuthiuron, the presence of the herbicide negated these benefits.

This underscores the significance of considering multiple parameters and employing complementary assessment methods to accurately evaluate the efficiency of phytoremediation practices, thereby facilitating sustainable and effective soil remediation strategies.

3.3. Production of C. juncea: Evaluating Ecotoxicity and Phytoremediation Efficiency

3.3.1. Growth Dynamics

The cultivation of *C. juncea* as a bioindicator provided valuable insights into the residual phytotoxicity of tebuthiuron and the effectiveness of the phytoremediation treatments. In the reference treatment (Ref), which involved soil without prior plant cultivation (*CE* or *MP*), no inoculants (BACT or FUNG), and no herbicide (TBT), *C. juncea* exhibited slow height development (Figure 4). This baseline serves as a control for comparing the effects of various treatments.

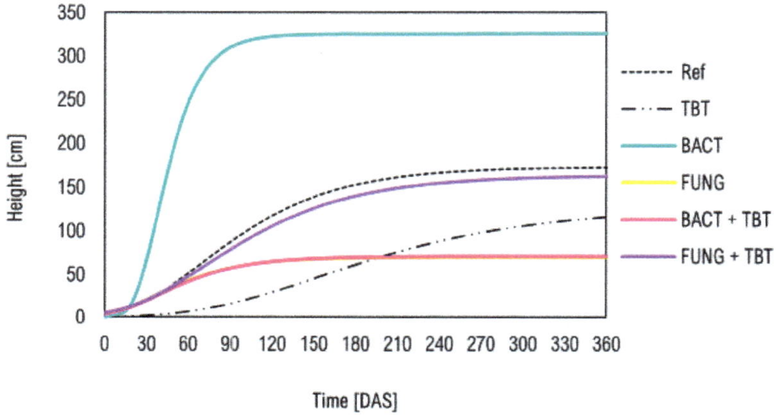

Figure 4. Kinetic growth of *C. juncea* height as bioindicator species in soil with tebuthiuron (TBT) and fungal (FUNG) or bacterial (BACT) inoculants from the Gompertz model. Ref—reference control soil without leguminous plants, inoculants, and tebuthiuron.

Interestingly, soil treated with the bacterial inoculant (BACT) without prior leguminous cultivation or herbicide application showed a faster increase in *C. juncea* height, with a higher specific growth rate (k = 0.0582) and greater maximum height (α = 324.99) around 60 DAS (Table 3). In contrast, the fungal inoculant (FUNG) had a minimal contribution to *C. juncea*'s height development, indicated by a lower growth rate (k = 0.0302) and lower maximum height (α = 69.95). This suggests that the bacterial inoculant may promote the growth of *C. juncea* in uncontaminated soils more effectively than the fungal inoculant.

The presence of tebuthiuron in the soil demonstrated a significant phytotoxic effect on *C. juncea*, evidenced by a lower specific growth rate (k = 0.0115) (Table 3 and Figure 5). Despite the slower growth rate, *C. juncea* in soil with tebuthiuron alone (without inoculants or prior cultivation) reached a maximum height exceeding 100 cm, which was higher than in treatments with FUNG or BACT + TBT. This indicates that while tebuthiuron adversely affects growth, *C. juncea* can still attain considerable height in its presence.

Remarkably, the phytotoxic effect of tebuthiuron was mitigated by the fungal inoculant in certain treatments, suggesting the involvement of microorganisms in bioremediation. Previous studies have associated microbial genera such as *Methylobacterium*, *Microbacterium*, *Paenibacillus*, and *Streptomyces* with tebuthiuron degradation [32,33]. The long-term presence of the fungal inoculant may contribute to the dissipation of tebuthiuron, reducing its toxicity to subsequent plantings.

Table 3. Parameters of the Gompertz kinetic models for the height of *C. juncea* as a bioindicator species in tebuthiuron (TBT) soil with *M. pruriens* (MP) and *C. ensiformis* (CE) and bacterial (BACT) or fungal (FUNG) inoculants. Ref—reference control soil without leguminous plants, inoculants, and tebuthiuron.

Treatments	Complexity			Adequacy		
	α	β	k	R^2	AIC	BIC
MP	156.35	5.94	0.01061	0.98	49.51	50.30
MP + TBT	86.91	3.29	0.03295	0.98	45.17	45.96
MP + BACT	181.96	3.76	0.01741	0.98	45.39	46.18
MP + FUNG	42.01	3.23	0.05290	0.98	41.15	41.94
MP + BACT + TBT	147.18	3.75	0.02018	0.99	41.67	42.46
MP + FUNG + TBT	248.45	8.52	0.06591	0.98	54.98	55.77
CE	178.13	3.82	0.01486	0.96	47.42	48.20
CE + TBT	115.67	3.34	0.02142	0.98	46.18	45.39
CE + BACT	78.74	3.00	0.02500	0.97	45.81	45.02
CE + FUNG	66.01	2.91	0.02667	0.96	47.20	47.98
CE + BACT + TBT	115.76	5.60	0.09303	0.97	47.70	48.49
CE + FUNG + TBT	199.88	4.11	0.01602	0.98	43.88	44.67
Ref	172.93	3.78	0.01897	0.98	44.72	45.51
TBT	324.99	8.91	0.05819	0.98	51.98	52.77
BACT	69.95	2.95	0.03017	0.97	48.00	47.21
FUNG	127.47	5.76	0.01148	0.99	44.82	45.61
BACT + TBT	70.50	3.08	0.02991	0.98	42.20	42.99
FUNG + TBT	163.15	3.47	0.01731	0.97	50.26	51.05

Parameters α, β, and k denote the superior asymmetry, the inflection point, and the exponential decay of the specific growth rate, respectively; $\beta = 1$ keeps the relative decrease with time constant; $\beta > 1$ accelerates the relative decrease with time; $\beta < 1$ slows the relative decrease with time; R^2: coefficient of determination; AIC: Akaike Information Criterion; BIC: Bayesian Information Criterion.

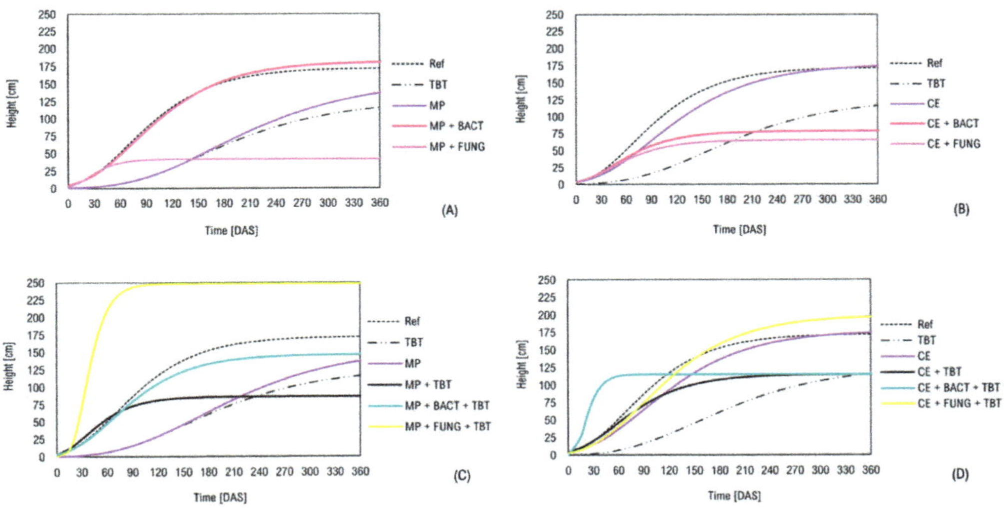

Figure 5. Kinetic growth of *C. juncea* height as bioindicator species in tebuthiuron (TBT) soil with *M. pruriens* (MP) and *C. ensiformis* (CE) and fungal (FUNG) or bacterial (BACT) inoculants. Ref—reference control soil without leguminous plants, inoculants, and tebuthiuron. (**A**,**B**) Treatments without TBT. (**C**,**D**) Treatments with TBT.

Prior cultivation of *M. pruriens* and *C. ensiformis* significantly influenced the growth of *C. juncea*. Soil previously cultivated with *C. ensiformis* had a more positive effect on *C. juncea*'s height compared to soil with *M. pruriens*. As shown in Table 3, the parameters

α and k for *C. ensiformis* were greater than those for *M. pruriens* ($\alpha = 156.35$ vs. 78.13; $k = 0.0149$ vs. 0.0106). While *M. pruriens* is beneficial for soil health due to nitrogen fixation and nutrient cycling, it may exhibit allelopathic effects that inhibit the growth of nearby plants through the production of bioactive compounds [34].

Interactions between the green manure species and microbial inoculants yielded distinct results in *C. juncea*'s growth. For instance, the negative effect of *M. pruriens* on *C. juncea* was alleviated when associated with the bacterial inoculant (MP + BACT), resulting in a higher specific growth rate ($k = 0.0174$) and maximum height ($\alpha = 181.96$ cm) compared to MP alone. Conversely, the association of *C. ensiformis* with the bacterial inoculant (CE + BACT) had an antagonistic effect on *C. juncea*'s growth, yielding a smaller growth curve compared to soil with *CE* alone. Similarly, the association of fungal inoculants with either phytoremediation species was generally detrimental to *C. juncea*, particularly in the MP + FUNG treatment, where the maximum height was the lowest ($\alpha = 42.01$ cm), despite a higher growth rate ($k = 0.0529$).

When analyzing treatments involving tebuthiuron and prior cultivation with *M. pruriens* or *C. ensiformis*, high phytotoxicity and severe limitations in *C. juncea* height were observed compared to the control treatments without the herbicide. Nevertheless, the height growth rates after prior cultivation with the herbicide remained higher than those in the control tests, ranging from $k = 0.0330$ for MP + TBT to $k = 0.0214$ for CE + TBT. The presence of phytotoxic compounds in the soil and the natural senescence of the phytoremediator and sentinel species may contribute to these adverse effects [7].

Notably, the presence of the fungal inoculant reduced the phytotoxic effect of tebuthiuron in certain treatments. In the MP + FUNG + TBT treatment, *C. juncea* exhibited a rapid height increase ($k = 0.0659$), resulting in a steep growth curve and an impressive maximum height ($\alpha = 248.45$ cm). Similar results were observed in the CE + FUNG + TBT treatment, where the maximum height ($\alpha = 199.88$ cm) was higher compared to other treatments with the same species. These findings indicate the potential of the fungal inoculant in mitigating tebuthiuron toxicity, possibly through microbial degradation of the herbicide.

The bacterial inoculant also exhibited a mitigating effect, though it was generally less pronounced than that of the fungal inoculant. In the CE + BACT + TBT treatment, *C. juncea*'s growth rate was higher ($k = 0.0930$), and in the MP + BACT + TBT treatment, the maximum height ($\alpha = 147.18$ cm) was higher compared to other treatments with the same species. This suggests that the bacterial inoculant can aid in tebuthiuron remediation in soil, albeit with some limitations [35].

Overall, these findings provide a detailed assessment of the growth dynamics of *C. juncea* as a bioindicator plant. The results elucidate the impact of microbial inoculants and the herbicide tebuthiuron on the development of the bioindicator, highlighting the potential of certain inoculants—particularly the fungal inoculant—in mitigating the herbicide's phytotoxic effects. Additionally, the efficacy of fungal inoculants in mitigating tebuthiuron toxicity may be attributed to their robust enzymatic systems capable of degrading complex organic pollutants. Fungi such as *Trichoderma* spp. produce a variety of extracellular enzymes, including laccases and peroxidases, which can oxidize and break down persistent herbicides [36]. These enzymes facilitate the cleavage of chemical bonds within the tebuthiuron molecule, transforming it into less toxic metabolites that are more amenable to further microbial degradation or assimilation by plants. The deployment of such fungi in bioaugmentation strategies not only enhances the degradation of recalcitrant compounds but also improves soil health by suppressing pathogenic microorganisms and promoting plant growth [37]. This contributes valuable knowledge to the field of phytoremediation, informing strategies for effectively addressing tebuthiuron-contaminated soils.

3.3.2. Phytomass Accumulation

The accumulation of dry biomass in *Crotalaria juncea* varied significantly across treatments, revealing critical insights into the interplay between microbial inoculants, herbicide presence, and phytoremediation efficacy (Figure 6). Notably, the treatment combining tebuthiuron with bacterial inoculants (TBT + BACT) resulted in the highest dry biomass production, reaching approximately 18 g. Despite being higher, there was no statistical significance compared to the other treatments, which suggested that the addition of specific bacterial strains effectively mitigated the herbicide's phytotoxic effects on *C. juncea*. The bacteria likely facilitated enhanced degradation or transformation of tebuthiuron, reducing its toxicity and promoting plant growth. This finding underscores the potential of bacterial inoculants as a pivotal component in phytoremediation strategies for tebuthiuron-contaminated soils.

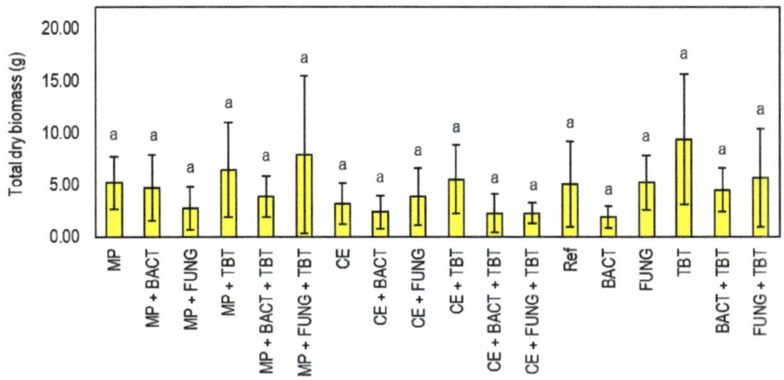

Figure 6. Production of total dry biomass of *C. juncea* in soil associated or not with tebuthiuron (TBT), bacterial (BACT), or fungal (FUNG) inoculants and/or the different plants *M. pruriens* (MP) and *C. ensiformis* (CE) after 70 DAS. Ref—reference control soil without leguminous plants, inoculants, and tebuthiuron. Same lowercase letters did not indicate statistical differences by the Tukey test ($p < 0.05$).

In stark contrast, the treatment with tebuthiuron alone (TBT) without any inoculants resulted in significantly lower biomass accumulation, around 7 g. This substantial reduction reflects the negative impact of the herbicide on the growth of *C. juncea*, confirming its phytotoxicity in the absence of bioremediation interventions. The persistence and toxicity of tebuthiuron in this treatment could be attributed to its chemical stability and the substantial organic matter content in the soil, which may enhance herbicide adsorption and reduce its bioavailability for degradation [30].

Fungal inoculants (FUNG), when applied in combination with tebuthiuron (TBT + FUNG), were less effective than bacterial inoculants in mitigating the herbicide's phytotoxicity, resulting in dry biomass of approximately 9 g. While fungi are known to play roles in biodegradation and plant growth promotion, their efficacy in this context was inferior to that of bacteria. This suggests that the specific fungal species used may not have possessed the necessary metabolic pathways to effectively degrade tebuthiuron or may have had less synergistic interactions with *C. juncea* compared to the bacterial strains.

The combinations involving prior cultivation of *M. pruriens* (MP) or *C. ensiformis* (CE) with microbial inoculants and tebuthiuron yielded variable biomass outcomes. Generally, these treatments did not achieve biomass values as high as the TBT + BACT treatment. Biomass production in these groups ranged between 5 and 12 g, indicating that the interactions between the leguminous plants, microbial inoculants, and tebuthiuron are complex. Possible antagonistic effects, such as competition for nutrients, allelopathic interactions, or microbial community shifts, may have limited the growth of *C. juncea* in these scenarios [7,34].

These findings align with previous studies emphasizing the critical role of microorganisms in enhancing plant tolerance to soil contaminants. The superior efficacy observed with bacterial inoculants may be attributed to their ability to metabolize toxic compounds, produce plant growth-promoting substances, or enhance nutrient availability, thereby creating more favorable conditions for plant development [10,32]. In contrast, the lower efficacy of fungal inoculants could be due to less efficient degradation pathways for tebuthiuron or weaker interactions with the phytoremediator and bioindicator plants.

Overall, the results highlight the importance of selecting appropriate microbial inoculants in phytoremediation strategies. The significant increase in biomass with bacterial inoculation demonstrates its potential application in agricultural practices to mitigate tebuthiuron contamination. The enhanced biomass production not only indicates improved plant health but also suggests a greater capacity for phytoremediation, as higher biomass is often correlated with increased pollutant uptake and degradation [27,28].

3.4. Bioassays with *L. sativa*: Validating Ecotoxicity and Phytoremediation Efficiency

Ecotoxicity testing with *L. sativa* is a pivotal tool for evaluating soil quality, particularly in environments potentially affected by herbicides like tebuthiuron. This indirect method not only detects the presence of herbicide but also verifies reductions in soil toxicity. Multiple researchers have conducted ecotoxicity tests using *L. sativa* following biological pesticide remediation experiments [7,19].

The germination index (GI) of *Lactuca sativa* served as a sensitive indicator of soil ecotoxicity and the effectiveness of phytoremediation treatments over time. In the uncultivated treatments (Figure 7), the reference soil (Ref) initially exhibited a significantly higher GI compared to other treatments without tebuthiuron from 0 to 40 DAS. However, by the end of the experimental period, the bacterial inoculant treatment (BACT) surpassed both the Ref and fungal inoculant (FUNG) treatments in final GI values (0.95 > 0.92 > 0.89, respectively) (Table 4). The higher growth rate (k value) observed for BACT indicates its positive influence on the germination and early development of *L. sativa* seeds, possibly through enhanced nutrient availability or the production of growth-promoting substances.

Table 4. Parameters of the Gompertz kinetic models for the germination index of *L. sativa* in ecotoxicity bioassays in tebuthiuron (TBT) soil with *M. pruriens* (MP) and *C. ensiformis* (CE) and bacterial (BACT) or fungal (FUNG) inoculants. Ref—reference control soil without leguminous plants, inoculants, and tebuthiuron.

Treatments	Complexity			Adequacy		
	α	β	k	R^2	AIC	BIC
MP	1.00	6.98	0.04774	0.98	−11.64	−12.47
MP + TBT	0.97	10.09	0.06319	0.98	−13.20	−14.04
MP + BACT	1.00	7.06	0.04656	0.98	−13.15	−13.99
MP + FUNG	0.97	10.09	0.06319	0.98	−13.20	−14.04
MP + BACT + TBT	1.00	6.81	0.04618	0.98	−11.94	−12.77
MP + FUNG + TBT	1.00	7.49	0.05091	0.98	−11.67	−12.51
CE	1.00	5.94	0.04001	0.98	−11.13	−11.97
CE + TBT	1.00	7.25	0.03861	0.99	−21.18	−22.02
CE + BACT	1.00	6.96	0.04861	0.98	−10.72	−11.56
CE + FUNG	1.00	7.03	0.04880	0.98	−10.94	−11.77
CE + BACT + TBT	1.00	6.46	0.04224	0.98	−12.94	−13.77
CE + FUNG + TBT	1.00	7.06	0.04797	0.98	−11.94	−12.77
Ref	1.00	3.27	0.02642	0.99	−19.10	−19.93
TBT	1.00	3.65	0.03053	0.99	−24.09	−24.92
BACT	1.00	2.84	0.02250	0.96	−8.90	−9.73
FUNG	1.00	3.01	0.03141	0.99	−15.20	−16.03
BACT + TBT	1.00	3.28	0.03499	0.99	−18.46	−19.29
FUNG + TBT	1.00	3.33	0.02641	0.99	−19.68	−20.51

Parameters α, β, and k denote the superior asymmetry, the inflection point, and the exponential decay of the specific growth rate, respectively; $\beta = 1$ keeps the relative decrease with time constant; $\beta > 1$ accelerates the relative decrease with time; $\beta < 1$ slows the relative decrease with time; R^2: coefficient of determination; AIC: Akaike Information Criterion; BIC: Bayesian Information Criterion.

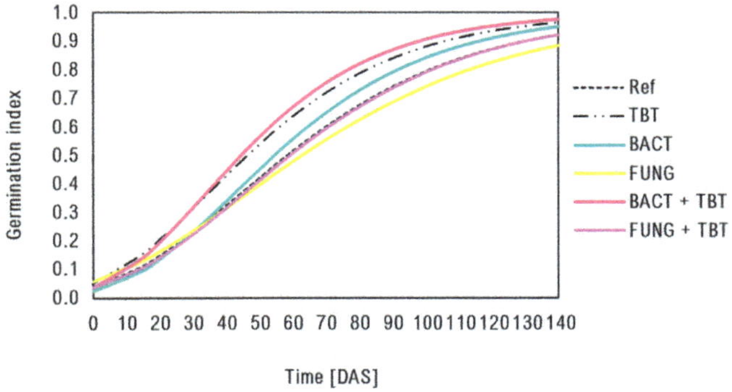

Figure 7. Kinetic evolution from the Gompertz model of the germination index of *L. sativa* in ecotoxicity bioassays in soil with tebuthiuron (TBT) and fungal (FUNG) or bacterial (BACT) inoculants from the Gompertz model. Ref—reference control soil without leguminous plants, inoculants, and tebuthiuron.

Intriguingly, the treatment containing only tebuthiuron (TBT) without any inoculants showed a higher final GI (0.96) and faster growth rate (k = 0.0314) compared to the Ref. This unexpected result suggests that natural attenuation processes were at play, whereby indigenous soil microorganisms gradually degraded the herbicide over time, reducing its phytotoxicity [38,39]. As tebuthiuron concentrations decreased, the inhibitory effects on seed germination diminished, allowing *L. sativa* to achieve higher GI values.

The impact of microbial inoculants in tebuthiuron-contaminated soil differed between bacteria and fungi. The BACT + TBT treatment exhibited the highest final GI (0.98) and growth rate (k = 0.0350), indicating that bacterial inoculants effectively enhanced the dissipation of tebuthiuron's toxic effects, facilitating seed germination and growth. In contrast, the FUNG + TBT treatment had a lower final GI (0.92) and growth rate (k = 0.0264), suggesting that the fungal inoculant was less effective in mitigating the herbicide's ecotoxicity.

The prior cultivation of *M. pruriens* (MP) and *C. ensiformis* (CE) also influenced the GI of *L. sativa* (Figure 8A,C). Initially, the GI in these treatments was lower than the Ref between 0 and 40 DAS, possibly due to residual allelopathic compounds from the leguminous plants or incomplete degradation of tebuthiuron. Over time, however, the GI increased, reaching 0.99 in the MP treatment (Table 4), indicating a reduction in soil toxicity. The higher growth rate (k = 0.0477) compared to the Ref (k = 0.0264) suggests that the prior cultivation of *M. pruriens* improved soil conditions, possibly through enhanced microbial activity, nutrient cycling, and the degradation of residual herbicides.

The combination of microbial inoculants with leguminous plants further affected the GI. In the MP + BACT treatment without tebuthiuron, the growth rate increased (k = 0.0632) compared to MP alone, although the final GI was slightly lower (0.97 < 0.99). This indicates that while bacterial inoculants can accelerate the reduction in ecotoxicity, they may also introduce competitive dynamics that slightly affect germination rates. In treatments with tebuthiuron, microbial inoculation appeared to reduce the negative ecotoxicological impact, as evidenced by higher GI values compared to treatments without inoculants.

Overall, the GI of *L. sativa* provided valuable insights into the temporal dynamics of soil ecotoxicity and the effectiveness of phytoremediation strategies. The results highlight the potential of combining leguminous plants with specific microbial inoculants to enhance the degradation of tebuthiuron and reduce its phytotoxic effects. However, the complexity of interactions among plants, microorganisms, and contaminants underscores the need for careful selection and optimization of phytoremediation components to achieve effective soil remediation [38,39].

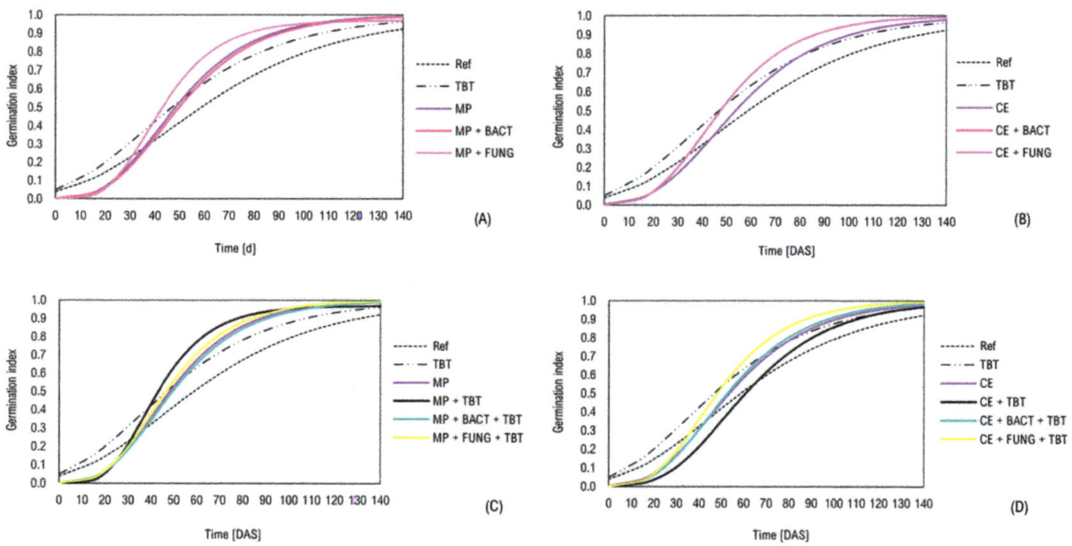

Figure 8. Kinetic evolution from the Gompertz model of the germination index of *L. sativa* in ecotoxicity bioassays in tebuthiuron (TBT) soil with *M. pruriens* (MP) and *C. ensiformis* (CE) and bacterial (BACT) or fungal (FUNG) inoculants. Ref—reference control soil without leguminous plants, inoculants, and tebuthiuron. (**A**,**B**) Treatments without TBT. (**C**,**D**) Treatments with TBT.

Furthermore, the physicochemical properties of the soil, such as pH, organic matter content, and cation exchange capacity, significantly influence the bioavailability and persistence of tebuthiuron [40]. Soils with high organic matter can adsorb greater amounts of herbicides, potentially reducing their immediate bioavailability to plants and microbes but also prolonging their environmental persistence [41]. Adjusting soil conditions through amendments like compost or biochar can enhance microbial activity and modify sorption characteristics, thereby improving degradation rates [42]. Tailoring phytoremediation strategies to account for these soil properties is essential for optimizing contaminant removal and ensuring the sustainability of remediation efforts [43].

3.5. The Impact of the Soil–Herbicide–Plant–Microbe Nexus on L. sativa's GI: A Deeper Understanding of Bioremediation

A multivariate response analysis using a mixed linear model (Figure 9) from 0 to 140 DAS elucidated the intricate interactions among tebuthiuron, leguminous plants, and microbial inoculants on the germination index of *L. sativa*. This analysis revealed that the combined effects of these factors played a crucial role in determining seed germination rates and seedling development.

When leguminous plants and microbial inoculants were combined, the treatments displayed a less steep slope in the model, indicating a strong interdependence between these factors. This synergistic interaction resulted in higher germination rates of *L. sativa*, even in the presence of tebuthiuron (accounted for as a random effect in the model). The combination of potential phytoremediators and bioaugmentation appeared to mitigate the phytotoxic effects of the herbicide more effectively than either factor alone.

In contrast, the individual effects of leguminous plants or microbial inoculants exhibited steeper slopes, suggesting a lower impact on reducing tebuthiuron toxicity when applied independently. This implies that microbial inoculants alone may not sufficiently alleviate the herbicide's phytotoxicity, and similarly, the cultivation of leguminous species without microbial augmentation may have limited efficacy.

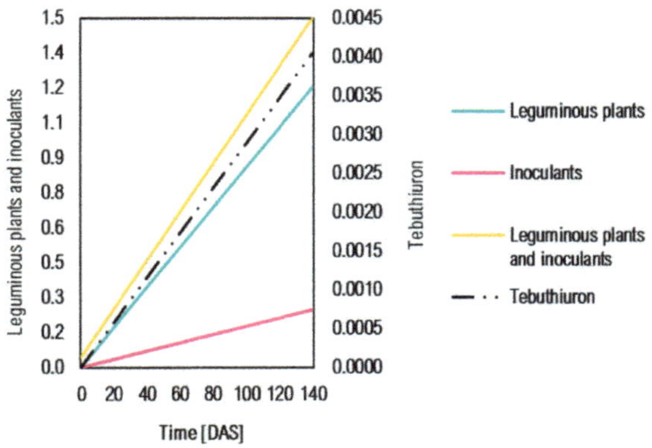

Figure 9. Dynamics of random (tebuthiuron) and fixed (leguminous plants—*M. pruriens* and *C. ensiformis*; and microbial inoculants—fungal and bacterial) effects on the specific rate of the germination index (GI) of *L. sativa* in ecotoxicity bioassays.

The enhanced performance of the combined treatments can be attributed to several mechanisms. The leguminous plants likely improved soil health by increasing organic matter content, enhancing nutrient availability through nitrogen fixation, and stimulating microbial activity [21,22]. The introduced microbial inoculants may have possessed specific degradative capabilities for tebuthiuron or facilitated the proliferation of indigenous degrader populations, leading to accelerated herbicide dissipation [9,10].

These findings underscore the importance of integrated phytoremediation strategies that leverage synergistic interactions between plants and microorganisms. By combining phytoremediators with bioaugmentation, it is possible to enhance the degradation of persistent contaminants like tebuthiuron, improve soil quality, and reduce ecotoxicity more effectively than with single-factor approaches.

3.6. Limitations and Directions to Improve the Credibility and Practicality of Phytoremediation

While this study demonstrates the potential of combining leguminous plants and microbial inoculants for the phytoremediation of tebuthiuron-contaminated soils, several limitations warrant consideration for future research and practical application.

The experiments were conducted under controlled greenhouse conditions, which may not fully capture the complexities of field environments. Factors such as soil heterogeneity, climatic variations, and ecological interactions can significantly influence phytoremediation outcomes. Field trials are essential to validate the effectiveness of the proposed strategies under real-world conditions, accounting for spatial and temporal variability.

Assessing the long-term sustainability of phytoremediation efforts is crucial. Continuous monitoring of contaminant levels, soil health indicators, and ecological impacts over extended periods will provide insights into the persistence of remediation effects and potential rebound of contaminant concentrations. In addition, agricultural soils are often contaminated with a mixture of pesticides and other pollutants. Future studies should investigate the efficacy of phytoremediation strategies in the context of multiple contaminants to develop comprehensive remediation approaches. Additionally, understanding how soil properties such as texture, organic matter content, and microbial diversity influence remediation processes will enable more tailored interventions. A deeper understanding of the microbial community dynamics is also essential. Metagenomic and metatranscriptomic analyses can identify key microbial taxa involved in contaminant degradation and elucidate functional pathways. This knowledge can inform the selection or engineering of more effective microbial consortia for bioaugmentation.

Economic analyses comparing phytoremediation to conventional remediation methods are necessary to assess cost-effectiveness. Factors such as the cost of microbial inoculants, plant cultivation, timeframes for remediation, and potential economic benefits from biomass utilization should be considered. Developing scalable and economically viable phytoremediation models will facilitate broader adoption in agricultural practices. The effective implementation of phytoremediation strategies requires supportive policies, regulatory frameworks, and stakeholder engagement. Educating farmers, land managers, and policymakers about the benefits and limitations of phytoremediation will promote its integration into sustainable land management practices.

3.7. Future Perspectives

Advancements in phytoremediation research hold promise for enhancing the efficiency and applicability of this eco-friendly remediation method. Future directions include (a) plant–microbe interactions, exploring novel symbiotic relationships, and co-cultivation techniques can optimize contaminant degradation. Genetic studies may reveal plant traits that enhance microbial colonization and activity; (b) genetic engineering, developing transgenic plants with enhanced metabolic capabilities to degrade specific contaminants offers potential, though ecological risks and ethical considerations must be addressed; (c) integrated remediation strategies, combining phytoremediation with other remediation technologies, such as biostimulation, chemical oxidation, or nanotechnology, may overcome limitations associated with single-method approaches; (d) urban and aquatic applications, extending phytoremediation to urban settings and aquatic environments requires adaptation to unique challenges, such as space constraints and pollutant dispersion dynamics; and (e) policy development and public awareness, establishing clear guidelines, safety protocols, and public education initiatives will facilitate acceptance and trust in phytoremediation practices.

As environmental concerns escalate globally, the practical application of phytoremediation on a larger scale becomes increasingly significant. By addressing the current limitations and capitalizing on emerging research, phytoremediation can evolve into a more robust and versatile tool for mitigating environmental pollution, promoting ecosystem health, and supporting sustainable agricultural practices.

4. Conclusions

This study has demonstrated that integrating leguminous plants with microbial inoculants offers a promising and sustainable strategy for the phytoremediation of tebuthiuron-contaminated agricultural soils. *M. pruriens* exhibited a faster growth rate than *C. ensiformis*, but both species experienced growth inhibition due to the herbicide's phytotoxic effects. The application of microbial inoculants significantly mitigated these negative impacts, with fungal inoculants particularly enhancing plant performance in *M. pruriens*. This suggests that specific fungi may possess metabolic pathways capable of degrading or transforming tebuthiuron into less toxic compounds, thereby promoting plant growth even in contaminated conditions. The use of *C. juncea* as a bioindicator validated the effectiveness of the remediation process. Treatments combining tebuthiuron with bacterial inoculants resulted in the highest biomass production of *C. juncea* and a marked reduction in herbicide toxicity, highlighting the potential of bacterial strains in enhancing tebuthiuron degradation. Ecotoxicological bioassays with *L. sativa* further confirmed that the combination of leguminous plants with microbial inoculants accelerated soil detoxification, achieving higher germination indices and growth rates compared to treatments without inoculants or prior cultivation. These findings accentuate the synergistic effects of plant–microbe interactions in enhancing the degradation of persistent herbicides, improving soil health, and reducing environmental toxicity.

While the results are promising, it is important to recognize that the experiments were conducted under controlled greenhouse conditions, which may not fully replicate the complexities of field environments. Field trials are essential to validate these phytoremediation strategies in real-world settings and to assess their long-term sustainability and effectiveness under diverse environmental conditions. Future research should focus on scaling up these approaches, analyzing microbial community dynamics to identify key degraders of tebuthiuron, optimizing plant–microbe combinations for specific soils and climates, evaluating economic feasibility, and developing supportive policies to facilitate the integration of phytoremediation into sustainable agricultural practices. In conclusion, the integration of leguminous plants with specific microbial inoculants presents a viable and eco-friendly approach for remediating tebuthiuron-contaminated soils. This strategy not only contributes to the restoration of soil quality and protection of environmental health but also advances sustainable land management practices, promoting healthier agricultural ecosystems and benefiting society as a whole.

Author Contributions: Conceptualization: V.H.C., V.L.d.S. and P.R.M.L.; methodology: V.H.C., E.B.d.M., L.G.d.V., E.P.P., R.N.M. and P.R.M.L.; investigation: V.H.C., T.S.V. and Y.A.F.; data curation: B.R.d.A.M., R.N.M. and P.R.M.L.; formal analysis: V.H.C. and B.R.d.A.M.; writing original draft: V.H.C., B.R.d.A.M., YAF, L.T., E.P.P. and P.R.M.L.; review and editing: V.H.C., V.L.d.S., E.B.d.M., L.G.d.V., L.T., E.P.P., R.N.M. and P.R.M.L.; funding acquisition, project administration, resources, supervision, and validation: P.R.M.L. All authors have read and agreed to the published version of the manuscript.

Funding: São Paulo Research Foundation (FAPESP-Brazil, 2021/01884-6); National Council for Scientific and Technological Development (CNPq-Brazil, 313530/2021-1), Fundação Agrisus—Brazil (PA 3740/24), Coordination for Improvement of Higher Education Personnel (CAPES-Brazil, financing code n. 001); and Pro-Rectory of Graduate Studies (Pró-Reitoria de Pós-Graduação) of São Paulo State University (Unesp-Brazil).

Data Availability Statement: The authors confirm that the data supporting the findings of this study are available in an online repository (UNESP-Brazil) and on request from the corresponding author (PRML).

Acknowledgments: The authors acknowledge the São Paulo Research Foundation (FAPESP-Brazil, 2021/01884-6); the National Council for Scientific and Technological Development (CNPq-Brazil, 313530/2021-1), the Fundação Agrisus—Brazil (PA 3740/24), the Coordination for Improvement of Higher Education Personnel (CAPES-Brazil, financing code n. 001); and Pro-Rectory of Graduate Studies (Pró-Reitoria de Pós-Graduação) of São Paulo State University (Unesp-Brazil). We would also like to thank the NETA group (FCAT/Unesp-Brazil), led by Evandro Pereira Prado, for helping with the application of pesticides.

Conflicts of Interest: The authors declare no conflicts of interest.

References

1. Qian, Y.; Matsumoto, H.; Liu, X.; Li, S.; Liang, X.; Liu, Y.; Zhu, G.; Wang, M. Dissipation, Occurrence and Risk Assessment of a Phenylurea Herbicide Tebuthiuron in Sugarcane and Aquatic Ecosystems in South China. *Environ. Pollut.* **2017**, *227*, 389–396. [CrossRef] [PubMed]
2. Herbicide Resistance Action Committee—Global Herbicide Classification Lookup—Tebuthiuron. Available online: https://hracglobal.com/tools/classification-lookup/?s=Tebuthiuron&mode=&letter=&number=#classificationLookup (accessed on 10 April 2024).
3. PPDB—Pesticide Properties Database: Tebuthiuron (Ref: EL 103), 2023. Available online: http://sitem.herts.ac.uk/aeru/iupac/Reports/614.htm (accessed on 1 July 2024).
4. Aransiola, S.A.; Maddela, N.R. *Phytoremediation in Food Safety: Risks and Prospects*; CRC Press: Boca Raton, FL, USA, 2024; ISBN 978-1-04-011546-6.
5. Sagrilo, E.; Leite, L.F.C.; Galvão, S.R.S.; Lima, E.F. *Manejo Agroecológico do Solo: Os Benefícios da Adubação Verde*; Documentos 193; Embrapa: Teresina, PI, USA, 2009; 22p.
6. Mendes, K.F.; Maset, B.A.; Mielke, K.C.; Sousa, R.N.; Martins, B.A.B.; Tornisielo, V.L. Phytoremediation of Quinclorac and Tebuthiuron-Polluted Soil by Green Manure Plants. *Int. J. Phytoremediation* **2021**, *23*, 474–481. [CrossRef] [PubMed]
7. Ferreira, L.C.; Moreira, B.R.A.; Montagnolli, R.N.; Prado, E.P.; Viana, R.S.; Tomaz, R.S.; Cruz, J.M.; Bidoia, E.D.; Frias, Y.A.; Lopes, P.R.M. Green Manure Species for Phytoremediation of Soil with Tebuthiuron and Vinasse. *Front. Bioeng. Biotechnol.* **2021**, *8*, 613642. [CrossRef] [PubMed]
8. Frias, Y.A.; Lima, E.W.; Aragão, M.B.; Nantes, L.S.; Moreira, B.R.A.; Cruz, V.H.; Tomaz, R.S.; Lopes, P.R.M. *Mucuna pruriens* Cannot Develop Phytoremediation of Tebuthiuron in Agricultural Soil with Vinasse: A Morphometrical and Ecotoxicological Analysis. *Front. Bioeng. Biotechnol.* **2023**, *11*, 1156751. [CrossRef] [PubMed]

9. Sarapirom, P.; Wiriyaampaiwong, P.; Rueangsan, K.; Plangklang, P.; Teerakun, M. The Combination Technique of Bioaugmentation and Phytoremediation on the Degradation of Paraquat in Contaminated Soil. *Asia-Pac. J. Sci. Technol.* **2022**, *27*, 9. [CrossRef]
10. Lopes, P.R.M.; Cruz, V.H.; Menezes, A.B.; Gadanhoto, B.P.; Moreira, B.R.A.; Mendes, C.R.; Mazzeo, D.E.C.; Dilarri, G.; Montagnolli, R.N. Microbial Bioremediation of Pesticides in Agricultural Soils: An Integrative Review on Natural Attenuation, Bioaugmentation and Biostimulation. *Rev. Environ. Sci. Biotechnol.* **2022**, *21*, 851–876. [CrossRef]
11. Khan, S.; Afzal, M.; Iqbal, S.; Khan, Q.M. Plant–Bacteria Partnerships for the Remediation of Hydrocarbon Contaminated Soils. *Chemosphere* **2013**, *90*, 1317–1332. [CrossRef]
12. Lima, E.W.; Brunaldi, B.P.; Frias, Y.A.; Moreira, B.R.A.; Alves, L.S.; Lopes, P.R.M. A Synergistic Bacterial Pool Decomposes Tebuthiuron in Soil. *Sci. Rep.* **2022**, *12*, 9225. [CrossRef]
13. Aparício, J.D.; Garcia-Velasco, N.; Urionabarrenetxea, E.; Soto, M.; Álvarez, A.; Polti, M.A. Evaluation of the Effectiveness of a Bioremediation Process in Experimental Soils Polluted with Chromium and Lindane. *Ecotoxicol. Environ. Saf.* **2019**, *181*, 255–263. [CrossRef]
14. Raimondo, E.E.; Saez, J.M.; Aparicio, J.D.; Fuentes, M.S.; Benimeli, C.S. Bioremediation of Lindane-Contaminated Soils by Combining Bioaugmentation and Biostimulation: Effective Scaling-Up from Microcosms to Mesocosms. *J. Environ. Manag.* **2020**, *276*, 111309. [CrossRef]
15. Pires, F.R.; Procópio, S.O.; dos Santos, J.B.; Souza, C.M.; Dias, R.R. Avaliação da Fitorremediação de Tebuthiuron Utilizando *Crotalaria juncea* como Planta Indicadora. *Rev. Ciênc. Agron.* **2008**, *39*, 245–250.
16. *NBR 10006*; Associação Brasileira De Normas Técnicas (ABNT). Procedimento para Obtenção de Extrato Solubilizado de Resíduos Sólidos. ABNT: Rio de Janeiro, Brazil, 2004; 20p.
17. Sobrero, M.C.; Ronco, A. Ensayo de Toxicidad Aguda con Semillas de Lechuga (*Lactuca sativa* L.). In *Ensayos Toxicológicos y Métodos de Evaluación de Calidad de Aguas: Standerización, Intercalibración, Resultados y Aplicaciones*; Morales, G.C., Ed.; IMTA: Mexico City, Mexico, 2004; pp. 71–79.
18. Labouriau, L.G.; Agudo, M. On the Physiology of Seed Germination in *Salvia hispanica* L. I. Temperature Effects. *An. Acad. Bras. Ciênc.* **1987**, *59*, 37–56.
19. Nantes, L.S.; Aragão, M.B.; Moreira, B.R.A.; Frias, Y.A.; Valério, T.S.; Lima, E.W.; Silva Viana, R.; Lopes, P.R.M. Synergism and Antagonism in Environmental Behavior of Tebuthiuron and Thiamethoxam in Soil with Vinasse by Natural Attenuation. *Int. J. Environ. Sci. Technol.* **2023**, *20*, 4883–4892. [CrossRef]
20. Santos, H.G.; Jacomine, P.K.T.; Anjos, L.H.C.; Oliveira, V.A.; Lumbreras, J.F.; Coelho, M.R.; Almeida, J.A.; Araújo Filho, J.C.; Oliveira, J.B.; Cunha, T.J.F. *Sistema Brasileiro de Classificação de Solos*, 5th ed.; Embrapa: Brasília, Brazil, 2018; 355p.
21. Echart, C.L.; Cavalli-Molina, S. Fitotoxicidade do Alumínio: Efeitos, Mecanismo de Tolerância e Seu Controle Genético. *Ciênc. Rural* **2001**, *31*, 531–541. [CrossRef]
22. Calegari, A. Plantas de Cobertura. In *Sistema Plantio Direto com Qualidade*; Casão Junior, R., Siqueira, R., Mehta, Y.R., Passini, J.J., Eds.; IAPAR: Londrina, Brazil, 2006; pp. 55–73.
23. Belo, A.F.; Santos, E.A.; Santos, J.B.; Ferreira, L.R.; Silva, A.A.; Cecon, P.R.; Silva, L.L. Efeito da Umidade do Solo sobre a Capacidade de *Canavalia ensiformis* e *Stizolobium aterrimum* em Remediar Solos Contaminados com Herbicidas. *Planta Daninha* **2007**, *25*, 239–249. [CrossRef]
24. Madalão, J.C.; Souza, M.F.; Silva, A.A.; Silva, D.V.; Jakelaitis, A.; Pereira, G.A.M. Action of *Canavalia ensiformis* in Remediation of Contaminated Soil with Sulfentrazone. *Bragantia* **2017**, *76*, 292–299. [CrossRef]
25. Zhang, N.; Guo, D.; Zhu, Y.; Wang, X.; Zhu, L.; Liu, F.; Teng, Y.; Christie, P.; Li, Z.; Luo, Y. Microbial Remediation of a Pentachloronitrobenzene-Contaminated Soil under *Panax notoginseng*: A Field Experiment. *Pedosphere* **2020**, *30*, 563–569. [CrossRef]
26. Madariaga-Navarrete, A.; Rodríguez-Pastrana, B.R.; Villagómez-Ibarra, J.R.; Acevedo-Sandoval, O.A.; Perry, G.; Islas-Pelcastre, M. Bioremediation Model for Atrazine Contaminated Agricultural Soils Using Phytoremediation (*Phaseolus vulgaris* L.) and a Locally Adapted Microbial Consortium. *J. Environ. Sci. Health B* **2017**, *52*, 367–375. [CrossRef]
27. Camargo, D.; Bispo, K.L.; Sene, L. Associação de *Rhizobium* sp. a Duas Leguminosas na Tolerância à Atrazina. *Rev. Ceres* **2011**, *58*, 425–431. [CrossRef]
28. Abdulrahman, A.A.; Oladele, F.A. Leaf Size and Transpiration Rates in *Agave americana* and *Aloe vera*. *Phytol. Balc.* **2017**, *23*, 95–100.
29. Kuiper, I.; Lagendijk, E.L.; Bloemberg, G.V.; Lugtenberg, B.J.J. Rhizoremediation: A Beneficial Plant-Microbe Interaction. *Mol. Plant-Microbe Interact.* **2004**, *17*, 6–15. [CrossRef] [PubMed]
30. Christoffoleti, P.J.; López-Ovejero, R.F.; Damin, V.; Carvalho, S.J.P.; Nicolai, M. *Comportamento dos Herbicidas Aplicados ao Solo na Cultura da Cana-de-Açúcar*, 1st ed.; CP: Piracicaba, SP, Brasil, 2009; 72p.
31. Oliveira, M.A.; Pires, F.R.; Ferraço, M.; Belo, A.F. Validation of an Analytical Method for Sulfentrazone Residue Determination in Soil Using Bacterial Chromatography and a Comparison of Chromatographic Sensitivity to Millet as a Bioindicator Species. *Molecules* **2014**, *19*, 10982–10997. [CrossRef]
32. Mostafa, F.I.Y.; Helling, C.S. Isolation and 16S DNA Characterization of Soil Microorganisms from Tropical Soils Capable of Utilizing the Herbicides Hexazinone and Tebuthiuron. *J. Environ. Sci. Health B* **2003**, *38*, 783–797. [CrossRef]
33. Sheldon, P.J.; Johnson, D.J.; August, P.R.; Liu, H.W.; Sherman, D.H. Characterization of a Mitomycin-Binding Drug Resistance Mechanism from the Producing Organism. *J. Bacteriol.* **1999**, *179*, 1796–1804. [CrossRef] [PubMed]

34. Erasmo, E.A.L.; Azevedo, W.R.; Sarmento, R.A.; Cunha, A.M.; Garcia, S.L.R. Potencial de Espécies Utilizadas como Adubo Verde no Manejo Integrado de Plantas Daninhas. *Planta Daninha* **2004**, *22*, 337–342. [CrossRef]
35. Melo, C.A.D.; Passos, A.B.R.J.; Madalão, J.C.; Silva, D.V.; Massenssini, A.M.; Silva, A.A.; Costa, M.D.; Souza, M.F. Bioaugmentation as an Associated Technology for Bioremediation of Soil Contaminated with Sulfentrazone. *Ecol. Indic.* **2019**, *99*, 343–348. [CrossRef]
36. Harman, G.E.; Howell, C.R.; Viterbo, A.; Chet, I.; Lorito, M. *Trichoderma* Species—Opportunistic, Avirulent Plant Symbionts. *Nat. Rev. Microbiol.* **2004**, *2*, 43–56. [CrossRef]
37. Wang, L.; Chi, X.-Q.; Zhang, J.-J.; Sun, D.-L.; Zhou, N.-Y. Bioaugmentation of a Methyl Parathion Contaminated Soil with *Pseudomonas* sp. Strain WBC-3. *Int. Biodeterior. Biodegrad.* **2014**, *87*, 116–121. [CrossRef]
38. Xie, J.; He, Z.; Liu, X.; Liu, X.; Van Nostrand, J.D.; Deng, Y.; Wu, L.; Zhou, J.; Qiu, G. GeoChip-Based Analysis of the Functional Gene Diversity and Metabolic Potential of Microbial Communities in Acid Mine Drainage. *Appl. Environ. Microbiol.* **2011**, *77*, 991–999. [CrossRef]
39. Wauchope, R.D.; Buttler, T.M.; Hornsby, A.G.; Augustijn-Beckers, P.W.M.; Burt, J.P. The SCS/ARS/CES Pesticide Properties Database for Environmental Decision-Making. *Rev. Environ. Contam. Toxicol.* **1992**, *123*, 1–155. [CrossRef]
40. Pignatello, J.J.; Xing, B. Mechanisms of Slow Sorption of Organic Chemicals to Natural Particles. *Environ. Sci. Technol.* **1996**, *30*, 1–11. [CrossRef]
41. Beesley, L.; Moreno-Jiménez, E.; Gomez-Eyles, J.L. Effects of Biochar and Greenwaste Compost Amendments on Mobility, Bioavailability, and Toxicity of Inorganic and Organic Contaminants in a Multi-Element Polluted Soil. *Environ. Pollut.* **2010**, *158*, 2282–2287. [CrossRef] [PubMed]
42. Dietz, A.C.; Schnoor, J.L. Advances in Phytoremediation. *Environ. Health Perspect.* **2001**, *109*, 163–168. [CrossRef] [PubMed]
43. Strong, P.J.; Claus, H. Laccase: A Review of Its Past and Its Future in Bioremediation. *Crit. Rev. Environ. Sci. Technol.* **2011**, *41*, 373–434. [CrossRef]

agronomy

MDPI

Article

Environmental Constraints Mitigation Directly Drove the Diversifications of Fungal Community and Functional Profile in Amended Coastal Salt-Affected Soils

Yunlong Li [1], Weifeng Chen [1], Yiwei Xu [1], Jie Yu [2], Wengang Zuo [1], Yuhua Shan [1,3] and Yanchao Bai [1,4,*]

1 College of Environmental Science and Engineering, Yangzhou University, Yangzhou 225127, China; liyunlong@yzu.edu.cn (Y.L.); weifengchen334@163.com (W.C.); 15868772091@163.com (Y.X.); 007615@yzu.edu.cn (W.Z.); shanyh@yzu.edu.cn (Y.S.)

2 Department of Agriculture and Rural Affairs of Jiangsu Province, Jiangsu Cultivated Land Quality and Agro-Environment Protection Station, Nanjing 210003, China; 18851119218@163.com

3 Jiangsu Collaborative Innovation Center for Solid Organic Waste Resource Utilization, Nanjing 210095, China

4 Key Laboratory of Arable Land Quality Monitoring and Evaluation, Key Laboratory of Saline-Alkali Soil Improvement and Utilization (Coastal Saline-Alkali Lands), Ministry of Agriculture and Rural Affairs, Yangzhou University, Yangzhou 225127, China

* Correspondence: ycbai@yzu.edu.cn

Abstract: Although organic amendment has been widely demonstrated to be capable of reassembling soil microbiomes in coastal salt-affected soils, quantitative characterization in respect to how abiotic and biotic components drive the diversifications of soil microbial community and function remains rudimentary. We investigated the effects of types and application rates of different exogenous organic ameliorants (sewage sludge, S; vermicompost, V) on the physicochemical properties, fungal community diversity, and fungal functional traits in coastal salt-affected soils. Results revealed that both S and V amendments exhibited significantly positive impacts on the alleviation of environmental constraints and the reassembly of fungal community and functional profile. Of note, efficacy of the two exogenous organic ameliorants was significantly influenced by the types of carbon sources utilized and the rates at which they are applied. More pronounced variations in soil physicochemical and microbiological properties were observed in soils amended by S and V at the application rate of 200 t ha^{-1}. Both S and V treatments positively boosted soil fungal community diversification with higher compositional diversities found in the majority of amended soils when compared to unamended soils. In addition, S and V applications favored the flourishment of functions linked to saprotrophic strategy with soil saprotrophs and wood saprotrophs predominating the functional profiles in soils amended by S and V, respectively. Results obtained from redundancy analysis and structural equation modeling revealed that pH, salinity, and SOC were the principle edaphic factors that significantly and directly affected the reassembly of fungal community. In addition, fluctuations in the prevalence of specific fungal genera, including *Stachybotrys*, *Trichoderma*, *Mortierella*, and *Acremonium*, emerged as a decisive biotic factor influencing diversifications of fungal communities and functional profiles. Taken together, this study not only highlighted the efficacy of S and V application on environmental constraints alleviation in coastal mudflats, but also qualitatively identified and quantitatively illustrated potential pathways and effectiveness of edaphic and biotic components driving the diversification of soil fungal microbiomes and functional profiles. These findings would enhance our understanding in respect to the microbial-mediated amelioration process of salt-affected soils within coastal mudflat ecosystem.

Keywords: coastal salt-affected soil; exogenous organic ameliorants; salinization; soil organic carbon; fungal community reassembly; fungal functional profile

1. Introduction

Coastal mudflats are extensively distributed globally, covering an estimated area of approximately 1.8 million km^2 [1]. Utilization of coastal mudflats has been recognized as an effective approach to address the imbalance between agricultural land supply and demand, which has been exacerbated by rapid population growth and urban expansion [2]. However, as a result of prolonged exposure to seawater and significant evapotranspiration, newly reclaimed lands in coastal mudflat regions have inherent limitation for subsequent cultivations due to numerous obstacle factors in soils. Specifically, salinization, soil organic carbon (SOC) shortage, nutritional deficiency (e.g., nitrogen, phosphorus, and potassium), and undiverse microbial community structure and function are commonly regarded as the main obstacle factors of coastal saline-alkali soils [3]. Consequently, overcoming these abiotic and biotic constraints mentioned above is essential for sustainable development of coastal salt-affected soils [4]. However, it is full of challenges as these abiotic and biotic constraints are concomitants and interact with each other in complex ways.

It is well-established that soil microbial communities, as well as their associated functionalities, play a vital role in enhancing soil fertility and sustainability, particularly in coastal regions affected by salinity [5–7]. Consequently, diversification of microbial community structure and function has been extensively regarded as one of the crucial links to land quality and soil fertility improvement in coastal mudflat lands. Numerous research investigations have demonstrated that soil microbiome assemblage is profoundly affected by environmental factors that are specific to particular habitats [8,9]. Generally, saline-alkali stress was found to be negatively related to community compositional and functional diversity, while improved nutritional status exhibited opposite impacts usually [10,11]. Of note, most findings mentioned above were reported in farmland soils, while knowledge regarding dominant drivers affecting soil microbial community assembly remains a topic of active debate in coastal highly salinized lands. For example, multiple studies found that fungal communities in saline soil exhibited greater vulnerability to soil saline-alkali stress than bacterial communities. In contrast, several studies revealed that fungal populations were more tolerant to soil salinity [12,13]. Furthermore, Zhang et al. [14] and Vanegas et al. [15] documented that soil pH and salinity were decisive drivers in soil microbiomes' reconstruction, while Rath et al. [13] demonstrated that soil microbial community was significantly influenced by the scarcity of SOC directly rather than the salinity levels. Consequently, elucidating microbial variations in response to soil environmental change and deciphering mechanisms underlying these changes will facilitate the targeted improvement of coastal saline soils.

Exogenous organic ameliorants have been increasingly applied for the quick elevation of SOC, which is commonly viewed as one essential prerequisite for successful amelioration of coastal salt-affected soils [16,17]. On one hand, organic matter-rich ameliorant applications can improve soil structure and, thereby, reduce the accumulation of salt in surface soil through capillary action, resulting in a significant mitigation of salt damages on soil microbes [18,19]. Moreover, ample carbon substrates, energy substances, and nutrients introduced by exogenous organic ameliorants play significant roles in the flourishment of soil microbiomes in coastal mudflats [20]. Recent investigations revealed that organic ameliorant amendments were conducive to the alleviation of obstacle factors and diversification of microbial communities and their functional capabilities in coastal saline soils [21,22]. Nevertheless, contrasting research findings were documented due to the different properties of soil and organic ameliorants (e.g., types of soil and organic matters, application rates and availabilities of organic matters, patterns and durations of organic amendments, etc.) [23–25].

Obviously, mechanisms governing the effects of organic ameliorant applications on the soil microbiome in coastal mudflat regions require further investigation. The current study carried out field experiments to assess the impacts of exogenous organic ameliorants with different types and application rates on soil physicochemical and microbiological properties in saline soils of coastal mudflats. We aimed to qualitatively identify what abiotic and biotic factors influence soil fungal microbiome reassembly and quantitatively evaluate the significance of the key agents driving the diversifications of fungal community and functionality. The corresponding outcomes will broaden our knowledge in respect to the fundamental mechanisms associated with microbial community assembly and provide practical guidance for sustainable exploitation and utilization of saline soils in coastal mudflat environments.

2. Materials and Methods

2.1. Experimental Site and Organic Ameliorants Used in This Study

The experiment site was located in Dongtai City (120°56′51″ E, 32°49′56″ N), Jiangsu Province, China. This area is characterized by a northern subtropical continental monsoon climate. The average annual air temperature is 16.5 °C, with the lowest and highest mean monthly temperatures occurring in January (2 °C) and July (32 °C), respectively. The mean annual precipitation is approximately 1417 mm, and the main period of rainfall is from June to September. The dominant soil type is categorized as Solonchaks [26] and the main planting system in this area is rice–wheat rotation. In this study, S and V were chosen as exogenous organic ameliorants for coastal salt-affected soil amendment, which were procured from a local fertilizer manufacturer. The main physicochemical characteristics of the initial soil and the two organic ameliorants are listed in Table 1 [27].

Table 1. Physicochemical properties of the soils and exogenous organic materials.

Treatments		pH	EC ($mS\ cm^{-1}$)	SOC ($g{\cdot}kg^{-1}$)	TN ($g{\cdot}kg^{-1}$)	C/N	TP ($g{\cdot}kg^{-1}$)	AN ($mg{\cdot}kg^{-1}$)	AP ($mg{\cdot}kg^{-1}$)
Mudflat soil		9.02	1.38	1.97	0.282	6.99	0.507	17.08	6.99
Sewage sludge		6.32	2.77	216.2	51.2	4.22	5.51	3440	813
Vermicompost		6.33	1.46	464.9	19.27	24.1	24.12	15.87	2467
CK		9.33 ± 0.09 a	1.55 ± 0.07 a	3.60 ± 0.32 c	0.33 ± 0.01 b	10.78 ± 3.19 a	0.4 ± 0.02 c	38.0 ± 0.3 a	21.4 ± 1.8 c
S50		9.11 ± 0.04 ab	1.25 ± 0.04 b	11.27 ± 1.13 b	0.67 ± 0.06 b	16.90 ± 4.60 a	0.61 ± 0.04 bc	107.1 ± 14.8 a	84.5 ± 11.4 bc
S100		8.78 ± 0.02 cd	0.82 ± 0.08 d	11.92 ± 1.83 b	0.70 ± 0.13 b	17.12 ± 6.35 a	0.68 ± 0.14 b	82.8 ± 11.4 a	94.8 ± 21.6 ab
S200		8.57 ± 0.20 d	0.72 ± 0.06 d	21.69 ± 3.71 a	1.33 ± 0.35 a	16.35 ± 6.08 a	1.24 ± 0.09 a	183.8 ± 64.0 a	159.0 ± 37.4 a
V50		9.04 ± 0.06 abc	1.34 ± 0.04 b	11.73 ± 1.32 b	0.68 ± 0.03 b	17.27 ± 5.16 a	0.72 ± 0.06 b	83.9 ± 7.4 a	85.0 ± 7.6 bc
V100		8.96 ± 0.08 bc	1.22 ± 0.03 b	13.59 ± 3.44 b	0.90 ± 0.25 ab	15.06 ± 5.56 a	0.82 ± 0.10 b	112.4 ± 23.4 a	108.2 ± 15.4 ab
V200		8.83 ± 0.08 bcd	0.98 ± 0.02 c	16.46 ± 2.12 ab	0.95 ± 0.18 ab	17.38 ± 5.35 a	0.83 ± 0.10b	128.3 ± 34.3 a	127.9 ± 27.5 ab
Two-way ANOVAs	T	ns	***	ns	ns	ns	ns	ns	ns
	R	**	***	**	**	ns	***	ns	*
	T × R	ns	*	ns	ns	ns	**	ns	ns

Notes: CK, control soil, "S" and "V" indicate coastal salt-affected soils amended by sewage sludge and vermicompost, respectively. Numbers "50", "100", and "200" adjoining S and V indicate the two organic ameliorants applied at amounts of 50 t ha^{-1}, 100 t ha^{-1}, and 200 t ha^{-1}, respectively. Values are means and standard errors (n = 3). Different letters in each row denote significant differences among treatments (Duncan's test, $p < 0.05$). T: types of organic amendments; R: application rates of organic amendments; T × R: interaction of types application rates. * $p < 0.05$, ** $p < 0.05$, *** $p < 0.001$, ns: p not significant (two-way ANOVA).

2.2. Experimental Layout and Soil Sampling

Field microplot experiments consisted of three treatment groups, namely, (1) coastal salt-affected soil without organic amendments (CK); (2–4) S-groups, soils amended by S; (5–7) V-groups, soils amended by V. The application rates of S and V were 50, 100, and 200 t ha^{-1}. Every treatment was comprised of three randomly distributed microplots (4 m × 4 m). S and V were once applicated into each microplot according to corresponding design application rates and mixed uniformly with the 0~20 cm soil layer using a rototiller in October 2019. Subsequently, two salt-enduring crops, barley (*Hordeum vulgare* L.) and maize (*Zea mays* L.), were cultivated in succession and rotated. Soil samples were collected in October 2020. In October 2020, a total of 21 soil samples were obtained from every microplot, with each soil sample comprised of five random soil cores (0~20 cm). Afterward, soil samples were homogenized and, respectively, stored at −80 °C and 4 °C for the following determinations.

2.3. Measurements of Soil Physico-Chemical Properties

Soil pH, electrical conductivity (EC), contents of SOC, TN, TP, AN, and AP, were determined following the methodology outlined by Lu [28]. Specifically, pH and EC were measured in soil–water extract suspension by using a PE-20 digital pH meter and a PE-30 EC meter, respectively. Determinations of SOC, TN, TP, AN, and AP of soil were conducted through $K_2Cr_2O_7$-H_2SO_4 digestion, Kjeldahl digestion, and Mo-Sb colorimetry, alkaline hydrolysis diffusion, and molybdenum blue methods, respectively.

2.4. Amplicon Sequencing and Bioinformatics Processing

Soil genomic DNA was isolated, quality-controlled, amplified, and sequenced following the established methods of Li et al. [29]. DADA2 plugin [30] in QIIME2 (version 2023.5) pipeline [31] was applied for sequence denoising. The Naïve Bayes consensus taxonomy classifier available in QIIME2 was performed for the taxonomic categorization of amplicon sequence variants (ASVs), and UNITE 8.0 database was employed for the classification of bacterial and fungal taxa, respectively. Afterward, fungal sequences across treatments were rarefied to 40,000 for subsequent community diversity analysis. Compositional diversity indices of soil fungal community were calculated to characterize the differentiations in community composition across treatments. Principal coordinates analysis (PCoA) accompanied by permutational multivariate analysis of variance (PERMANOVA) with 999 permutations were employed to assess the dissimilarities in fungal community structural diversity among treatments based on BC distance matrices. ASVs significantly enriched or depleted in each organic ameliorant treatments in comparison to control were, respectively, detected by using the R package DESeq2 (version 1.30.0). Functional prediction of fungal community was performed based on the Fungal Traits database [32].

2.5. Statistical Analysis

One-way analysis of variance (ANOVA) and multiple comparative analyses (Duncan) were performed for the difference significance test of soil physicochemical properties, community diversity indices (e.g., Shannon diversity, richness, evenness, and Bray–Curtis dissimilarity), community taxonomic compositions, and fungal functions across all treatments ($p < 0.05$). Two-way ANOVA was performed to test the effectiveness of types and application rates of exogenous organic materials on selected soil physicochemical properties and fungal community diversity indices. Principal component analysis (PCA) was employed to identify the primary fungal populations accounting for the variations in functions of fungal microbiomes across all treatments. Transformation-based redundancy discriminant analysis (tb-RDA) was performed to assess the explanatory powers of soil physicochemical characteristics on soil fungal community structure. Similarly, RDA was employed to examine the associations between fungal community functional profiles and dominant fungal genera (top 20). Procrustes analysis was applied to elucidate the correlations between edaphic variables, fungal microbiome, and functional traits. Structural equation modeling (SEM) analysis was performed to explore the relationships among soil physicochemical property, fungal microbiome, and fungal functioning using *lavaan* R package. The goodness of fit of the model was assessed through non-significant chi-square test (χ^2, $p > 0.05$), along with the goodness-of-fit index (GFI), and root mean square error of approximation (RMSEA).

3. Results

3.1. Soil Edaphic Properties

The utilization of organic ameliorants resulted in a statistically significant ($p < 0.05$) improvement in the soil edaphic variables with lower pH and salinity, and higher SOC and nutrient content found in soils amended by S and V in comparison to control soil (Table 1). pH and salinity significantly declined with the increasement of application rate of the two organic ameliorants, while C-, N-, and P-content showed opposite contrary trends. Particularly, S rather than V exhibited more profound effects on soil saline-alkali obstacle alleviation and nutrient status (C-, N-, P-content) improvement, with the minimum values of pH and salinity and maximum values of SOC, TN, TP, AN, and AP recorded in S200 treatment in comparison to those in other treatments. In addition, only soil salinity was significantly affected by organic ameliorants types, application rates, and their interaction effects concurrently.

3.2. Fungal Community Diversities

Organic amendments significantly ($p < 0.05$) altered soil fungal community diversities (Figure 1). Compositional diversities were all significantly affected by organic ameliorant types, application rates, and their interactions. Compositional diversity increased with the increasement of application rates of S and V (except for richness in V50) (Figure 1a–c). As for structural diversity, organic ameliorant types and application rates exhibited significant impacts on fungal community structure (Figure 1d). More pronounced Bray–Curtis dissimilarity was found in vermicompost-amended soils as compared to S-related soils. Results obtained from PCoA and hierarchical clusters showed that fungal community structures drastically ($p < 0.001$) altered as a result of S and V amendments. Apparently, S- and V-amended soils exhibited unique community structures that were distinctly different from those in unamended soil. Comparatively, fungal community structures in soils amended by high amounts of ameliorants (e.g., S100, S200, V100, and V200) showed larger differentiation as compared to those in soils amended by low amounts of S and V (S50 and V50) (Figure 1e,f).

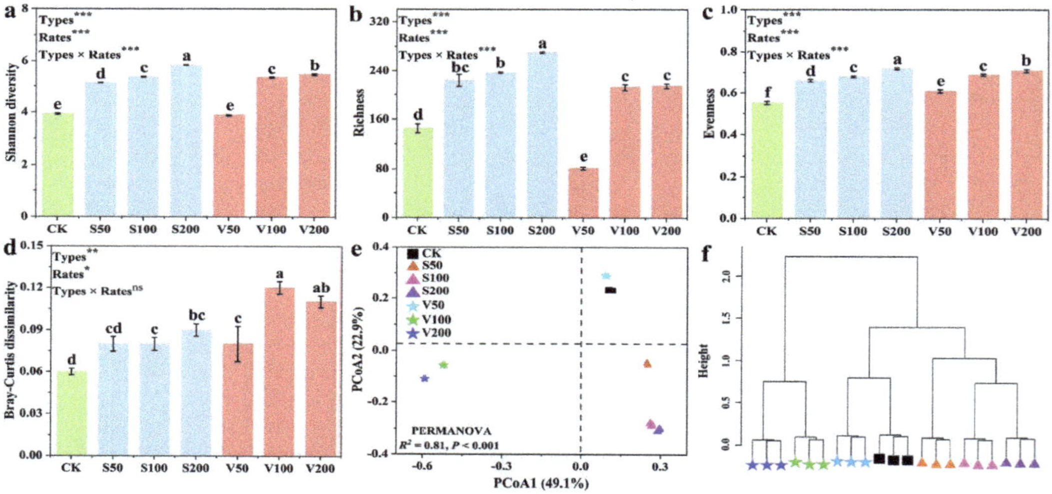

Figure 1. Soil fungal community diversities across treatments. (**a–c**) indicate fungal community compositional diversity indices, (**d**) indicates dissimilarities within fungal community structure, (**e**,**f**) indicate PCoA and hierarchical cluster analysis, respectively. *Error bars* denote standard errors, and different letters denote significant differences among treatments (Duncan's test, $p < 0.05$). T: types of organic amendments; R: application rates of organic amendments; T × R: interaction of types application rates. * $p < 0.05$, ** $p < 0.05$, *** $p < 0.001$, ns: p not significant (two-way ANOVA).

3.3. Fungal Community Taxanomic Compostion

Organic amendments exerted a significant influence ($p < 0.05$) on the composition of fungal communities across various taxonomic levels (Figure 2). At phylum level, *Ascomycota*, followed by *Basidiomycota*, *Mortierellomycota*, *Chytridiomycota*, and *Rozellomycota*, were dominant fungal phyla with accumulative relative abundances (RAs) larger than 96.0% across different treatments (Figure 2a). In particular, *Ascomycota* was the predominant fungal phylum among treatments, with RAs significantly decreased from 71.2% (CK) to 52.0~52.9% (S related treatments) and 49.4~65.6% (V related treatments), respectively (Figure 2a). At genus level, the predominant fungal genera of each treatment were *Coprinellus* (CK, 16.1%), *Coprinellus* (S50, 16.1%), *Mortierella* (S100, 20.2%), *Mortierella* (S200, 25.5%), *Fusarium* (V50, 15.9%), *Mortierella* (V100, 12.7%), and *Mortierella* (V200, 15.4%), respectively. Compared with CK, enriched *Mortierella*, *Trichoderma*, and *Phanerodontia* were observed in all organic-matter-amended soils, whereas genera *Stachybotrys*, *Cladorrhinum*, and *Alternaria*, showed opposite patterns. Particularly, *Preussia* and *Waitea* positively responded to the applications of sewage sludge; conversely, *Phialophora*, *Acremonium*, *Talaromyces*, and *Candolleomyces* genera positively responded to the V amendment with higher RAs only observed in V-related soil samples (Figure 2b). Additionally, soils amended with S (Figure 2c–e) and V (Figure 2f–h), respectively, harbored distinct fungal community compositions with specific sets of significantly responding ASVs as compared to CK. In total, 120~171 and 91~160 ASVs with significant variations in RAs were detected in soils amended by S and V in comparison to control, respectively. In particular, ASVs assigned to the *Mortierella* were significantly enriched in all amended soils, and ASVs belonging to the *Trichoderma* and *Leucothecium* were the top-responding fungal genera, with significant enrichment found in S-amended soils, whereas genus *Acremonium* was the top responding fungi, with significantly higher RAs in all V-amended soils in comparison to control soil.

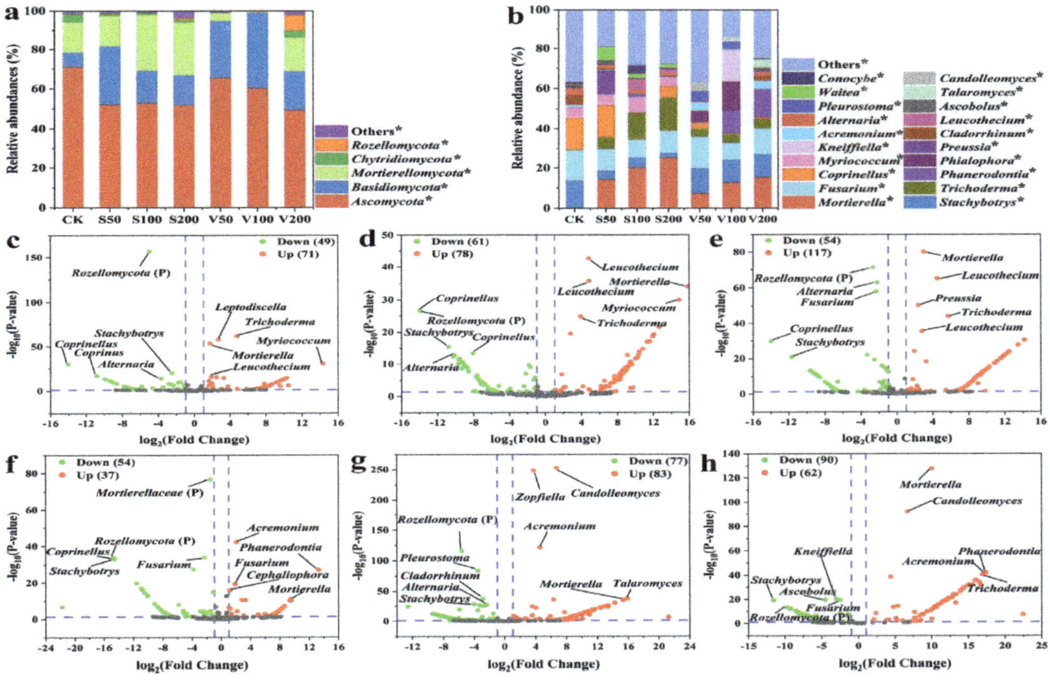

Figure 2. Fungal community compositions at different taxonomic levels. (**a**,**b**) indicate dominant fungal phyla (top 5) and genera (top 20) across different treatments, respectively; (**c**–**h**), ASVs with significant variations in RAs in soils amended by organic ameliorants in comparison to control soil, respectively ($p < 0.05$). "*" denotes significant differences among treatments (Duncan's test, $p < 0.05$).

3.4. Fungal Functioinal Traits

Organic amendments resulted in significant ($p < 0.05$) variations in functional profiles of fungal microbiome (Figure 3). Predicted functions across different treatments were classified into three groups, namely, saprotroph, followed by pathotroph and symbiotroph, which accounted for 75.8~89.4% of all functional traits. Particularly, saprotroph was the predominant fungal functionality among different treatments with RAs significantly increased from 52.6% (CK) to 58.9~61.2% (S related treatments), and 60.2~85.0% (V related treatments), respectively (Figure 3a). At the primary level, fungal trophic traits across different treatments were classified into 13 categories, namely, animal parasite, arbuscular mycorrhizal, dung saprotroph, epiphyte, foliar endophyte, litter saprotroph, mycoparasite, nectar/tap saprotroph, plant pathogen, pollen saprotroph, soil saprotroph, wood saprotroph, and unspecified saprotroph. The predominant fungal traits of each treatment were unspecified saprotroph (CK, 20.1%), soil saprotroph (S50, 33.2%), soil saprotroph (S100, 40.2%), soil saprotroph (S200, 34.8%), wood saprotroph (V50, 57.3%), wood saprotroph (V100, 59.2%), and wood saprotroph (V200, 35.4%), respectively (Figure 3b). Additionally, PCoA revealed that S and V amendments significantly ($p < 0.001$) altered the functional community structure of the fungal microbiome. Comparatively, functional community structures in soils amended by high amounts of ameliorants showed larger differentiation as compared to those in soils amended by low amounts of organic ameliorants (Figure 3c). PCA showed similar results that S- and V-related soils harbored apparently distinct functional profiles, which separated from the CK treatment. Additionally, arbuscular mycorrhizal, followed by plant pathogen, foliar endophyte, epiphyte, mycoparasite, soil saprotroph, and wood saprotroph, were dominant components accounting for the dissimilarities of organic amended soils in comparison to control soil (Figure 3d).

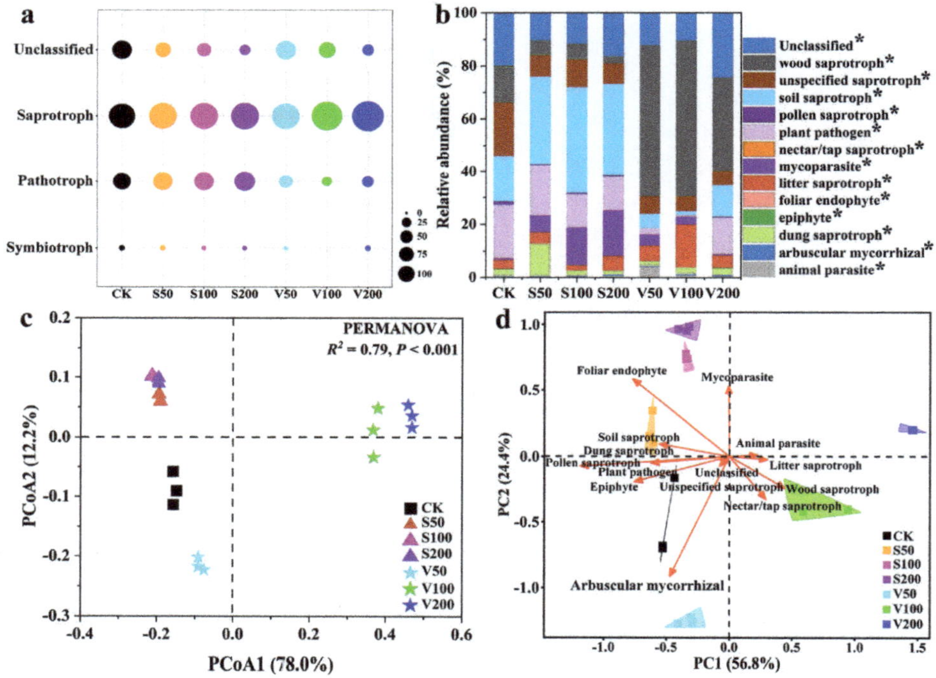

Figure 3. Fungal functionalities across treatments based on the Fungal Traits database. (**a**,**b**) indicate variations in different categorical traits of fungal community, (**c**), PCoA of community functional structure at genus level, (**d**) primary functional traits obtained from PCA. "*" denotes significant differences among treatments (Duncan's test, $p < 0.05$).

3.5. Linking Soil Physicochemical Property, Fungal Microbiome, and Functional Profile

Procrustes analysis revealed that the ordination of soil fungal microbiome significantly correlated with soil physicochemical properties ($M^2 = 0.7644$, $p = 0.039$) (Figure 4a). RDA results showed that six environmental factors, namely, salinity, accompanied by SOC, pH, AP, C/N, and AN, were strongly related to the obvious partitioning of taxonomic structures of fungal microbiome in coastal salt-affected soils (Figure 4c). These physicochemical attributes explained 35.9% ($p = 0.036$) of the soil fungal microbiome variations in total, with the explanatory capabilities of the initial two axes accounting for 31.3% (axe1, $p_{adj} = 0.029$) and 4.6% (axe2, $p_{adj} = 0.611$), respectively. As for functional profiles, results derived from Procrustes analysis showed that soil fungal functions significantly correlated with community composition ($M^2 = 0.4599$, $p < 0.001$) (Figure 4b). RDA results revealed that dominant fungal genera, such as *Trichoderma*, *Mortierella*, *Acremonium*, *Phanerodontia*, *Talaromyces*, *Candolleomyces*, *Stachybotrys*, and *Leucothecium*, were main variables leading to the functional variance of soil fungal microbiome (Figure 4c). These fungal populations explained 70.8% ($p = 0.016$) of the soil fungal microbiome variations in total, with the explanatory capabilities of the initial two axes accounting for 58.9% (axe1, $p_{adj} = 0.017$) and 11.9% (axe2, $p_{adj} = 0.236$), respectively (Figure 4d).

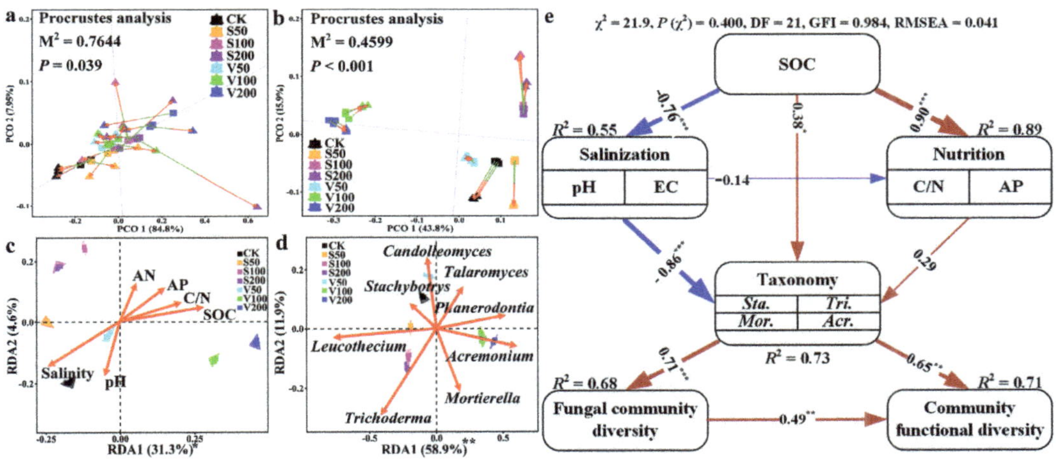

Figure 4. Correlations among physicochemical properties, fungal community, and functional traits. a and b indicate Procrustes analysis of associations between fungal community structure and soil physicochemical property (**a**) and functional profile and community composition (**b**), respectively; c and d indicate RDA of abiotic drivers in variations of soil fungal community (**c**) and biotic drivers (genera level) in variations of soil functional structure (**d**), respectively. (**e**) Structural equation models (SEMs) revealing the relationships among soil physicochemical properties, fungal microbiome, and community functioning. Red and blue arrows signify positive and negative correlations, respectively. Arrow width corresponds to magnitude of the association. Values adjacent to the arrows denote standardized path coefficients. R^2 denotes explained variance for induced variables. *Sta.*, *Tri.*, *Mor.*, and *Acr.* indicate relative abundances of genera *Stachybotrys*, *Trichoderma*, *Mortierella*, and *Acremonium*, respectively. "*", "**", and "***" denote significant correlations at $p < 0.05$, $p < 0.01$, and $p < 0.001$, respectively.

SEM analysis exhibited a good fit for the physicochemical properties, fungal microbiomes, and community functions in coastal salt-affected soils ($\chi^2 = 21.9$, $p = 0.400$, GFI = 0.984, RMSEA < 0.041) (Figure 4e). We observed SOC was closely and negatively related to soil salinization (standardized path coefficient = −0.76, $p < 0.001$), whereas significantly and positively correlated with soil nutrition (standardize path coefficient = 0.90, $p < 0.001$). Furthermore, SOC (standardized path coefficient = 0.38, $p < 0.01$) and salinization (standardized path coefficient = −0.86, $p < 0.001$) jointly played strong roles in directly controlling variations in taxonomic compositions, including genera of *Stachybotrys*, *Trichoderma*, *Mortierella*, and *Acremonium*. Moreover, taxonomic composition of fungal microbiome was significantly and positively (standardized path coefficient = 0.71, $p < 0.001$) related to fungal community diversity (PCO1). Overall, taxonomic composition (standardized path coefficient = 0.65, $p < 0.01$) accompanied by fungal community diversity (standardized path coefficient = 0.49, $p < 0.01$) jointly drove the diversification of community functional diversity (PC1), which, in total, explained 71% of the variations in the fungal community function.

4. Discussion

It is well documented that applying exogenous organic ameliorants is favorable for obstacle factor alleviation in coastal salt-affected soils [29,33]. In the current study, significantly alleviated soil limiting factors observed in amended soils were consistent with previous studies [34]. Specifically, reduced pH in organic amended soils may be partially ascribed to the neutralizing effects of lower-pH S (pH = 6.32) and V (pH = 6.33). Furthermore, the release of acidic compounds, such as humic acids and fulvic acids, during the degradation of S and V may have also contributed to the observed decrease in soil pH [35]. Consistent with other studies, increased carbon storage and nutrient contents observed in amended soils might be primarily attributed to the abundant organic substances, nitrogen, and phosphorus, being brought into soils by applying S and V [36,37]. Our results also revealed that SOC elevation significantly and negatively related to the reduction of soil salinity, suggesting that organic matter inputs can not only enhance SOC sequestration in coastal salt-affected soils, but also facilitate soil structure improvement with more noncapillary pores and eventually promote the downward leaching of salt in the plough layer [18,19]. Comparatively, noticeable differences in soil physicochemical properties were detected in soils amended by different organic ameliorants (S vs. V) and application rates (50 vs. 100 vs. 200 t ha^{-1}), which implied that effects of exogenous organic ameliorants on environmental constraint alleviation in coastal saline soils were significantly influenced by the types of carbon sources utilized and the rates at which they were applied. It aligned with those reported by Li et al. [27], who also noted that more pronounced performance of S and higher-amount amendments on obstacle factor alleviation as compared to V and lower-amount treatments, respectively. Overall, these findings indicated that the application of S and V was feasible and an effective agricultural strategy for mitigating environmental constraints in coastal salinized soils.

In addition to the amelioration of soil physicochemical attributes, organic amendments have been extensively demonstrated to facilitate the reassembly of soil microbial communities [21,22]. In the present study, we found organic materials drastically modulated soil fungal microbiomes. Moreover, significant correlations were found among soil abiotic factors (e.g., pH, salinity, SOC, C/N, AN) and fungal characteristics (e.g., community composition and structure), implying that reassembly of fungal microbiome in coastal saline soils was dominated by the specific microhabitat constructed by different exogenous organic ameliorant applications [38,39]. Generally, compositional diversities were higher in organic-matter-amended soils, possibly on the basis that organic-matter-induced microhabitat modification favored the colonization of more diverse microbial populations, which eventually contributed to the formation of fungal community diversification [40,41]. Soil pH, salinity, SOC, nutrient, and C: N ratios were widely considered as key determinants influencing microbial community assembly among multitudinous edaphic variables, particularly in saline-alkali soils [42]. The present study indicated that mitigated saline-alkali stress in organic-material-amended soils could promote the colonization of more microbial populations, as only a limited number of microorganisms exhibit tolerance to fluctuations in pH and salinity levels [43,44]. In contrast, enhanced supply of carbon substrates, energy substances, and nutrients are also responsible for the flourishment of more divergent microorganisms, as they are essential for microbial growth and proliferation [20]. The C: N ratio has traditionally been regarded as an indicator of soil nutrient availability. Fluctuations in soil C: N ratio can disrupt the equilibrium of fungal stoichiometry, subsequently influencing fungal anabolic processes and nutrient acquisition strategies [14]. Typically, the optimal C: N ratio for soil fungi ranges from approximately 15:1 to 25:1 [45,46]. In the present study, the observed increase in C: N ratios may serve as a significant factor contributing to the notable correlation identified between soil C: N ratios and the taxonomic compositions of the fungal communities.

In this study, we observed that soils amended by different treatments harbored differentiated fungal microbiomes. It implied that effects of exogenous organic ameliorants on soil fungal community assembly communities were significantly influenced by the types and quantities of these organic materials utilized [47]. On one hand, it may be associated with the selective promotion of specific fungal groups by distinct types of carbon sources [48]. Additionally, differentiation in environmental constraints alleviation degrees mediated by different organic amendments might also be responsible for the distinctive responses of specific fungal species to different treatments [49]. In particular, a marked reduction in fungal richness was identified in soil treated by V50, corroborating the observations of Jiang et al. [50] and Zheng et al. [51], who also reported that compositional diversity of soil microbial communities exhibited negative responses to the organic fertilizations. This outcome might be attributed to the overall effects of relatively weak stimulation of specific fungal populations due to inadequate V applications, coupled with insufficient mitigation of environmental constraints. This hypothesis was further substantiated by the recorded lower variability in the relative abundances of dominant fungal phyla, as well as the diminished numbers of enriched and depleted ASVs in soils treated with V50. Together, results mentioned above revealed that sewage sludge and vermicompost application can drive fungal microbiome diversifications with higher compositional diversity and distinguishable community structure, mainly through microhabitat modification in coastal mudflat salt-affected soils. Furthermore, efficacy of exogenous organic ameliorants on soil fungal microbiome was found to be dependent on both the types of applied rates of organic materials utilized.

Soil microbial community functions have been increasingly considered as significant ecological bio-indicators. These functions are essential for a thorough understanding of the mechanisms underlying the assembly of microbial community functional profiles in response to various environmental constraints [52,53]. As reported in previous investigations, organic amendments were capable of the assembling of fungal population links to saprotrophs, resulting in enhancement functions involved in decomposing organic matters and overcoming nutrient limitation [54]. In this study, proliferation of soil saprotrophs and wood saprotrophs in S- and V-amended soils indicated that organic amendment favored the establishment of fungal seedbanks characterized by diverse functionalities related to the cycles of soil organic matter and nutrients [55,56]. As is well known, soil functionalities are deeply dependent on microbial microbiomes and, correspondingly, fungal functional profiles might be influenced as a result of the variations in fungal communities [52]. The positive relationships between fungal consortia (*Stachybotrys*, *Trichoderma*, *Mortierella*, and *Acremonium*) and fungal functional diversity seen in this study were supported by the notion that fungal microbiomes serve as important predictors and contributors to microbial functioning across various soil types [57]. On one hand, fungal genera, including *Trichoderma*, *Mortierella*, and *Acremonium*, are commonly recognized as decomposers, directly affecting the persistence and mineralization of soil organic matters [58–60]. Additionally, competitions existing among different fungal groups possessing differentiated adaptive capacity to the environment, as well as substrate utilization patterns, can lead to variations in fungal community diversity, and, thereby, indirectly result in distinct trophic strategies in regard to soil microbial functional profiles [61]. These outcomes reflect the fact that environmental constraint alleviation in organic amended soils were conductive to the reassembly of soil fungal profiles possessing abundant functional annotations linked to saprotrophic traits. Particularly, enriched fungal genera of *Trichoderma*, *Mortierella*, and *Acremonium* may serve as crucial pioneer species in the diversification of soil microbial functions, as species within these genera are implicated in various processes that contribute to soil health [62]. It is important to highlight that the findings only presented the status quo one year after soil amendment, and dynamics-based analyses should be incorporated in the subsequent studies to thoroughly elucidate the mechanisms governing the responses of fungal communities and their functions to the organic amendments in coastal salt-affected soils.

5. Conclusions

In conclusion, this study suggested that soil pH, salinity, SOC, and fungal consortia composed of *Stachybotrys*, *Trichoderma*, *Mortierella*, and *Acremonium* were key abiotic and biotic components driving the diversifications of fungal communities and functional profiles in coastal salt-affected soils. In particular, exogenous organic ameliorant amendments exhibited outstanding effects on soil environmental constraint alleviation as well as reassembly of fungal communities and functions, and effectiveness tended to be type-dependent and dosage-dependent. Fungal community reassembly in amended soils was strongly associated with variations in in edaphic factors, especially soil pH, salinity, and SOC. Fungal functional profiles in amended soils were dominated by functionalities linking to saprotrophs and tightly related to the potential saprotrophic fungal groups affiliated to genera *Trichoderma*, *Mortierella*, and *Acremonium*. Collectively, this study provided references for identification and assessment of key abiotic and biotic agents driving the diversifications of fungal community and functional profiles in coastal salt-affected soils amended by exogenous organic matter. The dynamic evolution characteristics of soil edaphic and microbiological parameters, as well as the role of potential saprotrophic fungal populations, should be further investigated, which will be beneficial for us to holistically understand the underlying mechanisms of the microbial-driven soil quality improvement process in coastal mudflat lands.

Author Contributions: Conceptualization, Y.S. and Y.B.; methodology, J.Y.; software, W.C. and Y.X.; validation, W.Z.; formal analysis, J.Y.; investigation, W.C. and Y.X.; data curation, W.C. and Y.X.; writing—original draft preparation, Y.L.; writing—review and editing, Y.L.; visualization, Y.L.; supervision, W.Z.; funding acquisition, Y.S. and Y.B. All authors have read and agreed to the published version of the manuscript.

Funding: This research was supported by the Strategic Priority Research Program of the Chinese Academy of Sciences (No. XDA0440402), Research Fund for Agricultural Science and Technology of Jiangsu Province (CX(23)1019), the Key Research and Development Project of Jiangsu Coast Development Group Co., Ltd. (2022YHTDJB02), and Research Fund for Jiangsu Agricultural Industry Technology System (JATS[2023]317 and JATS[2023]318) for Yanchao Bai.

Data Availability Statement: Data are contained within the article.

Conflicts of Interest: The authors declare no conflicts of interest.

References

1. Dissanayake, N.G.; Frid, C.L.; Drylie, T.P.; Caswell, B.A. Ecological functioning of mudflats: Global analysis reveals both regional differences and widespread conservation of functioning. *Mar. Ecol. Prog. Ser.* **2018**, *604*, 1–20. [CrossRef]
2. Li, J.; Liu, D.Y.; Gong, H.R.; Liu, Z.; Zhang, Y.T. Crop yield increments will enhance soil carbon sequestration in coastal arable lands by 2100. *J. Clean. Prod.* **2023**, *432*, 139800. [CrossRef]
3. Su, X.S.; Wang, Y.Q.; Wang, G.X.; Zhang, Y.; Gong, X.L.; Yu, J.; Gou, F.G.; Lyu, H. Assessment and prediction of coastal saline soil improvement effects combining substrate amendments and salt barrier materials in typical region of the Yangtze River Delta. *Soil Tillage Res.* **2022**, *223*, 105483. [CrossRef]
4. Nikalje, G.C.; Srivastava, A.K.; Pandey, G.K.; Suprasanna, P. Halophytes in biosaline agriculture: Mechanism, utilization, and value addition. *Land Degrad. Dev.* **2018**, *29*, 1081–1095. [CrossRef]
5. Zhang, W.W.; Chong, W.; Xue, R.; Wang, L.J. Effects of salinity on the soil microbial community and soil fertility. *J. Integr. Agric.* **2019**, *18*, 1360–1368. [CrossRef]
6. Zhu, Y.; Guo, B.; Liu, C.; Lin, Y.C.; Fu, Q.L.; Li, N.Y.; Li, H. Soil fertility, enzyme activity, and microbial community structure diversity among different soil textures under different land use types in coastal saline soil. *J. Soils Sediments* **2021**, *21*, 2240–2252. [CrossRef]
7. Zhang, G.L.; Bai, J.H.; Zhai, Y.J.; Jia, J.; Zhao, Q.Q.; Wang, W.; Hu, X.Y. Microbial diversity and functions in saline soils: A review from a biogeochemical perspective. *J. Adv. Res.* **2024**, *59*, 129–140. [CrossRef]
8. Ding, Y.A.; Geng, Y.C.; Zhou, W.C.; Li, D.H. Habitat-specific environmental factors regulate the spatial variability of biological soil crust microbial communities on the Qinghai-Tibet Plateau. *Sci. Total Environ.* **2023**, *901*, 165937. [CrossRef]

9. Li, Z.Z.; Yang, Y.Z.; Zheng, H.; Hu, B.A.; Dai, X.H.; Meng, N.; Zhu, J.Y.; Yan, D.N. Environmental changes drive soil microbial community assembly across arid alpine grasslands on the Qinghai-Tibetan Plateau, China. *Catena* **2023**, *228*, 107175. [CrossRef]
10. Hicks, L.C.; Frey, B.; Kjøller, R.; Lukac, M.; Moora, M.; Weedon, J.T.; Rousk, J. Toward a function-first framework to make soil microbial ecology predictive. *Ecology* **2022**, *103*, e03594. [CrossRef]
11. Chen, L.F.; He, Z.B.; Zhao, W.Z.; Liu, J.L.; Zhou, H.; Li, J.; Meng, Y.Y.; Wang, L.S. Soil structure and nutrient supply drive changes in soil microbial communities during conversion of virgin desert soil to irrigated cropland. *Eur. J. Soil Sci.* **2020**, *71*, 768–781. [CrossRef]
12. Liu, B.S.; Hu, Y.H.; Wang, Y.; Xue, H.H.; Li, Z.H.; Li, M. Effects of saline-alkali stress on bacterial and fungal community diversity in *Leymus chinensis* rhizosphere soil. *Environ. Sci. Pollut. Res.* **2022**, *29*, 70000–70013. [CrossRef] [PubMed]
13. Rath, K.M.; Murphy, D.N.; Rousk, J. The microbial community size, structure, and process rates along natural gradients of soil salinity. *Soil Biol. Biochem.* **2019**, *138*, 107607. [CrossRef]
14. Zhang, G.L.; Bai, J.H.; Tebbe, C.C.; Huang, L.B.; Jia, J.; Wang, W.; Wang, X.; Yu, L.; Zhao, Q.Q. *Spartina alterniflora* invasions reduce soil fungal diversity and simplify co-occurrence networks in a salt marsh ecosystem. *Sci. Total Environ.* **2021**, *758*, 143667. [CrossRef]
15. Vanegas, J.; Munoz-Garcia, A.; Perez-Parra, K.A.; Figueroa-Galvis, I.; Mestanza, O.; Polania, J. Effect of salinity on fungal diversity in the rhizosphere of the halophyte *Avicennia germinans* from a semi-arid mangrove. *Fungal Ecol.* **2019**, *42*, 100855. [CrossRef]
16. Somenahally, A.C.; McLawrence, J.; Chaganti, V.N.; Ganjegunte, G.K.; Obayomi, O.; Brady, J.A. Response of soil microbial communities, inorganic and organic soil carbon pools in arid saline soils to alternative land use practices. *Ecol. Indic.* **2023**, *150*, 110227. [CrossRef]
17. Xiao, M.; Chen, C.; Yao, R.J.; Wang, X.P.; Liu, G.M. Response of soil fungal community in coastal saline soil to short-term water management combined with bio-organic fertilizer. *Agronomy* **2024**, *14*, 1441. [CrossRef]
18. Cong, P.F.; Ouyang, Z.; Hou, R.X.; Han, D.R. Effects of application of microbial fertilizer on aggregation and aggregate-associated carbon in saline soils. *Soil Tillage Res.* **2017**, *168*, 33–41. [CrossRef]
19. Amini, S.; Ghadiri, H.; Chen, C.R.; Marschner, P. Salt-affected soils, reclamation, carbon dynamics, and biochar: A review. *J. Soils Sediments* **2016**, *16*, 939–953. [CrossRef]
20. Guo, L.L.; Nie, Z.Y.; Zhou, J.; Zhang, S.X.; An, F.H.; Zhang, L.; Tóth, T.; Yang, F.; Wang, Z.C. Effects of different organic amendments on soil improvement, bacterial composition, and functional diversity in saline–sodic soil. *Agronomy* **2022**, *12*, 2294. [CrossRef]
21. Wang, X.; Li, Z.; Cheng, Y.D.; Yao, H.; Li, H.; You, X.W.; Zhang, C.S.; Li, Y.Q. Wheat straw hydrochar induced negative priming effect on carbon decomposition in a coastal soil. *IMeta* **2023**, *2*, e134. [CrossRef] [PubMed]
22. Wang, K.; Wang, S.; Zhang, X.; Wang, W.Y.; Wang, X.Y.; Kong, F.L.; Xi, M. The amelioration and improvement effects of modified biochar derived from *Spartina alterniflora* on coastal wetland soil and *Suaeda salsa* growth. *Environ. Res.* **2024**, *240*, 117426. [CrossRef] [PubMed]
23. Zhang, X.Y.; Qi, L.L.; Li, W.M.; Hu, B.X.; Dai, Z.X. Bacterial community variations with salinity in the saltwater-intruded estuarine aquifer. *Sci. Total Environ.* **2021**, *755*, 142423. [CrossRef] [PubMed]
24. Lopes, B.C.; Figueiredo, R.S.; Araújo, J.C.; Matos, A.T. Bacterial community dynamics in tropical soil after sewage sludge amendment. *Water Sci. Technol.* **2020**, *82*, 2937–2947. [CrossRef] [PubMed]
25. Mattana, S.; Petrovičová, B.; Landi, L.; Gelsomino, A.; Cortés, P.; Ortiz, O.; Renella, G. Sewage sludge processing determines its impact on soil microbial community structure and function. *Appl. Soil Ecol.* **2014**, *75*, 150–161. [CrossRef]
26. IUSS Working Group WRB (International Union of Soil Science Working Group on World Reference Base). *World Reference Base for Soil Resources 2006 (First Update 2007)*; World Soil Resources Reports No. 103; FAO: Rome, Italy, 2007. [CrossRef]
27. Li, Y.L.; Wang, Y.M.; Gu, C.H.; Shen, C.; Xu, L.; Zhao, Y.L.; Yi, S.Q.; Zuo, W.G.; Shan, Y.H.; Zhang, Z.Q.; et al. Differential effects of organic ameliorants on the reassembly of bacterial communities in newly amended coastal mudflat salt-affected soil. *Agronomy* **2022**, *12*, 2525. [CrossRef]
28. Lu, R.K. *Soil Agricultural Chemical Analysis Method*; China Agricultural Science Technology Press: Beijing, China, 2000.
29. Li, Y.L.; Shen, C.; Wang, Y.M.; Xu, L.; Zhao, Y.L.; Yi, S.Q.; Zuo, W.G.; Yao, R.J.; Zhang, X.; Gu, C.H.; et al. Alleviated environmental constraints and restructured fungal microbiome facilitate aggregate formation and stabilization in coastal mudflat saline soil amended by sewage sludge. *Land Degrad. Dev.* **2023**, *34*, 3064–3075. [CrossRef]
30. Callahan, B.J.; McMurdie, P.J.; Rosen, M.J.; Han, A.W.; Johnson, A.J.A.; Holmes, S.P. DADA2: High-resolution sample inference from Illumina amplicon data. *Nat. Methods* **2016**, *13*, 581–583. [CrossRef]
31. Bolyen, E.; Rideout, J.R.; Dillon, M.R.; Bokulich, N.A.; Abnet, C.C.; Al-Ghalith, G.A.; Alexander, H.; Alm, E.J.; Arumugam, M.; Asnicar, F.J.N. Reproducible, interactive, scalable and extensible microbiome data science using QIIME 2. *Nat. Biotechnol.* **2019**, *37*, 852–857. [CrossRef]
32. Polme, S.; Abarenkov, K.; Henrik Nilsson, R.; Lindahl, B.D.; Clemmensen, K.E.; Kauserud, H.; Nguyen, N.; Kjoller, R.; Bates, S.T.; Baldrian, P. FungalTraits: A user-friendly traits database of fungi and fungus-like stramenopiles. *Fungal Divers.* **2020**, *105*, 1–16. [CrossRef]
33. Shao, Q.; Xia, X.J.; Li, G.H.; Li, H.; Lin, J.T.; Lou, Y.H.; Yang, Q.G.; Wang, H.; Yang, Z.C.; Pan, H.; et al. Effects of flue gas desulfurization gypsum and clover planting on qualities of soil and winter jujube in coastal saline-alkali orchard of north China. *Soil Ecol. Lett.* **2024**, *6*, 230185. [CrossRef]

34. Luo, X.X.; Chen, W.J.; Liu, Q.; Wang, X.; Miao, J.; Liu, L.Q.; Zheng, H.; Liu, R.H.; Li, F. Corn straw biochar addition elevated phosphorus availability in a coastal salt-affected soil under the conditions of different halophyte litter input and moisture contents. *Sci. Total Environ.* **2024**, *908*, 168355. [CrossRef] [PubMed]
35. Rashid, I.; Murtaza, G.; Zahir, Z.A.; Farooq, M. Effect of humic and fulvic acid transformation on cadmium availability to wheat cultivars in sewage sludge amended soil. *Environ. Sci. Pollut. Res.* **2018**, *25*, 16071–16079. [CrossRef] [PubMed]
36. Kacprzak, M.; Neczaj, E.; Fijakowski, K.; Grobelak, A.; Grosser, A.; Worwag, M.; Rorat, A.; Brattebo, H.; Almås, Å.; Singh, B.R. Sewage sludge disposal strategies for sustainable development. *Environ. Res.* **2017**, *156*, 39–46. [CrossRef]
37. Florentino, A.L.; Ferraz, A.V.; Gonçalves, J.L.M.; Asensiod, V.; Muraoka, T.; Dias, C.T.S.; Nogueira, T.A.R.; Capra, G.F.; Abreu-Junior, C.H. Long-term effects of residual sewage sludge application in tropical soils under *Eucalyptus* plantations. *J. Clean. Prod.* **2019**, *220*, 177–187. [CrossRef]
38. Guevara, R.; Ikenaga, M.; Dean, A.L.; Pisani, C.; Boyer, J.N. Changes in sediment bacterial community in response to long-term nutrient enrichment in a subtropical seagrass-dominated estuary. *Microb. Ecol.* **2014**, *68*, 427–440. [CrossRef]
39. Jousset, A.; Bienhold, C.; Chatzinotas, A.; Gallien, L.; Gobet, A.; Kurm, V.; Küsel, K.; Rillig, M.C.; Rivett, D.W.; Salles, J.F. Where less may be more: How the rare biosphere pulls ecosystems strings. *ISME J.* **2017**, *4*, 853–862. [CrossRef]
40. Curci, M.; Lavecchia, A.; Cucci, G.; Lacolla, G.; Corato, U.D.; Crecchio, C. Short-term effects of sewage sludge compost amendment on semiarid soil. *Soil Syst.* **2020**, *4*, 48. [CrossRef]
41. Liu, X.; Guo, K.L.; Huang, L.; Ji, Z.; Jiang, H.M.; Li, H.; Zhang, J.F. Responses of absolute and specific enzyme activity to consecutive application of composted sewage sludge in a Fluventic Ustochrept. *PLoS ONE* **2017**, *12*, e0177796. [CrossRef]
42. Fang, J.; Deng, Y.C.; Che, R.X.; Han, C.; Zhong, W.H. Bacterial community composition in soils covered by different vegetation types in the Yancheng tidal marsh. *Environ. Sci. Pollut. Res.* **2020**, *27*, 21517–21532. [CrossRef]
43. Rousk, J.; Bååth, E.; Brookes, P.C.; Lauber, C.L.; Lozupone, C.; Caporaso, J.G.; Knight, R.; Fierer, N. Soil bacterial and fungal communities across a pH gradient in an arable soil. *ISME J.* **2010**, *4*, 1340–1351. [CrossRef] [PubMed]
44. Rath, K.M.; Fierer, N.; Murphy, D.V.; Rousk, J. Linking bacterial community composition to soil salinity along environmental gradients. *ISME J.* **2019**, *13*, 836–846. [CrossRef] [PubMed]
45. Li, L.J.; Zhu-Barker, X.; Ye, R.; Doane, T.A.; Horwath, W.R. Soil microbial biomass size and soil carbon influence the priming effect from carbon inputs depending on nitrogen availability. *Soil Biol. Biochem.* **2018**, *119*, 41–49. [CrossRef]
46. Li, Q.R.; Tian, Y.Q.; Zhang, X.Y.; Xu, X.L.; Wang, H.M.; Kuzyakov, Y. Labile carbon and nitrogen additions affect soil organic matter decomposition more strongly than temperature. *Appl. Soil Ecol.* **2017**, *114*, 152–160. [CrossRef]
47. Calleja-Cervantes, M.E.; Aparicio-Tejo, P.M.; Villadas, P.J.; Irigoyen, I.; Irañeta, J.; Fernández-González, A.J.; Fernández-López, M.; Menéndez, S. Rational application of treated sewage sludge with urea increases GHG mitigation opportunities in Mediterranean soils. *Agric. Ecosyst. Environ.* **2017**, *238*, 114–127. [CrossRef]
48. Huang, X.Q.; Zhao, J.; Zhou, X.; Zhang, J.B.; Cai, Z.C. Differential responses of soil bacterial community and functional diversity to reductive soil disinfestation and chemical soil disinfestation. *Geoderma* **2019**, *348*, 124–134. [CrossRef]
49. Hollister, E.B.; Engledow, A.S.; Hammett, A.J.M.; Provin, T.L.; Wilkinson, H.H.; Gentry, T.J. Shifts in microbial community structure along an ecological gradient of hypersaline soils and sediments. *ISME J.* **2010**, *4*, 829–838. [CrossRef]
50. Jiang, S.Q.; Yu, Y.N.; Gao, R.W.; Wang, H.; Zhang, J.; Li, R.; Long, X.H.; Shen, Q.R.; Chen, W.; Cai, F. High-throughput absolute quantification sequencing reveals the effect of different fertilizer applications on bacterial community in a tomato cultivated coastal saline soil. *Sci. Total Environ.* **2019**, *687*, 601–609. [CrossRef]
51. Zheng, J.Y.; Wang, L.; Zhao, J.S.; Niu, Y.H.; Xiao, H.B.; Wang, Z.; Yu, S.X.; Shi, Z.H. Forty-year-old orchards promote carbon storage by changing aggregate-associated enzyme activities and microbial communities. *Catena* **2022**, *213*, 106195. [CrossRef]
52. Díaz, S.; Lavorel, S.; Chapin, F.S.; Tecco, P.A.; Gurvich, D.E.; Grigulis, K. Functional diversity-at the crossroads between ecosystem functioning and environmental filters. In *Terrestrial Ecosystems in a Changing World*; Global Change-The IGBP Series; Springer: Berlin/Heidelberg, Germany, 2007; pp. 81–91. [CrossRef]
53. Preston-Mafham, J.; Boddy, L.; Randerson, P.F. Analysis of microbial community functional diversity using sole-carbon-source utilisation profiles—A critique. *FEMS Microbiol. Ecol.* **2002**, *42*, 1–14. [CrossRef]
54. Kyaschenko, J.; Clemmensen, K.E.; Hagenbo, A.; Karltun, E.; Lindahl, B.D. Shift in fungal communities and associated enzyme activities along an age gradient of managed Pinus sylvestris stands. *ISME J.* **2010**, *11*, 863–874. [CrossRef] [PubMed]
55. Clocchiatti, A.; Hannula, S.E.; Berg, M.v.d; Korthals, G.; Boer, W.D. The hidden potential of saprotrophic fungi in arable soil: Patterns of short-term stimulation by organic amendments. *Appl. Soil Ecol.* **2020**, *147*, 103434. [CrossRef]
56. Dai, Z.M.; Enders, A.; Rodrigues Jorge, L.M.; Hanley, K.M.; Brookers, P.C.; Xu, J.M.; Lehmann, J. Soil fungal taxonomic and functional community composition as affected by biochar properties. *Soil Biol. Biochem.* **2018**, *126*, 159–167. [CrossRef]
57. Garcia-Franco, N.; Martínez-Mena, M.; Goberna, M.; Albaladejo, J. Changes in soil aggregation and microbial community structure control carbon sequestration after afforestation of semiarid shrublands. *Soil Biol. Biochem.* **2015**, *87*, 110–121. [CrossRef]
58. Rohrich, C.R.; Jaklitsch, W.M.; Voglmayr, H.; Iversen, A.; Vilcinskas, A.; Nielsen, K.F.; Thrane, U.; Dohren, H.V.; Bruckner, H.; Degenkolb, T.J. Front line defenders of the ecological niche! Screening the structural diversity of peptaibiotics from saprotrophic and fungicolous *Trichoderma/Hypocrea* species. *Fungal Divers.* **2014**, *69*, 117–146. [CrossRef]
59. Li, F.; Zhang, S.Q.; Wang, Y.; Li, Y.; Li, P.P.; Chen, L.; Jie, X.L.; Hu, D.S.; Feng, B.; Yue, K.; et al. Rare fungus, *Mortierella capitata* promotes crop growth by stimulating primary metabolisms related genes and reshaping rhizosphere bacterial community. *Soil Biol. Biochem.* **2020**, *151*, 108017. [CrossRef]

60. Baldrian, P.; Voriskova, J.; Dobiasova, P.; Merhautova, V.; Lisa, L.; Valaskova, V. Production of extracellular enzymes and degradation of biopolymers by saprotrophic microfungi from the upper layers of forest soil. *Plant Soil* **2011**, *338*, 111–125. [CrossRef]
61. Egidi, E.; Delgado-Baquerizo, M.; Plett, J.M.; Wang, J.T.; Eldridge, D.J.; Bardgett, R.D.; Maestre, F.T.; Singh, B.K. A few *Ascomycota* taxa dominate soil fungal communities worldwide. *Nat. Commun.* **2019**, *10*, 2369. [CrossRef]
62. Li, F.; Chen, L.; Redmile-Gordon, M.; Zhang, J.B.; Zhang, C.Z.; Ning, Q.; Li, W. *Mortierella elongata*'s roles in organic agriculture and crop growth promotion in a mineral soil. *Land Degrad. Dev.* **2018**, *29*, 1642–1651. [CrossRef]

agronomy

MDPI

Article

Arbuscular Mycorrhizae Affect Soil Nitrogen Fertilizer Utilization, Denitrification Functional Genes, and N_2O Emissions During Biochar Amendment

Yanfang Wang [1], Jing Liu [1], Xuxian Deng [2], Yuyang Li [1], Jiakai Gao [1] and Ling Liu [1,*]

[1] College of Agriculture, Henan University of Science and Technology, Luoyang 471023, China
[2] Luoyang Seed Industry Development Center, Luoyang 471023, China; hnlydxx@126.com
* Correspondence: liuling1978@haust.edu.cn

Abstract: Arbuscular mycorrhizal fungi (AMF) can form symbionts with plant roots, acquire soil nitrogen, and affect nitrous oxide (N_2O) production. Biochar, as a soil additive for the management of agricultural soil, affects soil nitrogen (N) utilization and plant growth. However, how AMF regulates soil N unitization, the denitrification process, and N_2O emissions in plant–soil systems remains largely unknown, particularly under the biochar amendment. In this study, a microcosm experiment was conducted to investigate the impacts of different mycorrhizal treatments (CK: neither AMF hyphae nor plant roots; AMF: only AMF hyphae; AMF + R: AMF hyphae and plant roots) on plant growth, soil N fertilizer utilization, N_2O production and consumption, functional gene abundance, and N_2O emission at two biochar addition levels (B0: no biochar; B1: biochar addition rate of 10 $g{\cdot}kg^{-1}$ soil) in a maize planting soil system. The results revealed that AMF alone and AMF with plant root treatments enhanced the fresh weight of maize plants by 10.15% and 19.23% and decreased soil inorganic N contents by 33.28% and 75.56%, respectively. The combination of biochar, AMF, and plant roots showed the largest increase in maize plant biomass. The AMF and AMF with root treatments all significantly decreased the *nir*S + *nir*K/*nos*Z ratio and N_2O emissions at two biochar levels. The presence of AMF and plant roots during biochar amendment showed the smallest *nir*S + *nir*K/*nos*Z ratio and N_2O emissions. The AMF combined with biochar and AMF and plant roots combined with biochar treatments increased *nir*S + *nir*K/*nos*Z by 24.32% and 26.90% and decreased N_2O accumulation emission by 21.12% and 38.13%, respectively. The results imply that biochar, AMF, and plant roots reduced N_2O emissions directly by reducing soil N and increasing soil N unitization efficiency and indirectly by shifting the N_2O production and consumption gene abundance in agroecosystems. These findings suggest that the addition of biochar and AMF and/or the presence of plant roots can interact to alleviate soil N_2O emissions by manipulating plant inorganic N acquisition and the soil denitrification process.

Keywords: arbuscular mycorrhizal fungi; denitrification; plant biomass; nitrogen uptake; nitrogen retention

1. Introduction

Nitrous oxide (N_2O) is a potent greenhouse gas that can destroy the ozone layer in the stratosphere and cause climate warming [1,2]. Due to the large amount of nitrogen (N) fertilizer applied, agricultural soils have become the main emission source of N_2O, accounting for about 60% of the global anthropogenic N_2O emission [3]. The production of soil N_2O mainly comes from soil nitrification and denitrification processes [4,5]. In the process of soil nitrification, microorganisms in the soil oxidize the ammonium salt (NH_4^+) to nitrite (NO_2^-) or nitrate (NO_3^-), and N_2O is produced mainly by the oxidation of hydroxylamine (NH_2OH) to NO_2. The microorganisms involved in this process are the ammonia-oxidized archaea (AOA) and ammonia-oxidizing bacteria (AOB), which are encoded by the ammonia mono oxygen-enzyme gene (*amo*A). Denitrification is the main N_2O emission source in agricultural soils [6]. In the denitrification process, nitrogen oxides (NO_3^- and NO_2^-) are reduced to N_2O and N_2 by various reductases, among which nitrite reductase and nitrous oxide reductase are key, and rate-limiting enzymes in soil denitrification [6,7]. The *nir*S and *nir*K genes encode nitrite reductases, which produce N_2O, and *nos*Z genes encode N_2O reductases that reduce N_2O to N_2 [8]. Therefore, the ratio of *nir*S + *nir*K/*nos*Z has been used as an indicator for N_2O emission potential [9].

Arbuscular mycorrhizal fungi (AMF) are a kind of beneficial soil microorganisms, which are widespread and can form symbionts with the plant roots of over 80% of the land plants [10]. AMF can help plant roots obtain soil nutrients and absorb photosynthates from host plants [11]. Plant roots combined with AMF play a vital role in the soil N cycles and significantly affect the absorption and mineralization of N [12,13]. Yet, only a few researchers have studied the impacts of plant roots and AMF on soil N_2O emission, and they have produced different results. For instance, Zhang et al. [14] revealed that plant roots combined with AMF suppressed soil N_2O emission by changing the N cycling microbial community. Bender et al. [15] also revealed that the AMF significantly reduced soil N_2O emission by shifting the key gene abundance of N_2O emission. Qiu et al. [9] reported that AMF combined with plant roots reduced the *nir*K gene abundance and the (*nir*K + *nir*S)/*nos*Z ratio, decreasing soil N_2O emission. Changes in soil denitrifiers play a major controlling role in the emission of N_2O. However, some studies indicated that plant roots and AMF may increase soil N_2O emission by producing more root exudates and labile carbon from plant roots, which supply high carbon for denitrifiers to stimulate denitrification [16,17]. Therefore, the impacts of plant roots and AMF on the denitrification process and shifting soil N_2O emission need further investigation.

Biochar is a carbon-rich solid material produced through the high-temperature cracking of crop straw, firewood, weeds, and feces in the absence of oxygen conditions [18]. In recent years, biochar has been used as a soil additive for the management of agricultural soil, which produced important impacts on soil quality, crop yield, carbon sequestration, and greenhouse gas emissions [2,14,19,20]. Previous studies have indicated that biochar addition can regulate the nitrification and denitrification process by facilitating biofilm formation and retaining soil moisture [21,22]. Biochar application can reduce N leaching from soil and activate N fixation by increasing soil carbon sequestration [23,24]. Moreover, biochar amendment can increase the soil cation exchange and enhance the adsorption of soil NH_4^+-N, thus reducing soil inorganic N content and inhibiting the nitrification process [25,26]. In addition, biochar could promote soil O_2 content to restrain the denitrification process and reduce N_2O emission [27,28]. However, few studies have explicitly examined the effects of AMF and plant roots on denitrifying genes and N_2O emissions, particularly in biochar amendment. Some recent studies have shown that biochar has a special effect on soil microbes, such as AMF [29–32]. For example, a crop straw-derived biochar application can improve soil physicochemical characteristics and facilitate the germination and growth of AMF spores [33,34], and biochar can regulate plant fungal signal substances that affect the germination and hyphal branching of AMF [35,36]. Zhang et al. [37] also showed that biochar may indirectly influence AMF by affecting the activity of phosphorus bacteria in soil. Yet, it remains largely unexplored how plant roots associated with AMF affect N utilization and N_2O emission under biochar application. Therefore, the microcosm experiment was conducted to investigate the impacts of AMF, plant roots, and biochar on plant growth, soil N utilization, soil denitrification, and N_2O emission in maize planting soil systems. We hypothesized that (1) the biochar addition would enhance the growth of AMF mycelial, increase the activity of AMF mycelial microorganisms, and promote the absorption and transfer of effective nitrogen to plants to reduce the substrate concentration of denitrification microorganisms and then affect the soil denitrification process and N_2O emissions, and (2) the combined of biochar, AMF, and plant roots would reduce N_2O-producing gene abundance and increase soil N utilization, corresponding to decreases in N_2O emission.

2. Materials and Methods

2.1. Biochar, AMF Inocula, Soil, and Plant Preparation

The biochar was produced from wheat straw pyrolyzed under 500 °C in anaerobic conditions for 2 h, and the heating rate was 20 °C min^{-1} (Sanli New Energy Co. Ltd., Shangqiu, China). The organic C and N contents were measured using a Perkin-Elmer 2400 CHN elemental analyzer (Perk-Elmer Co. Ltd., Norwalk, CT, USA). The total K content was determined using acid digestion and elemental analysis through atomic adsorption spectroscopy (Thermo Fisher Scientific Co. Ltd., Waltham, MA, USA). The available P was extracted using 0.5 mol L^{-1} $NaHCO_3$ and measured colorimetrically (Beijing Puxi General Instrument Co. LTD., Beijing, China). The pH was measured for a 1:5 (w/v) suspension of biochar and deionized water with a compound glass electrode. The surface area and pore size were measured using a TriStar II 3020 automated surface area and pore size analyzer (Micromeritics Co. Ltd., Atlanta, GA, USA). The basic biochar properties were surface area at 8.9 m^2 g^{-1}, average pore size at 3.14 nm, total pore volume at 0.015 cm^3 g^{-1}, organic C content at 46.76%, total N content at 0.59%, available P content at 0.89 g kg^{-1}, K content at 2.72%, and pH at 10.4.

The AMF inoculum, *Rhizophagus intraradices* BJ09, and *Funneliformis mosseae* BJ04A were purchased from the Beijing Academy of Agriculture and Forestry Sciences. The inocula were pot-cultured with Sudangrass [Sorghum sudanese (Piper) Stapf.] in the autoclaved (121 °C for 2 h) mixture of zeolite and river sand at 4:1 (w/w) in the greenhouse for four months. The ratio of the substrate and AMF inoculum was 100:1 (w/w). The temperature was maintained at 16–25 °C, with regular watering to maintain plant growth. The AMF inocula contains spores (approximately 30 spores per g^{-1} of substrate), mycelium, and fine root segments.

Surface horizon soil (0–20 cm) was collected from agricultural soil at Henan University of Science and Technology (34°38″ N, 112°28″ E), Luoyang, China, on 5 May 2020. The soil was air-dried and passed through a 5 mm sieve for the microcosm experiment. The soil had a pH of 7.2, 14.09 g kg^{-1} of organic matter, 53.76 mg kg^{-1} of available N, 10.18 mg kg^{-1} of available P, and 112.74 mg kg^{-1} of available K. The planted crop was maize (*Zea mays* L. var. Zhengdan-958).

2.2. Experimental Design

2.2.1. Microcosm Experiment

The microcosm experiment was established as a completely randomized trial with two biochar levels and three mycorrhizal levels, making a total of six treatments. The two levels of biochar were no biochar application (B0) and biochar addition rate of 10 g·kg^{-1} soil (B1). To insulate the impact of AMF hyphae alone or AMF hyphae with roots on N_2O emission, the plexiglass mesocosms (45 cm length, 15 cm width, 18 cm depth) were constructed and separated into three identical compartments by different mesh membrane sizes to create different mycorrhizal treatments as follows: 0.45 μm pore size nylon mesh (neither AMF nor plant roots, CK) (Figure 1A), 20 μm pore size nylon mesh (only AMF hyphae can grow through, AMF) (Figure 1B), and 1.6 mm pore size nylon mesh (AMF hyphae and plant roots can penetrate through, AMF + R) (Figure 1C). The three meshes used in this experiment all allow water, soluble C, and N produced by plant roots to pass through. Each compartment of half of the plexiglass mesocosms was filled with 3.0 kg of soil evenly mixed with 100 g of AMF inoculum. Meanwhile, each compartment of the other half of the plexiglass mesocosms was filled with 3.0 kg of an even mixture of soil and the prepared biochar at a rate of 10 g·kg^{-1}, as well as 100 g AMF of inoculum. Each treatment had four replicates, and there was a total of 24 plexiglass mesocosms. The two compartments at the left and right of the plexiglass mesocosm were named as host compartments, in which maize was planted, and the compartment in the middle of the plexiglass mesocosm was designated as the test compartment, in which N_2O emission was measured (Figure 1).

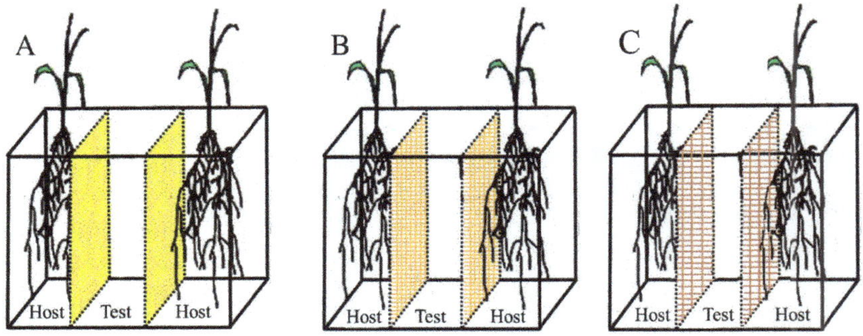

Figure 1. Schematic diagram of root compartments. The plexiglass mesocosms (45 cm length, 15 cm width, 18 cm depth) were constructed and separated into three identical compartments by different mesh membrane sizes to create different mycorrhizal treatments: (**A**) 0.45 µm pore size nylon mesh (neither AMF nor plant roots); (**B**) 20 µm pore size nylon mesh (only AMF hyphae can grow through); (**C**) 1.6 mm pore size nylon mesh (AMF hyphae and plant roots can penetrate through). Host compartments with growing maize and test compartments with N_2O collection.

2.2.2. Plant Growth Conditions

The experiment was conducted in the greenhouse at Henan University of Science and Technology from 10 May to 3 August 2020. The air temperature in the greenhouse ranged between 24.2 °C and 36.4 °C. Two maize seeds were sown in the host compartment at the left and right of the plexiglass mesocosms on 10 May 2020 and thinned to one plant on the seventh day after germination. At the jointing stage of maize, the 10.21% abundance of ^{15}N labeled urea fertilizer was hole-applied into each test compartment of plexiglass mesocosms to achieve a rate of 180 kg N ha^{-1} soil. The plexiglass mesocosms were placed randomly; deionized water was periodically watered for host compartments with plants, as needed. The test compartments without plants were also watered with a small amount of deionized water and usually were observed to ensure that plant roots and AMF could pass through the mesh.

2.3. *Measurements*

2.3.1. Measurement of Plant Biomass, N Contents, and Root AMF Colonization

The maize plants were harvested after 85 days of growth. The aboveground parts and roots were sampled as described by Zhang et al. [14]. The fresh weight of the maize above-ground parts and roots was determined with an electronic balance. Then, the aboveground parts were dried to obtain a stable weight and ground to measure the N contents. The total N contents were determined using the micro-Kjeldahl method after digestion with H_2SO_4-H_2O_2. The ^{15}N abundance of the maize plant was determined using the isotope mass spectrometer (Elementar Analysensystems GmbH Inc., Langenselbold, Germany). Fresh fibrous roots (≤2 mm in diameter) were cut into 1 cm root segments, which were put into the test tube, added 10% KOH(w/v) for digestion, and then stained with acid fuchsin. The root AMF colonization was observed under a microscope (100×) using the gridline intersection method [38].

2.3.2. Measurement of Soil N_2O Emissions

The N_2O gas samples were collected from the test compartment using the static chamber method [39]. Gas samples were collected from 1 day after fertilization (DAF) and discontinuously collected at 1–5, 7, 9, 11, 13, 18, and 21–28 DAF. The concentrations of N_2O were measured using gas chromatography (Agilent 7890A, Agilent Technologies Co. Ltd., Santa Clara, CA, USA). The calculations of N_2O fluxes and cumulative emissions by the research of Gao et al. [40].

2.3.3. Soil Sampling and Analyses

After the maize plants were harvested, soil samples of the test compartment were collected, and the visible residues and plant roots were carefully removed. Some of the soil samples were stored in a −80 °C refrigerator for soil DNA extraction, and the remainder of the soil was stored in a 4 °C refrigerator for soil total nitrogen, soil ^{15}N abundance, NH_4^+-N, and NO_3^--N measurements. The soil total nitrogen, NH_4^+-N, and NO_3^--N contents were measured using a continuous flow analyzer (SEAL Analytic, Norderstedt, Germany). The soil ^{15}N abundance was determined in a similar manner as that described for the plants' ^{15}N mensurate.

2.3.4. Soil DNA Extracted and Denitrification Functional Gene Abundances

A total of 0.5 g of the soil samples stored in the −80 °C refrigerator to extract soil DNA were taken according to the method and procedure of the Fast DNA SPIN Kit for Soil (MP Biomedicals Co. Ltd., Santa Ana, CA, USA). After extraction, the integrity of the proposed DNA fragments was measured using 1% agarose gel electrophoresis. The extracted DNA purity and concentration were measured using an ultraviolet spectrophotometer (Thermo Fisher Scientific Co. Ltd., Waltham, MA, USA). The abundances of the *nir*S, *nir*K, and *nos*Z genes were measured through quantitative real-time polymerase chain reaction (qPCR) (Applied Biosystems, Waltham, MA, USA). The qPCR primers are given in Table 1.

Table 1. Primers and qPCR cycle conditions.

Target Gene	Primer	Sequence	Reaction Conditions of qPCR	Reference
nirS	*nirS*Cd3AF *nirS*R3cd	GTSAACGTSAAGGARACSGG GASTTCGGRTGSGTCTTGA	Procedure 1	[41]
nirK	*nirK*1F *nirK*5R	GGMATGGTKCCSTGGCA GCCTCGATCAGRTTRTGG	Procedure 2	[42]
nosZ	*nosZ*-1F *nosZ*-1622R	CGYTGTTCMTCGACAGCCAG CGSACCTTSTTGCCSTYGCG	Procedure 1	[43]

Procedure 1: 5 min—95 °C, 30 s—95 °C, 30 s—58 °C annealing, 35 cycles, 1 min—72 °C, 5 min—72 °C, Procedure 2: 5 min—95 °C, 30 s—95 °C, 30 s—55 °C annealing, 35 cycles, 1 min—72 °C.

2.4. Calculation of Nitrogen Utilization

The ^{15}N utilization efficiency by plants (NUE), ^{15}N retention rate by soil (NRR), and ^{15}N loss rate (NLR) were calculated according to the following formulas used by Robinson [44].

$$\text{NUE or NRR\%} = \times 100 \quad (1)$$

$$\text{NLR\%} = 100 - \text{NUE} - \text{NRR} \quad (2)$$

$$\text{N1} = \text{N0} \times \text{A} \quad (3)$$

$$\text{A\%} = \frac{\text{APC}(^{15}\text{N})}{\text{C}} \times 100 \quad (4)$$

where NUE is the ^{15}N utilization efficiency for maize plants, NRR is the ^{15}N retention rate for soil, and NLR is the ^{15}N loss rate, which is calculated according to Equations (3) and (4). N1 is the ^{15}N contents from ^{15}N urea in the plant or soil ($g \cdot kg^{-1}$), N2 is the total ^{15}N contents from ^{15}N urea ($g \cdot kg^{-1}$), N0 refers to total N contents in the plant or soil ($g\ kg^{-1}$), A is the percentages of ^{15}N amount from ^{15}N urea of total N in the plant or soil (%), APC (^{15}N) is the ^{15}N-atom percent excess of the plant or soil (%), and C is the ^{15}N-atom percent excess of labeled ^{15}N urea (%).

2.5. Statistical Analysis

All the data analyses were performed using SPSS 25.0 software. The significance was tested using analysis of variance (ANOVA) and LSD for multiple comparisons. Two-way ANOVA was used to analyze the significance of biochar and AMF effects. Correlations among plant biomass, AMF, soil NH_4^+-N and NO_3^--N, abundance of denitrification functional genes, and N_2O emission were analyzed by using Pearson's correlation.

3. Results

3.1. Maize Plant Biomass, Soil Inorganic N Contents, and AMF Colonization of Roots

Both biochar application and AMF significantly affected maize biomass (Tables 2 and 3). Compared with no biochar input, the biochar application significantly increased maize biomass under the same treatment. Compared with the CK, the AMF and AMF + R treatments all significantly enhanced the growth traits of fresh weight ($p < 0.05$) under two biochar application levels (Table 1).

Table 2. Plant biomass, root colonization rate, and soil inorganic N contents with CK, AMF, and AMF + R treatments under different biochar application levels.

Biochar Levels	Treatment	Fresh Weight of the Root (g plant^{-1})	Fresh Weight of the Shoot (g plant^{-1})	Soil NH_4^+-N (mg N kg^{-1})	Soil NO_3^--N (mg N kg^{-1})	AMF Colonization (%)
	CK	110.78 ± 10.23 e	243.12 ± 20.34 d	1.02 ± 0.24 a	45.38 ± 5.12 a	41.17 ± 4.78 c
B0	AMF	132.07 ± 15.36 cd	246.08 ± 22.67 d	0.91 ± 0.25 b	27.70 ± 3.24 c	40.27 ± 5.28 d
	AMF + R	136.63 ± 11.25 c	266.47 ± 30.89 c	0.87 ± 0.26 b	9.65 ± 1.23 d	38.23 ± 4.17 fde
	CK	132.96 ± 16.39 cd	260.46± 28.67 c	0.95 ± 0.14 b	32.20 ± 3.78 b	45.14 ± 4.96 a
B1	AMF	141.66 ± 15.56 b	304.61 ± 32.78 b	0.73 ± 0.15 c	23.06 ± 2.79 c	44.67 ± 5.02 b
	AMF + R	163.63 ± 21.15 a	338.62 ± 37.59 a	0.68 ± 0.12 c	7.33 ± 1.11 e	43.46 ± 4.79 bc

Values are mean ± SE (n = 4). B0, no biochar application. B1, biochar application, Different letters among the treatments indicated the significance according to Duncan's LSD test at 0.05 level.

Table 3. Results (F-values) of two-way ANOVA for the effects of biochar, AMF, and their interaction on plant and soil-related indices.

Factors	Biochar(B)	AMF	B × AMF
Shoot biomass (g plant^{-1})	16.7 **	42.5 **	12.1 *
Root biomass (g plant^{-1})	14.2 **	5.7	2.7
AMF colonization (%)	29.6 **	2.4	1.5
Soil NH_4^+-N (mg N kg^{-1})	47.9 ***	113.6 ***	16.8 **
Soil NO_3^--N (mg N kg^{-1})	20.78 **	109.7 ***	8.9
N_2O emission flux (μg·m^{-2}·h^{-1})	116.2 ***	79.4 ***	25.7 **
nirK (copies g^{-1} soil)	35.4 ***	78.5 ***	13.7 **
nirS(copies g^{-1} soil)	56.2 ***	82.3 ***	17.5 **
nosZ(copies g^{-1} soil)	38.9 ***	78.5 ***	24.7 ***
nirS + *nirK*/*nosZ*	87.2 ***	70.3 ***	19.4 **

Significant effects: * $p \leq 0.05$, ** $p \leq 0.01$, *** $p \leq 0.001$.

Compared with the CK, the AMF and AMF + R treatments all decreased the AMF colonization rate of maize roots under both biochar application levels (Tables 2 and 3). The AMF and AMF + R treatments reduced the average AMF colonization rate by 1.87% and 5.68%, respectively. There was no significant difference in the AMF colonization rate between the AMF and AMF + R treatments (Table 2). Compared with no biochar input, the biochar application showed a trend of increasing root AMF colonization. Compared with no biochar input, the CK, AMF, and AMF + R treatments enhanced the AMF colonization rate by 9.64%, 10.93%, and 13.68%, respectively, at the B1 input level.

Compared to B0 addition, B1 application significantly reduced the average soil NH_4^+-N and NO_3^--N contents by 16.43% and 22.62% (Table 2). Compared with the CK, the AMF and AMF + R treatments significantly decreased soil NH_4^+-N and NO_3^--N contents at two biochar input levels. The AMF and AMF + R treatments at the B0 level reduced the soil NH_4^+-N contents by 10.78% and 14.71%, respectively, and the NO_3^--N contents by 38.96% and 78.73%, respectively, and at the B1 level, they reduced the soil NH_4^+-N contents by

23.16% and 30.53%, respectively, and NO_3^--N contents by 28.38% and 77.23%, respectively, compared with the CK treatment (Table 2).

3.2. Soil N_2O Emissions

The N_2O emission rates showed a variation trend of increasing first and then decreasing during the experimental period at both biochar levels (Figure 2), and the N_2O emission peak occurred on the third to fourth days after fertilization. Compared with the CK, the AMF and AMF + R treatments all significantly decreased the N_2O emission peak by 40.62% and 53.23%, respectively, during the B1 application and 27.63% and 34.67%, respectively, during the B0 application. The result showed that biochar application significantly decreased soil N_2O emission rates (Figure 2). Both biochar input and AMF presence significantly affected soil N_2O emission flux (Table 3).

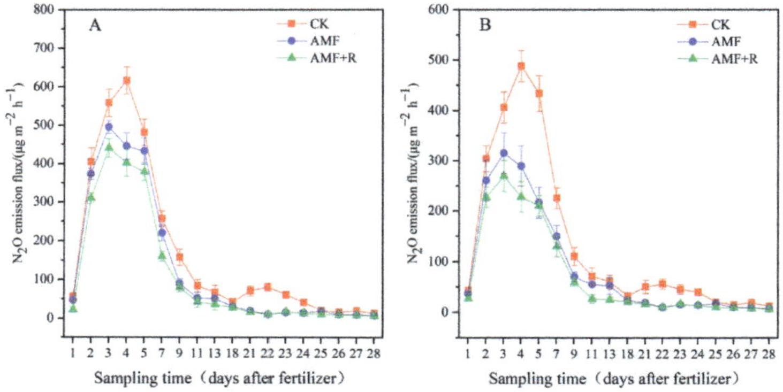

Figure 2. Effects of biochar input, AMF, and roots on soil N_2O emission flux. Values are means ± SE ($n = 4$) at any given time point. (**A**) No biochar application; (**B**) biochar application.

The soil N_2O accumulation emission during different treatments showed the trend of CK > AMF > AMF + R at both biochar levels (Figure 3). Compared with the CK, the AMF and AMF + R treatments significantly decreased N_2O accumulation emission by 14.41% and 25.02%, respectively, during the B0 application and 21.12% and 38.13%, respectively, during the B1 application. Overall, the soil N_2O cumulative emissions were significantly decreased during AMF and AMF + R treatments compared with CK ($p < 0.05$), and the AMF + R treatment had the best effect in reducing the soil N_2O cumulative emissions.

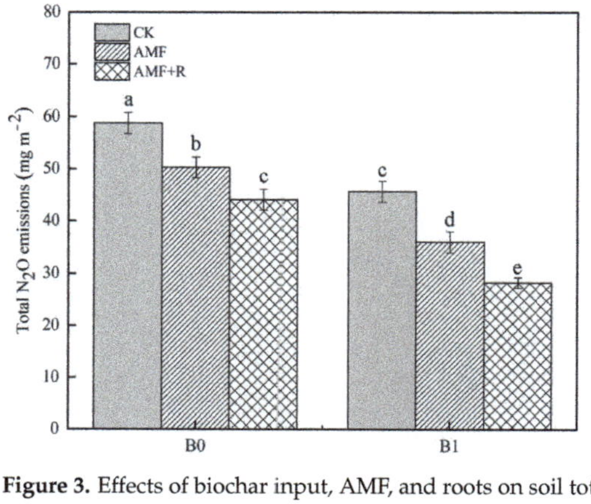

Figure 3. Effects of biochar input, AMF, and roots on soil total N_2O emission. Values are mean ± SE ($n = 4$). B0, no biochar application; B1, biochar application. Different letters among the treatments indicated the significance according to Duncan's LSD test at 0.05 level.

3.3. Abundances of Denitrification Genes

The AMF and AMF + R treatments enhanced the *nir*S, *nir*K, and *nos*Z gene copy numbers at both biochar application levels (Figure 4). During B1 application, the AMF and AMF + R treatments increased the *nir*S gene copy numbers by 23.83% and 49.10%, respectively, the *nir*K gene copy numbers by 68.01% and 113.75%, respectively, and the *nos*Z gene copy numbers by 110.86% and 175.51%, respectively, compared with CK. During B0 application, the AMF and AMF + R treatments increased the *nir*S gene copy numbers by 27.02% and 45.86%, respectively, the *nir*K gene copy numbers by 62.55% and 127.18%, respectively, and the *nos*Z gene copy numbers by 66.27% and 118.71%, respectively.

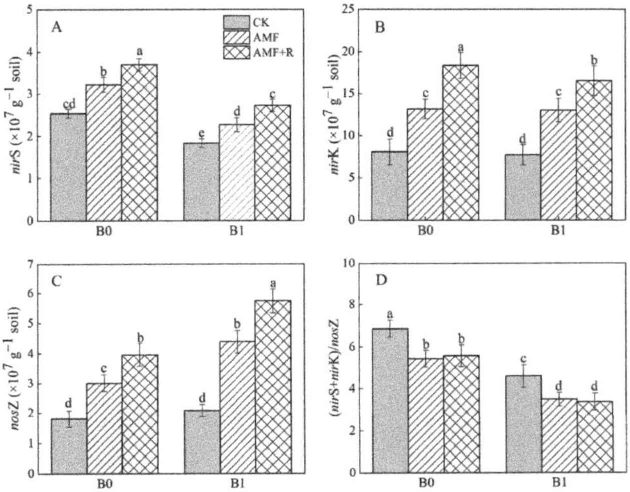

Figure 4. Effects of biochar input, AMF, and roots on soil *nir*K (**A**), *nir*S (**B**), and *nos*Z (**C**) gene copy numbers and *nir*S + *nir*K/*nos*Z ratio (**D**). Values are mean ± SE (n = 4). B0, no biochar application; B1, biochar application. Different letters among the treatments indicated the significance according to Duncan's LSD test at 0.05 level.

The AMF and AMF + R treatments significantly decreased *nir*S + *nir*K/*nos*Z at both biochar levels. The AMF and AMF + R treatments increased *nir*S + *nir*K/*nos*Z by 24.32% and 26.90%, respectively, at the B1 level and 20.83% and 18.84%, respectively, at the B0 level. The results showed that biochar input significantly decreased *nir*S + *nir*K/*nos*Z compared with no biochar addition (Figure 4). Both biochar input and AMF or root presence significantly affected the abundances of the denitrification genes (Table 3).

3.4. Nitrogen Content and Nitrogen Use Efficiency of Maize Plants

The biochar, AMF, and AMF with roots all had significant ($p < 0.05$) effects on the uptake and utilization of N nutrients for maize plants (Table 4). Compared with the CK, the AMF and AMF + R treatments all significantly increased the ^{15}N absorption and ^{15}N utilization efficiency of maize under both biochar addition levels. The AMF and AMF + R treatments increased N utilization efficiency by 18.03% and 25.17%, respectively, at the B0 level and 4.46% and 20.15%, respectively, at the B1 level. Compared with no biochar input, the biochar application significantly increased the N uptake and absorption, ^{15}N content, and nitrogen use efficiency of maize during the same treatment.

The soil ^{15}N retention rate during different treatments showed the trend of AMF + R > AMF > CK at the same biochar levels. The largest value of the soil ^{15}N retention rate was 31.09% during the AMF +R treatment with biochar application (Table 4). Contrarily, the soil ^{15}N loss rate during different treatments showed the trend of CK > AMF > AMF + R at the same biochar levels. Compared with CK, the AMF and AMF + R treatments decreased the soil ^{15}N loss rate by 10.33% and 14.42%, respectively, at the B0 level and 9.06% and 19.16%, respectively, at the B1 level.

Table 4. Effects of biochar input, AMF, and roots on nitrogen uptake of maize and the fate of ^{15}N-labeled urea N.

Biochar Levels	Treatment	Nitrogen Content of Maize (g kg^{-1})	^{15}N Absorption of Maize (g kg^{-1})	^{15}N Utilization Efficiency (%)	Soil ^{15}N Retention Rate (%)	Soil ^{15}N Loss Rate (%)
	CK	16.79 ± 1.04 e	3.35 ±8.93 d	20.02 ± 2.25 d	26.79 ± 3.04 e	53.19 ± 4.28 a
B0	AMF	22.04 ± 1.23 c	3.79 ±9.23 c	23.63 ±1.40 c	27.78 ± 3.23 d	48.59 ± 4.28 c
	AMF + R	27.56 ± 2.18 a	4.13 ± 8.59 b	25.06 ± 2.09 b	28.56 ± 14.18 c	46.38 ± 4.28 d
	CK	18.29 ± 2.27 d	3.75 ±3.10 c	23.72 ± 6.24 c	27.29 ± 2.27 d	48.99 ± 4.28 b
B1	AMF	24.80 ± 2.47 b	4.36 ±3.43 b	24.74 ± 1.63 b	29.80 ± 1.47 b	45.46 ± 4.28 d
	AMF + R	28.09 ± 1.77 a	4.98 ±7.07 a	28.50 ± 3.37 a	31.09 ± 3.77 a	40.41 ± 4.28 e

Values are mean ± SE (n = 4). B0, no biochar application; B1, biochar application. Different letters among the treatments indicated the significance according to Duncan's LSD test at 0.05 level.

3.5. Factors Affecting Soil N_2O Emissions

N_2O emission was correlated positively with the soil NH_4^+-N, NO_3^--N, and (*nir*K + *nir*S)/*nos*Z ratio. The *nir*K, *nir*S, and *nos*Z gene abundance, AMF colonization, and shoot biomass were correlated negatively with N_2O emission (Figure 5).

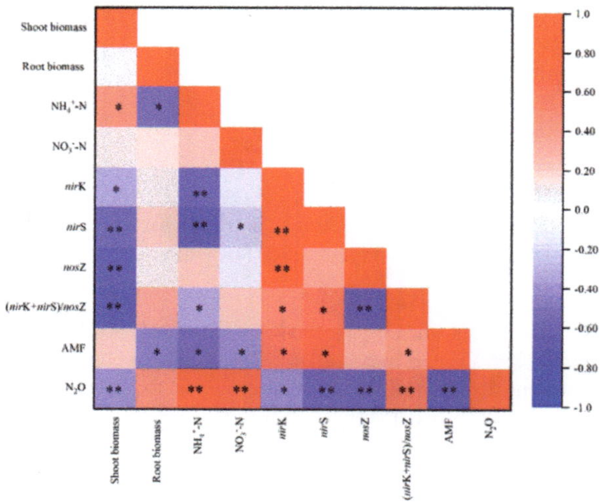

Figure 5. Relationships among plant biomass, soil inorganic N, denitrification gene abundance, AMF, and soil N_2O emission. ** $p < 0.01$ and * $p < 0.05$.

4. Discussion

4.1. Effects of Biochar, AMF, and Roots on Maize Growth

The studies revealed that the AMF alone or AMF with plant roots increased maize biomass (Table 2). The positive effects on maize growth are likely because AMF and plant roots promote nutrient acquisition and water uptake [30,34], which are beneficial to plant growth. Consistent with some previous studies [32,37], in this study, biochar application also promoted maize biomass. This may be because biochar could supply valuable macronutrients (e.g., N and P) and metal ions (e.g., Ca^{2+} and Mg^{2+}), which directly promote plant growth [45]. The second reason may be that biochar has the characteristics of a porous structure and large surface area, which can increase nutrient retention and reduce nutrient loss, thus indirectly promoting plant growth [46,47]. The results showed that biochar combined with AMF and/or plant roots had the highest shoot and root biomass. There was likely an interaction between biochar and AMF, which is consistent with previous studies [48,49]. In this study, biochar application enhanced AMF colonization, which increased the acquisition of soil nutrients and water and improved maize growth.

4.2. Effects of Biochar, AMF, and Roots on Soil N Utilization

Using the ^{15}N labeling technique, this study revealed that biochar input decreased soil NH_4^+-N and NO_3^--N concentrations and increased the ^{15}N utilization efficiency of maize and ^{15}N retention rate in soil. The first reason may be that biochar improved the soil's physical and chemical properties, which enhanced the inorganic N uptake of plants, ^{15}N utilization efficiency, and ^{15}N retention rate and resulted in a reduction of soil inorganic N content [50–53]. Additionally, previous studies have shown that biochar addition could increase the ion exchange capacity of soil and improve the adsorption and fixation of soil NH_4^+-N and NO_3^--N. These inorganic N were absorbed by biochar and could be released again for the absorption and utilization of plants, thus improving N utilization efficiency [32,54,55]. The third possible explanation for the increase in ^{15}N utilization efficiency is that biochar can increase nutrient availability, stimulate beneficial soil microbial activity, and reduce nutrient leaching [56–58]. Some studies have indicated that biochar addition increased soil pH and supplied a more suitable environment for the growth of nitrifying microbes, which could improve the nitrification process and decrease soil NH_4^+-N and NO_3^--N contents [59,60]. The results also showed that AMF and AMF + R treatments significantly decreased soil NH_4^+-N and NO_3^--N contents (Table 2). The AMF and/or plant roots play a vital role in soil N utilization and plant N uptake [15,61,62]. The extraradical mycelia of AMF and plant roots can absorb inorganic N from soil and transport them to host plants [63]. Moreover, AMF can promote the degradation of nitrogen by soil microorganisms, thus improving N uptake in plants and reducing soil N content [64,65]. Therefore, the AMF significantly reduced soil inorganic N contents and enhanced N utilization efficiency in this study.

The biochar combined with AMF alone or AMF with plant roots all significantly decreased soil NH_4^+-N and NO_3^--N contents and increased the ^{15}N utilization efficiency of maize and the ^{15}N retention rate of the soil, compared with no biochar input or biochar input with no AMF treatments. There are some reasons for the abovementioned results. Firstly, biochar application can improve the physiochemical properties of soil and promote the germination and growth of AMF spores. In addition, biochar has the properties of a porous structure, providing a favorable habitat for AMF [35,66]. Secondly, biochar may affect the germination and branching of AMF by regulating plant fungal signal substances [58,67]. Thirdly, the biochar indirectly affects AMF by affecting the activity of phosphorus bacteria in the soil [67]. Therefore, AMF can improve plants to absorb more soil inorganic N with biochar application.

4.3. Effects of Biochar, AMF, and Roots on N_2O Emissions

In the present study, the application of AMF alone or AMF with plant roots significantly decreased soil N_2O emission, compared with CK treatment (Figures 2 and 3). The application of biochar and AMF with plant roots showed the largest decrease in soil N_2O emission, which was consistent with the study by Hao et al. [36]. There are several reasons that biochar, AMF, and plant roots suppress soil N_2O emissions. On the one hand, biochar, AMF, and plant roots have inherent characteristics that help plants acquire nutrients from the soil [20,68]. In this study, biochar, AMF, and plant roots all reduced soil NH_4^+-N and NO_3^--N concentrations and increased the utilization of soil N. On the other hand, biochar application facilitates the germination of AMF spores and the extension of extraradical mycelial, which enhance N uptake in the host plant, reduce the N substrate of soil nitrifiers and denitrifiers, and decrease N_2O emissions [8,35].

In agricultural soil, N_2O is mainly produced through the participation of soil nitrifiers and denitrifiers. The denitrification process, in general, plays a dominant role in soil N_2O

emissions [8,69]. In the process of denitrification, the denitrifiers acquire energy from organic matter and convert NO_3^- to N_2O. In this study, the abundance of denitrifying functional genes was measured, which was helpful in estimating whether the change in soil N_2O emission was related to biochar-, AMF-, and plant root-induced changes in the abundance of denitrifying functional genes (Figure 4). This research showed that AMF and AMF + R treatments all enhanced *nir*S, *nir*K, and *nos*Z abundance and reduced the ratio of *nir*K + *nir*S/*nos*Z at both biochar application levels (Figure 4). Zhang et al. [14] showed that AMF alone or AMF and plant root treatments all enhanced *nir*S abundance and reduced the *nirk*/*nir*S ratio at high N and normal N input levels with conventionally managed soil. The present research showed that AMF with plant roots decreased N_2O emission directly by reducing soil N and indirectly by changing the abundance of N_2O-producing functional genes in the agroecosystem. Qiu et al. [9] reported that AMF with plant roots reduced *nir*K abundance and the (*nir*K + *nir*S)/*nos*Z ratio, leading to a reduction in soil N_2O emissions. In summary, the effect of AMF and plant roots on N_2O emission was mainly exhibited through regulating denitrifying functional genes and the denitrification process. This study showed that biochar addition decreased *nir*S and *nir*K abundance and enhanced *nos*Z abundance during AMF and AMF + R treatments (Figure 4). The positive correlation between the soil's N_2O emission and the (*nir*K + *nir*S)/*nos*Z ratio (Figure 5) was consistent with the results from other agricultural soils [9,36]. For instance, Hao et al. [36] reported that biochar addition and/or AMF inoculation decreased N_2O emissions in wheat-planted soil, compared with soil in the absence of biochar and AMF addition. This decrease in soil emission was mainly associated with the diminished abundance of N_2O-producing functional genes (*nir*K and *nir*S) and the increased abundance of N_2O-consuming functional genes (*nos*Z), and the (*nir*K + *nir*S)/*nos*Z ratio was the strongest driver for soil N_2O. In brief, the results suggest that the application of biochar, AMF, and plant roots may alter N_2O emission by modifying the denitrifying functional gene abundance. Furthermore, the detailed mechanisms of the interaction of biochar, AMF, and plant roots on soil denitrifying microbes and functional genes are still unclear and need further research.

5. Conclusions

The biochar application and presence of AMF or AMF with plant roots reduced soil N_2O emissions by increasing soil N unitization efficiency and decreasing the (*nir*K + *nir*S)/*nos*Z ratio in maize-planted soils. The presence of AMF and plant roots during biochar amendment showed the highest decrease in N_2O emission and an increase in soil N unitization efficiency. The biochar, AMF, and plant roots may interact to shift plant growth, N utilization, denitrifying functional genes, and N_2O emissions. These results suggest that biochar, in combination with AMF, may be considered an effective method for mediating N cycling and reducing N_2O emissions in agroecosystems. This study was performed under microcosm conditions, which were different from the field environment (fertilization, precipitation, and tillage practices). Therefore, it is necessary to conduct field experiments to study the effects and mechanisms of biochar and AMF on soil N cycle and N_2O emission.

Author Contributions: Conceptualization, Y.W.; Methodology, Y.L. and J.G.; Software, J.L. and J.G.; Validation, L.L.; Formal analysis, Y.L. and J.G.; Investigation, Y.W. and J.L.; Resources, X.D. and L.L.; Writing—original draft, Y.W.; Writing—review & editing, X.D. and L.L. All authors have read and agreed to the published version of the manuscript.

Funding: This work was supported by the National Natural Science Foundation of China (31700367) and the Natural Science Foundation of Henan Province (242300420598).

Data Availability Statement: Data is contained within the article.

Conflicts of Interest: The authors declare no conflict of interest.

References

1. Ravishankara, A.R.; Daniel, J.S.; Portmann, R.W. Nitrous oxide (N_2O): The dominant ozone-depleting substance emitted in the 21st century. *Science* **2009**, *326*, 123–125. [CrossRef]
2. Wu, P.; Xie, M.H.; Clough, T.J.; Yuan, D.; Wu, S.H.; He, X.D.; Hu, C.S.; Zhou, S.G.; Qin, S.P. Biochar-derived persistent free radicals and reactive oxygen species reduce the potential of biochar to mitigate soil N_2O emissions by inhibiting *nosZ*. *Soil Biol. Biochem.* **2023**, *178*, 108970. [CrossRef]
3. Luo, Z.; Lam, S.K.; Fu, H.; Hu, S.; Chen, D. Temporal and spatial evolution of nitrous oxide emissions in China: Assessment, strategy and recommendation. *J. Clean Prod.* **2019**, *223*, 360–367. [CrossRef]
4. Butterbach-Bahl, K.; Baggs, E.M.; Dannenmann, M.; Kiese, R.; Zechmeister-Boltenstern, S. Nitrous oxide emissions from soils: How well do we understand the processes and their controls? *Phil. Trans. R. Soc. B* **2013**, *368*, 20130122. [CrossRef]
5. Firestone, M.K.; Firestone, R.B.; Tiedje, J.M. Nitrous oxide from soil denitrification: Factors controlling its biological production. *Science* **1980**, *208*, 749–751. [CrossRef]
6. Jones, C.M.; Putz, M.; Tiemann, M.; Hallin, S. Reactive nitrogen restructures and weakens microbial controls of soil N_2O emissions. *Commun. Biol.* **2022**, *5*, 273. [CrossRef]
7. Hallin, S.; Philippot, L.; Löffler, F.E.; Sanford, R.A.; Jones, C.M. Genomics and ecology of novel N_2O-reducing microorganisms. *Trends Microbiol.* **2018**, *26*, 43–55. [CrossRef]
8. Jones, C.M.; Spor, A.; Brennan, F.P.; Breuil, M.C.; Bru, D.; Lemanceau, P.; Griffiths, B.; Hallin, S.; Philippot, L. Recently identified microbial guild mediates soil N_2O sink capacity. *Nat. Clim. Change* **2014**, *4*, 801–805. [CrossRef]
9. Qiu, Y.P.; Jiang, Y.; Guo, L.J.; Zhang, L.; Burkey, K.O.; Zobel, R.W.; Reberg-Horton, S.C.; Shew, H.D.; Hu, S.J. Shifts in the composition and activities of denitrifiers dominate CO_2 stimulation of N_2O emissions. *Environ. Sci. Technol.* **2019**, *53*, 11204–11213. [CrossRef]
10. Kiers, E.T.; Duhamel, M.; Beesetty, Y.; Mensah, J.A.; Franken, O.; Verbruggen, E.; Fellbaum, C.R.; Kowalchuk, G.A.; Hart, M.M.; Bago, A.; et al. Reciprocal rewards stabilize cooperation in the mycorrhizal symbiosis. *Science* **2011**, *333*, 880–882. [CrossRef]
11. Smith, S.E.; Smith, F.A. Roles of arbuscular mycorrhizas in plant nutrition and growth: New paradigms from cellular to ecosystem scales. *Annu. Rev. Plant Biol.* **2011**, *62*, 227–250. [CrossRef]
12. Storer, K.; Coggan, A.; Ineson, P.; Hodge, A. Arbuscular mycorrhizal fungi reduce nitrous oxide emissions from N_2O hotspots. *New Phytol.* **2018**, *220*, 1285–1295. [CrossRef]
13. Veresoglou, S.D.; Chen, B.D.; Rillig, M.C. Arbuscular mycorrhiza and soil nitrogen cycling. *Soil Biol. Biochem.* **2012**, *46*, 53–62. [CrossRef]
14. Zhang, X.L.; Qiu, Y.P.; Gilliam, F.S.; Gillespie, C.J.; Tu, C.; Reberg-Horton, S.C.; Hu, S.J. Arbuscular mycorrhizae shift community composition of N-cycling microbes and suppress soil N_2O emission. *Environ. Sci. Technol.* **2022**, *56*, 13461–13472. [CrossRef]
15. Bender, S.F.; Plantenga, F.; Neftel, A.; Jocher, M.; Oberholzer, H.R.; Köhl, L.; Giles, M.; Daniell, T.J.; van der Heijden, M.G.A. Symbiotic relationships between soil fungi and plants reduce N_2O emissions from soil. *ISME J.* **2014**, *8*, 1336–1345. [CrossRef]
16. Ai, C.; Liang, G.Q.; Wang, X.B.; Sun, J.W.; He, P.; Zhou, W. A distinctive root-inhabiting denitrifying community with high $N_2O/(N_2O+N_2)$ product ratio. *Soil Biol. Biochem.* **2017**, *109*, 118–123. [CrossRef]
17. Henry, S.; Texier, S.; Hallet, S.; Bru, D.; Dambreville, C.; Chèneby, D.; Bizouard, F.; Germon, J.C.; Philippot, L. Disentangling the rhizosphere effect on nitrate reducers and denitrifiers: Insight into the role of root exudates. *Environ. Microbiol.* **2008**, *10*, 3082–3092. [CrossRef]
18. Manya, J.J. Pyrolysis for biochar purposes: A review to establish current knowledge gaps and research needs. *Environ. Sci. Technol.* **2012**, *46*, 7939–7954. [CrossRef]
19. Dong, W.X.; Walkiewicz, A.; Bieganowski, A.; Oenema, O.; Nosalewicz, M.; He, C.H.; Zhang, Y.M.; Hu, C.S. Biochar promotes the reduction of N_2O to N_2 and concurrently suppresses the production of N_2O in calcareous soil. *Geoderma* **2020**, *362*, 114091. [CrossRef]
20. Yuan, H.J.; Zhang, Z.J.; Li, M.Y.; Clough, T.; Wrage-Mönnig, N.; Qin, S.P.; Ge, T.D.; Liao, H.P.; Zhou, S.G. Biochar's role as an electron shuttle for mediating soil N_2O emissions. *Soil Biol. Biochem.* **2019**, *133*, 94–96. [CrossRef]
21. He, P.J.; Zhang, H.H.; Duan, H.W.; Shao, L.M.; Lü, F. Continuity of biochar-associated biofilm in anaerobic digestion. *Chem. Eng. J.* **2020**, *390*, 124605. [CrossRef]
22. Liu, H.Y.; Li, H.B.; Zhang, A.P.; Rahaman, M.A.; Yang, Z.L. Inhibited effect of biochar application on N_2O emissions is amount and time-dependent by regulating denitrification in a wheat-maize rotation system in North China. *Sci. Total Environ.* **2020**, *721*, 137636. [CrossRef]
23. Clough, T.; Condron, L.; Kammann, C.; Müller, C. A review of biochar and soil nitrogen dynamics. *Agronomy* **2013**, *3*, 275–293. [CrossRef]
24. Luo, X.; Chen, L.; Zheng, H.; Chang, J.; Wang, H.; Wang, Z.; Xing, B. Biochar addition reduced net N mineralization of a coastal wetland soil in the Yellow River Delta, China. *Geoderma* **2016**, *282*, 120–128. [CrossRef]
25. Yang, F.; Cao, X.D.; Gao, B.; Zhao, L.; Li, F.Y. Short-term effects of rice straw biochar on sorption, emission, and transformation of soil NH_4^+-N. *Environ. Sci. Pollut. R.* **2015**, *22*, 9184–9192. [CrossRef]
26. Zhang, H.; Voroney, R.P.; Price, G.W. Effects of temperature and processing conditions on biochar chemical properties and their influence on soil C and N transformations. *Soil Biol. Biochem.* **2015**, *83*, 19–28. [CrossRef]

27. Case, S.D.C.; McNamara, N.P.; Reay, D.S.; Stott, A.W.; Grant, H.K.; Whitaker, J. Biochar suppresses N_2O emissions while maintaining N availability in a sandy loam soil. *Soil Biol. Biochem.* **2015**, *81*, 178–185. [CrossRef]
28. Niu, Y.H.; Chen, Z.M.; Müller, C.; Zaman, M.; Kim, D.G.; Yu, H.Y.; Ding, W.X. Yield-scaled N_2O emissions were effectively reduced by biochar amendment of sandy loam soil under maize—Wheat rotation in the North China plain. *Atmos. Environ.* **2017**, *170*, 58–70. [CrossRef]
29. Ibrahim, M.M.; Tong, C.X.; Hu, K.; Zhou, B.Q.; Xing, S.H.; Mao, Y.L. Biochar-fertilizer interaction modifies N-sorption, enzyme activities and microbial functional abundance regulating nitrogen retention in rhizosphere soil. *Sci. Total Environ.* **2020**, *739*, 140065. [CrossRef]
30. Jabborova, D.; Annapurna, K.; Choudhary, R.; Bhowmik, S.N.; Desouky, S.E.; Selim, S.; Azab, I.H.E.; Hamada, M.M.A.; Nahhas, N.E.; Elkelish, A. Interactive impact of biochar and arbuscular mycorrhizal on root morphology; physiological properties of fenugreek (*Trigonella foenum-graecum* L.) and soil enzymatic activities. *Agronomy* **2021**, *11*, 2341. [CrossRef]
31. Lehmann, J.; Rillig, M.C.; Thies, J.; Masiello, C.A.; Hockaday, W.C.; Crowley, D. Biochar effects on soil biota—A review. *Soil Biol. Biochem.* **2011**, *43*, 1812–1836. [CrossRef]
32. Liu, L.; Li, J.W.; Yue, F.X.; Yan, X.W.; Wang, F.Y.; Bloszies, S.; Wang, Y.F. Effects of arbuscular mycorrhizal inoculation and biochar amendment on maize growth, cadmium uptake and soil cadmium speciation in Cd-contaminated soil. *Chemosphere* **2018**, *194*, 495–503. [CrossRef]
33. Gomez, J.D.; Denef, K.; Stewart, C.E.; Zheng, J.; Cotrufo, M.F. Biochar addition rate influences soil microbial abundance and activity in temperate soils. *Eur. J. Soil Sci.* **2014**, *65*, 28–39. [CrossRef]
34. Liang, J.F.; An, J.; Gao, J.Q.; Zhang, X.Y.; Song, M.H.; Yu, F.H. Interactive effects of biochar and AMF on plant growth and greenhouse gas emissions from wetland microcosms. *Geoderma* **2019**, *346*, 11–17. [CrossRef]
35. Hammer, E.C.; Forstreuter, M.; Rillig, M.C.; Kohler, J. Biochar increases arbuscular mycorrhizal plant growth enhancement and ameliorates salinity stress. *Appl. Soil Ecol.* **2015**, *96*, 114–121. [CrossRef]
36. Hao, Z.; Dong, Z.J.; Han, S.; Zhang, A.F. Effects of biochar and arbuscular mycorrhizal fungi on winter wheat growth and soil N_2O emissions in different phosphorus environments. *Front. Plant Sci.* **2022**, *13*, 1069627. [CrossRef]
37. Zhang, L.; Qiu, Y.P.; Cheng, L.; Wang, Y.; Liu, L.L.; Tu, C.; Bowman, D.C.; Burkey, K.O.; Bian, X.M.; Zhang, W.J.; et al. Atmospheric CO_2 enrichment and reactive nitrogen inputs interactively stimulate soil cation losses and acidification. *Environ. Sci. Technol.* **2018**, *52*, 6895–6902. [CrossRef]
38. Giovannetti, M.; Mosse, B. Evaluation of techniques for measuring vesicular arbuscular mycorrhizal infection in roots. *New Phytol.* **1980**, *84*, 489–500. [CrossRef]
39. Shi, Y.L.; Liu, X.R.; Zhang, Q.W. Effects of combined biochar and organic fertilizer on nitrous oxide fluxes and the related nitrifier and denitrifier communities in a salinealkali soil. *Sci. Total Environ.* **2019**, *686*, 199–211. [CrossRef]
40. Gao, J.; Duan, M.; Zhang, X.; Li, Q.; Yu, F. Effects of frequency and intensity of drying-rewetting cycles on hydrocotyle vulgaris growth and greenhouse gas emissions from wetland microcosms. *Catena* **2018**, *164*, 44–49. [CrossRef]
41. Guo, G.X.; Deng, H.; Qiao, M.; Mu, Y.J.; Zhu, Y.G. Effect of pyrene on denitrification activity and abundance and composition of denitrifying community in an agricultural soil. *Environ Pollut.* **2011**, *159*, 1886–1895. [CrossRef]
42. Braker, G.; Fesefeldt, A.; Witzel, K.P. Development of PCR primer systems for amplification of nitrite reductase genes (*nirK* and *nirS*) to detect denitrifying bacteria in environmental samples. *Appl. Environ. Microb.* **1998**, *64*, 3769–3775. [CrossRef]
43. Throbäck, I.N.; Enwall, K.; Jarvis, Å.; Hallin, S. Reassessing PCR primers targeting *nirS*; *nirK* and *nosZ* genes for community surveys of denitrifying bacteria with DGGE. *FEMS Microbiol. Ecol.* **2004**, *49*, 401–417. [CrossRef]
44. Robinson, D. $\delta^{15}N$ as an integrator of the nitrogen cycle. *Trends Ecol. Evol.* **2001**, *16*, 153–162. [CrossRef]
45. Ghorbani, M.; Amirahmadi, E. Biochar and soil contributions to crop lodging and yield performance—A meta-analysis. *Plant Physiol. Bioch.* **2024**, *215*, 109053. [CrossRef]
46. Ghorbani, M.; Amirahmadi, E.; Bernas, J.; Konvalina, P. Testing biochar's ability to moderate extremely acidic soils in tea-growing areas. *Agronomy* **2024**, *14*, 533. [CrossRef]
47. Guo, R.; Qian, R.; Yang, L.; Khaliq, A.; Han, F.; Hussain, S.; Zhang, P.; Cai, T.; Jia, Z.K.; Chen, X.L.; et al. Interactive effects of maize straw-derived biochar and N fertilization on soil bulk density and porosity, maize productivity and nitrogen use efficiency in arid areas. *J. Soil Sci. Plant Nutr.* **2022**, *22*, 4566–4586. [CrossRef]
48. Liu, L.; Wang, Y.F.; Yan, X.W.; Li, J.W.; Jiao, N.Y.; Hu, S.J. Biochar amendments increase the yield advantage of legume-based inter cropping systems over monoculture. *Agri. Ecosyst. Environ.* **2017**, *237*, 16–23. [CrossRef]
49. Mickan, B.S.; Abbott, L.K.; Stefanova, K.; Solaiman, Z.M. Interactions between biochar and mycorrhizal fungi in a water-stressed agricultural soil. *Mycorrhiza* **2016**, *26*, 565–574. [CrossRef]
50. Liao, J.; Liu, X.; Hu, A.; Song, H.; Chen, X.; Zhang, Z. Effects of biochar-based controlled release nitrogen fertilizer on nitrogen-use efficiency of oilseed rape (*Brassica napus* L.). *Sci. Rep.* **2020**, *10*, 11063. [CrossRef]
51. Ma, R.; Guan, S.; Dou, S.; Wu, D.; Xie, S.; Ndzelu, B.S. Different rates of biochar application change ^{15}N retention in soil and 15 N utilization by maize. *Soil Use Manag.* **2020**, *36*, 773–782. [CrossRef]
52. Zhu, Q.H.; Peng, X.H.; Huang, T.Q.; Xie, Z.B.; Holden, N.M. Effect of biochar addition on maize growth and nitrogen use efficiency in acidic red soils. *Pedosphere* **2014**, *24*, 699–708. [CrossRef]

53. Fiorentino, N.; Sánchez-Monedero, M.A.; Lehmann, J.; Enders, A.; Fagnano, M.; Cayuela, M.L. Interactive priming of soil N transformations from combining biochar and urea inputs: A ^{15}N isotope tracer study. *Soil Biol. Biochem.* **2019**, *131*, 166–175. [CrossRef]
54. Zhang, X.; Zhang, J.Q.; Song, M.X.; Dong, Y.B.; Xiong, Z.Q. N_2O and NO production and functional microbes responding to biochar aging process in an intensified vegetable soil. *Environ. Pollut.* **2022**, *307*, 119491. [CrossRef] [PubMed]
55. Wang, Z.; Wang, Z.; Luo, Y.; Zhan, Y.N.; Meng, Y.L.; Zhou, Z.G. Biochar increases ^{15}N fertilizer retention and indigenous soil N uptake in a cotton-barley rotation system. *Geoderma* **2020**, *357*, 113944. [CrossRef]
56. Abel, S.; Peters, A.; Trinks, S.; Schonsky, H.; Facklam, M.; Wessolek, G. Impact of biochar and hydrochar addition on water retention and water repellency of sandy soil. *Geoderma* **2013**, *202*, 183–191. [CrossRef]
57. Huang, M.; Yang, L.; Qin, H.; Jiang, L.; Zou, Y. Quantifying the effect of biochar amendment on soil quality and crop productivity in Chinese rice paddies. *Field Crop Res.* **2013**, *154*, 172–177. [CrossRef]
58. Atkinson, C.J.; Fitzgerald, J.D.; Hipps, N.A. Potential mechanisms for achieving agricultural benefits from biochar application to temperate soils: A review. *Plant Soil.* **2010**, *337*, 1–18. [CrossRef]
59. Ulyett, J.; Sakrabani, R.; Kibblewhite, M.; Hann, M. Impact of biochar addition on water retention; nitrification and carbon dioxide evolution from two sandy loam soils. *Eur. J. Soil Sci.* **2014**, *65*, 96–104. [CrossRef]
60. Mandal, S.; Thangarajan, R.; Bolan, N.S.; Sarkar, B.; Khan, N.; Ok, Y.S.; Naidu, R. Biochar-induced concomitant decrease in ammonia volatilization and increase in nitrogen use efficiency by wheat. *Chemosphere* **2016**, *142*, 120–127. [CrossRef]
61. Hodge, A.; Storer, K. Arbuscular mycorrhiza and nitrogen: Implications for individual plants through to ecosystems. *Plant Soil.* **2015**, *386*, 1–19. [CrossRef]
62. Verzeaux, J.; Hirel, B.; Dubois, F.; Lea, P.J.; Tetu, T. Agricultural practices to improve nitrogen use efficiency through the use of arbuscular mycorrhizae: Basic and agronomic aspects. *Plant Sci.* **2017**, *264*, 48–56. [CrossRef] [PubMed]
63. Hodge, A.; Fitter, A.H. Substantial nitrogen acquisition by arbuscular mycorrhizal fungi from organic material has implications for N cycling. *Proc. Natl. Acad. Sci. USA* **2010**, *107*, 13754–13759. [CrossRef] [PubMed]
64. Cavagnaro, T.R.; Bender, S.F.; Asghari, H.R.; Heijden, M. The role of arbuscular mycorrhizas in reducing soil nutrient loss. *Trends Plant Sci.* **2015**, *20*, 283–290. [CrossRef] [PubMed]
65. Zhang, S.J.; Lehmann, A.; Zheng, W.S.; You, Z.Y.; Rillig, M.C. Arbuscular mycorrhizal fungi increase grain yields: A meta-analysis. *New Phytol.* **2019**, *222*, 543–555. [CrossRef]
66. Ezawa, T.; Yamamoto, K.; Yoshida, S. Enhancement of the effectiveness of indigenous arbuscular mycorrhizal fungi by inorganic soil amendments. *Soil Sci. Plant Nutr.* **2002**, *48*, 897–900. [CrossRef]
67. Warnock, D.D.; Lehmann, J.; Kuyper, T.W.; Rillig, M.C. Mycorrhizal responses to biochar in soil—Concepts and mechanisms. *Plant Soil.* **2007**, *300*, 9–20. [CrossRef]
68. Nguyen, T.T.N.; Xu, C.Y.; Tahmasbian, I.; Che, R.X.; Xu, Z.H.; Zhou, X.H.; Wallace, H.M.; Bai, S.H. Effects of biochar on soil available inorganic nitrogen: A review and meta-analysis. *Geoderma* **2017**, *288*, 79–96. [CrossRef]
69. Liang, D.; Robertson, G.P. Nitrification is a minor source of nitrous oxide (N_2O) in an agricultural landscape and declines with increasing management intensity. *Glob. Change Biol.* **2021**, *27*, 5599–5613. [CrossRef]

agronomy

MDPI

Article

Isolation, Characterization, and Optimization of Culture Medium for Local Straw-Degrading Bacteria from Northeastern Black Soils of China

Lei Zhang [1,2], Tianyu Liu [3], Chengzhao Yan [4], Jinpeng Zhang [1], Rui Yu [1,*], Nana Luo [1] and Yong Yu [1]

1 Key Laboratory of Wetland Ecology and Environment, State Key Laboratory of Black Soils Conservation and Utilization, Northeast Institute of Geography and Agroecology, Chinese Academy of Sciences, Changchun 130102, China
2 University of Chinese Academy of Sciences, Beijing 100049, China
3 College of Horticulture, Jilin Agricultural University, Changchun 130118, China
4 College of Resources and Environment, Jilin Agricultural University, Changchun 130118, China
* Correspondence: yurui@iga.ac.cn

Abstract: In order to solve the problem of low and poor straw degradation in typical black soil areas of Northeast China, the present study was carried out to screen the potential of in situ strains with cellulose degradation ability from black soils of Northeast China to play a role in the resourceful utilization of straw and the development of sustainable agriculture. The straw degradation potential of the strains was evaluated by combining sodium carboxymethyl cellulose plate screening and cellulase viability assay; the species identification of the strains was carried out by morphology, physiology, biochemistry, and molecular biology; and the basic medium formulation of the strains was optimized by Box–Behnken response surface methodology. Ten cellulose-degrading strains were identified: ZL-5, ZL-69, ZL-88, ZL-95, ZL-111, ZL-137, ZL-139, ZL-140, ZL-187, and ZL-216, of which ZL-139 had the highest cellulase production capacity, with a cellulase secretion of 7.8781 U/mL in the enzyme-producing medium. ZL-139 was identified as *Bacillus cereus*; the optimized best formulation was glucose—4.284 g/L, yeast extract—1.454 g/L, $MgSO_4$—0.417 g/L, KH_2PO_4—0.5 g/L, KH_2PO_4—0.5 g/L, K_2HPO_4—1.5 g/L, and NaCl—1.0 g/L. In conclusion, strain ZL-139 has good potential for crop straw degradation and can be a candidate strain for a straw-rotting agent in northeast China, with promising prospects for development and utilization.

Keywords: black soil; straw resource utilization; cellulase; *Bacillus cereus*; Box–Behnken response surface methodology

1. Introduction

With the continuous development of global agriculture, discarded straw seriously affects the sustainability of agriculture and jeopardizes the health level of the environment. China is one of the world's largest producers of crop straw, generating about 700 million tons of straw annually; however, the lack of feasible treatment and utilization technologies [1] has resulted in more than 200 million tons of this being disposed of by incineration, leading to serious environmental pollution and the wasting of resources [2,3]. Burning one ton in an open field releases 1168 kg of CO_2, 1.0 kg of CH_4, and 0.06 kg of N_2O as greenhouse gasses and emits large amounts of particulate matter [4]; meanwhile, the open burning of straw evaporates soil moisture, destabilizes soil aggregates, reduces organic matter, and lowers the particle size of soil aggregates [5,6]. The Chinese government has formulated a series of relevant policies and regulations, but, in general, the dissemination of policies banning straw burning in China's rural communities has had mixed results [7], and farmers are not sufficiently cognizant of the policy to ban straw burning, which has led to the occurrence of straw burning from time to time. In recent years, as the straw problem has become more and more prominent, how to deal with the large amount of excess straw has become a major challenge for "One Health [8]", but straw has a complex structure and a low rate of natural degradation [9] , so adopting a reasonable method for the resource utilization of straw can help to achieve low-carbon agriculture.

Currently, there are physical, chemical, and biological methods for straw resource utilization. Compared with physical and chemical treatments, biological methods for corn stover consume low energy and are easy to implement, energy-efficient, and environmentally friendly, while rapidly promoting the in situ degradation of stover [10,11]. Previous studies have shown that fungi have a strong ability to degrade straw [12,13]. However, compared to fungi, bacteria have the advantages of a small size, a short reproduction cycle, a simple structure, high resistance, and wide distribution in nature, which enable them to be cultivated and produced on a large scale [14]. They can enter the plant cells in large quantities to degrade the cellulose therein [15]; at the same time, the organic acids they produce can also promote cellulose degradation [16]. For example, Dong [17] et al. found that inoculation with *Brevibacillus borstelensis*, *Bacillus paralicheniformis*, *Fusobacterium*, *Bacillus subtilis*, and *Providencia vermicola* significantly accelerated the straw composting humification process of straw compost; a correlation analysis by Sun [18] et al. showed a significant positive correlation between the rate of straw degradation and the bacterial population during the first 30 d of straw degradation; and Zhang [19] et al. isolated *Bacillus cereus* and *Ochrobactrum pseudintermedium* from the rumen of buffaloes, which were finally proven to be able to degrade agricultural wastes after the preliminary screening and verification of enzyme activity. However, in practical applications, the local soil microflora will largely affect the role of exogenous microorganisms [20] and reduce the straw reduction efficiency of exogenous microorganisms. For example, Zhao [21] et al. found that there was no significant change in the chao index of the soil bacterial community when exogenous strains were inoculated, and Jia [22] et al. concluded that the reason why there was no significant remediation effect of inoculation with exogenous strains during soil remediation might be that the soil properties affected the microbial remediation of exogenous degradative strains. So, the development of straw-degrading bacteria from local soil has important application value.

Northeastern black soils are rich in organic matter and nutrient content, as well as having an excellent water-holding capacity, making them a valuable resource for high-yield agriculture and playing a vital role in global agricultural production [23]. At the same time, black soil possesses rich microbial resources, and the isolation of straw-degrading bacteria from black soil can promote the in situ degradation of straw to a greater extent. The aim of this study was to provide strain resources for straw resource utilization. The focus of this study was on (1) the in situ isolation and screening of straw-degrading strains from the black soil region of northeastern China; (2) the morphological, physiological, biochemical, and molecular biological identification of the strains with the highest straw degradation ability; and (3) the optimization of the basic medium for the best strains. We hope that this study will provide strain resources and a theoretical basis for the preparation of straw-degrading microbial agents and the resource utilization of crop straw.

2. Materials and Methods

2.1. Sample Collection

The experimental soil samples were collected from the black soil of Gongzhuling Demonstration Base (43°66′70″ N, 125°00′06″ E) in Changchun City, Jilin Province. The Gongzhuling Demonstration Base has a temperate continental monsoon climate, with an average annual temperature of 5.9 °C, an annual precipitation of 600 mm, and a type of black soil with a good granular structure, usually with an organic matter content of between 3 and 6 per cent and a neutral or slightly alkaline pH. The top 5 cm of soil was removed, and about 1000 g of soil was shoveled at a depth of 5–40 cm; the impurities such as plant leaves, rhizomes, and stones were removed and were then sealed in a bag, numbered, recorded, and stored at room temperature. Soil from 5 to 40 cm was selected. The top layer of soil

(0–5 cm) was selected due to it being light and dry, whereas the deeper layer of soil (>40 cm) was chosen due to nutrient deprivation; therefore, this depth (5–40 cm) is best suited for the survival of soil bacteria and has the maximum soil bacterial abundance. Corn stover was collected from Changchun Comprehensive Agricultural Experiment Station of Northeast Institute of Geography and Agroecology, Chinese Academy of Sciences (43°00′11″ N, 125°40′34″ E), the climate characteristics of the research institute are similar to those of the Gongzhuling Demonstration Base, and the surface dust of corn stover was washed with sterile water, dried at 80 °C for 12 h, and then cut into 1–2 cm sections for storage. Corn is one of the most important cash crops in Northeast China, and so it is of practical significance to use corn stover as the research object in this experiment. The cleaning and drying of corn stover is carried out to avoid experimental errors due to impurities and moisture.

2.2. Screening of Cellulose-Degrading Bacteria

Here, 5 g of soil sample was weighed and added to a triangular flask containing 95 mL of sterile water and was incubated for 72 h at 35 °C with 150 rpm-min-1 shaker. After standing for 5 h to allow the solution to stratify, 2 mL of the supernatant was aspirated and inoculated into a liquid enrichment medium containing 10 g/L tryptone, 5 g/L yeast extract, 10 g/L NaCl, and 30 g/L corn straw. The enrichment was incubated for 7 d at 35 °C, 150 rpm·min^{-1}, during which time the triangular flasks were labeled with a graduated scale and the water level was replenished at 12 h intervals. The water was diluted to 10^{-2}, 10^{-3}, 10^{-4}, 10^{-5}, and 10^{-6} concentrations according to the gradient dilution method [24]. After mixing, 100 μL of diluted solutions of different concentrations were applied to the Luria–Bertani (LB) medium using a pipette gun [25]. The use of 100 μL of dilution solution for spreading was to ensure uniform spreading at different dilution concentrations to the extent that single colonies appeared, which were then inverted and cultured at 35 °C in the dark for 24 h. Strains with different colony morphologies were picked using an inoculation loop and isolated and purified by streaking on an LB medium [26] , numbered, and preserved. The morphological characteristics of colonies formed by various bacteria on culture media have certain stability and specificity; therefore, colonies can be used as a preliminary basis for identifying different bacterial species.

The already purified strains were inoculated in a screening medium containing 20.0 g/L CMC-Na, 2.5 g/L Na_2HPO_4, 1.5 g/L KH_2PO_4, 2.5 g/L peptone, 0.5 g/L yeast extract, and 20 g/L agar [27]. The colonies were grown in dark culture at 35 °C for 8 h. The colonies were stained using 1 g/L Congo red coloring solution for 20 min, and the Congo red coloring solution was then poured off. Add 1 mol/L NaCl solution to the screening medium and rinse the surface of the plate 2–3 times; observe the production of a hydrolysis circle and use vernier calipers to accurately measure the diameter of the hydrolysis circle (D) and the diameter of the colony (d) on the plate; and calculate the ratio of the diameter of the hydrolysis circle to the diameter of the colony [28–30]. The preliminary determinations of the high and low cellulase activity of the isolated strains were carried out in three replicates for each strain.

2.3. Measurement of Cellulase Activity

The experimental strains were inoculated with a sodium carboxymethyl cellulose liquid enzyme-producing medium containing 10.0 g/L tryptone, 10.0 g/L beef extract, 10.0 g/L CMC-Na, 1.5 g/L NaCl, 1.0 g/L KH_2PO_4, and 0.3 g/L $MgSO_4{\cdot}7H_2O$, and were then placed in a constant-temperature shaker for 5 d at 35 °C and 150 rpm·min^{-1}. Furthermore, 5 mL of the sample was centrifuged at 10,000 rpm min^{-1} for 3 min, and the supernatant was taken as the crude enzyme solution, with three replicates for each strain.

The test tube was filled with 200 μL of the crude enzyme solution consisting of 1% CMC-Na solution and 200 μL of crude enzyme solution (the control group was prepared by adding an equal amount of inactivated crude enzyme solution, which was sterilized in an autoclave at 121 °C for 30 min), mixed homogeneously, and then water-bathed for 30 min in a water bath at 50 °C. Then, 600 μL of DNS reagent (DNS reagent (3,5-dinitrosalicylic acid reagent) is a commonly used chemical reagent, which is widely used for the determination of reducing sugar contents) was added [31,32] , mixed well, boiled in a water bath for 10 min, cooled rapidly in an ice-water bath, and then fixed to 5 mL with distilled water. Its absorbance value at 540 nm was determined, and cellulase activity was calculated from the plotted glucose standard curve ($Y = 0.1324X - 0.1045$, $R^2 = 0.9959$) with three replicates per strain. The enzyme activity was calculated as follows [33] :

$$\text{cellulase activity (U/mL)} = \frac{W \times \Delta f \times 10^3}{180.16 \times 30}$$

where w is the amount of glucose produced by the enzymatic reaction obtained from the glucose standard curve, g; Δf is the dilution, i.e., the ratio of the amount of test-tube solution to the volume of the enzyme solution; 10^3 indicates the unit conversion; 180.16 is the relative molecular mass of glucose, $g \cdot mol^{-1}$; 30 is the enzyme reaction time, min; and the amount of enzyme activity required to catalyze the production of 1 μmol of glucose in 1 min is one enzyme activity unit of enzyme activity.

2.4. Identification of Bacterial Strains

The experimental strains were inoculated in a solid enrichment medium and a liquid LB medium and were incubated overnight at 35 °C. For relevant morphological [34] and physiological and biochemical characterization, refer to the *Handbook of Systematic Identification of Common Bacteria* [35]. Scanning electron microscopy observation was performed by centrifuging the culture solution at 8000 $rpm \cdot min^{-1}$ for 5 min, discarding the supernatant, fixing the precipitated organisms with 2.5% glutaraldehyde at 4 °C overnight, and washing them three times in 0.2 M PBS buffer (pH = 6.8) for 15–20 min each time. Dehydration was carried out with ethanol gradient and then put into a freeze dryer for freeze drying. The dried bacterial samples were treated with gold spraying [36].

Total DNA of the strain was extracted for PCR amplification, and primers were used as universal primers for bacterial 16S rDNA sequences [37] 27F: 5′-AGAGTTTGATCCTGGCTCAG-3′ and 1492R: 5′-GGTTACCTTGTTACGACTT-3′. The amplification system is as follows: 2X Tap PCR Master Mix—25 μL, upstream and downstream primers—1 μL, cDNA—2.5 μL, ddH_2O—20.1 μL. PCR reaction conditions: 95 °C for 3 min, 95 °C for 30 s, 55 °C for 30 s, 72 °C for 1 min, 34 cycles; 72 °C extension for 5 min. The PCR products were sent to Bioengineering (Songjiang, Shanghai, China) Co., Ltd. for sequencing after 1% agarose gel electrophoresis, the sequencing results were analyzed by BLAST comparison in the NCBI (https://www.ncbi.nlm.nih.gov/) database on 1 April 2024, and the sequences of the strains with the highest homology were selected to construct a phylogenetic tree [38].

2.5. Optimization of Optimal Medium

For a clear description of the Box–Behnken response surface design approach, the overall optimization process is described in Figure S1. Keeping the contents of KH_2PO_4, Na_2HPO_4, NaCl, and $MgSO_4$ constant in the basic medium formulated as 5 g/L glucose, 1.0 g/L NH_4NO_3, 0.5 g/L KH_2PO_4, 1.5 g/L K_2HPO_4, 1.0 g/L NaCl, 0.5g/L $MgSO_4$, and by controlling the variables sequentially, the formulation of the unique carbon source and the unique nitrogen source were screened one-way to determine the best formulation for response surface optimization.

Using glucose addition as factor A, yeast extract addition as factor B, $MgSO_4$ addition as factor C, and OD_{600} as the response variable, a three-factor, three-level response surface design was carried out to optimize the optimal medium formulation using the Box–Behnken experimental design. The Box–Behnken response surface is a statistical method used to study the interactions between variables in a multivariate system. The Box–Behnken response surface test factors and level design of the optimal medium for ZL-139 are shown in Table 1.

Table 1. Box–Behnken response surface test factors and levels (g/L).

Levels	Factors		
	A Glucose	B Yeast Extract	C $MgSO_4$
−1	3.0	0.5	0.3
0	5.0	1.0	0.5
1	7.0	1.5	0.7

2.6. Statistical Analysis

Excel2013 software was used to organize the data, the phylogenetic tree was constructed using MEGA11.0 software to construct the Neighbor-Joining (NJ) Kimura's 2-parameter model, and the confidence level of the branch points of the evolutionary tree was evaluated by the resampling method (bootstrap test). The bootstrap test is a statistical resampling technique that is widely used to assess the branch support and reliability of phylogenetic trees. Because of the random variation in the data, sampling errors, or imperfections in the model assumptions, some branches of a phylogenetic tree may not be stable or reliable. The bootstrap test estimates these uncertainties by resampling the original data multiple times, thus helping to assess the reliability of the phylogenetic tree. The Shapiro–Wilk test, one-way ANOVA statistical analysis, and LSD multiple comparisons were performed on one-way data using the agricolae and ggplot2 packages in Rstudio 4.3.1. The agricolae package has a powerful role to play in the design of agricultural experiments and statistical analyses and was utilized to perform the LSD multiple comparisons in this experiment; ggplot2 allows for the visualization of data as it provides a graphical syntax-based plotting system, and this was utilized to generate bar graphs in this experiment.

3. Results

3.1. Screening of Straw-Degrading Bacterial Strains and Comparison of Enzyme Activity

Ten strains of cellulose-degrading bacteria were screened using a screening medium as shown in Figure 1A, where ZL-5, ZL-69, ZL-88, ZL-95, ZL-111, ZL-137, ZL-139, ZL-140, ZL-187, and ZL-216 are the effects of the cellulose degradation ability, CK-1 is an effect without cellulose degradation ability, and CK-2 is the screening medium plane. The white circle on the medium plane is the colony level and the yellow hydrolysis circle is the cellulose degradation ability level.

Figure 1B shows the ratio (D/d) of the diameter of the hydrolysis circle (D) to the diameter of the colony (d) in the plane of the medium measured by a vernier caliper, which responds to the cellulose degradation ability of the strains. The D/d of cellulose-degrading bacteria ZL-5, ZL-69, ZL-88, ZL-95, ZL-111, ZL-137, ZL-139, ZL-140, ZL-187, and ZL-216 were 1.7188, 1.4258, 1.6698, 1.5601, 1.4615, 1.6277, 1.8072, 1.2832, 1.4028, and 1.6125, respectively, and strain ZL-139 had the largest D/d at 1.8072, which was significantly higher than that of the other nine strains, with a 40.83% enhancement over ZL-140 and a 28.83% enhancement over ZL-187.

The cellulase activity (U/mL) of different strains was determined using the DNS method (Figure 2); the cellulase activity (U/mL) of strains ZL-5, ZL-69, ZL-88, ZL-95, ZL-111, ZL-137, ZL-139, ZL-140, ZL-187, and ZL-216 were 6.7601, 6.8300, 7.0745, 7.7208, 6.5680, 6.3758, 7.8781, 6.2884, 7.2492, and 6.6553, and the cellulase activity of strain ZL-139 was higher than that of the other nine strains, which was 7.8781 U/mL, 25.28% higher than the enzyme activity of ZL-140 and 23.56% higher than that of ZL-139, and this value can directly reflect the strain's cellulose degradation ability.

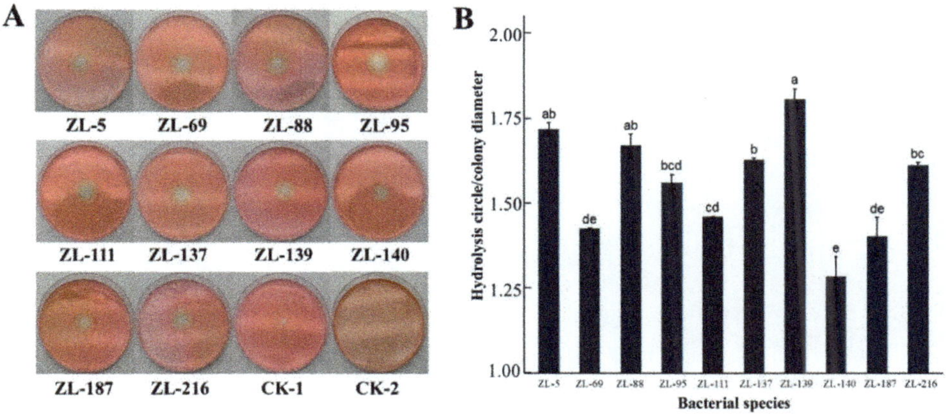

Figure 1. Schematic diagram of cellulose degradation ability. (**A**): ZL-5, ZL-69, ZL-88, ZL-95, ZL-111, ZL-137, ZL-139, ZL-140, ZL-187, and ZL-216 are the effects of cellulose degradation ability, CK-1 is an effect without cellulose degradation ability, CK-2 is the screening medium plane, the white circle on the medium plane is the colony level, and the yellow hydrolysis circle is the cellulose degradation ability level. (**B**): Ratio of hydrolysis circle (D)/colony diameter (d) for different bacterial strains. The different lowercase letters indicate significant differences among treatments ($p < 0.05$).

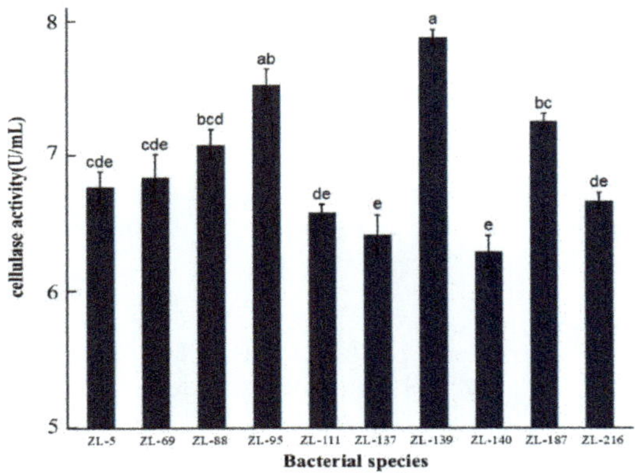

Figure 2. Comparison of cellulase activity among different bacterial strains. The different lowercase letters indicate significant differences among treatments ($p < 0.05$).

3.2. Morphological, Physiological, Biochemical, and Molecular Biology Identification of Bacterial Strains

According to plate scribing on the LB medium, ZL-139 single colonies were raised, smooth, moist, and sticky, and the colonies were white to yellowish, round or oval, with a soft texture and waxy droplets (Figure 3A); as observed by scanning electron microscopy (SEM), the strains appear as elongated rod-shaped cells arranged singly, in pairs, or in short or long chains, measuring (1.2–1.8) μm × (3.0–5.0) μm, with smooth or slightly wrinkled surfaces. Flagella can be seen on some of the cells, and these flagella protrude from one or both ends of the cell, presenting a free state; some strains may have a capsule, which is manifested as a layer of hyaline or translucent material covering the cell surface. In addition, the endospores of the strains can be observed, which are located at one end of the cell, with a clear demarcation line between them and the mother cell (Figure 3B).

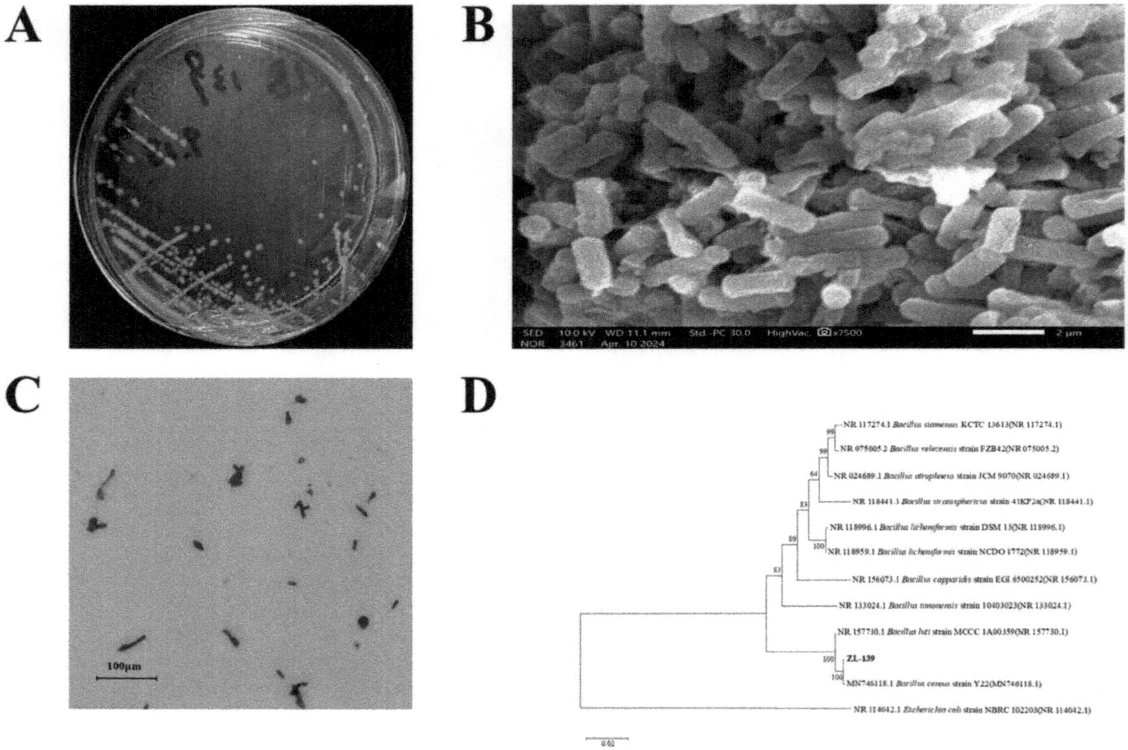

Figure 3. Identification of strain ZL-139. (**A**): Growth of strain ZL-139 on LB medium. (**B**): SEM observation of ZL-139 (X7500). (**C**): Galanz staining diagram of ZL-139. (**D**): Phylogenetic tree established using N-J method based on 16S rDNA sequences of ZL-139 and related strains.

Table 2 shows the statistics of the physiological and biochemical indices of ZL-139, which showed that ZL-139 was an aerobic strain with positive Gram staining (Figure 3C), a positive methyl red (M.R) reaction, a starch hydrolysis reaction, a gelatin hydrolysis reaction, and a negative contact enzyme and diacetyl (V-P) reaction.

Table 2. Physiological and biochemical characteristics of strain ZL-139 and *Escherichia coli* (CK). Note: "+" indicates a positive reaction, "−" indicates a negative reaction.

Tests	Results	
Name	**ZL-139**	***Escherichia coli* (CK)**
Catalase test	−	+
V-P experiment	−	−
M.R experiment	+	+
Starch hydrolysis	+	−
Gelatin hydrolysis	+	−
Aerobic test	aerobic	facultative anaerobe
Gram's stain	+	−

The 16s rDNA sequence was obtained after extracting the total DNA of ZL-139 by PCR amplification, and the sequence was uploaded on the Gen Bank database of National Center for Biotechnology Information (NCBI) for BLAST 16S rDNA sequence homology comparison. Bacillus cereus was the closest relative with 100% BLAST similarity (Figure 3D).

The combined morphological observation, physiological and biochemical characterization, and molecular identification showed that ZL-139 was *Bacillus cereus* (SUB14404564).

3.3. Optimization of Optimum Medium Based on Box–Behnken Response Surface Methodology

The one-factor test with different carbon sources was conducted on strain ZL-139, as shown in Figure 4A, and the carbon source variables were maltose, glucose, fructose, starch, and sucrose in that order. There were significant differences in the OD_{600} of the basic medium bacterial broth after incubation, as determined by an enzyme labeling instrument, and its OD_{600} was 1.0380, 1.1657, 1.0490, 1.0137, and 0.9503, respectively, in which, compared with fructose, sucrose, maltose, and starch, ZL-139 was more likely to absorb glucose, and its OD_{600} was 1.1657 after 8 h of incubation, which was 22.66% higher than sucrose, i.e., in the basic medium, the most suitable carbon source was glucose. Figure 4B shows a one-way test with different nitrogen sources performed by strain ZL-139; the nitrogen source variables were beef paste, urea, NH_4NO_3, tryptone, and yeast extract in that order, and the OD_{600} was measured to be 0.9977, 0.4353, 0.3257, 0.9097, and 1.1687, respectively, after 8h of incubation. ZL-139 grew best when the added nitrogen source was yeast extract, and its OD_{600} was 1.1687. The worst growth was observed when the added nitrogen source was NH_4NO_3, with an OD_{600} of 0.3257; yeast extract was significantly more effective than beef paste, urea, ammonium nitrate, and tryptone, with an increase in its OD_{600} of 168.5% in comparison to the nitrogen source of urea and of 258.9% in comparison to the nitrogen source of NH_4NO_3, i.e., in the basal medium, the most suitable nitrogen source was yeast extract.

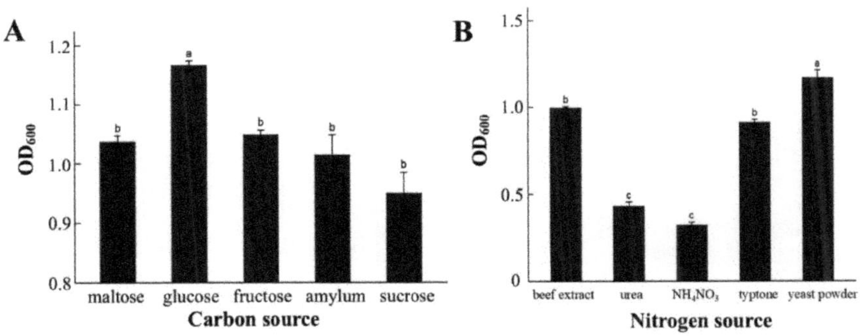

Figure 4. Effects of different carbon and nitrogen sources on the growth of strain ZL-139. The different lowercase letters indicate significant differences among treatments ($p < 0.05$). Effect of different nutrients on the growth of strain ZL-139. (**A**) Effect of carbon source on the growth of strain ZL-139. (**B**) Effect of nitrogen source on the growth of strain ZL-139. Different lower case letters indicate significant differences between treatments ($p < 0.05$).

The optimization of the optimal medium formulation was accomplished by performing a three-factor, three-level response surface design using a Box–Behnken experimental design with glucose addition as factor A, yeast extract addition as factor B, $MgSO_4$ addition as factor C, and OD_{600} as the response variable. Table 3 shows the Box–Behnken experimental design and results, and the analysis of variance of the regression equations of the experimental results using the Box–Behnken response surface method is shown in Table 4. The following multivariate quadratic regression mathematical model was established by fitting the data to Y using OD_{600} (Y) as the response surface and A, B, and C as the independent variables:

$$Y = -0.568167 + 0.245875A + 0.983667B + 1.98667C + 0.001250AB + 0.012992AC - 0.265833BC - 0.024906A^2 - 0.166500B^2 - 1.73646C^2$$

where Y is the OD_{600} value, A is the glucose addition (g/L), B is the yeast extract addition (g/L), and C is the MgSO4 addition (g/L).

Table 3. Box–Behnken experimental design and results.

Test Number	A	B	C	OD_{600}
1	−1	−1	0	0.916
2	1	−1	0	0.927333
3	−1	1	0	1.425667
4	1	1	0	1.442
5	−1	0	−1	1.14
6	1	0	−1	1.15
7	−1	0	1	1.14
8	1	0	1	1.169667
9	0	−1	−1	0.9
10	0	1	−1	1.489
11	0	−1	1	0.98
12	0	1	1	1.462667
13	0	0	0	1.208
14	0	0	0	1.4
15	0	0	0	1.33
16	0	0	0	1.307
17	0	0	0	1.35

Table 4. Box–Behnken experiment regression equation and analysis of variance.

Source	Sum of Squares	df	Mean Square	F-Value	*p*-Value
Model	0.6297	9	0.07	23.83	0.0002
A	0.0006	1	0.0006	0.193	0.6736
B	0.5492	1	0.5492	187.05	<0.0001
C	0.0007	1	0.0007	0.229	0.6469
AB	6.25×10^{-6}	1	6.25×10^{-6}	0.0021	0.9645
AC	0.0001	1	0.0001	0.0329	0.8611
BC	0.0028	1	0.0028	0.9628	0.3592
A^2	0.0418	1	0.0418	14.23	0.007
B^2	0.0073	1	0.0073	2.48	0.1589
C^2	0.0203	1	0.0203	6.92	0.0339
Residual	0.0206	7	0.0029		
Lack of Fit	0.0004	3	0.0001	0.0293	0.9922
Pure Error	0.0201	4	0.005		
Cor Total	0.6502	16			

As can be seen from Table 4, the *p*-value of the model used in this experiment is 0.0002 ($p < 0.01$), indicating that the model is highly significant; at the same time, the *p*-value of the disfitting term is 0.9922 ($p > 0.05$), indicating that the disfitting term of this model is not significant, indicating that there is no significant difference between the model used in this experiment and the results of the fitting by the experiment and that, therefore, the model can be used for the components of the basic medium for ZL-139 optimization. The correlation coefficient R^2Adj is 0.9278, indicating that there is a variability of 7.22% in the model, 92.78% of the experimental data can be explained by using the polynomial regression equation obtained in this experiment, and the coefficient of determination of the model of this polynomial regression equation is 96.84%, indicating that the model established in this experiment has a better fit.

In order to show the intensity of the interactions among factors more intuitively, the two-by-two interactions of factors A (glucose), B (yeast extract), and C ($MgSO_4$) on the OD_{600} of ZL-139 Bacillus cereus were plotted as three-dimensional response surface and contour plots (Figure 5), in which A, B, C, D, E, and F denote the three-dimensional response surface and contour plots of the effects of glucose and yeast extract interactions, yeast extract and $MgSO_4$ interactions, and glucose and $MgSO_4$ interactions on ZL-139 OD_{600}, respectively.

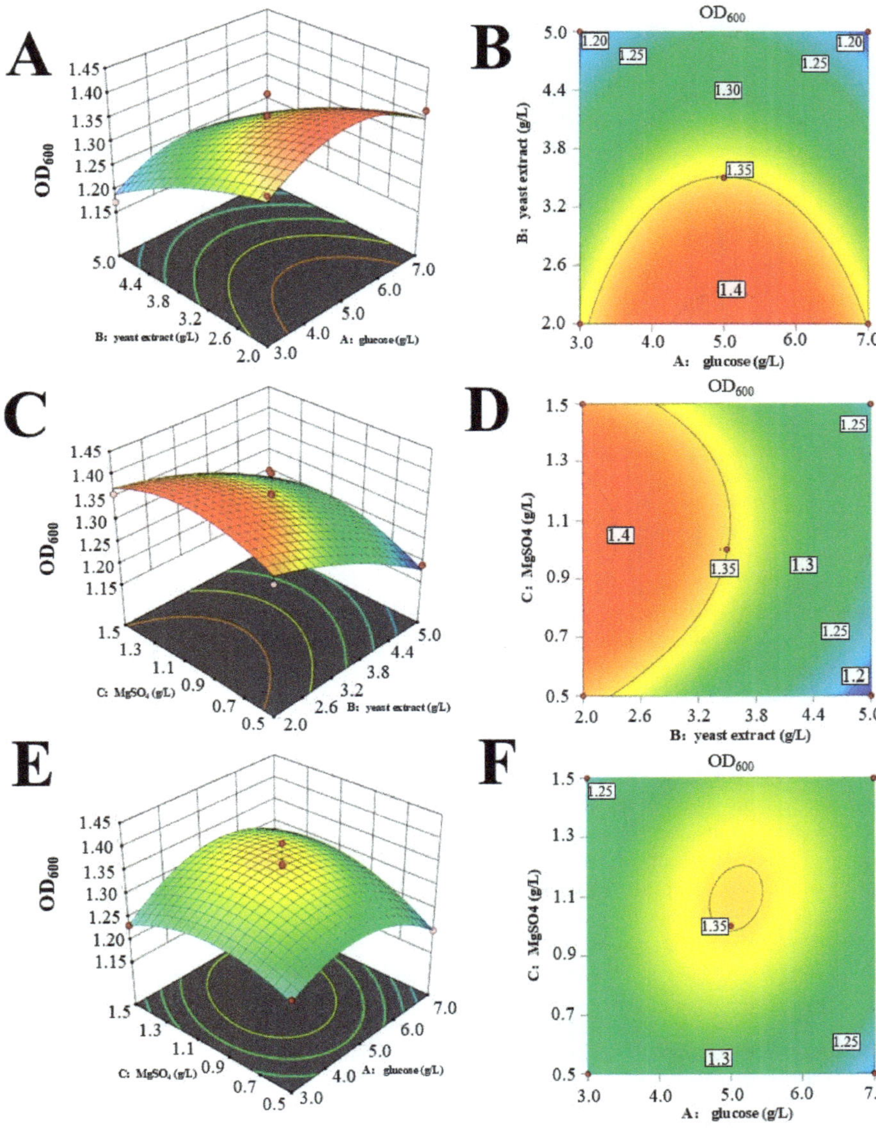

Figure 5. Three-dimensional response surface and contour maps of the ZL-139 basic culture medium designed based on Box–Behnken response surface methodology. (**A**,**B**): three-dimensional response surface and contour plots of the effect of glucose and yeast extract on the biomass of ZL-139 OD_{600}, respectively; (**C**,**D**): three-dimensional response surface and contour plots of the effect of yeast extract and $MgSO_4$ on the biomass of ZL-139 OD_{600}, respectively; (**E**,**F**): three-dimensional response surface and contour plots of the effect of glucose and $MgSO_4$ on the biomass of ZL-139 OD_{600}, respectively; (**E**,**F**): three-dimensional response surface and contour plot of the effects of glucose and $MgSO_4$ on ZL-139 OD_{600}, respectively.

The regression model was used to analyze the relationship between the three factors and the OD_{600} measured by ZL-139, and, at the same time, to predict the optimal conditions for each factor, which showed that the optimal culture conditions for strain ZL-139 were as follows: glucose addition—4.284 g/L, yeast extract addition—1.454 g/L, $MgSO_4$ addition—0.417 g/L, KH_2PO_4—0.5 g/L, K_2HPO_4—1.5 g/L, NaCl—1.0 g/L, and the maximum theoretical value of OD_{600} was 1.501 per 20 mL of the liquid medium under this condition.

According to the optimal culture conditions of ZL-139 obtained after the fitting analysis, the medium indexes were adjusted to 4.3 g/L for glucose addition, 1.45 g/L for yeast extract addition, 0.42 g/L for $MgSO_4$ addition, 0.5 g/L for KH_2PO_4, 1.5 g/L for K_2HPO_4, and 1.0 g/L for NaCl in the actual cultivation operation, and the OD_{600} average was measured to be 1.473 ± 0.017 with the values predicted by three parallel tests. The components of the optimal medium for ZL-139 were verified, and the mean value of OD_{600} of ZL-139 was measured as 1.473 ± 0.017 in three parallel experiments, with a difference of 4.203% from the predicted value.

4. Discussion

The screening of bacteria for cellulose degradation capacity using a CMC-Na-rich screening medium and Congo red staining method has been widely used for targeted screening because Congo red and cellulose interact in the medium, and when bacteria secrete cellulase, cellulose is degraded to cellulose ditangles, glucose, and organic acids, and the pH around the colony decreases, leading to the formation of a hyaline hydrolysis circle [28,39], which can be used to detect the cellulose degradation capacity of bacteria. Therefore, the ability of bacteria to degrade cellulose can be initially detected by the ratio of the diameter of the hydrolysis circle (D) to the diameter of the colony (d) [40,41]. Because cellulase activity is directly proportional to D/d, Yu [42] et al. isolated five bacterial species with straw degrading ability, named z-5, 4-1, 3-1, 2-2, and m-2, from vegetable soil, small grass inter-root soil, silkworm dung, and forest soil, where strain 2-2 had the largest D/d, at 1.76, and strain 3-1 had the smallest, at 1.22, which were smaller than the strains isolated from black soil in the present experiment; Yuan [43] et al. isolated 160 strains of straw-degrading bacteria from straw-returned soil, in which the largest hydrolysis circle diameter was XJ-132 with a hydrolysis circle diameter of 31.0 mm, which was smaller than that of the isolate ZL-139 in the present experiment. Therefore, the isolated isolate from the black soil with rich humus content in the present experiment has a greater potential for application.

Suitable culture conditions can promote bacterial multiplication and enhance vigor, which is also important for bacterial cellulase activity. Furthermore, 30–40 °C and 120–180 $rpm \cdot min^{-1}$ are the optimal culture condition intervals for straw-degrading bacteria; for example, Li [44] et al. chose 30 °C and 180 $rpm \cdot min^{-1}$ as the culture conditions when investigating the cellulase activity of the bacterial strain TH-21; Guo [45] et al. chose 37 °C as the incubation condition when screening and evaluating the cellulase activity of *Cellumonas xylaniitica* and *Escherichia coli JM109*; and Singh [46] et al. chose 32 °C and 120 $rpm \cdot min^{-1}$ as the incubation condition when studying the enzyme activity of the seven isolates isolated in their experiment. Therefore, 35 °C and 150 $rpm \cdot min^{-1}$ were chosen as the incubation conditions, which would be helpful for the study of cellulase activity in this experiment.

The DNS method [47,48] is based on the principle that reducing sugars produced by the cellulase hydrolysis of cellulose can reduce yellow 3,5-dinitrosalicylic acid (DNS) to brownish-red 3-amino-5-nitrosalicylic acid under an alkaline environment, and the color shade of this reduction product is proportional to the content of reducing sugars within a certain concentration range. Wan [49] et al. measured the cellulase content of screened *Bacillus subtilis*, *Bacillus licheniformis*, *Bacillus megaterium*, *Canadian krusei*, *Trametes versicolor*, and *Trametes versicolor* for cellulase activity, with the highest cellulase activity being 7.26 U/mL and 6.24 U/mL, lower than that of ZL-139 isolated in this experiment. Meanwhile, as a bacterial strain isolated in situ from the black soil of Northeast China, it may play a better role in efficiently degrading the straw in the black soil of Northeast China, so the best cellulase-producing bacterial strain, ZL-139, isolated in this experiment can be used as a high-quality bacterial strain for degrading the straw.

Discovering straw-degrading bacteria that can play a more effective has a strong practical significance; for example, in soil fertility, straw-degrading bacteria, through the decomposition of straw, help the formation of soil granular structure and improve soil aeration and water retention. At the same time, the decomposition of the nutrients can release nitrogen, phosphorus, potassium, and other nutrients, which can be absorbed and used by the crop, thus reducing the use of chemical fertilizers. Therefore, the application of straw-degrading bacteria can significantly improve the efficiency and environmental friendliness of agricultural production, providing strong support for future agricultural practices and environmental protection.

SEM is a routine and very standard imaging and microanalysis technique [50] that provides high-resolution images that enable researchers to observe in detail the surface microstructure and morphological stereoscopic features of bacteria. This high-resolution imaging is very important for identifying bacterial species and understanding their surface characteristics, such as observing the shape, size, and surface texture of the bacteria, as well as subtle structures such as flagella, hyphae, and pods. ZL-139, under a scanning electron microscope, has typical morphological characteristics of the genus Bacillus. Its elongated rod-like structure can enhance its surface area to volume ratio, which helps to improve nutrient uptake efficiency. The unique arrangement can enhance the interaction between bacteria and the stability of the community. The cell size of (1.2–1.8) μm × (3.0–5.0) μm enables the bacteria to pass through tiny spaces and gaps, and, at the same time, helps them to multiply and spread rapidly in different environments. The smoothness and folds on the surface of ZL-139 help nutrient adsorption and metabolite release. The flagellum is the main locomotor organ of ZL-139 and enables better attachment of ZL-139 to hosts or other surfaces. Capsules are a polysaccharide or protein layer that provide protection, adhesion, and other abilities for ZL-139. Endospores are one of the most important features of the genus Bacillus that ensure the survival and reproduction of ZL-139.

In the phylogenetic tree established by the N-J method based on the 16S rDNA sequence of ZL-139, the similarity between ZL-139 and *Bacillus cereus* is 100%. *Bacillus cereus* is a Gram-positive, spore-producing bacterium that is widely found in nature, including soil, water, and food. This bacterium is highly viable due to its ability to form heat-resistant spores under unfavorable conditions. *Bacillus cereus* plays an important role in agriculture and has received increasing attention due to its properties such as disease resistance, growth promotion, and high stress tolerance [51] , but *Bacillus cereus* has also been associated with a variety of diseases, such as food poisoning, and therefore, the utilization of *Bacillus cereus* requires special attention in terms of its potential applications and personal safety. Verónica [52] et al. found that *Bacillus cereus* MH778713 is a plant growth-promoting bacterium (PGPB) that stimulates plant growth and confers tolerance to biotic and abiotic stresses. For example, *Bacillus cereus* MH778713 stimulated the growth of tomato seedlings in greenhouse experiments and protected tomato plants from Fusarium wilt. Wang [53] et al. found that *Bacillus cereus* BL4 could promote the bioremediation of Cd-contaminated soil by *Miscanthus floridulus* through its effects on plant physicochemical properties and soil microbiology. Zhao [54] et al. found that *Bacillus cereus* significantly altered the fungal composition of the compost and increased microbial co-operation in the composting process, thereby affecting the transformation of composted material when added to the compost of chicken manure and wood chips.

Medium composition is one of the most important parameters affecting bacterial cell growth and enzyme production [55]. Figure 4 demonstrates the effect of different carbon and nitrogen sources of medium growth rates of ZL-139, which can be incubated economically and efficiently by varying the carbon and nitrogen sources in a single factor to understand the preferences of the strains for different carbon and nitrogen sources. Glucose has an extremely wide range of applications, such as in fermentation production; glucose is the main carbon source for the growth of bacteria and product synthesis, and its content directly determines the proliferation and metabolism level of the bacterium. Using glucose as the optimal carbon source is significantly more effective than the other kinds of carbon sources in the carbon source medium of ZL-139. El-dein [56] et al. optimized the conditions for extracellular polysaccharide (EPS) production by *Pleurotus pulmonarius* and found that glucose was the best stimulant for EPS production, which is also in agreement with the findings obtained by Xu [57] et al., where they optimized the conditions for the growth of *Paecilomyces tenuipes C240* using a one-way test—glucose was the best carbon source. The phenomenon of catabolite repression is a mechanism for efficiently utilizing resources in the presence of nutrients to promote growth, development, and survival [58]. For example, *E. coli*, under unrestricted growth conditions with multiple carbohydrates, will usually preferentially take up and utilize one of them [59] , although other carbon sources can also promote *E. coli* growth. At the molecular level, *Bacillus subtilis*, for example, will produce a transcriptional repressor protein that represses promoters of genes that catabolize non-preferred carbon sources during glucose availability [60]. Thus, moderate amounts of glucose can provide the energy required for microbial growth and support essential cellular activities. If the concentration of glucose is high, it will inhibit the expression of cellulase genes in the strain through the carbon source repression effect, and if the concentration of glucose is low, the strain may not be able to obtain enough energy to support cellulase synthesis.

Cryptococcus gattii is an important yeast that may be one of the raw materials for yeast extracts. Moodley [61] et al. sequenced a gene, WM276, from *Cryptococcus gattii* and annotated it as a putative cellulase. Therefore, yeast extracts may be closely related to cellulase synthesis; for example, certain peptides can act as signaling molecules that trigger the expression of cellulase genes. Yeast extract is commonly used as a high-quality nitrogen source, which is mainly naturally reproduced by a variety of yeasts and is rich in nutrients such as amino acids, vitamins, and growth factors, making it a purely natural, non-polluting, and healthy source of nutrients [62]. In addition to supplying organic nitrogen, it can also provide amino acids and B vitamins [63,64] in order to support the growth of the cultures. ZL-139 was poorly receptive to inorganic nitrogen sources, with an OD_{600} of only 0.3257. He [65] et al. found that inorganic nitrogen sources had little or no effect on its growth when incubated with different nitrogen sources in *Valsa mali Mayabe et Yamada*, but Cheng [66] et al. found that in *Ustilaginoidea virens (Cooke) Tak*, using $CaNO_3$ as a nitrogen source had the most pronounced growth-promoting effect, which may be related to the microorganism's own energy preference.

The role of $MgSO_4$ in the culture medium mainly consists of providing magnesium, participating in the activity of various enzymes, and promoting microbial growth. For example, Junillon [67] et al., in their study of *Salmonella spp.*, found that the addition of magnesium to the medium significantly increased its growth rate; Dutta [68] et al., in their study of cellulase from NCEL03 isolated from Gnathang valley of East Sikkim, found that as the concentration of added MgN increased, the viability of cellulase also increases gradually until it reached 12.5 μg/mL; Ji et al [69]., in their study of *7α-Hydroxysteroid dehydrogenase,* found that magnesium ions significantly enhance its activity and thermal stability. Therefore, by combining the results of the one-way test and the analysis of substance properties, glucose, yeast extract, and $MgSO_4$ were selected for the experimental design of the Box–Behnken response surface methodology in this experiment.

Box–Behnken design is a design of experiment methodology commonly used in response surface methodology (RSM), where the aim is to determine the relationship between multiple variables and find the best settings to optimize the process. This method is particularly suitable for exploring the effects of changes in factor levels on response variables and can be used to construct quadratic polynomial models. Whereas bacterial media usually consist of multiple components, each of which may have a concentration that affects microbial growth, the Box–Behnken design is able to effectively deal with these multifactorial problems. Meanwhile, the Box–Behnken design can identify the optimal parameter combinations in a small number of experiments, avoiding the extreme conditions that may be harmful; for example, some extreme concentrations may inhibit the growth of bacteria or even lead to death, so it is very suitable for the optimization of culture media. Also, many studies have shown that the response surface experimental design is an effective method for optimizing media and culture conditions [70]. Ghazanfari [71] et al. optimized the culture conditions of natural lactic acid bacteria (LAB) to obtain a maximum yield of 500 mg/mL of L-glutamic acid, which reduced the production cost to a greater extent; Patel [72] et al. designed a new medium for response surface methodology to enhance the production of tacrolimus by *Streptomyces tsukubensis*, and the optimized tacrolimus yield increased by 2.94-fold with a maximum yield of 574 mg/L; *Bacillus thuringiensis* (Bt) is an entomopathogenic bacterium, and Saberi [73] et al. used a response surface methodology to optimize the culture of Bt as an effective biocide for Coleoptera very well, which can play an important role in sustainable pest control management; Rodrigues [74] et al. used response surface methodology to optimize the best culture medium for *Komagataeibacter xylinus BPR 2001* (*K. xylinus*), resulting in an increase in bacterial nanocellulose (BNC) yield and productivity to 6.4 $\pm$ 0.54 g/L and 0.74 $\pm$ 0.079 g/L/d. In this experiment, a Box–Behnken experimental design was carried out with glucose addition as factor A, yeast extract addition as factor B, $MgSO_4$ addition as factor C, and OD_{600} as the response variable. As can be seen in Figure 5, the response surfaces of the interaction of the factors on the effect of OD_{600} were all parabolic, indicating that there is a maximum value in the fitted regression model and that ZL-139 reached a maximum OD_{600} of 1.501 per 20 mL in the basic medium, which is favorable for the development and utilization of strain ZL-139 and which is also consistent with the findings of Beeler [75] et al. ZL-139 is an efficient straw-degrading bacterium that can be used in biomass conversion, enzyme production, biofuel production, environmental protection, and other fields. After optimizing the basic culture of ZL-139, its OD_{600} can reach 1.501 per 20 mL, which will help to improve the cultivation efficiency and reduce the cultivation cost in the laboratory culture. At the same time, due to the increase in the biomass of ZL-139 in the basic medium, we can enhance the scope of its industrial or agricultural applications to a greater extent; for example, it can shorten the fermentation cycle of ZL-139, and it lays a good theoretical foundation for the production of large quantities of straw-degrading bacterial agents for the purpose of straw resource utilization.

5. Conclusions

In this study, we successfully isolated a strain, ZL-139, identified as *Bacillus cereus*, which exhibits strong straw degradation ability. This cellulase activity of 7.8781 U/mL is indicative of ZL-139's potential in agricultural waste management compared to previously reported values. The optimum medium for ZL-139 was optimized by one-way tests, a Box–Behnken experimental design, and model analysis, and the results showed that the optimum culture conditions for ZL-139 in the basic medium were glucose–4.284 g/L, yeast extract—1.454 g/L, $MgSO_4$—0.417 g/L, KH_2PO_4—0.5 g/L, K_2HPO_4—1.5 g/L, and NaCl—1.0 g/L, at which time the maximum theoretical OD_{600} value per 20 mL of the liquid medium was 1.501. These components were selected based on their roles in supporting bacterial growth and maximizing enzyme production. However, due to the phenomenon of catabolite repression, only the biomass of ZL-139 in the optimized medium was determined in this study, which was improved but not necessarily able to promote the cellulase activity, and we will continue to optimize the medium in the later stage in order to produce a better effect. At the same time, the screening and medium optimization of ZL-139 in the laboratory conditions in the present study may not be completely consistent with the results of the indoor experiment. Therefore, it is necessary to conduct field experiments on this basis of further clarifying the straw degradation effect of ZL-139 in the future, and future field trials should focus on parameters such as degradation rate, soil health improvements, and impacts on crop yield to comprehensively assess the effectiveness of ZL-139 in natural environments. Finally, as discussed, *Bacillus cereus* has some pathogenicity, such as food poisoning, so those using Bacillus cereus need to pay special attention to its potential application value and personal safety. In summary, the findings from this study not only contribute to the understanding of straw degradation mechanisms, but also highlight the potential applications of ZL-139 in enhancing agricultural sustainability and developing innovative biotechnological solutions.

Supplementary Materials: The following supporting information can be downloaded at: https://www.mdpi.com/article/10.3390/agronomy14112591/s1, Figure S1: Overall optimization process for Box-Behnken response surface design.

Author Contributions: Conceptualization, L.Z. and R.Y.; methodology, L.Z.; software, T.L. and C.Y.; validation, J.Z.; investigation, N.L.; data curation, L.Z. and Y.Y.; writing—original draft preparation, L.Z.; writing—review and editing, R.Y.; supervision, R.Y. All authors have read and agreed to the published version of the manuscript.

Funding: This work was funded by the Strategic Priority Research Program of the National Natural Science Foundation of China (Grant No. 42301134), Natural Science Foundation of Jilin Province (Grant Nos. 20210101109JC and YDZJ202301ZYTS232), Jilin Province and Chinese Academy of Sciences cooperation in high-tech industrialization of special projects (Grant No. 2023SYHZ0052), the National Natural Science Foundation of China (Grant No. 42307187), and the Young Scientist Group Project of Northeast Institute of Geography and Agroecology, Chinese Academy of Sciences (Grant No. 2023QNXZ05).

Data Availability Statement: The datasets used and/or analyzed during the current study are available from the corresponding author on reasonable request.

Conflicts of Interest: The authors declare no conflicts of interest.

References

1. Liu, B.; Xia, H.; Jiang, C.; Riaz, M.; Yang, L.; Chen, Y.; Fan, X.; Xia, X. 14 year applications of chemical fertilizers and crop straw effects on soil labile organic carbon fractions, enzyme activities and microbial community in rice-wheat rotation of middle China. *Sci. Total Environ.* **2022**, *841*, 156608. [CrossRef] [PubMed]
2. Meng, X.; Ran, C.; Liu, B.; Zhao, Z.; Bai, T.; Zhao, M.; Cheng, Z.; Chen, G.; Geng, Y. Effect of straw return with nitrogen fertilizer on photosynthetic characteristics and yield of rice in soda saline–alkali rice paddy fields. *Cereal Res. Commun.* **2023**, *51*, 509–526. [CrossRef]
3. Hong, J.; Ren, L.; Hong, J.; Xu, C. Environmental impact assessment of corn straw utilization in China. *J. Clean. Prod.* **2016**, *112*, 1700–1708. [CrossRef]
4. Soam, S.; Borjesson, P.; Sharma, P.K.; Gupta, R.P.; Tuli, D.K.; Kumar, R. Life cycle assessment of rice straw utilization practices in India. *Bioresour. Technol.* **2017**, *228*, 89–98. [CrossRef] [PubMed]
5. Zhang, T.; Wooster, M.J.; Green, D.C.; Main, B. New field-based agricultural biomass burning trace gas, PM2.5, and black carbon emission ratios and factors measured in situ at crop residue fires in Eastern China. *Atmos. Environ.* **2015**, *121*, 22–34. [CrossRef]
6. Halder, M.; Ahmad, S.J.; Rahman, T.; Joardar, J.C.; Siddique, M.A.B.; Islam, M.S.; Islam, M.U.; Liu, S.; Rabbi, S.; Peng, X. Effects of straw incorporation and straw-burning on aggregate stability and soil organic carbon in a clay soil of Bangladesh. *Geoderma Reg.* **2023**, *32*, e00620. [CrossRef]
7. Liu, X.; Liu, J.; Liu, Z. Developing a model of propagating the straw burning prohibition policy in Chinese rural communities and exploring its countermeasures. *Energy Rep.* **2024**, *11*, 2556–2564. [CrossRef]
8. Gray, G.C.; Nguyen-Tien, T. Threatened by many complex food security problems—Agriculture, academic, and government professionals seek new one health research partnerships. *One Health* **2024**, *19*, 100890. [CrossRef]
9. Cai, J.; Wang, Y.; Liu, J.; Zhang, X.; Li, F. Pretreatment enhanced structural disruption, enzymatic hydrolysis, fermentative hydrogen production from rice straw. *Int. J. Hydrogen Energy* **2022**, *47*, 11778–11786. [CrossRef]
10. Zhang, W.; Wu, S.; Cai, L.; Liu, X.; Wu, H.; Xin, F.; Zhang, M.; Jiang, M. Improved Treatment and Utilization of Rice Straw by *Coprinopsis cinerea*. *Appl. Biochem. Biotechnol. Part A Enzym. Eng. Biotechnol.* **2018**, *184*, 616–629. [CrossRef]
11. Feng, H.; Sun, Y.; Zhi, Y.; Mao, L.; Luo, Y.; Wei, X.; Zhou, P. Lignocellulose degradation by the isolate of Streptomyces griseorubens JSD-1. *Funct. Integr. Genom.* **2015**, *15*, 163–173. [CrossRef] [PubMed]
12. Kersten, P.; Cullen, D. Extracellular oxidative systems of the lignin-degrading Basidiomycete Phanerochaete chrysosporium. *Fungal Genet. Biol.* **2007**, *44*, 77–87. [CrossRef] [PubMed]
13. Venkatesagowda, B. Enzymatic demethylation of lignin for potential biobased polymer applications. *Fungal Biol. Rev.* **2019**, *33*, 190–224. [CrossRef]
14. Xu, C.; Su, X.; Wang, J.; Zhang, F.; Shen, G.; Yuan, Y.; Yan, L.; Tang, H.; Song, F.; Wang, W. Characteristics and functional bacteria in a microbial consortium for rice straw lignin-degrading. *Bioresour. Technol.* **2021**, *331*, 125066. [CrossRef]

15. Brunecky, R.; Donohoe, B.S.; Yarbrough, J.M.; Mittal, A.; Scott, B.R.; Ding, H.; Taylor Ii, L.E.; Russell, J.F.; Chung, D.; Westpheling, J. The Multi Domain Caldicellulosiruptor bescii CelA Cellulase Excels at the Hydrolysis of Crystalline Cellulose. *Sci. Rep.* **2017**, *7*, 9622. [CrossRef] [PubMed]
16. Khan, A.W.; Murray, W.D. Influence of Clostridium saccharolyticum on cellulose degradation by *Acetivibrio cellulolyticus*. *J. Appl. Microbiol.* **2008**, *53*, 379–383. [CrossRef]
17. Dong, S.; Wei, Y.; Yu, Q.; Gao, Y.; Chen, H.; Zhou, K.; Cheng, M.; Wang, B.; Wei, Y.; Hu, X. Inoculating functional bacteria improved the humification process by regulating microbial networks and key genera in straw composting by adding different nitrogen sources. *Bioresour. Technol.* **2024**, *393*, 130022. [CrossRef]
18. Sun, C.; Chen, L.; Chen, J.; Guo, L.; Ao, J.; Wang, Z. Effect of Microbial Preparation on the Degradation and Number of Microbiology of the Straw. *J. Mirobiol.* **2023**, *43*, 84–89.
19. Zhang, K.; Li, J.; Wang, Z.; Xie, B.; Xiong, Z.; Li, H.; Ahmed, M.; Fang, F.; Li, J.; Li, X. Cloning, expression and application of a novel laccase derived from water buffalo ruminal lignin-degrading bacteria. *Int. J. Biol. Macromol.* **2024**, *266*, 131109. [CrossRef]
20. Wu, D.; Wei, Z.; Gao, X.; Wu, J.; Chen, X.; Zhao, Y.; Jia, L.; Wen, D. Reconstruction of core microbes based on producing lignocellulolytic enzymes causing by bacterial inoculation during rice straw composting. *Bioresour. Technol.* **2020**, *315*, 123849. [CrossRef]
21. Zhao, B.; Dong, W.; Chen, Z.; Zhao, X.; Cai, Z.; Feng, J.; Li, S.; Sun, X. Microbial inoculation accelerates rice straw decomposition by reshaping structure and function of lignocellulose-degrading microbial consortia in paddy fields. *Bioresour. Technol.* **2024**, *413*, 131545. [CrossRef] [PubMed]
22. Jia, W.; Ye, Q.; Shen, D.; Yu, K.; Zheng, Y.; Liu, M.; Jiang, J.; Wang, W. Enhanced mineralization of chlorpyrifos bound residues in soil through inoculation of two synergistic degrading strains. *J. Hazard. Mater.* **2021**, *412*, 125116. [CrossRef] [PubMed]
23. Yang, X.; Zhao, H.; Yang, J.; Ma, Y.; Liu, Z.; Li, C.; Wang, T.; Yan, Z.; Du, N. MiR-150-5p regulates melanoma proliferation, invasion and metastasis via SIX1-mediated Warburg Effect. *Biochem. Biophys. Res. Commun.* **2019**, *515*, 85–91. [CrossRef]
24. Gao, Z.; Li, P.; Li, C.; Tang, R.; Wang, M.; Chen, J.; Yang, Y.; He, Z.; Xiao, Z.; Ma, Y.; et al. Identification, functional annotation, and isolation of phosphorus-solubilizing bacteria in the rhizosphere soil of Swida wilsoniana (Wanger) Sojak. *Appl. Soil Ecol.* **2024**, *194*, 105207. [CrossRef]
25. Nakamura, M.; Akada, R. Blending of selected yeast extract and peptone for inducible and constitutive protein production in Escherichia coli using the pET system. *J. Biosci. Bioeng.* **2024**. [CrossRef]
26. Siddiqui, I.; Sankhwar, R.; Kumar, A.; Gupta, R.K. GC-MS analysis of hospital waste soil and isolation of antimicrobial agent producing bacteria. *Microbe* **2024**, *3*, 100079. [CrossRef]
27. Sharker, B.; Islam, M.A.; Hossain, M.A.A.; Ahmad, I.; Al Mamun, A.; Ghosh, S.; Rahman, A.; Hossain, M.S.; Ashik, M.A.; Hoque, M.R.; et al. Characterization of lignin and hemicellulose degrading bacteria isolated from cow rumen and forest soil: Unveiling a novel enzymatic model for rice straw deconstruction. *Sci. Total Environ.* **2023**, *904*, 166704. [CrossRef] [PubMed]
28. Li, F.; Xie, Y.; Gao, X.; Shan, M.; Sun, C.; Niu, Y.D.; Shan, A. Screening of cellulose degradation bacteria from Min pigs and optimization of its cellulase production. *Electron. J. Biotechnol.* **2020**, *48*, 29–35. [CrossRef]
29. Dar, M.A.; Pawar, K.D.; Pandit, R.S. Prospecting the gut fluid of giant African land snail, Achatina fulica for cellulose degrading bacteria. *Int. Biodeterior. Biodegrad.* **2018**, *126*, 103–111. [CrossRef]
30. Hussain, A.A.; Abdel-Salam, M.S.; Abo-Ghalia, H.H.; Hegazy, W.K.; Hafez, S.S. Optimization and molecular identification of novel cellulose degrading bacteria isolated from Egyptian environment. *J. Genet. Eng. Biotechnol.* **2017**, *15*, 77–85. [CrossRef]
31. Zhu, N.; Zhu, Y.; Li, B.; Jin, H.; Dong, Y. Increased enzyme activities and fungal degraders by Gloeophyllum trabeum inoculation improve lignocellulose degradation efficiency during manure-straw composting. *Bioresour. Technol.* **2021**, 337, 125427. [CrossRef] [PubMed]
32. Chen, L.; Tang, T.; Wang, Z.; Zhao, N.; Wu, S.; Liu, Y. A novel fungal and bacterial consortium promotes the degradation of rice straw: Conditions optimization and degradation properties. *Int. Biodeterior. Biodegrad.* **2024**, *194*, 105875. [CrossRef]
33. Bai, D.; Hong, Z.; Guo, J.; Li, J.; Wang, J. Screening and Enzyme Activity Determination of Rice Straw Degradation Bacteria in Xiaozhan. *Mod. Agric. Sci. Technol.* **2024**, *03*, 114–118.
34. Xu, J.; Jiang, J.; Song, Z.; Hong, J.; Zhao, T.; Wu, K.; Guo, X.; Chen, Z.; Guan, X.; Pan, X. The isolation, identification and efficacy of Bacillus velezensis XF-8 in tomato bacterial wilt control. *Adv. Agrochem* **2024**. [CrossRef]
35. Dong, X.Z.; Cai, M.Y. *The Manual of Systematic Determinative Bacteriology*; Science Press: Beijing, China, 2001.
36. El-Ouahmani, N.; Amri, O.; Chahid, A.; Yacoubi, B.; Zekhnini, A. *Ononis natrix* L.: A natural biocide against marine biofouling-forming bacteria. *J. Indian Chem. Soc.* **2024**, *101*, 101332. [CrossRef]
37. Bing, H.; Qi, C.; Gu, J.; Zhao, T.; Yu, X.; Cai, Y.; Zhang, Y.; Li, A.; Wang, X.; Zhao, J.; et al. Isolation and identification of NEAU-CP5: A seed-endophytic strain of B. velezensis that controls tomato bacterial wilt. *Microb. Pathog.* **2024**, *192*, 106707. [CrossRef]
38. Koichiro, T.; Glen, S.; Sudhir, K. MEGA11: Molecular Evolutionary Genetics Analysis version 11. *Mol. Biol. Evol.* **2021**, *38*, 3022–3027.
39. Raza, A.; Bashir, S.; Tabassum, R. Evaluation of Cellulases and Xylanases Production from *Bacillus* spp. Isolated from Buffalo Digestive System. *Kafkas Univ. Vet. Fak. Derg.* **2019**, *25*, 39–46.
40. Samanta, A.K.; Kolte, A.P.; Senani, S.; Sridhar, M.; Jayapal, N. A simple and efficient diffusion technique for assay of endo β-1,4-xylanase activity. *Braz. J. Microbiol.* **2011**, *42*, 1349–1353. [CrossRef]

41. Pratima, G.; Kalpana, S.; Avinash, S. Isolation of cellulose-degrading bacteria and determination of their cellulolytic potential. *Int. J. Microbiol.* **2012**, *2012*, 578925.
42. Yu, H.; Guo, X. Screening of Straw-degrading Bacteria and Study on Their Cellulosedegrading Performances. *Biotechnol. Bull.* **2019**, *35*, 58–63.
43. Yuan, H.; Qian, D.; Ye, W.; Lu, H.; Zhang, G. Isolation and Identification of A Growth Promoting Straw Degradable Bacterial Strain. *J. Mirobiol.* **2023**, *43*, 77–84.
44. Li, F.; Pan, S.; Yang, L.; Liao, S.; Yang, D.; Wang, Q.; Huang, S. Efficient production 2,3-butanediol from biomass-derived sugars by Raoultella ornithinolytica TH-21, a newly isolated lignocellulose-degrading bacterium. *Ind. Crops Prod.* **2024**, *217*, 118841. [CrossRef]
45. Guo, H.; Chen, H.; Fan, L.; Linklater, A.; Zheng, B.; Jiang, D.; Qin, W. Enzymes produced by biomass-degrading bacteria can efficiently hydrolyze algal cell walls and facilitate lipid extraction. *Renew. Energy* **2017**, *109*, 195–201. [CrossRef]
46. Singh, A.; Kumar, R.; Maurya, A.; Chowdhary, P.; Raj, A. Isolation of functional ligninolytic Bacillus aryabhattai from paper mill sludge and its lignin degradation potential. *Biotechnol. Rep.* **2022**, *35*, e00755. [CrossRef]
47. Seesatat, A.; Rattanasuk, S.; Bunnakit, K.; Maneechot, P.; Sriprapakhan, P.; Artkla, R. Biological degradation of rice straw with thermophilic lignocellulolytic bacterial isolates and biogas production from total broth by rumen microorganisms. *J. Environ. Chem. Eng.* **2021**, *9*, 104499. [CrossRef]
48. Miller, G.L. Use of Dinitrosalicylic Acid Reagent for Determination of Reducing Sugar. *Anal. Chem.* **1959**, *31*, 426–428. [CrossRef]
49. Wan, C.Y.; Liu, R.; Huang, F.H.; Huang, Q. Study on Screening of Complex Microbial Community in Straw Degrading of Rapeseed. *J. Chin. Biotechnol.* **2011**, *31*, 70–74.
50. Huang, L.; Gan, Y. A review on SEM imaging of graphene layers. *Micron* **2024**, *187*, 103716. [CrossRef]
51. Huang, Y.; Yang, L.; Pan, K.; Yang, Z.; Yang, H.; Liu, J.; Zhong, G.; Lu, Q. Heavy metal-tolerant bacteria *Bacillus cereus* BCS1 degrades pyrethroid in a soil–plant system. *J. Hazard. Mater.* **2024**, *461*, 132594. [CrossRef]
52. Verónica, R.; Javier, M.; María, D.R.B.; Dolores, C.; JoséAntonio, M.; Antonino, B. *Bacillus cereus* MH778713 elicits tomato plant protection against Fusarium oxysporum. *J. Appl. Microbiol.* **2021**, *132*, 470–482.
53. Wang, X.; Luo, S.; Chen, Y.; Zhang, R.; Lei, L.; Lin, K.; Qiu, C.; Xu, H. Potential of Miscanthus floridulus associated with endophytic bacterium *Bacillus cereus* BL4 to remediate cadmium contaminated soil. *Sci. Total Environ.* **2023**, *857*, 159384. [CrossRef] [PubMed]
54. Zhao, H.; Li, S.; Pu, J.; Wang, H.; Dou, X. Effects of Bacillus-based inoculum on odor emissions co-regulation, nutrient element transformations and microbial community tropological structures during chicken manure and sawdust composting. *J. Environ. Manag.* **2024**, *354*, 120328. [CrossRef]
55. Limkar, M.B.; Pawar, S.V.; Rathod, V.K. Statistical optimization of xylanase and alkaline protease co-production by *Bacillus* spp using Box-Behnken Design under submerged fermentation using wheat bran as a substrate. *Biocatal. Agric. Biotechnol.* **2019**, *17*, 455–464. [CrossRef]
56. El-Dein, M.M.N.; El-Fallal, A.A.; Toson, E.-S.A.; Hereher, F.E. Exopolysaccharides Production by *Pleurotus pulmonarius*: Factors Affecting Formation and Their Structures. *Pak. J. Biol. Sci.* **2004**, *7*, 1078.
57. Xu, C.-P.; Kim, S.-W.; Hwang, H.-J.; Choi, J.-W.; Yun, J.-W. Optimization of submerged culture conditions for mycelial growth and exo-biopolymer production by Paecilomyces tenuipes C240. *Process Biochem.* **2003**, *38*, 1025–1030. [CrossRef]
58. Nair, A.; Sarma, S.J. The impact of carbon and nitrogen catabolite repression in microorganisms. *Microbiol. Res.* **2021**, *251*, 126831. [CrossRef]
59. Bettenbrock, K.; Fischer, S.; Kremling, A.; Jahreis, K.; Sauter, T.; Gilles, E.-D. A Quantitative Approach to Catabolite Repression in *Escherichia coli*. *J. Biol. Chem.* **2006**, *281*, 2578–2584. [CrossRef] [PubMed]
60. Reuß, D.R.; Rath, H.; Thürmer, A.; Benda, M.; Daniel, R.; Völker, U.; Mäder, U.; Commichau, F.M.; Stülke, J. Changes of DNA topology affect the global transcription landscape and allow rapid growth of a Bacillus subtilis mutant lacking carbon catabolite repression. *Metab. Eng.* **2018**, *45*, 171–179. [CrossRef]
61. Moodley, D.; Botes, A. A carboxymethyl cellulase from the yeast Cryptococcus gattii WM276: Expression, purification and characterisation. *Protein Expr. Purif.* **2025**, *225*, 106594. [CrossRef]
62. Jiang, L. Optimization of fermentation conditions for pullulan production by Aureobasidium pullulan using response surface methodology. *Carbohydr. Polym. Sci. Technol. Asp. Ind. Important Polysacch.* **2010**, *79*, 414–417. [CrossRef]
63. Shen, J.-W.; Shi, C.-W.; Xu, C.-P. Exopolysaccharides from Pleurotus pulmonarius: Fermentation Optimization, Characterization and Antioxidant Activity. *Food Technol. Biotechnol.* **2013**, *51*, 520–527.
64. Pokhrel, C.P.; Ohga, S. Submerged culture conditions for mycelial yield and polysaccharides production by Lyophyllum decastes. *Food Chem.* **2007**, *105*, 641–646. [CrossRef]
65. He, Y.; Yu, Z.; Wang, H.; Huang, L.; Wang, H. Effects of different carbon and nitrogen sources on activity of extracellular pectinase of Valsa mali var. mali. *J. Northwest A F University. Nat. Sci. Ed.* **2014**, *42*, 81–86.
66. Cheng, Y.; Cao, Y.; Zhang, J.; Chen, W. Optimization on the Culture Conditions of Ustilaginoidea virens(Cooke) Tak. *J. Henan Agric. Sci.* **2011**, *40*, 120–123.
67. Junillon, T.; Morand, L.; Flandrois, J.P. Enhanced tetrazolium violet reduction of Salmonella spp. by magnesium addition to the culture media. *Food Microbiol.* **2014**, *42*, 132–135. [CrossRef]

68. Dutta, N.; Biswas, S.; Saha, M.K. Biophysical characterization and activity analysis of nano-magnesium supplemented cellulase obtained from a psychrobacterium following graphene oxide immobilization. *Enzym. Microb. Technol.* **2016**, *95*, 248–258. [CrossRef]
69. Ji, S.; Pan, Y.; Zhu, L.; Tan, J.; Tang, S.; Yang, Q.; Zhang, Z.; Lou, D.; Wang, B. A novel 7α-hydroxysteroid dehydrogenase: Magnesium ion significantly enhances its activity and thermostability. *Int. J. Biol. Macromol.* **2021**, *177*, 111–118. [CrossRef]
70. Kumaran, J.; Poulose, S.; Joseph, V.; Bright Singh, I.S. Enhanced biomass production and proximate composition of marine microalga Nannochloropsis oceanica by optimization of medium composition and culture conditions using response surface methodology. *Anim. Feed Sci. Technol.* **2021**, *271*, 114761. [CrossRef]
71. Ghazanfari, N.; Fallah, S.; Vasiee, A.; Tabatabaei Yazdi, F. Optimization of fermentation culture medium containing food waste for l-glutamate production using native lactic acid bacteria and comparison with industrial strain. *LWT* **2023**, *184*, 114871. [CrossRef]
72. Patel, G.; Khobragade, T.P.; Avaghade, S.R.; Patil, M.D.; Nile, S.H.; Kai, G.; Banerjee, U.C. Optimization of media and culture conditions for the production of tacrolimus by Streptomyces tsukubaensis in shake flask and fermenter level. *Biocatal. Agric. Biotechnol.* **2020**, *29*, 101803. [CrossRef]
73. Saberi, F.; Marzban, R.; Ardjmand, M.; Pajoum Shariati, F.; Tavakoli, O. Optimization of culture media to enhance the ability of local Bacillus thuringiensis var. tenebrionis. *J. Saudi Soc. Agric. Sci.* **2020**, *19*, 468–475. [CrossRef]
74. Rodrigues, A.C.; Fontão, A.I.; Coelho, A.; Leal, M.; Silva, F.A.G.S.d.; Wan, Y.; Dourado, F.; Gama, M. Response surface statistical optimization of bacterial nanocellulose fermentation in static culture using a low-cost medium. *New Biotechnol.* **2019**, *49*, 19–27. [CrossRef] [PubMed]
75. Beeler, N.; Hühn, T.; Rohn, S.; Colombi, R. Development and optimization of a cocoa extraction treatment by means of the response surface methodology (RSM) and artificial neural networks (ANN). *Ind. Crops Prod.* **2024**, *222*, 119610. [CrossRef]

agronomy

MDPI

Article

Combined Application of High-Throughput Sequencing and Metabolomics to Evaluate the Microbial Mechanisms of Plant-Growth-Promoting Bacteria in Enhancing the Remediation of Cd-Contaminated Soil by Hybrid Pennisetum

Shan-Shan Gao [1], Ying-Jun Zhang [1], Yang Shao [1], B. Larry Li [2], Han Liu [1], Yu-Ying Li [2], Xue-Min Ren [1,*] and Zhao-Jin Chen [2,*]

1 College of Water Resource and Modern Agriculture, Nanyang Normal University, Nanyang 473061, China; 19540340615@nynu.edu.cn (S.-S.G.)
2 Overseas Expertise Introduction Center for Discipline Innovation of Watershed Ecological Security in the Water Source Area of the Mid-Line Project of South-to-North Water Diversion, Nanyang 473061, China
* Correspondence: 20131092@nynu.edu.cn (X.-M.R.); zhaojin_chen@163.com or zjchen@nynu.edu.cn (Z.-J.C.)

Abstract: The contamination of soil with the heavy metal cadmium (Cd) is increasingly prominent and severely threatens food security in China. Owing to its low cost, suitable efficacy, and ability to address the shortcomings of plant remediation by enhancing the ability of plants to take up Cd, plant–microbe combination remediation technology has become a research hotspot in heavy metal pollution remediation. A pot experiment was performed to examine the effects of inoculation with the plant-growth-promoting bacterium *Brevibacillus* sp. SR-9 on the biomass, Cd accumulation, and soil nutrients of hybrid Pennisetum. The purpose of this study was to determine how *Brevibacillus* sp. SR-9 alleviates stress caused by heavy metal contamination. High-throughput sequencing and metabolomics were used to determine the effects of inoculation on the soil bacterial community composition and microbial metabolic functions associated with hybrid Pennisetum. The results suggest that mutation of *Brevibacillus* sp. SR-9 effectively alleviates Cd pollution stress, leading to increased biomass and accumulation of Cd in hybrid Pennisetum. The aboveground biomass and the root weight increased by 12.08% and 27.03%, respectively. Additionally, the accumulation of Cd in the aboveground sections and roots increased by 21.16% and 15.50%, respectively. Measurements of the physicochemical properties of the soil revealed that the strain *Brevibacillus* sp. SR-9 slightly increased the levels of available phosphorus, total nitrogen, total phosphorus, and available potassium. High-throughput DNA sequencing revealed that *Brevibacillus* sp. SR-9 implantation modified the composition of the soil bacterial community by increasing the average number of Actinobacteria and *Bacillus*. The total nitrogen content of the soil was positively correlated with the Actinobacteria abundance, total phosphorus level, and available phosphorus level. Metabolomic analysis revealed that inoculation affected the abundance of soil metabolites, and 59 differentially abundant metabolites were identified ($p < 0.05$). Among these, 14 metabolites presented increased abundance, whereas 45 metabolites presented decreased abundance. Fourteen metabolic pathways were enriched in these metabolites: the folate resistance pathway, the ABC transporter pathway, D-glutamine and D-glutamic acid metabolism, purine metabolism, and pyrimidine metabolism. The abundance of the metabolites was positively correlated with the levels of available phosphorus, total potassium, total phosphorus, and total nitrogen. According to correlation analyses, the development of hybrid Pennisetum and the accumulation of Cd are strongly associated with differentially abundant metabolites, which also impact the abundance of certain bacterial populations. This work revealed that by altering the makeup of microbial communities and their metabolic processes, bacteria that promote plant development can mitigate the stress caused by Cd. These findings reveal the microbiological mechanisms through which these bacteria increase the ability of hybrid Pennisetum to take up the Cd present in contaminated soils.

Keywords: cadmium; plant–microbe remediation; hybrid Pennisetum; bacterial community; metabolomics

1. Introduction

Soil heavy metal contamination has worsened due to economic factors, urbanization, and the rapid growth of agricultural inputs [1–3]. In China, the total area of arable land affected by heavy metal pollution has reached 20 million hectares [4]. Cadmium (Cd) is one of the most hazardous heavy metals on the planet [5] and interferes with the absorption and transport of essential nutrients in plants, disrupts plant metabolism, and inhibits plant growth [6,7]. Cd treatment dramatically decreased the root tip number, surface area, and root length of chili peppers [8]. Additionally, Cd can enter animals and humans through the food chain, and long-term exposure can lead to kidney deposition, ultimately causing kidney failure, lung disease, and bone fragility [9]. Thus, it is critical to create economical and ecologically sound techniques for remediating soils polluted with Cd [10].

Currently, in soil heavy metal pollution remediation research, phytoremediation is regarded as a very promising green technology because of its cost-effectiveness, simplicity, and environmentally friendly nature [11]. Hybrid Pennisetum species have strong stress resistance, wide adaptability, and high biomass yield. Additionally, recent studies have shown that hybrid Pennisetum species have strong Cd tolerance and moderate Cd accumulation capacity and can extract a considerable total amount of Cd from the soil [12,13]. Additionally, hybrid Pennisetum has application value in anaerobic digestion for biogas production, direct combustion for power generation, pulp and paper production, and the manufacture of regenerated cellulose films [14]. However, in practical remediation, issues such as slow plant growth, low biomass, long remediation cycles, poor heavy metal tolerance, and difficulty in recycling contaminated plants significantly limit the effectiveness of remediation [15].

Plant growth promoting bacteria (PGPB) are bacteria that can directly or indirectly increase plant growth [16]. In the field of heavy metal remediation, by increasing metal accumulation and improving heavy metal tolerance, PGPB can increase the ability of plants to adapt to their surroundings and, therefore, increase phytoremediation effectiveness [17]. Kong et al. [18] and Mohammadzadeh et al. [19] reported that PGPB could affect the phytotoxic effects of heavy metals and enhance the absorption of heavy metals by plants. Soil microorganisms are essential for the soil ecosystem and strongly affect plant development in soils polluted with heavy metals. They also play a vital role in the biogeochemical cycling of heavy metals [20,21]. The rhizosphere is the site of interactions among plants, microbes, and soil, the rhizosphere microbial community structure and diversity are regarded as important indicators of soil health [22]. Studies have shown that inoculation with PGPB can alter the rhizosphere microbial community structure to promote plant growth [23]. Research has also shown that under heavy metal stress, PGPB can increase the complexity and connectivity of rhizosphere bacterial community network modules, suggesting that PGPB may improve network stability and thereby promote plant growth [24]. Similarly, rhizosphere metabolites—comprising root exudates and small molecules produced by microbial metabolism in the rhizosphere—play a vital role in the collaborative resistance of plants and microbes to heavy metal stress [25]. Zuluaga et al. [26] and Faten et al. [27] reported that inoculation with PGPB changed the metabolism and function of the plant rhizosphere, improved plant resistance to stress, and increased plant biomass.

High-throughput sequencing technology is now commonly utilized in soil, rhizosphere, and environmental studies [28–30]. High-throughput sequencing can be employed to investigate the impact of heavy metals on plants, offering useful knowledge for the remediation of heavy metal contamination. Using high-throughput technologies, Sheng et al. [31] investigated how chromium (Cr) affects the bacterial population of Iris species, illuminating the mechanisms underlying the stress response of Iris plants to Cr heavy metal stress. Metabolomic techniques are now widely used to elucidate the physiological effects of various pollutants on exposed organisms. For example, Navarro-Reig et al. [32] used untargeted metabolomics to perform a comprehensive LC–MS analysis of metabolites extracted from the aboveground tissues of rice plants under Cd and copper (Cu) stress, identifying 97 different metabolites. Compared with Cu, Cd had a more significant effect on rice, primarily affecting secondary metabolism and amino acid, purine, and carbon and glycerolipid metabolism pathways. Ghate et al. [33] studied the oxidative stress response of rice to arsenic (As) and potassium salt exposure using both untargeted and targeted metabolomics via LC–MS. The results highlighted two key metabolites identified as response metabolites to As. Currently, there is limited research on how PGPB alter the microbial composition and metabolism of renewable energy plants such as hybrid Pennisetum for the remediation of soil heavy metal contamination. The following questions were addressed in this study: 1. Can the plant-growth-promoting bacterium *Brevibacillus* sp. SR-9 promote the growth of hybrid Pennisetum and increase the heavy metal content in Cd-contaminated soils, thereby improving the phytoremediation efficiency? 2. Does the plant-growth-promoting bacterium *Brevibacillus* sp. SR-9 promote hybrid Pennisetum growth by altering its microbial community composition and metabolism? Is there an interaction between the makeup of the microbial population in the rhizosphere and the metabolism of hybrid Pennisetum that influences the phytoremediation efficiency?

2. Materials and Methods

2.1. Experimental Materials

The tested soil was collected from a garden near a pomegranate orchard at Nanyang Normal University. The material was filtered to eliminate contaminants such as leaves and stones. Afterward, it was air-dried and subsequently passed through a sieve with a mesh size of 20 for further analysis. The seeds of hybrid Pennisetum used in the experiments were purchased from Shuyang County Tuojing Horticulture Co., L (Suqian, China). The bacteria, which were chosen for their ability to promote plant development and withstand heavy metals, were identified as members of the *Brevibacillus* genus and given the name *Brevibacillus* sp. SR-9. The strain *Brevibacillus* sp. SR-9 (SUB11374180) [34] was obtained from Cd-contaminated soil used for farming in Nanyang, China. *Brevibacillus* sp. SR-9 bacteria are highly tolerant to heavy metals and are proficient in the production of iron carriers and indole-3-acetic acid (IAA) and the solubilization of phosphorus and potassium.

2.2. Experimental Methods

The *Brevibacillus* sp. SR-9 strain was introduced into LB liquid medium and incubated at 30 °C with agitation at a speed of 180 revolutions per minute for 24 h to prepare the *Brevibacillus* sp. SR-9 bacterial suspension. After incubation, the bacterial cells were collected via centrifugation and washed with deionized water. The suspension was then diluted to achieve an optical density (OD) value of 1.0–1.5 for further use.

Pot experiment: We measured and portioned the soil and then added analytical grade $CdSO_4 \cdot 8H_2O$ to attain a concentration of 20 milligrams per kilogram (mg/kg) of Cd ions in the soil. The combination was completely blended and allowed to reach a state of equilibrium for 30 days. The dirt that had been prepared was placed into pots, with each pot containing 0.75 kg of soil. The experiment included two treatments: a control group (CK) without bacterial inoculation with a Cd concentration of 20 mg/kg and an experimental group inoculated with *Brevibacillus* sp. SR-9 at the same Cd concentration of 20 mg/kg. Hybrid Pennisetum seeds were evenly sown in plastic pots and grown outdoors. The pots were kept under adequate light, and their positions were rotated randomly every 4 days to provide uniform and stable growth conditions. Thinning was carried out after the hybrid Pennisetum seedlings grew to a height of 5 cm, with only 3 seedlings remaining in each pot. The test bacterium was inoculated into the experimental group at 30, 50, and 70 days after germination. The control group was administered an equal amount of tap water at the same intervals. The plants were grown for 95 days before harvesting and sample collection.

2.3. Soil Cd Content and Physicochemical Property Measurements

Digestion with hydrochloric acid, nitric acid, hydrofluoric acid, and perchloric acid was performed, and the Cd content in different plant tissues was determined via an inductively coupled plasma–optical emission spectrometer (ICP–OES) (Optima 2100 DV, Perkin Elmer, USA) after digestion. The soil pH was determined via a potentiometric approach with a soil-to-water ratio of 1:5. The indophenol blue colorimetric technique was employed to determine the total nitrogen (TN) concentration in the soil samples. The total phosphorus (TP) content in the soil samples was determined via the sodium hydroxide fusion–molybdenum antimony colorimetric method. The flame photometry technique was used to assess the available potassium (AK) concentration in the soil samples. The 0.5 $mol \cdot L^{-1}$ $NaHCO_3$ extraction–molybdenum antimony colorimetric technique was used to evaluate the available phosphorus (AP) concentration in the soil samples.

2.4. High-Throughput Sequencing

For the purpose of harvesting and sequencing DNA, the soil samples were transported to Maggi Biomedical Technology Co., Ltd. (Shanghai, China) The purity of the isolated genomic DNA was assessed via 1% agarose gel electrophoresis after the extraction process. PCR detection was performed via a TransGen AP 221-02 and TransStart Fastpfu DNA polymerase (MP Biomedicals, Irvine, CA, USA) in a 20 μL reaction. PCR was conducted with an ABI GeneAmp® 9700 instrument (Illumina Inc., San Diego, CA, USA), which utilizes particular primers that are synthesized with barcodes corresponding to the specified sequencing areas. All the materials were processed via conventional experimental methods, with three replicates per sample. The PCR products obtained from identical samples were combined and subjected to analysis via 2% agarose gel electrophoresis. Following Tris-HCl elution and the use of an AxyPrep DNA Gel Extraction Kit (AXYGE, Silicon Valley, CA, USA), the PCR products were recovered and reassessed via 2% agarose gel electrophoresis. A QuantiFluor™-ST Blue Fluorescence Quantification System (Promega, Shanghai, China) was used to quantify the PCR products on the basis of the electrophoresis results. The samples were subsequently mixed at the necessary ratios on the basis of the sequencing volume specifications before Illumina sequencing was conducted. Illumina sequencing produces paired-end (PE) reads, which are initially assembled via overlap relationships. The quality of each sequence was subsequently evaluated and refined. Operational taxonomic unit (OTU) clustering analysis and taxonomic categorization were carried out following sample separation. Numerous diversity indices were examined, and the sequencing depth was evaluated via the OTU data. In addition, the community structure was subjected to statistical analysis at various taxonomic levels via categorization information. On the basis of the aforementioned investigations, numerous comprehensive statistical and visual analyses, multivariate analyses, and significance tests were carried out on the phylogenetic data and community compositions of many samples.

2.5. Metabolomic Analysis

One gram of each soil sample was weighed, and 20 microliters of internal standard and 1 milliliter of a solution containing 50% methanol were added. A centrifuge was used to separate and collect the remaining liquid after sedimentation. The liquid above the sediment was passed through a filter with a pore size of 0.22 μm via a sterile syringe. All samples were stored at −80 °C for LC–MS analysis. The untargeted metabolomics analysis in this experiment was carried out by employing an ultrahigh-performance liquid chromatography system (UHPLC-TOF MS) (Agilent Technologies Inc., CA, USA) for separation. Instrument calibration was conducted as needed to ensure accurate mass spectrometry data. The appropriate injection mode and injection volume were selected on the basis of the sample properties and analytical requirements. The analysis was initiated, and the instrument was controlled according to the preset conditions. After sample separation, MS analysis was performed using an Agilent 6538 Ultra-High-Resolution Q-TOF (Jungkap Park, Santa Clara, CA, USA) mass spectrometer in both positive and negative ion modes.

The raw data were processed to extract metabolite information and convert it into an analyzable format through baseline correction, peak alignment, peak identification, missing value imputation, data filtering, and normalization. The techniques of partial least squares discriminant analysis (PLS-DA) and orthogonal partial least squares discriminant analysis (OPLS-DA) were employed to obtain insights into the variations in metabolites across different groups. Metabolites that exhibited notable differences across groups were chosen on the basis of variable importance in projection (VIP) values obtained from the OPLS-DA in conjunction with a *t*-test ($p < 0.05$). The identified differentially abundant metabolites were mapped to corresponding metabolic pathways via the KEGG database.

2.6. Data Analysis

SPSS 17.0 (IBM, Armonk, NY, USA) was used to analyze the heavy metal concentrations, high-throughput sequencing, and soil physicochemical parameter data. To assess group differences, one-way ANOVA and *t*-tests were employed. The transfer factor (TF) was calculated as the aboveground Cd content/root Cd content [35].

3. Results

3.1. Effects of Brevibacillus sp. SR-9 on Hybrid Pennisetum Growth, Cd Content, and Cd Accumulation under Cd-Induced Stress

As shown in Table 1, the biomass results of hybrid Pennisetum indicate that, compared with those of the control group (CK), the biomass of the aerial parts increased by 12.08%, and the root dry weight significantly increased by 27.03% ($p < 0.01$) in the group treated with *Brevibacillus* sp. SR-9. These findings suggest that the introduction of *Brevibacillus* sp. SR-9 increased the ability of hybrid Pennisetum roots to withstand the presence of Cd. Compared with that in the control group, which was not inoculated with *Brevibacillus* sp. SR-9, the accumulation of Cd in the aerial portions and roots of hybrid Pennisetum increased by 21.16% and 15.50%, respectively, in the group treated with *Brevibacillus* sp. SR-9. Thus, inoculation with *Brevibacillus* sp. SR-9 can increase Cd accumulation in various tissues of hybrid Pennisetum. The transport coefficient increased from 0.14 to 0.16, representing a 19.94% increase compared with that of the CK group. These findings indicate that inoculation with *Brevibacillus* sp. SR-9 enhances the ability of the plant to transport heavy metals from the soil into the plant, thereby increasing the remediation capacity of the plant.

Table 1. Effects of inoculation with *Brevibacillus* sp. SR-9 on biomass, Cd content, Cd accumulation, and translocation factor (TF) in different parts of hybrid Pennisetum.

Group	Aboveground Biomass (g)	Root Dry Weight (g)	Aboveground Cd Content (mg/kg)	Root Cd Content (mg/kg)	Aboveground Cd Accumulation (ug)	Root Cd Accumulation (ug)	TF
CK	5.38 ± 0.76	0.74 ± 0.19	2.38 ± 0.32	17.48 ± 2.98	13.23 ± 2.76	12.84 ± 4.30	0.14
SR-9	6.03 ± 0.50	0.94 ± 0.16	2.55 ± 0.61	15.68 ± 2.69	16.03 ± 2.88	14.83 ± 1.62	0.16

3.2. Effects of Brevibacillus sp. SR-9 on the Physicochemical Properties of Soil Contaminated with Cd

Table 2 shows that, compared with those in the control group, the total nitrogen content, total phosphorus content, available phosphorus content, and available potassium content in the microbial treatment group significantly increased ($p < 0.05$) by 6.25%, 3.46%, 12.00%, and 2.49%, respectively, with the increase in available phosphorus being particularly notable. Under the influence of *Brevibacillus* sp. SR-9, soil nutrients increased, leading to increased growth of hybrid Pennisetum.

3.3. Impact of Brevibacillus sp. SR-9 on Soil Bacterial Communities

At the OTU level, diversity index analysis revealed that (Figure 1), compared with the CK, the inclusion of *Brevibacillus* sp. SR-9 resulted in decreases in the ACE index, Chao1 index, and Sobs index of 3.1%, 0.65%, and 1.4%, respectively. The application of microbial therapy resulted in a decrease in the overall abundance of bacteria and caused changes in the composition of the bacterial population. Nevertheless, there were no substantial alterations in the Shannon or Simpson indices.

Table 2. Effects of inoculation with *Brevibacillus* sp. SR-9 on the physicochemical properties of hybrid Pennisetum rhizosphere soil.

Group	pH	Total Nitrogen Content (mg/kg)	Total Phosphorus Content (mg/kg)	Available Phosphorus Content (mg/kg)	Available Potassium Content (mg/kg)
CK	7.12 ± 0.10	1.76 ± 0.09	242.07 ± 8.08	0.025 ± 0.001	107.33 ± 2.17
SR-9	7.05 ± 0.05	1.87 ± 0.11	250.44 ± 15.04	0.028 ± 0.003	111.00 ± 2.74

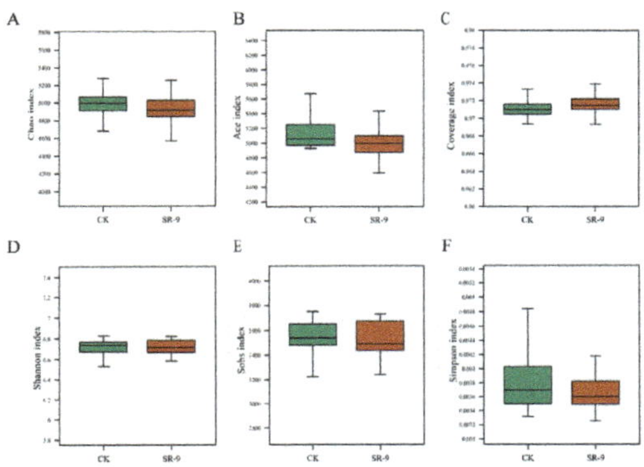

Figure 1. Assessment of the bacterial community diversity in different samples. (**A**) Comparison of the Chao index between groups, (**B**) comparison of the ACE index between groups, (**C**) comparison of the coverage index between groups, (**D**) comparison of the Shannon index between groups, (**E**) comparison of the Sobs index between groups, and (**F**) comparison of the Simpson index between groups. Green indicates the CK group, and orange indicates the SR-9 group.

Venn diagram analysis revealed the differences in the OTU compositions and unique species of rhizosphere soil bacterial communities under the different treatments. A total of 15,123 bacterial OTUs were identified across both treatments, with 6545 OTUs shared between them. The CK group had 1134 unique OTUs, accounting for 7.50% of the total OTUs, whereas the microbial treatment group had 899 unique OTUs, representing 5.95% of the total OTUs. As shown in Figure 2, PLS-DA analysis at the OTU level revealed that the nonmicrobial treatment group and the microbial treatment group were positioned at opposite sides of the plot, clustering separately. These findings indicate that there were differences between the two treatment groups and the *Brevibacillus* sp. SR-9 treatment altered the composition of the soil bacterial community.

On the basis of high-throughput sequencing of the soil bacterial 16S rDNA gene and clustering at 97% similarity, a total of 39 phyla and 958 genera were identified. On the basis of the phylum-level analysis (Figure 3A) of the microbial populations residing in the rhizosphere of hybrid Pennisetum under various therapeutic regimens, the dominant phyla were Actinobacteria, Proteobacteria, Acidobacteria, Chloroflexi, and Firmicutes. Together, these phyla accounted for more than 80% of the total bacterial abundance in the samples. In the CK treatment group, the relative abundances of the dominant phyla were 26.7% for Actinobacteria, 24.9% for Proteobacteria, 15.5% for Acidobacteria, 11.3% for Chloroflexi, and 4.5% for Firmicutes. In the microbial treatment group, the relative abundances were 27.7% for Actinobacteria, 24.4% for Proteobacteria, 15.2% for Acidobacteria, 11.2% for Chloroflexi, and 4.7% for Firmicutes. Compared with the CK group, the microbial treatment group presented a 1% increase in the relative abundance of Actinobacteria, whereas the relative abundance of Acidobacteria decreased by 0.3%. Furthermore, the abundance of Proteobacteria was greater in the CK group than in the microbial treatment group. As shown in Figure 3B, at the genus level, the four most dominant bacterial genera based on relative abundance were *Sphingomonas*, *RB41*, *Bacillus*, and *Gaiella*. Compared with that in the nonmicrobial treatment group, the relative abundance of the *Bacillus* genus in hybrid Pennisetum rhizosphere soil increased by 0.3% with *Brevibacillus* sp. SR-9 treatment. Compared with the microbial treatment group, the CK group presented a substantial increase in the relative abundance of *Sphingomonas*, although the relative abundances of *RB41* and *Gaiella* did not significantly differ between the two groups. The differences and changes in the phylum- and genus-level bacterial communities between the two treatment groups indicate that inoculation with *Brevibacillus* sp. SR-9 affects the rhizosphere microbial communities.

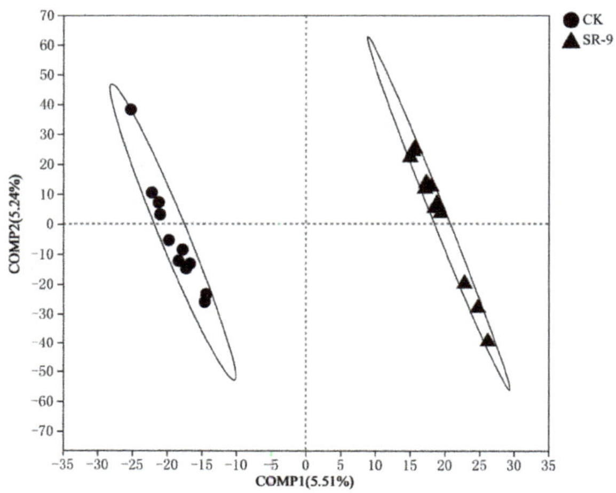

Figure 2. PLS-DA of the bacterial community composition.

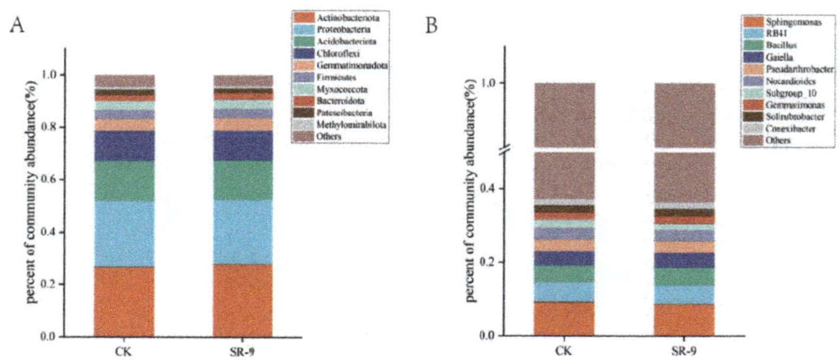

Figure 3. The relative abundance of communities in different treatment groups at the phylum (**A**) and genus (**B**) levels.

The symbiotic relationships between bacteria under the two different treatments were explored via co-occurrence network analysis, resulting in a phylum-level co-occurrence network diagram of the rhizosphere soil (Figure 4). The results reveal that the bacterial co-occurrence networks differed between treatments. The CK group had 387 nodes and 5879 edges, whereas the microbial treatment group had 378 nodes and 9110 edges. The average degree, network density, and clustering coefficient in the microbial treatment group were greater than those in the CK group. Nevertheless, the microbial treatment group presented a reduced average network distance. Additionally, the microbial treatment group had a lower modularity (0.277) than the CK group (0.353) (Table 3). These findings indicate that the microbial treatment group had a more complex and compact bacterial network than the CK group.

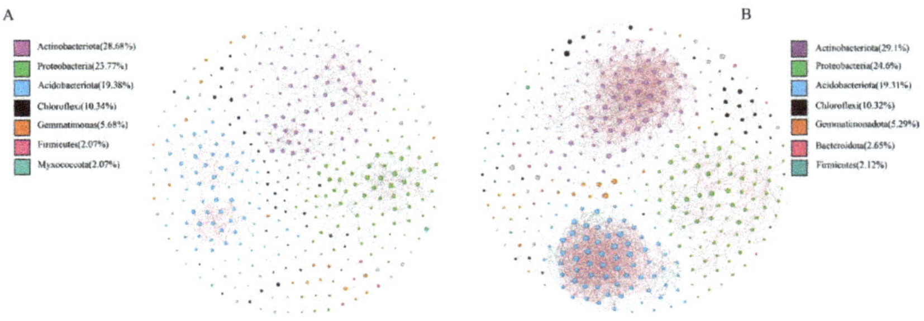

Figure 4. Co-occurrence patterns of bacterial community networks in the CK (**A**) and SR-9 (**B**) groups. The figure is divided into 7 sections comprising the bacterial community networks under each treatment group. Within each square, the groups and the top 7 significant bacterial communities of the co-occurrence network are enumerated.

Table 3. Topological parameters of the microbial co-occurrence networks.

Group	CK	SR-9
Nodes	387	378
Edges	5879	9110
Positive correlation	3325	5248
Negative correlation	2554	3862
Average degree	30.3824	48.2011
Average weight degree	21.2292	33.9104
Average path length	1.5166	1.3958
Network diameter	2.6302	3.1889
Network density	0.0787	0.1279
Clustering coefficient	0.406	0.4906
Betweenness centralization	0.0079	0.0063

Redundancy analysis (RDA) was used to examine the relationships between the physicochemical parameters of the soil in hybrid Pennisetum and the composition of the bacterial population at the phylum level in the rhizosphere. As shown in Figure 5, the results reveal that the soil physicochemical parameters accounted for 70.2% of the variance in the organization of the bacterial population. The initial two principal components explained 54.66% and 15.54% of the total variation, respectively. Among these factors, pH, total nitrogen, available potassium, and available phosphorus were identified as important factors influencing disparities in the hybrid Pennisetum microbial populations present in the soil. Among the five most abundant microbial phyla in hybrid Pennisetum soil, Acidobacteria was directly related to soil pH, while the remaining three phyla were inversely related to the pH of the soil. Moreover, the presence of Actinobacteria was positively correlated with total nitrogen, total phosphorus, and available phosphorus in the soil. At the genus level, the soil physicochemical properties explained 54.86% of the variation, with the first two canonical axes accounting for 38.49% and 16.37% of the variance, respectively. The abundance of the *Bacillus* genus was directly related to the available phosphorus, available potassium, and pH, but was negatively correlated with the total nitrogen and total phosphorus. In contrast, *Sphingomonas* exhibited the opposite pattern, with an inverse relationship with AP, AK, and pH and a positive correlation with TN and TP.

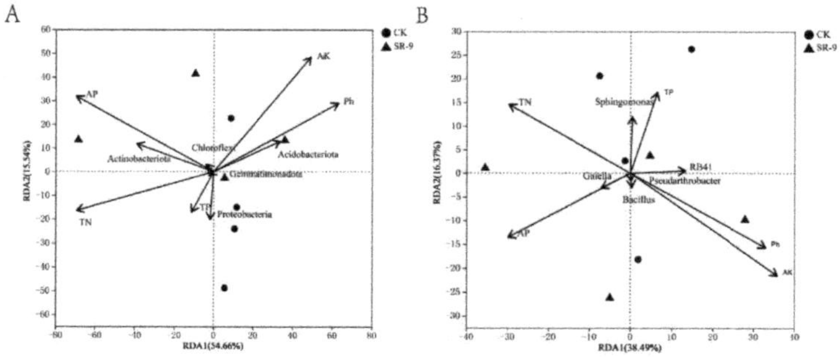

Figure 5. RDA ordination biplot between bacterial communities and environmental factors at the phylum (**A**) and genus (**B**) levels.

3.4. Impact of Brevibacillus sp. SR-9 on Metabolic Functions

To investigate the impact of *Brevibacillus* sp. SR-9 inoculation on the metabolic profiles of hybrid Pennisetum rhizosphere soil under Cd stress, PLS-DA and OPLS-DA score analyses were performed, with 200-cycle cross-validation to assess the model stability. As shown in Figure 6, in both the cationic and anionic modes, the soil samples from the microbial treatment group and the nonmicrobial treatment group were well separated, indicating that there were differences in the soil metabolites between the two treatments. These findings suggest that *Brevibacillus* sp. SR-9 inoculation affects the rhizosphere soil metabolite profile. Selected sample cross-validation revealed that the R^2Y values of the original model in both the cationic and anionic modes were close to 1, indicating that the model accurately represented the sample data. The Q^2 value of approximately 0.5 suggests a clear differentiation between the two sample groups. Additionally, the intercept of the Q regression line with the *y*-axis below zero reflects the model's stability, as shown in Figure 7.

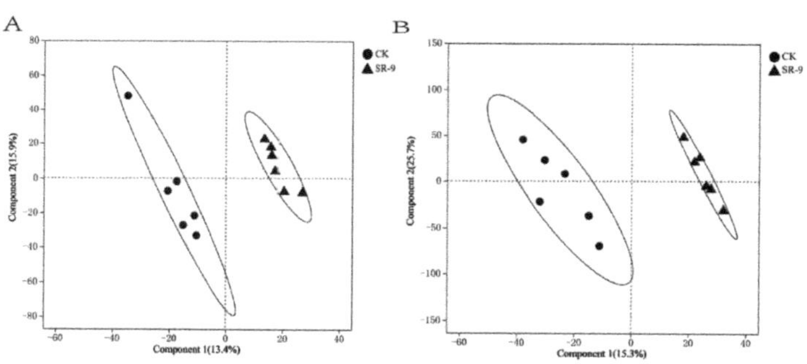

Figure 6. PLS-DA analysis chart of cationic (**A**) and anionic (**B**) metabolites in the rhizosphere soil of hybrid Pennisetum.

A total of 613 metabolites were detected across both sample groups, with 299 metabolites common to both groups. The CK samples contained 306 metabolites, whereas the microbial treatment samples contained 307 metabolites. A volcano plot of the 613 metabolites was created (Figure 8). In the plot, each point represents a metabolite, with larger scatter points indicating higher VIP values. This means that metabolites with larger scatter points are considered more reliably differentially abundant. To further identify differences in metabolites between the microbial treatment and control soils, OPLS-DA was performed with the criteria of $p < 0.05$ and VIP > 1. The results revealed 59 differentially abundant metabolites ($p < 0.05$), with the abundance of 14 metabolites increasing and the abundance of 45 metabolites decreasing in the microbial treatment group. These differentially abundant metabolites include petasitenine, 8,13-dihydroxy-9,11-octadecadienoic acid,

3-hydroxy-10′-apo-b,y-carotenal, and hypoxanthine. The metabolites whose abundance increased included 3-hydroxy-10′-apo-b, y-carotenal, petasitenine, and margaroylglycine. This indicates that inoculation with *Brevibacillus* sp. SR-9 affects the differential abundance of metabolites present in the rhizosphere soil of hybrid Pennisetum under the influence of Cd-induced stress.

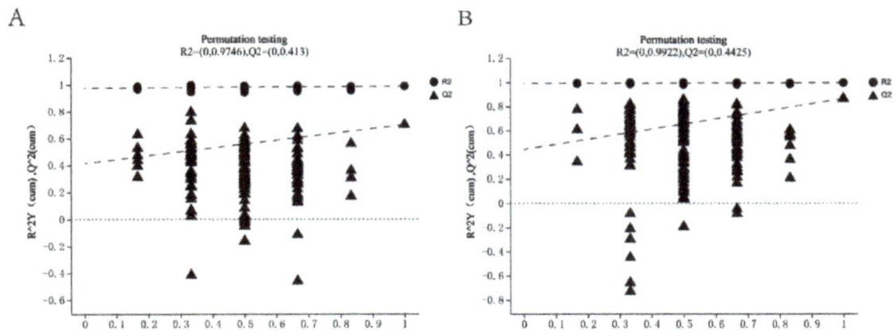

Figure 7. Permutation test of OPLS-DA models in the CK group (**A**) and *Brevibacillus* sp. SR-9 (**B**) group. The abscissa represents the displacement retention of the displacement test, the ordinate represents the values of the R2 (dot) and Q2 (triangular) displacement tests, and the two dotted lines represent the regression lines of R2 and Q2.

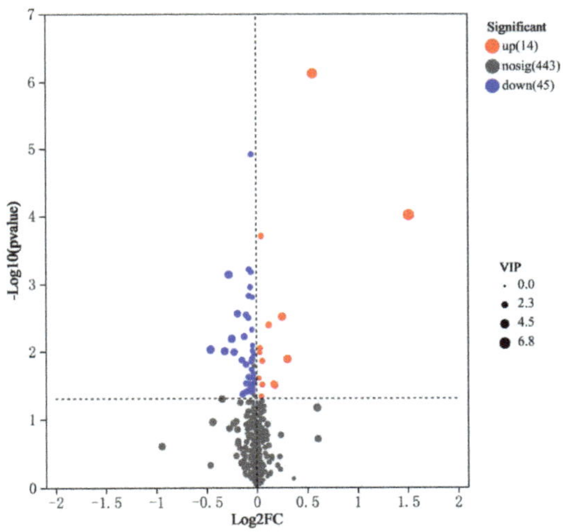

Figure 8. Volcano plot showing differentially abundant metabolites between groups. Each dot in the graph represents a specific metabolite, and the size of the dot represents the VIP value. The default red dots indicate the metabolites with significantly increased abundance, whereas the blue dots indicate the metabolites with significantly decreased abundance. The gray dots indicate nonsignificantly differentially abundant metabolites.

To further understand the effects of *Brevibacillus* sp. SR-9 on metabolites in hybrid Pennisetum soil and explore the mechanisms behind metabolite changes, a KEGG pathway enrichment analysis was conducted on the selected differentially abundant metabolites. On the basis of the KEGG database, the KEGG IDs of the differentially abundant metabolites were used to obtain pathway enrichment results. Pathways significantly enriched among the differentially expressed metabolites were identified via hypergeometric testing. This technique facilitated the investigation of metabolic pathway enrichment for the differentially abundant metabolites. The analysis identified three main pathways: environmental information processing, human diseases, and metabolism. Among these, the metabolism category had the most enriched pathways. The main pathways included lipid metabolism, metabolism of cofactors and vitamins, metabolism of other amino acids, nucleotide metabolism, and terpenoid and polyketide metabolism. The metabolic pathway bubble plot (Figure 9) revealed that the pathways significantly enriched in differentially abundant metabolites were the folate resistance pathway, the ABC transporter pathway, d-glutamine and d-glutamic acid metabolism, purine metabolism, and pyrimidine metabolism ($p < 0.05$). Notably, the purine metabolism and pyrimidine metabolism pathways were enriched in a greater number of metabolites. These pathways play important physiological roles in plant stress responses.

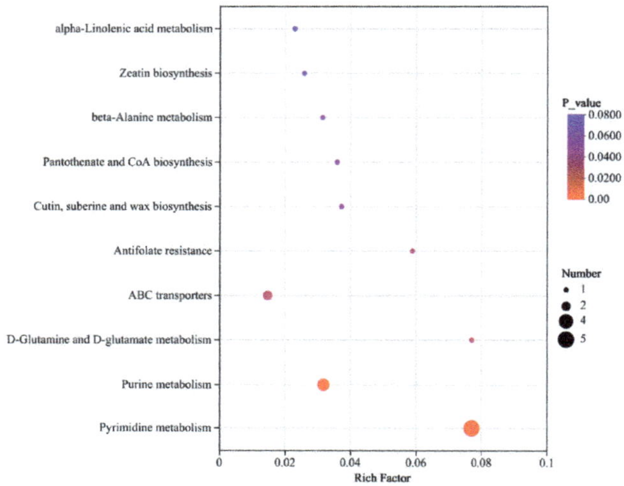

Figure 9. Bubble map of the KEGG enrichment analysis results. The size of the bubble in the figure represents the number of differentially abundant metabolites that were enriched in that pathway, and the color of the bubble represents the size of the *p*-value of different enriched pathways.

To investigate the correlation between differentially abundant metabolites and soil nutrients, a correlation analysis was conducted between the physicochemical properties of hybrid Pennisetum rhizosphere soil and the metabolites whose abundances increased in both groups (Figure 10). The results indicate that the abundances of MG (0:0/16:0/0:0) and Cis-4-hydroxycyclohexylacetic acid were positively correlated with the available potassium. Moreover, the abundance of methyl dihydrojasmonate was significantly positively correlated with the available phosphorus. The abundance of propofol glucuronide and kiwiionoside was significantly negatively correlated with soil pH.

3.5. Correlation Analysis

Brevibacillus sp. SR-9 altered both the composition and concentration of metabolites in the rhizosphere and affected the abundances of certain rhizosphere bacterial species. Correlation analysis between the differentially abundant metabolites and the abundance of rhizosphere bacteria revealed which bacterial populations were influenced by the differentially abundant metabolites. Spearman correlation analysis was used to examine the relationships between bacterial communities and differentially abundant metabolites (Figure 11), and the findings revealed a substantial positive correlation between Subgroup_10 and C17 sphinganine ($p = 0.036$), with a correlation coefficient of 0.74. Sphingomonas was also significantly positively correlated with 3,4,5-trihydroxy-6-(8-hydroxy-2-oxo-2H-furo[2,3-h] chromen-4-yl}oxy)oxane-2-carboxylic acid ($p = 0.02$), with a 0.79 correlation coefficient.

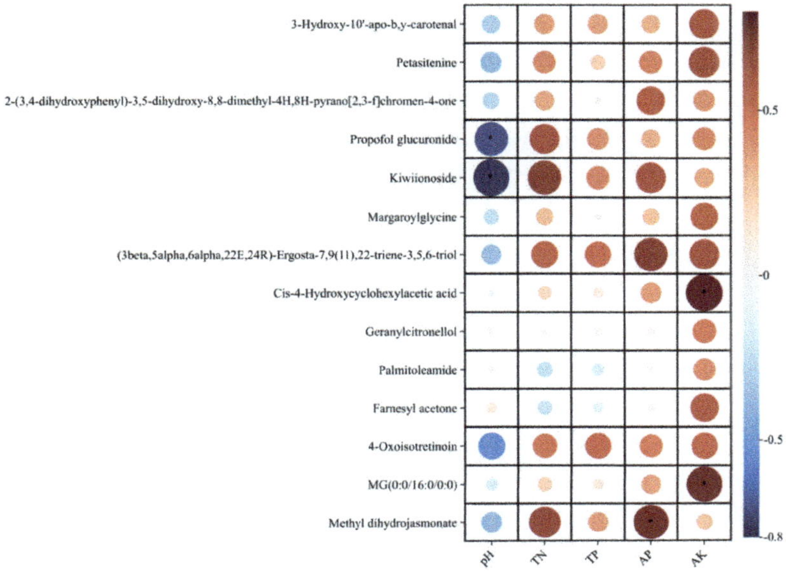

Figure 10. Correlation analysis diagram of the differentially abundant metabolites and soil physicochemical properties of hybrid Pennisetum. The abscissa shows the physical and chemical properties of the soil, the ordinate shows the metabolites whose abundances increased, and a larger circle size and deeper color indicate higher values. * Indicates significance.

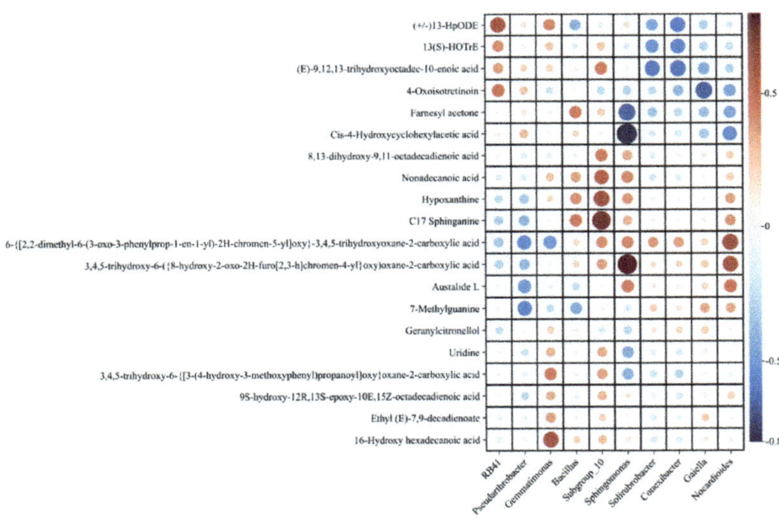

Figure 11. Correlation analysis of the differentially abundant metabolites and bacterial communities in hybrid Pennisetum. The abscissa represents the bacterial community at the genus level, and the ordinate represents the top 20 differentially abundant metabolites, and a larger circle size and deeper color indicate higher values.

Sphingomonas was significantly negatively correlated with Cis-4-hydroxycyclohexylacetic acid ($p = 0.01$), with a correlation coefficient of -0.83. At the genus level, *Bacillus* was primarily positively correlated with farnesyl acetone, hypoxanthine, and C17 sphinganine. Additionally, the effects of various bacterial populations and metabolites on hybrid

Pennisetum growth indices, Cd accumulation, and soil physicochemical parameters were analyzed via linear regression and Mantel testing (Figure 12). The findings indicate that the abundance of *Sphingomonas* is positively correlated with the Cd content in the aboveground part (r = 0.52, p = 0.05) and with the DTPA-Cd content (r = 0.91, p = 0.03). The abundance of 3-hydroxy-10′-apo-b,y-carotenal was positively correlated with the belowground dry weight (r = 0.42, p = 0.01), the Cd content in the aboveground part(r = 0.76, p = 0.04), and the DTPA-Cd content (r = 0.59, p = 0.05). The abundance of 8,13-dihydroxy-9,11-octadecadienoic acid was positively correlated with the Cd content in the aboveground part (r = 0.56, p = 0.03) and exhibited a substantial positive association with the DTPA-Cd content (r = 0.97, p = 0.008).

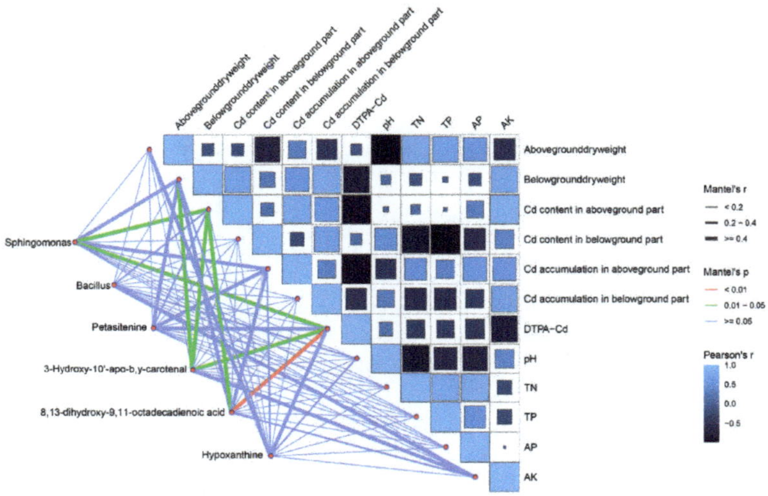

Figure 12. Relationships between hybrid Pennisetum growth indices and soil physicochemical properties, bacterial communities, and differentially abundant metabolites. Hybrid Pennisetum growth indices are shown in a butterfly plot, and the physicochemical properties of the soil and bacterial communities and differentially abundant metabolites are shown on the right.

4. Discussion

Hybrid Pennisetum is a nonfood energy crop with high biomass potential that shows promise for soil heavy metal pollution remediation in China [11]. Nevertheless, when subjected to intense heavy metal pressure, the biomass of hybrid Pennisetum diminishes, resulting in decelerated development and diminished effectiveness in remediation [36]. PGPB significantly affect soil remediation by secreting metabolites such as IAA, siderophores, enzymes, and organic acids. These substances promote plant growth and increase the ability of plants to accumulate heavy metals [37]. *Bacillus* is an important PGPB. Shah [38] reported that treating tomatoes with *Bacillus* can improve seed germination rates, seedling vigor, and several growth characteristics, including plant height, root length, and fresh weight. In this study, *Brevibacillus* sp. SR-9 was inoculated into soil with a Cd concentration of 20 mg/kg, and a pot cultivation experiment was conducted with hybrid Pennisetum. The findings demonstrate that the introduction of *Brevibacillus* sp. SR-9 resulted in a 12.08% increase in the aboveground biomass of hybrid Pennisetum, thus facilitating its development. Studies such as showing that *Pseudomonas aeruginosa* CPSB1 promotes wheat growth under Cd stress [37] and showing that SaMR12 increases Cd accumulation in *Sedum alfredii* report similar findings to those of this experiment [39]. This result may be related to the IAA and siderophores secreted by *Brevibacillus* sp. SR-9. For example, siderophores produced by *Streptomyces tendae* significantly increase the growth of sunflowers (*Helianthus annuus*) and their uptake of Fe and Cd [40]. Ma et al. [41] reported that siderophores produced by *Bacillus pumilus* E2S2, which was isolated from *Sedum plumbizincicola*, increased the growth of the host plant. Chen et al. [42] demonstrated that under Cd treatment, inoculation with SaMR12 enhanced the tolerance of *Sedum alfredii* to Cd, mitigated Cd-induced oxidative stress, and thus promoted plant growth and increased Cd absorption. These conclusions align with the outcomes of this investigation. The concentrations of Cd in the aboveground portions and roots increased by 21.16% and 15.50%, respectively. The translocation factor increased by 19.94% compared with that of the noninoculated group. These findings indicate that inoculation with *Brevibacillus* sp. SR-9 promotes the proliferation of hybrid Pennisetum, effectively alleviates Cd stress, and enhances plant remediation efficiency. According to Wu et al. [13], the presence of *Bacillus* BM18-2 resulted in an 83.1% increase in the biomass of hybrid Pennisetum and a 28.6% increase in Cd accumulation. Moreira et al. [43] reported that inoculation with *Ralstonia eutropha* and *Chryseobacterium humi* increased Cd accumulation in maize (*Zea mays*) roots by 18.6%. Moreover, heavy metal accumulation in soil leads to alterations in its physicochemical characteristics, which in turn affect the forms of heavy metals. These changes in metal forms determine their mobility and bioavailability [44]. Sofia Houida et al. [45] reported that introducing PGPB into soil had a substantial positive effect on plant development and resulted in increased levels of nitrogen, phosphate, and potassium. These findings suggested that these bacteria increase the availability of nutrients in the soil. Similarly, this study revealed that the introduction of *Brevibacillus* sp. SR-9 resulted in a 6.25% increase in the overall nitrogen content of the soil, a 3.46% increase in its overall phosphorus content, a 12.00% increase in its available phosphorus content, and a 2.49% increase in its available potassium content. These findings suggest that the *Brevibacillus* sp. SR-9 strain can increase plant nutrient uptake by secreting beneficial substances or converting less accessible soil components into forms that plants can absorb, thereby increasing the utilization of diverse nutrients [46,47].

The introduction of PGPB into the rhizosphere has dual effects: it directly influences plants and has an impact on the microbial population in the soil around roots [48]. The introduction of PGPB by inoculation can increase plant growth and enhance the effectiveness of remediation by changing the bacterial composition in the rhizosphere [49]. α diversity is a key indicator reflecting the diversity and abundance of bacterial populations [50]. In this study, the richness of rhizosphere soil bacteria decreased after inoculation with PGPB. This outcome aligns with the findings of Pongsilp [51], possibly because inoculation with plant growth-promoting microorganisms lowered the soil pH and increased the total nitrogen content [3]. This study employed partial least squares discriminant analysis (PLS-DA) at the operational taxonomic unit (OTU) level and revealed that the introduction of *Brevibacillus* sp. SR-9 affected the structure of the soil microbial community. This result is similar to those of Pongsilp et al. [51] and may be due to the properties of *Brevibacillus* sp. SR-9, such as the ability to secrete IAA, siderophores, and phosphate solubilizers, along with its heavy metal resistance, which allows for it to effectively colonize the rhizosphere and impact the composition of microorganisms in the rhizosphere of hybrid Pennisetum. It is also possible that soil microorganisms compete with exogenous PGPB for nutrients and ecological niches, leading to changes in the bacterial community structure. Habibollahi et al. [52] reported that the dominant phyla within soils polluted with high concentrations of metallic elements included Proteobacteria, Actinobacteria, Acidobacteria, and Bacteroidetes. Similarly, our high-throughput sequencing results revealed that at the phylum level, the predominant bacterial phyla in the rhizosphere soil of hybrid Pennisetum were Actinobacteria, Proteobacteria, and Acidobacteria. Notably, after SR-9 inoculation, the relative abundance of Proteobacteria significantly increased. Proteobacteria, a dominant bacterial phylum, includes groups such as the chemolithotrophic Acidithiobacillus species involved in the soil iron and sulfur cycles. These groups can survive under heavy metal contamination and enhance Cd chelation and precipitation through processes such as iron oxidation, thereby reducing the movement of heavy metals in soil and reducing the harmful effects of Cd on plants [53]. Additionally, Proteobacteria are closely linked to carbon and nitrogen cycles, which increase soil fertility in the plant rhizosphere and provide plants with essential nutrients. Acidobacteria are associated mainly with soil metal speciation and mineral weathering. This group, which is often acidophilic, can lead to soil acidification and aggregation issues, reflecting its role in altering soil conditions [54]. In this study, the reduction in Acidobacteria by 0.3% after *Brevibacillus* sp. SR-9 inoculation suggested that the strain may lower the bioavailability of Cd in the soil. Indeed, Acidobacteria may serve as an important indicator of Cd bioavailability in Cd-contaminated soils of hybrid Pennisetum. Studies have also confirmed that Actinobacteria, among other phyla, are typically among the most abundant phyla in soil and are also known to be resistant to heavy metals [55]. At the genus level, the most prevalent genera were *Sphingomonas*, *RB41*, *Bacillus*, and *Gaiella*. Chen et al. reported that *Sphingomonas* SaMR12 significantly increased the uptake of Cd by *Sedum alfredii* [56]. Additionally, *Bacillus* spp. produce various antibiotics, including bacillomycin, bacilysin, and mycosubtilin, which inhibit pathogens and are well documented for their heavy metal resistance, typically present in soils that are polluted with various heavy metals [57,58]. Co-occurrence network analysis of microbial community members reveals the interactions between different taxa, elucidates network characteristics, and identifies key taxa that play significant roles in community interactions and organization. This approach has been widely used to explore complex interaction networks and ecological rules in microbial communities across various environments [59,60]. In a co-occurrence network, the clustering coefficient measures the level of clustering within a network's nodes, reflecting how nodes tend to form clusters within the network. The average path length is the count of edges in the shortest path connecting any two nodes in the network, indicating the level of distance between nodes [61]. Modularity measures the degree of modularity within a network, indicating that species within the same module interact more frequently with each other than with species in different modules [62,63]. This study revealed that although the modularity index decreased after treatment, the mean network distance decreased, and the clustering coefficient increased, suggesting that bacterial injection improved the diversity of the soil microbial community. Therefore, the arrangement of bacterial communities in rhizosphere soil is characterized by distinct modules rather than being haphazard [64]. Soil physicochemical conditions significantly influence microbial community composition, with the adaptability of microorganisms to these factors determining their community structure [65]. This study revealed that pH, total nitrogen, available potassium, and available phosphorus were important factors influencing the differences in the bacterial community composition in the soil of *Leymus chinensis*. The Actinobacteria phylum and *Bacillus* genus are related to soil physicochemical properties. PGPB stimulate the activation and increase the availability of soil mineral nutrients, thus facilitating plant development.

Root rhizosphere metabolomics is a recent high-throughput omics screening technique focused on rhizosphere metabolites. This study explored the relationships between rhizosphere metabolites and crop physiology–soil nutrient changes, revealing the physiological and ecological functions within the system. This approach has been widely applied in microbiological research and other fields [66]. Rhizosphere soil metabolites are a range of organic and inorganic compounds, as well as signaling molecules, that are secreted by both plant roots and soil microorganisms during plant–soil interactions. These metabolites include primary and secondary metabolites derived from various photosynthetic products [67]. LC–MS analysis was conducted to investigate the differential changes in rhizosphere soil metabolites of *Artemisia argyi* following treatment with *Brevibacillus* sp. SR-9. OPLS-DA revealed 59 differentially abundant metabolites, with 14 showing increased abundance and 45 showing decreased abundance. These results suggest that the introduction of the PGPB *Brevibacillus* sp. SR-9 affected the composition of metabolites in the rhizosphere soil, which led to changes in microbial activity. This, in turn, alleviated Cd stress and promoted the growth of *Artemisia argyi*. This outcome aligns with the discoveries made by Zuluaga et al. [26]. According to Wang et al. [68], Cd treatment has a considerable effect on the expression levels of many metabolites associated with amino acid, lipid, carbohydrate, and nucleotide metabolism. These changes are attributed to the plant's response to Cd stress, in which it produces and regulates key metabolites such as carbohydrates, amino acids, and lipids to maintain cellular homeostasis. Similarly, in this study, after *Brevibacillus* sp. SR-9 treatment, the metabolites whose abundance increased included mainly lipids, amino acids, carbohydrates, organic acids, flavonoids, and other metabolites. Among these, carbohydrates serve as mediators of the mutualistic association between crops and PGPB, playing a role in the regulatory processes of PGPBin plants [69]. Flavonoids derived from plants have the ability to impede free radical reactions and diminish oxidative stress [70]. Lipid compounds may delay leaf senescence in plants [71]. Tan et al. [72] reported that in the presence of Cd stress, amino acids and their derivatives have the ability to create compounds with heavy metal ions by utilizing their carboxyl and amino groups. This process ultimately enhances the ability of plants to withstand the harmful effects of heavy metals. In addition, organic acids can modify the chemical structure of heavy metals, hence decreasing their level of toxicity [73]. The KEGG enrichment analysis identified many pathways that showed substantial enrichment in the differentially abundant metabolites. These pathways include the folate resistance pathway, the ABC transporter pathway, d-glutamine and d-glutamic acid metabolism, purine metabolism, and pyrimidine metabolism. Plants rely on purine and pyrimidine nucleotides for a wide range of metabolic reactions. They play crucial roles in nucleic acid production, function as energy sources, and serve as building blocks for the production of fundamental substances such as sucrose, polysaccharides, and phospholipids, as well as secondary compounds [74]. Hence, the process of creating and breaking down nucleotides is vital for the growth and development of plants. Research has confirmed that ABC transporters play essential roles in plant heavy metal detoxification. For example, ABC transporters can increase the resistance of black wheat by transporting PC-Cd or GSH-Cd from the cytoplasm to the vacuole, thereby facilitating detoxification. Additionally, ABC transporters can export intracellular Cd across membranes to the extracellular space, thus increasing plant tolerance to Cd [75]. The study also revealed that differentially abundant metabolites such as alkaloids, organic acids, lipids, and lipid-like molecules were positively correlated with the available potassium, available phosphorus, total phosphorus, and total nitrogen. This may be because the root-secreted metabolites affect the soil nutrient composition, which in turn impacts plant growth. The composition and concentration of rhizosphere metabolites can alter the structure of the rhizosphere microbial community. Changes in microbial community structure imply shifts in microbial community functions, which can have an impact on the development of plants. These findings indicate that after immunization with *Brevibacillus* sp. SR-9, soil metabolites not only directly or indirectly provide nutrients to hybrid Pennisetum but also indirectly govern the community and organization of microorganisms in the rhizosphere. This regulation, in turn, affects hybrid Pennisetum growth and ultimately alleviates Cd stress. Soil metabolites serve as a reflection of changes in soil microbial communities, as alterations at the organismal and enzymatic levels manifest as modified metabolite profiles [76]. Under external pressures, plants drive the adjustment of microbial distribution to ensure optimal metabolic activity [77]. Therefore, metabolic communication is a crucial mechanism for plant–microbe interactions [68].

5. Conclusions

This work revealed that inoculating hybrid Pennisetum with the PGPB *Brevibacillus* sp. SR-9 greatly altered the bacterial composition and absorption of Cd and reduced the content of Cd in the soil. The contents of total nitrogen, total phosphorus, available phosphorus, and available potassium in the soil increased, alleviating Cd stress and thereby promoting the formation and proliferation of hybrid Pennisetum biomass to a certain degree. Moreover, the *Brevibacillus* sp. SR-9 strain altered the composition of the bacterial population in the root soil under Cd stress, affecting the proportions of Actinobacteria, Proteobacteria, Acidobacteria, Planctomycetes, and *Bacilli* and increasing the relative abundance of bacterial genera in the root soil of hybrid Pennisetum. Furthermore, following the introduction of the *Brevibacillus* sp. SR-9 strain, there was a noticeable alteration in the proportion of metabolites present in the soil samples. This shift had an impact on many metabolic pathways, including the folate resistance pathway, the ABC transporter pathway, d-glutamine and d-glutamate metabolism, purine metabolism, and pyrimidine metabolism. This, in turn, increased the ability of hybrid Pennisetum to respond to Cd stress. This study deepens our understanding of the biochemical processes occurring in the rhizosphere and presents novel opportunities for the using microorganisms to assist plants in restoring soils polluted with heavy metals.

Author Contributions: Z.-J.C. and X.-M.R. designed the experiments; Z.-J.C., X.-M.R. and S.-S.G. participated in writing the paper; S.-S.G., Y.-J.Z., Y.S. and H.L. performed the experiments and analyzed the data; Z.-J.C., X.-M.R., B.L.L. and Y.-Y.L. reviewed the manuscript. All authors have read and agreed to the published version of the manuscript.

Funding: This research was funded by the National Natural Science Foundation of China (Grant No. U2004145), the Program for Science & Technology Innovation Talents in the Universities of Henan Province (No. 23HASTIT018), the Key Research and Development Projects of Henan Province (Grant Nos. 221111520600 and 231111113000), and the Key Scientific and Technological Project of Henan Province (Grant Nos. 242102521067 and 232102320252).

Data Availability Statement: The data that support the findings of this study are available upon request from the corresponding author.

Conflicts of Interest: The authors declare that they have no conflicts of interest.

References

1. Kim, J.J.; Kim, Y.S.; Kumar, V. Heavy metal toxicity: An update of chelating therapeutic strategies. *J. Trace Elem. Med. Biol.* **2019**, *54*, 226–231. [CrossRef] [PubMed]
2. Vardhan, K.H.; Kumar, P.S.; Panda, R.C. A review on heavy metal pollution, toxicity and remedial measures: Current trends and future perspectives. *J. Mol. Liq.* **2019**, *290*, 111197. [CrossRef]
3. Deng, J.; Yin, Y.; Zhu, W.; Zhou, Y. Variations in Soil Bacterial Community Diversity and Structures Among Different Revegetation Types in the Baishilazi Nature Reserve. *Front. Microbiol.* **2018**, *9*, 2874. [CrossRef] [PubMed]
4. Hu, H.; Jin, Q.; Kavan, P. A study of Heavy Metal Pollution in China: Current status, Pollution-Control Policies and Countermeasures. *Sustainability* **2014**, *6*, 5820–5838. [CrossRef]
5. Rai, P.K.; Lee, S.S.; Zhang, M.; Tsang, Y.F.; Kim, K.H. Heavy metals in food crops: Health risks, fate, mechanisms, and management. *Environ. Int.* **2019**, *125*, 365–385. [CrossRef]
6. Chen, D.; Chen, D.; Xue, R.; Long, J.; Lin, X.; Lin, Y.; Jia, L.; Zeng, R.; Song, Y. Effects of boron, silicon and their interactions on cadmium accumulation and toxicity in rice plants. *J. Hazard. Mater.* **2019**, *367*, 447–455. [CrossRef]
7. Qin, S.; Liu, H.; Nie, Z.; Rengel, Z.; Gao, W.; Li, C.; Zhao, P. Toxicity of cadmium and its competition with mineral nutrients for uptake by plants: A review. *Pedosphere* **2020**, *30*, 168–180. [CrossRef]
8. Huang, B.; Xin, J.; Dai, H.; Liu, A.; Zhou, W.; Yi, Y.; Liao, K. Root morphological responses of three hot pepper cultivars to Cd exposure and their correlations with Cd accumulation. *Environ. Sci. Pollut. Res. Int.* **2015**, *22*, 1151–1159. [CrossRef]
9. Genchi, G.; Sinicropi, M.S.; Lauria, G.; Carocci, A.; Catalano, A. The Effects of Cadmium Toxicity. *Int. J. Environ. Res. Public Health* **2020**, *17*, 3782. [CrossRef]
10. Jin, J.; Wang, C.; Liu, R.; Gong, J.; Wang, J.; Niu, X.; Zheng, R.; Tang, Z.; Malik, K.; Li, C. Soil microbial community compositions and metabolite profiles of *Achnatherum inebrians* affect phytoremediation potential in Cd contaminated soil. *J. Hazard. Mater.* **2023**, *459*, 132280. [CrossRef]
11. Xie, Y.; Bu, H.; Feng, Q.; Wassie, M.; Amee, M.; Jiang, Y.; Bi, Y.; Hu, L.; Chen, L. Identification of Cd-resistant microorganisms from heavy metal-contaminated soil and its potential in promoting the growth and Cd accumulation of bermudagrass. *Environ. Res.* **2021**, *200*, 111730. [CrossRef] [PubMed]
12. Zhang, X.; Xia, H.; Li, Z.; Zhuang, P.; Gao, B. Potential of four forage grasses in remediation of Cd and Zn contaminated soils. *Bioresour. Technol.* **2010**, *101*, 2063–2066. [CrossRef]
13. Wu, J.; Kamal, N.; Hao, H.; Qian, C.; Liu, Z.; Shao, Y.; Zhong, X.; Xu, B. Endophytic *Bacillus megaterium* BM18-2 mutated for cadmium accumulation and improving plant growth in Hybrid *pennisetum*. *Biotechnol. Rep.* **2019**, *24*, e00374. [CrossRef] [PubMed]
14. Jia, X.; Zhang, B.; Chen, W.; Feng, B.; Guo, P. Development of phytoremediator screening strategy and exploration of Pennisetum aided chromium phytoremediation mechanisms in soil. *Chemosphere* **2022**, *289*, 133160. [CrossRef]
15. Zhang, Y.; Zhao, S.Y.; Zhang, R.H.; Li, B.L.; Li, Y.Y.; Han, H.; Duan, P.F.; Chen, Z.J. Screening of plant growth-promoting rhizobacteria helps alleviate the joint toxicity of PVC+Cd pollution in sorghum plants. *Environ. Pollut.* **2024**, *355*, 124201. [CrossRef]
16. Glick, B.R. Plant growth-promoting bacteria: Mechanisms and applications. *Scientifica* **2012**, *2012*, 963401. [CrossRef]

17. Sarwar, N.; Imran, M.; Shaheen, M.R.; Ishaque, W.; Kamran, M.A.; Matloob, A.; Rehim, A.; Hussain, S. Phytoremediation strategies for soils contaminated with heavy metals: Modifications and future perspectives. *Chemosphere* **2017**, *171*, 710–721. [CrossRef] [PubMed]
18. Kong, Z.; Glick, B.R. The Role of Plant Growth-Promoting Bacteria in Metal Phytoremediation. *Adv. Microb. Physiol.* **2017**, *71*, 97–132.
19. Mohammadzadeh, A.; Tavakoli, M.; Motesharezadeh, B.; Chaichi, M.R. Effects of plant growth-promoting bacteria on the phytoremediation of cadmium-contaminated soil by sunflower. *Arch. Agron. Soil Sci.* **2017**, *63*, 807–816. [CrossRef]
20. Verma, S.; Kuila, A. Bioremediation of heavy metals by microbial process. *Environ. Technol. Innov.* **2019**, *14*, 100369. [CrossRef]
21. Teng, Y.; Wang, X.; Li, L.; Li, Z.; Luo, Y. Rhizobia and their bio-partners as novel drivers for functional remediation in contaminated soils. *Front. Plant Sci.* **2015**, *6*, 32. [CrossRef] [PubMed]
22. Gardner, T.; Acosta-Martinez, V.; Senwo, Z.; Dowd, S.E. Soil Rhizosphere Microbial Communities and Enzyme Activities under Organic Farming in Alabama. *Diversity* **2011**, *3*, 308–328. [CrossRef]
23. Roesti, D.; Gaur, R.; Johri, B.N.; Imfeld, G.; Sharma, S.; Kawaljeet, K.; Aragno, M. Plant growth stage, fertiliser management and bio-inoculation of arbuscular mycorrhizal fungi and plant growth promoting rhizobacteria affect the rhizobacterial community structure in rain-fed wheat fields. *Soil Biol. Biochem.* **2006**, *38*, 1111–1120. [CrossRef]
24. Kong, Z.; Wu, Z.; Glick, B.R.; He, S.; Huang, C.; Wu, L. Co-occurrence patterns of microbial communities affected by inoculants of plant growth-promoting bacteria during phytoremediation of heavy metal-contaminated soils. *Ecotoxicol. Environ. Saf.* **2019**, *183*, 109504. [CrossRef]
25. Korenblum, E.; Massalha, H.; Aharoni, A. Plant-microbe interactions in the rhizosphere via a circular metabolic economy. *Plant Cell* **2022**, *34*, 3168–3182. [CrossRef]
26. Alzate Zuluaga, M.Y.; Milani, K.M.L.; Miras-Moreno, M.B.; Lucini, L.; Valentinuzzi, F.; Mimmo, T.; Pii, Y.; Cesco, S.; Rodrigues, E.P.; de Oliveira, A.L.M. The adaptive metabolomic profile and functional activity of tomato rhizosphere are revealed upon PGPB inoculation under saline stress. *Environ. Exp. Bot.* **2021**, *189*, 104552. [CrossRef]
27. Dhawi, F.; Datta, R.; Ramakrishna, W. Mycorrhiza and PGPB modulate maize biomass, nutrient uptake and metabolic pathways in maize grown in mining-impacted soil. *Plant Physiol. Biochem. PPB* **2015**, *97*, 390–399. [CrossRef]
28. Juszczuk-Kubiak, E.; Dekowska, A.; Sokołowska, B.; Połaska, M.; Lendzion, K. Evaluation of the Spoilage-Related Bacterial Profiles of Vcuum-Packaged Chilled Ostrich Meat by Next-Generation DNA Sequencing Approach. *Processes* **2021**, *9*, 803. [CrossRef]
29. Vezzulli, L.; Stagnaro, L.; Grande, C.; Tassistro, G.; Canesi, L.; Pruzzo, C. Comparative 16SrDNA Gene-Based Microbiota Profiles of the Pacific Oyster (*Crassostrea gigas*) and the Mediterranean Mussel (*Mytilus galloprovincialis*) from a Shellfish Farm (ligurian sea, Italy). *Microb. Ecol.* **2018**, *75*, 495–504. [CrossRef]
30. Zhang, M.; Zeng, G.; Liang, D.; Xu, Y.; Li, Y.; Huang, X.; Ma, Y.; Wang, F.; Liao, C.; Tang, C.; et al. An Analysis of the Colony Structure of Prokaryotes in the Jialing River Waters in Chongqing. *Int. J. Environ. Res. Public Health* **2022**, *19*, 5525. [CrossRef]
31. Sheng, L.; Zhao, W.; Yang, X.; Mao, H.; Zhu, S. Response characteristics of rhizosphere microbial community and metabolites of *Iris tectorum* to Cr stress. *Ecotoxicol. Environ. Saf.* **2023**, *263*, 115218. [CrossRef] [PubMed]
32. Navarro-Reig, M.; Jaumot, J.; Piña, B.; Moyano, E.; Galceran, M.T.; Tauler, R. Metabolomic analysis of the effects of cadmium and copper treatment in *Oryza sativa* L. Using untargeted liquid chromatography coupled to high resolution mass spectrometry and all-ion fragmentation. *Metallomics* **2017**, *9*, 660–675. [CrossRef]
33. Ghate, T.; Soneji, K.; Barvkar, V.; Ramakrishnan, P.; Prusty, D.; Islam, S.R.; Manna, S.K.; Srivastava, A.K. Thiourea mediated Ros-metabolites reprogramming restores root system architecture under arsenic stress in rice. *J. Hazard. Mater.* **2022**, *435*, 129020. [CrossRef]
34. Liu, Y.Q.; Chen, Y.; Ren, X.M.; Li, Y.Y.; Zhang, Y.J.; Zhang, H.; Han, H.; Chen, Z.J. Plant growth-promoting bacteria modulate gene expression and induce antioxidant tolerance to alleviate synergistic toxicity from combined microplastic and Cd pollution in sorghum. *Ecotoxicol. Environ. Saf.* **2023**, *264*, 115439. [CrossRef] [PubMed]
35. Ju, W.; Liu, L.; Fang, L.; Cui, Y.; Duan, C.; Wu, H. Impact of co-inoculation with plant-growth-promoting rhizobacteria and rhizobium on the biochemical responses of alfalfa-soil system in copper contaminated soil. *Ecotoxicol. Environ. Saf.* **2019**, *167*, 218–226. [CrossRef] [PubMed]
36. Kamal, N.; Liu, Z.; Qian, C.; Wu, J.; Zhong, X. Improving hybrid *Pennisetum* growth and cadmium phytoremediation potential by using *Bacillus megaterium* BM18-2 spores as biofertilizer. *Microbiol. Res.* **2021**, *242*, 126594. [CrossRef]
37. Rizvi, A.; Khan, M.S. Biotoxic impact of heavy metals on growth, oxidative stress and morphological changes in root structure of wheat (*Triticum aestivum L.*) and stress alleviation by *Pseudomonas aeruginosa* strain CPSB1. *Chemosphere* **2017**, *185*, 942–952. [CrossRef]
38. Shah, R.; Amaresan, N.; Patel, P.; Jinal, H.N.; Krishnamurthy, R. Isolation and characterization of *Bacillus* spp. Endowed with multifarious plant growth-promoting traits and their potential effect on tomato (*Lycopersicon esculentum*) seedlings. *Arab. J. Sci. Eng.* **2020**, *45*, 4579–4587. [CrossRef]
39. Pan, F.; Meng, Q.; Wang, Q.; Luo, S.; Chen, B.; Khan, K.Y.; Yang, X.; Feng, Y. Endophytic bacterium *Sphingomonas* SaMR12 promotes cadmium accumulation by increasing glutathione biosynthesis in *Sedum alfredii* Hance. *Chemosphere* **2016**, *154*, 358–366. [CrossRef]

40. Dimkpa, C.O.; Merten, D.; Svatos, A.; Büchel, G.; Kothe, E. Siderophores mediate reduced and increased uptake of cadmium by *Streptomyces tendae* f4 and sunflower (*Helianthus annuus*), respectively. *J. Appl. Microbiol.* **2009**, *107*, 1687–1696. [CrossRef]
41. Ma, Y.; Oliveira, R.S.; Wu, L.; Luo, Y.; Rajkumar, M.; Rocha, I.; Freitas, H. Inoculation with Metal-Mobilizing Plant-Growth-Promoting Rhizobacterium *Bacillus* sp. SC_2b and Its Role in Rhizoremediation. *J. Toxicol. Environ. Health. Part A* **2015**, *78*, 931–944. [CrossRef] [PubMed]
42. Chen, B.; Shen, J.; Zhang, X.; Pan, F.; Yang, X.; Feng, Y. The Endophytic Bacterium, *Sphingomonas SaMR12*, Improves the Potential for Zinc Phytoremediation by Its Host, *Sedum alfredii*. *PLoS ONE* **2014**, *9*, e106826. [CrossRef]
43. Moreira, H.; Marques, A.P.; Franco, A.R.; Rangel, A.O.; Castro, P.M. Phytomanagement of Cd-contaminated soils using maize (*Zea mays L.*) assisted by plant growth-promoting rhizobacteria. *Environ. Sci. Pollut. Res. Int.* **2014**, *21*, 9742–9753. [CrossRef]
44. Marques, A.P.; Moreira, H.; Franco, A.R.; Rangel, A.O.; Castro, P.M. Inoculating *Helianthus annuus* (sunflower) grown in zinc and cadmium contaminated soils with plant growth promoting bacteria--Effects on phytoremediation strategies. *Chemosphere* **2013**, *92*, 74–83. [CrossRef]
45. Houida, S.; Yakkou, L.; Kaya, L.O.; Bilen, S.; Raouane, M.; El Harti, A.; Amghar, S. Plant growth-promoting bacteria isolated from earthworms enhance spinach growth and its phytoremediation potential in metal-contaminated soils. *Int. Microbiol.* **2024**, *27*, 545–558. [CrossRef]
46. Ju, W.; Liu, L.; Jin, X.; Duan, C.; Cui, Y.; Wang, J.; Ma, D.; Zhao, W.; Wang, Y.; Fang, L. Co-inoculation effect of plant-growth-promoting rhizobacteria and rhizobium on EDDS assisted phytoremediation of Cu contaminated soils. *Chemosphere* **2020**, *254*, 126724. [CrossRef]
47. Naqqash, T.; Hameed, S.; Imran, A.; Hanif, M.K.; Majeed, A.; van Elsas, J.D. Differential Response of Potato Toward Inoculation with Taxonomically Diverse Plant Growth Promoting Rhizobacteria. *Front. Plant Sci.* **2016**, *7*, 144. [CrossRef] [PubMed]
48. Wang, W.; Liu, A.; Zheng, X.; Chen, X.; Fu, W.; Wang, G.; Ji, J.; Pei, S.; Guan, C. Combined application of a PGPR strain *Methylophilus* sp. FP-6 and adenine to enhance the phytoremediation potential of tobacco: Oxybenzone remediation, plant response and rhizosphere bacterial community adaptation. *Plant Soil* **2023**, *482*, 313–332. [CrossRef]
49. Chen, L.; Hao, Z.; Li, K.; Sha, Y.; Wang, E.; Sui, X.; Mi, G.; Tian, C.; Chen, W. Effectsof growth-promoting rhizobacteria on maize growth and rhizosphere microbial community under conservation tillage in Northeast china. *Microb. Biotechnol.* **2021**, *14*, 535–550. [CrossRef]
50. Chao, A.; Yang, M.C.K. Stopping rules and estimation for recapture debugging with unequal failure rates. *Biometrika* **1993**, *80*, 193–201. [CrossRef]
51. Pongsilp, N.; Nimnoi, P. Inoculation of *Ensifer fredii* strain LP2/20 immobilized in agar results in growth promotion and alteration of bacterial community structure of Chinese kale planted soil. *Sci. Rep.* **2020**, *10*, 15857. [CrossRef] [PubMed]
52. Habibollahi, M.H.; Baghizadeh, A.; Sabokbar, A.; Sharafi, K. Isolation and characterization of copper and cadmium resistant bacteria from industrial wastewaters and evaluating the biosorption of selected bacteria. *Desalination Water Treat.* **2017**, *93*, 139–144. [CrossRef]
53. Wang, X.; Zhang, X.; Liu, X.; Huang, Z.; Niu, S.; Xu, T.; Zeng, J.; Li, H.; Wang, T.; Gao, Y.; et al. Physiological, biochemical and proteomic insight into integrated strategies of an endophytic bacterium *Burkholderia cenocepacia* strain YG-3 response to cadmium stress. *Metallomics* **2019**, *11*, 1252–1264. [CrossRef]
54. Kalam, S.; Basu, A.; Ahmad, I.; Sayyed, R.Z.; El-Enshasy, H.A.; Dailin, D.J.; Suriani, N.L. Recent Understanding of Soil Acidobacteria and Their Ecological Significance: A Critical Review. *Front. Microbiol.* **2020**, *11*, 580024. [CrossRef]
55. Alotaibi, B.S.; Khan, M.; Shamim, S. Unraveling the Underlying Heavy Metal Detoxification Mechanisms of *Bacillus* Species. *Microorganisms* **2021**, *9*, 1628. [CrossRef] [PubMed]
56. Chen, B.; Zhang, Y.; Rafiq, M.T.; Khan, K.Y.; Pan, F.; Yang, X.; Feng, Y. Improvement of cadmium uptake and accumulation in *Sedum alfredii* by endophytic bacteria *Sphingomonas* SaMR12: Effects on plant growth and root exudates. *Chemosphere* **2014**, *117*, 367–373. [CrossRef] [PubMed]
57. Goswami, S.; Das, S. A Study on Cadmium Phytoremediation Potential of Indian Mustard, Brassica juncea. *Int. J. Phytoremediat.* **2015**, *17*, 583–588. [CrossRef]
58. Gupta, R.; Khan, F.; Alqahtani, F.M.; Hashem, M.; Ahmad, F. Plant Growth-Promoting Rhizobacteria (PGPR) Assisted Bioremediation of Heavy Metal Toxicity. *Appl. Biochem. Biotechnol.* **2024**, *196*, 2928–2956. [CrossRef]
59. Barberán, A.; Bates, S.T.; Casamayor, E.O.; Fierer, N. Using network analysis to explore co-occurrence patterns in soil microbial communities. *ISME J.* **2012**, *6*, 343–351. [CrossRef]
60. Qiu, L.; Zhang, Q.; Zhu, H.; Reich, P.B.; Banerjee, S.; van der Heijden, M.G.A.; Sadowsky, M.J.; Ishii, S.; Jia, X.; Shao, M.; et al. Erosion reduces soil microbial diversity, network complexity and multifunctionality. *ISME J.* **2021**, *15*, 2474–2489. [CrossRef]
61. Faust, K.; Raes, J. Microbial interactions: From networks to models. *Nat. Rev. Microbiol.* **2012**, *10*, 538–550. [CrossRef] [PubMed]
62. Bascompte, J.; Stouffer, D.B. The assembly and disassembly of ecological networks. *Philos. Trans. R. Soc. Lond. Ser. B Biol. Sci.* **2009**, *364*, 1781–1787. [CrossRef] [PubMed]
63. Freilich, M.A.; Wieters, E.; Broitman, B.R.; Marquet, P.A.; Navarrete, S.A. Species co-occurrence networks: Can they reveal trophic and non-trophic interactions in ecological communities? *Ecology* **2018**, *99*, 690–699. [CrossRef]
64. Liu, Y.Q.; Chen, Y.; Li, Y.Y.; Ding, C.Y.; Li, B.L.; Han, H.; Chen, Z.J. Plant growth-promoting bacteria improve the Cd phytoremediation efficiency of soils contaminated with PE-Cd complex pollution by influencing the rhizosphere microbiome of sorghum. *J. Hazard. Mater.* **2024**, *469*, 134085. [CrossRef]

65. Zheng, Q.; Hu, Y.; Zhang, S.; Noll, L.; Böckle, T.; Dietrich, M.; Herbold, C.W.; Eichorst, S.A.; Woebken, D.; Richter, A.; et al. Soil multifunctionality is affected by the soil environment and by microbial community composition and diversity. *Soil Biol. Biochem.* **2019**, *136*, 107521. [CrossRef]
66. Mhlongo, M.I.; Piater, L.A.; Madala, N.E.; Labuschagne, N.; Dubery, I.A. The chemistry of plant-microbe interactions in the rhizosphere and the potential for metabolomics to reveal signaling related to defense priming and induced systemic resistance. *Front. Plant Sci.* **2018**, *9*, 112. [CrossRef] [PubMed]
67. Bi, B.; Yuan, Y.; Zhang, H.; Wu, Z.; Wang, Y.; Han, F. Rhizosphere soil metabolites mediated microbial community changes of *Pinus sylvestris* var. *mongolica* across stand ages in the Mu Us Desert. *Appl. Soil Ecol.* **2022**, *169*, 104222. [CrossRef]
68. Wang, Y.; Ren, W.; Li, Y.; Xu, Y.; Teng, Y.; Christie, P.; Luo, Y. Nontargeted metabolomic analysis to unravel the impact of di (2-ethylhexyl) phthalate stress on root exudates of alfalfa (*Medicago sativa*). *Sci. Total Environ.* **2019**, *646*, 212–219. [CrossRef]
69. Fang, W.; St Leger, R.J. *Mrt*, a gene unique to fungi, encodes an oligosaccharide transporter and facilitates rhizosphere competency in *Metarhizium robertsii*. *Plant Physiol.* **2010**, *154*, 1549–1557. [CrossRef]
70. Tan, C.; Wang, Z.; Feng, X.; Pan, B.; Irfan, M.; Liu, C. Transcriptomic and metabolomics of flavonoid compounds in *Actinidia arguta* var. Arguta. *J. King Saud. Univ.—Sci.* **2021**, *33*, 101605. [CrossRef]
71. Ma, D.; Xu, B.; Feng, J.; Hu, H.; Tang, J.; Yin, G.; Xie, Y.; Wang, C. Dynamic metabolomics and transcriptomics analyses for characterization of phenolic compounds and their biosynthetic characteristics in wheat grain. *Front. Nutr.* **2022**, *9*, 844337. [CrossRef] [PubMed]
72. Tan, P.; Zeng, C.; Wan, C.; Liu, Z.; Dong, X.; Peng, J.; Lin, H.; Li, M.; Liu, Z.; Yan, M. Metabolic profiles of *Brassica juncea* roots in response to cadmium stress. *Metabolites* **2021**, *11*, 383. [CrossRef] [PubMed]
73. Ulbrich, T.C.; Rivas-Ubach, A.; Tiemann, L.K.; Friesen, M.L.; Evans, S.E. Plant root exudates and rhizosphere bacterial communities shift with neighbor context. *Soil. Biol. Biochem.* **2022**, *172*, 108753. [CrossRef]
74. Stasolla, C.; Katahira, R.; Thorpe, T.A.; Ashihara, H. Purine and pyrimidine nucleotide metabolism in higher plants. *J. Plant Physiol.* **2003**, *160*, 1271–1295. [CrossRef]
75. Cheng, Y.; Bao, Y.; Chen, X.; Yao, Q.; Wang, C.; Chai, S.; Zeng, J.; Fan, X.; Kang, H.; Sha, L.; et al. Different nitrogen forms differentially affect Cd uptake and accumulation in dwarf Polish wheat (*Triticum polonicum* L.) seedlings. *J. Hazard. Mater.* **2020**, *400*, 123209. [CrossRef]
76. Zhang, H.; Huang, M.; Zhang, W.; Gardea-Torresdey, J.L.; White, J.C.; Ji, R.; Zhao, L. Silver nanoparticles alter soil microbial community compositions and metabolite profiles in unplanted and cucumber-planted soils. *Environ. Sci. Technol.* **2020**, *54*, 3334–3342. [CrossRef]
77. Korenblum, E.; Dong, Y.; Szymanski, J.; Panda, S.; Jozwiak, A.; Massalha, H.; Meir, S.; Rogachev, I.; Aharoni, A. Rhizosphere microbiome mediates systemic root metabolite exudation by root-to-root signaling. *Proc. Natl. Acad. Sci. USA* **2020**, *117*, 3874–3883. [CrossRef]

agronomy

MDPI

Article

Responses of Methane Emission and Bacterial Community to Fertilizer Reduction Plus Organic Materials over the Course of an 85-Day Leaching Experiment

Jiakai Gao [1,2,*], Zhenyi Ma [1], Ling Liu [1], Zhaoyong Shi [1] and Jialong Lv [3]

[1] College of Agriculture, Henan University of Science and Technology, Luoyang 471023, China; mzymzy12345@126.com (Z.M.); liulinghenan@126.com (L.L.); shizy1116@126.com (Z.S.)

[2] State Key Laboratory Soil Eros & Dryland Farming Loess Pl, Institute Soil & Water Conservat, Northwest A&F University, Xianyang 712100, China

[3] College of Natural Resources and Environment, Northwest A&F University, Xianyang 712100, China; ljlll@nwsuaf.edu.cn

* Correspondence: jkgao2016@163.com

Abstract: Methane produced from paddy fields has a negative impact on global climate change. However, the role of soil bacterial community composition in mediating methane (CH_4) emission from waterlogged paddy soil using the column experiment is poorly known. In the present study, various fertilization treatments were adopted to investigate the effects of fertilizer reduction combined with organic materials (CK: control; CF: conventional fertilization; RF: 20% fertilizer reduction; RFWS: RF plus wheat straw amendment; RFRS: RF plus rapeseed shell amendment; RFAS: RF plus astragalus smicus amendment) on CH_4 emission and soil bacterial community during an 85-day leaching experiment. We hypothesized that the fertilizer reduction plus the organic materials could enrich the bacterial communities and increase CH_4 emission. The average CH_4 flux varied from 0.03 μg m^{-2} h^{-1} to 76.19 μg m^{-2} h^{-1} among all treatments in the nine sampling times, which may account for the experimental conditions such as air temperature, moisture, and anthropogenic factors. In addition, high-throughput sequencing was utilized to investigate the alteration of the soil bacterial community structure. It was revealed that the diversity and composition of the bacterial community in the topsoil amended with organic materials underwent significant shifts after the 85-day leaching experiment. *Proteobacteria* was identified as the dominant phylum of the soil bacteria, with an average proportion of 35.2%. For *Firmicutes*, the proportion of RFRS (11%) was higher than that in the CK (8%), RF (8%), RFWS (7%), RFAS (6%), and CF (5%) treatments. Additionally, *Gammaproteobacteria* and *Alphaproteobateria* were supposed to be the major class bacterial communities, with average proportions of 12.8% and 12.2%, respectively. For the RFWS treatment, the contribution of *Alphaproteobateria* was the highest among all the bacterial relative abundance. According to the correlation heatmap analysis, the top ten bacterial communities were positively related to soil microbial biomass carbon (MBC) and ammonia nitrogen (NH_4^+-N) ($p < 0.01$). The findings also indicated that the RFRS treatment was the favorable management to alleviate CH_4 emission during an 85-day leaching experiment or possibly in paddy production. Collectively, these results predict that the impacts of different treatments on CH_4 production are strongly driven by soil microbial communities and soil properties, with soil bacteria being more prone to the crop residue degradation stage and more sensitive to soil properties. The discoveries presented in this study will be useful for assessing the efficacy and mechanisms of organic material amendments on CH_4 emissions in paddy soil.

Keywords: fertilizer reduction; organic materials; methane; bacterial community; leaching

1. Introduction

Rice (*Oryza saliva*) is one of the most important grain crops in the world and is considered a staple food by 40% of the world's population. Meanwhile, previous investigations reported that the demand for rice will be expected to increase by 24% over the next 20 years [1,2]. As is well known, CH_4 ranks second to carbon dioxide (CO_2) in terms of global warming potential, with a 25-fold higher warming potential over a 100-year time span [3]. CH_4 emissions in agricultural activities in China amounted to 22,245 Gg in 2014, of which 40.1% originated from paddy fields [4]. Hence, to mitigate climate change, alternative agricultural practices need to be developed to curtail CH_4 emissions from rice paddies. Excessive utilization of chemical fertilizers and intensive cultivation can result in the loss of soil organic carbon (SOC) and the decline of nitrogen (N) use efficiency in croplands [5]. In recent years, scientists have utilized a series of agricultural practices such as irrigation regime, straw mulching and fertilizer reduction in rice production to mitigate methane emissions [6–8]. Crop residue incorporation, which is a potential alternative to agricultural management, has great positive environmental benefits, such as increasing soil fertility and crop yield [9], and decreasing chemical fertilizer application and air pollution from crop residue burning [10]. Additionally, straw incorporation combined with reduced chemical fertilizer can enhance N immobilization and decrease the soil oxygen and OC substrate sources for methanogens, resulting in an increase in CH_4 emission in rice paddies [11–13]. Previous observations indicated that the impact of straw return on soil microbial communities may also be attributed to differences in straw application rates, durations, fertilizer management, and other factors [14,15]. The chemical structure of different crop residues changes during decomposition, which in turn affects the decomposability of the remaining residues [16]. Therefore, the origins, types, and amounts of organic materials can play various roles in regulating soil characteristics and bacterial communities, thereby resulting in the difference in CH_4 emission rates in paddy soil. Consequently, further analysis is needed to understand how fertilizer reduction combined with organic materials on CH_4 emission and soil bacterial effects in long-term leaching experiments.

Generally, CH_4 emission from paddy soil involves three components: production, oxidation and transportation, which are dominated by soil microbial processes, including its production in flooded soil anaerobic conditions by *methanogens* and consumption (oxidation) in aerobic microsites by *methanotrophs* [17,18]. Previous studies also indicated *methanogens* are distributed among the seven orders of *Euryarchaeota*: *Methanosarcinales*, *Methanobacteriales*, *Methanomicrobiales*, *Methanococcales*, *Methanopyrales*, *Methanocellales* and *Methanoplasmatales* [19]. Therefore, key factors such as soil C/N and oxygen concentration, which regulate the ratio of *methanogens* to *methanotrophs* will ultimately determine the CH_4 emissions typically passing through a particular pathway from production in aerobic soil layers into the atmosphere [20]. Investigations indicated that CH_4 is produced by the decomposition of organic carbon in anaerobic soils or sediments [21]. The areas of bulk density, specific surface area and microhabitat of microbial organisms may be improved when straw returned to farmlands, resulting in the variations of CH_4 emissions fluxes. Straw incorporation with reduced chemical fertilizers also regulated the *methanogenic* and *methanotrophic* community structure. Previous investigations showed that straw retention and chemical fertilizer reduction changed *methanogenic* and *methanotrophic* community structure, meanwhile, increasing the relative abundance of *Methyloaracoccus* (a dominant genus of methanotrophs) from 0.02% to 61.8% [8,22]. Moreover, straw incorporation might accelerate the propagation of *methanotrophs* and improve CH_4 oxidation potential by influencing soil aeration and soil microorganism habitat [23]. Researchers also revealed that various carbon sources from organic fertilizer application might regulate CH_4 production by affecting the abundance of *Methanosarcinales* and *Methanobacteriales* [24,25]. Hereof, the effect of soil bacterial community on CH_4 production should be clarified in waterlogged paddy soil with fertilizer reduction plus organic material amendment.

However, the influences of diverse organic materials on soil bacterial communities and CH_4 emissions via soil column leaching simulation tests have received scant coverage. Herein, we postulated that the changes in soil physicochemical conditions (e.g., soil nutrients, aeration, and texture conditions) resulting from straw amendments might mediate the methane oxidation potential and community structure of methanotrophs in paddy field ecosystems. A long-term column leaching experiment was performed under fertilizer reduction in combination with various organic matters. The main objectives of the present study were to address the following questions: (1) Do fertilizer regimes stimulate flux? (2) How do fertilizer regimes change soil (0–10 cm) bacterial community and structure? (3) What are the potential microbial drivers of CH_4 emission over the course of an 85-day leaching experiment? The results obtained in the presented study may provide a new cognition on the effects of optimizing fertilization regimes on CH_4 emissions mechanisms and management strategies in rice production.

2. Materials and Methods

2.1. Design of Soil Column Leaching Experiment

Soil used in this study was taken from Xinji County, Hanzhong City, Shaanxi Province (33°0′16″ N, 108°48′44″ E), with a type of loggotrophic paddy soil and a texture of clayey soil. The conditions of irrigation and drainage were controlled by 10~30 mm thin water layer on the surface in the greening period, and the soil moisture in the root layer accounted for 60~80% of the field water capacity in the other growth periods. Organic materials including WS (Wheat Straw), RS (Rapeseed Shell) and AS (Astragalus Smicus) were also sampled, and then air-dried and crushed for further use. The total C, N, and P proportions of WS, RS, and AS were measured by Element Analyzer (Unicube, Elementar, Hanau, Germany) in dry weight (Table S1). Moreover, the materials and functional groups analysis of WS, RS and AS were employed by flourier transform infrared spectrometer (FTIR). The organic materials were dried to a constant weight at 60 °C and subsequently crushed using a 100-mesh sieve. Take a small amount (about 1–2 mg) of the sample and grind it into a fine powder using an agate mortar and pestle. Weigh out about 100 mg of potassium bromide (KBr) powder, which is infrared transparent and grind this in a separate agate mortar and pestle, and finally subjected to scanning at the specified wavelength of 4000–400 cm^{-1}. A resolution between 4 cm^{-1} and 8 cm^{-1} is sufficient for most applications and the number of scans can vary from 16 to 128, depending on the noise level that can be tolerated and the time available. The main functional groups, such as oxhydryl, carbonyl, alkyl, and sulfonyl were generally identified by the absorbance peaks at 3200–3600 cm^{-1}, 1650–1820 cm^{-1}, 2850–2960 cm^{-1}, and 1000–1400 cm^{-1}, respectively [26].

Soil column leaching experiments were adopted to explore the environmental implication of organic materials amendment in waterlogged paddy soil, in the Yangling, Shaanxi Province (34°15′ N, 108°34′ E). Soils in columns were collected in 0–40 cm soil layers every 10 cm (0–10, 10–20, 20–30, and 30–40 cm, respectively), which were dominated by a cultivation system of wheat–rice regime. The physicochemical characteristics of soils in the present study are shown in Table 1. SOC contents were determined by the $K_2Cr_2O_7$ oxidation-$FeSO_4$ titration method. Soil samples were extracted by 1 M NH_4Ac four times, and then titrated with standard solution of hydrochloric acid (HCl), calculating the CEC value basied on dosage of HCl finally. Soil pH was determined with a pH meter (PB-10, Sartorius, Gottingen, Germany) in a soil:water solution (1:2.5, *w*:*v*). Total N was measured by an alkaline hydrolysis diffusion approach, available P was determined by a molybdenum blue colorimetric technique, and available K was quantified by flame photometry. The soil was extracted in 1 M NH_4Ac (pH 7.0), in accordance with the methods detailed in our previous study [27]. Additionally, the dissolved organic carbon (DOC) released from soil (0–10 cm) was determined by TOC analyzer (TOC-L CPH, Shimadzu, Kyoto, Japan) (Table S4) after leaching experiment. The excitation-emission matrices (EEMs) analysis of soil DOC samples was simultaneously processed by a fluorescent spectrometer (Lengguang Tech., F97, Shanghai, China), with emission (Em) wavelength 250–550 nm, excitation (Ex) wavelength 200–500 nm and increment of 5 nm, respectively. The bandwidths of Ex and Em were 10 nm, and the scanning speed was 6000 nm min^{-1}. Additionally, the fluorescent indices, including FI (the ratio of the fluorescence intensity at Em 450 nm and 500 nm, and at Ex 370 nm) [28], HIX (the ratio of the integral area over Em 435–480 nm to 300–345 nm, at Ex 254 nm) [29] and BIX (the ratio of fluorescence intensity at Em 380 nm and 430 nm, and at Ex 310 nm) [30] were also calculated in our work.

Table 1. Physicochemical properties of soils used vertically (0–40 cm) in this study.

Depth (cm)	pH	SOM (g kg^{-1})	AK (mg kg^{-1})	AP (mg kg^{-1})	CEC (cmol kg^{-1})	TN (g kg^{-1})
10	7.38 ± 0.21 b	32.22 ± 0.44 a	130.40 ± 1.21 a	20.12 ± 0.61 a	18.60 ± 0.22 a	1.21 ± 0.10 a
20	7.59 ± 0.11 ab	24.58 ± 0.51 b	59.10 ± 4.21 b	13.80 ± 1.81 b	20.18 ± 2.81 a	1.24 ± 0.17 a
30	7.63 ± 0.02 a	10.97 ± 1.09 c	39.61 ± 1.10 c	7.55 ± 0.37 c	19.87 ± 3.61 a	0.39 ± 0.05 b
40	7.72 ± 0.03 a	6.53 ± 0.41 d	38.23 ± 1.11 c	5.66 ± 0.32 c	18.31 ± 1.31 a	0.26 ± 0.03 b

Notes: SOM: Soil organic matter; AP: Available potassium; AP: Available phosphorus; CEC: cation exchange capacity; TN: Total nitrogen. Data (means ± SD, n = 3) followed by different letters indicate significant differences among soil depths at $p < 0.05$.

According to the bulk density of soil in the four layers, about 1000 g air-dried soil was filled in soil every 10 cm column (Figure 1B). Moreover, 0–10 cm soil layers were identified as different treatments (i.e., CK: control treatment without organic material and fertilizer; CF: conventional fertilization; RF: 20% fertilizer reduction compared with CF; RFWS: RF plus wheat straw amendment; RFRS: RF plus rapeseed shell amendment; RFAS: RF plus astragalus smicus amendment). Each treatment was repeated three times and eighteen soil columns were conducted in this study. The inputs of fertilizer application and organic materials were consistent with our previous study [31]. The amounts of WS, RS, and AS applications were 3.93 g, 2.36 g, and 2.36 g per column in topsoil (0–10 cm), which were converted by the amount of field fertilizer (Table S2). Approximately 1.5 L deionized water was used to stabilize column for three days and saturate soil before leaching experiment. The leaching experiment begins when the soil water content was 70% of field water capacity (FWC), and the approximate 1 cm high overlying water maintained over the course of a 70-day leaching experiment. The leachate of the four depths was artificially collected by latex tubing (10 cm length) and plastic bottle (25 mL) under natural conditions. Additionally, the DOC, nitrate nitrogen (NO_3-N), NH_4^+-N, and total phosphorus (TP) concentrations have significant differences in different sampling depths. The surface water and leachate samples were collected after gas collection and the amounts of samples were approximately 15 mL for further detection every sampling time.

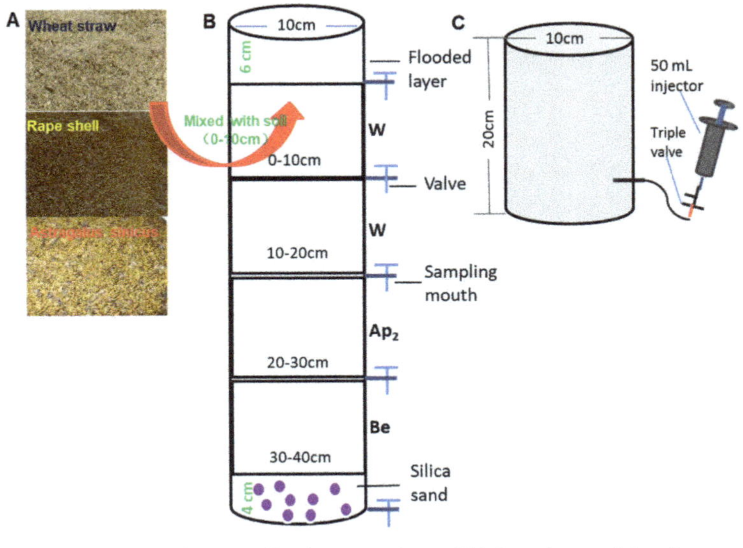

Figure 1. Schematic diagram of leaching experiment ((**A**) Organic materials: wheat straw, rapeseed shell, astragalus smicus; (**B**) Leaching column; (**C**) Column-shaped static chamber).

2.2. Collections and Detections of the CH_4 Fluxes

To facilitate the collection of gas samples, a cylindrical sampling chamber (20 cm × 10 cm) was utilized to sample methane gas on the 1st, 3rd, 6th, 10th, 15th, 25th, 40th, 60th, and 85th days (Figure 1C). Prior to gas sampling, the column was capped with the chamber, and the seam crossing was sealed with tape to avoid air leakage. A 50 mL syringe was employed to repeatedly mix the collected air for the zero-time sample. The other three gas samples were collected in the same manner every 10 min. Finally, the mixed gas samples were conserved in gas collection vessel for further detection. Gas chromatography (GC; Agilent 7890, Santa Clara, CA, USA), equipped with a flame ionization detector (FID) and and thermal conductivity detector (TCD), was employed to detect CH_4 concentration. The temperatures of the injection port, detector and oven were 250, 250 and 60 °C, respectively. Additionally, CH_4 fluxes were calculated by CH_4 accumulation from zero time to 1 h later air sampled. Beijing ZG Special Gases Science and Technology Corporation (Beijing, China) supplied the CH_4 standard gas with a concentration of 1.8 mg L^{-1}, used to calculate the integral area of gas volume fraction to sampled time. The GHG emission rates were calculated from the change in gas concentrations with time after the chamber was closed. The calculation of CH_4 fluxes followed the method described in the previous study [32]:

$$J = \frac{dc}{dt} * \frac{M}{V_0} * \frac{P}{P_0} * \frac{T_0}{T} * H$$

where J represents the CH_4 (μg m^{-2} h^{-1}) emission flux; dc/dt represents the slope of gas volume fraction to t (mm^3 (m^3 h)$^{-1}$); M is V is CH_4 molar mass; P is the atmospheric pressure of chamber (m^2); T is the absolute temperature in the chamber (K); V_0, P_0, T_0 are the molar volume of gas (mL mol^{-1}), atmospheric pressure (Pa), absolute temperature (K), respectively, at standard condition; H is the height of chamber (m).

2.3. Topsoil DNA Extraction and High throughput Sequencing

The polymerase of TransGen TransStart Fastpfu DNA Polymerase AP221-02 was operated to amplify bacterial 16S rRNA genes following the primers [338F (ACTCC-TACGGGAGGCAGCAG) and 806R (GGACTACHVGGGTWTCTAAT)] in V3-V4 hypervariable region [33]. QIA Quick Gel Extraction Kit (QIAGEN, Dusseldorf, Germany) and Real-Time PCR (ABI GeneAmp® 9700, New York, NY, USA) were used to purify and quantify PCR products, respectively. Additionally, High throughput sequencing was performed on Illumina's MiSeq-PE250 platform at the Allwegene Company (Beijing, China). The dataset was analyzed using QIIME and the sequences were clustered into operational taxonomic units (OTUs) at a similarity level of 97%. The Ribosomal Database Project (RDP) Classifier tool (NBTIC, Bethesda, ML, USA) was also employed to classify all the sequences into different taxonomic groups, which can calculate the abundances of OTUs and the parameters of diversities [34]. The raw sequences were deposited into the NCBI Sequence Read Archive (SRA) database with accession number PRJNA1065809.

2.4. Statistical Analysis

The original data were calculated by Microsoft Excel (Microsoft Excel 2019, Microsoft, Albuquerque, NM, USA). One-way analysis of variance (ANOVA) was regarded as being statistically significant in using SPSS 21.0 (SPSS Inc., Chicago, IL, USA). DUCAN test was used to determine significant differences among all treatments ($p < 0.05$). The correlations between various indices were assessed by Pearson's correlation analysis ($p < 0.05$, $p < 0.01$ and $p < 0.0001$). Beta diversity of bacterial samples in different treatments was analyzed by principal Component Analysis (PCA)and Principal Coordinate Analysis (PCoA) which were conducted by R language packages coupled with broken-stick model based on Bray–Curits distances at OTUs level (Supplementary Materials). All figures in this study were drawn by Origin Pro (Origin 2021, Northampton, MA, USA).

3. Results

3.1. Characteristics of Soil and Organic Materials Used in This Study

The physicochemical properties of the soils used vertically (0–40 cm) in the present study are shown in Table 1. The SOM content obviously decreased with an increase in soil depth, showing a range of 6.53 g kg^{-1} to 32.22 g kg^{-1} ($p < 0.05$) (Table 1). Similarly, diverse origins of organic materials led to significant variations in the properties of WS, RS, and AS, encompassing nutrient contents, phase compositions, and main functional groups. The total C, N, and P proportions of WS, RS, and AS are shown in Supplementary Table S1, which indicates the C/N order of WS (47.87) > RS (43.07) > AS (24.83). The three organic materials mainly consist of carbon components rather than nitrogen and phosphorus, which can provide more carbon sources for microbial activity and regulate CH_4 emissions.

As a whole, the order of absorption peak intensity was WS > RS > AS at the same wavenumber. FTIR characteristics of AS, RS, and WS were shown in Figure 2, which indicated a similar trend in the main absorption peaks of 858 cm^{-1}, 1300 cm^{-1}, 1539 cm^{-1}, 1715 cm^{-1}, and 3009 cm^{-1}. The absorption peak at 858 cm^{-1} pointed to the stretching vibration of -CHO in AS lower than that in WS and AS. The bending vibration O-H at 1300 cm^{-1} showed the compounds of phenols, alcohols or carboxylic acids. The bending vibration N-H at 1539 cm^{-1} illustrated the amine-group compound. The stretching vibration C=O at 1715 cm^{-1} forecasted the compounds of ketones, aldehydes or carboxylic acids. The stretching vibration C-H at 1715 cm^{-1} forecasted the alkene compound. The aromatic ketones and aromatic carboxylic acid compounds (benzophenone, benzoic) were considered the main substances in the three materials, which may provide nutrients for soil microorganisms during the decomposition of a long-term leaching experiment. Then, the activities and abundances of relative microorganisms can mediate the CH_4 emissions rates.

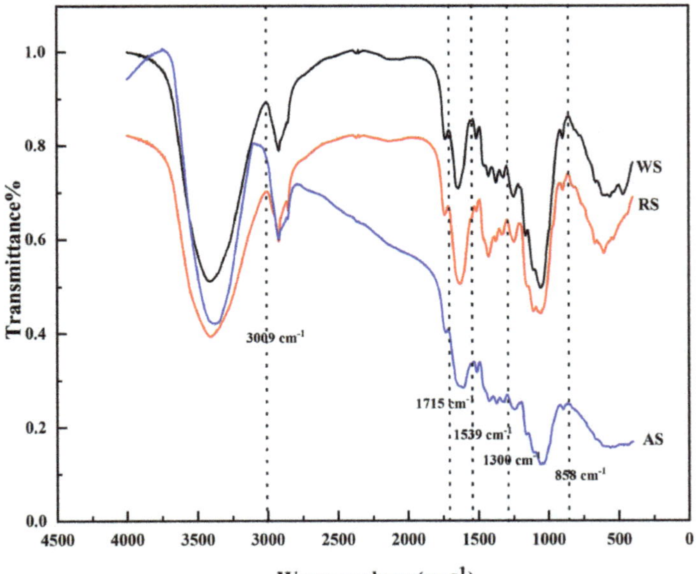

Figure 2. FTIR characteristics of the three organic materials (WS, RS and AS represent wheat straw, rapeseed shell and astragalus smicus, respectively) used in this study.

3.2. Changes of CH_4 Emissions over the Course of an 85-Day Leaching Experiment

The dynamic changes and the average emissions of the CH_4 fluxes are described in Figure 3. The emission of the CH_4 flux changed with a scope of 0.001 μg m^{-2} h^{-1}~320.51 μg m^{-2} h^{-1} and the average CH_4 emission flux varied from 0.03 μg m^{-2} h^{-1} to 76.19 μg m^{-2} h^{-1} during the leaching experiment in the waterlogged paddy soil. One-way ANOVA and the DUCAN test ($p < 0.05$) were used to analyze the difference in CH_4 emission fluxes among all treatments over the course of an 85-day leaching experiment. A significant ($p < 0.05$) difference in the CH_4 emission flux occurred in RFWS (320.51 μg m^{-2} h^{-1}) on 2 May, meanwhile, the peak of average CH_4 emission flux also occurred here (76.19 μg m^{-2} h^{-1}), which may be attributed to the abundant carbon released from WS during the leaching experiment. However, the CH_4 flux rate in the RFWS treatment (81.80 μg m^{-2} h^{-1}) was lower than that in the RFAS treatment (121.01 μg m^{-2} h^{-1}) on 2 July, which may account for the reduction of soil water or the destruction of anaerobic conditions.

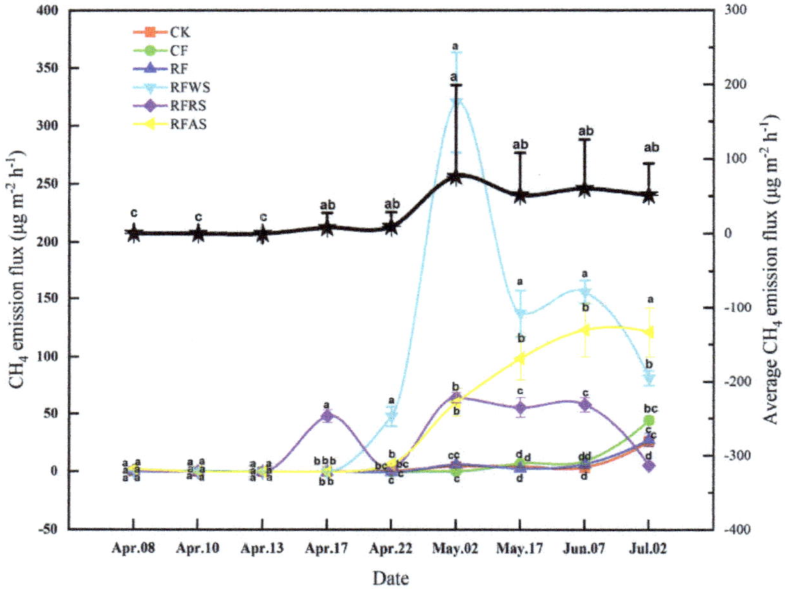

Figure 3. Dynamic changes of CH_4 flux in various treatments, and average CH4 emission flux over the course of an 85-day leaching experiment among all treatments (i.e., CK: control treatment without organic material and fertilizer; CF: conventional fertilization; RF:20% fertilizer reduction compared with CF; RFWS: RF plus wheat straw amendment; RFRS: RF plus rapeseed shell amendment; RFAS: RF plus astragalus smicus amendment). Data (means ± SD, n = 3) followed by different letters indicate significant differences among treatments at $p < 0.05$.

3.3. Variations of the Soil Bacterial Community Structure

To reveal the variations of the bacterial community and structure in the topsoil plus organic materials and fertilizer after a long-term leaching test, high throughput sequencing was employed to distinguish the variations of the soil bacterial structure (Figure 4). Chao, ACE and Shanon analyses showed that organic material amendments with fertilizer reduction enriched the soil bacterial community diversities, which attributed to the favorable habitat for soil organisms. The Chao index is directly related to species richness, for which it provides a higher estimate compatible with the sample abundance data, and the Shannon index is a species diversity metric. For alpha diversity indices of Chao, the values of bacterial diversities showed that the RFWS (3398.27) treatment was higher than CK (2940.77), CF (3174.30), RFRS (3036.43), and RFAS (3153.34) ($p < 0.05$). The ACE values displayed an order of RFWS (3413.34) > RF (3297.59) > CF (3165.81) > RFAS (3153.79) > RFRS (3051.42) > CK (2941.83). When the Shannon analysis was adopted, the diversity values were RFWS (10.04) > RFAS (10.01) > RF (9.939) > CF (9.81) > RFRS (9.77) > CK (9.70). However, no difference existed in Simpson analysis among all treatments (Table S3).

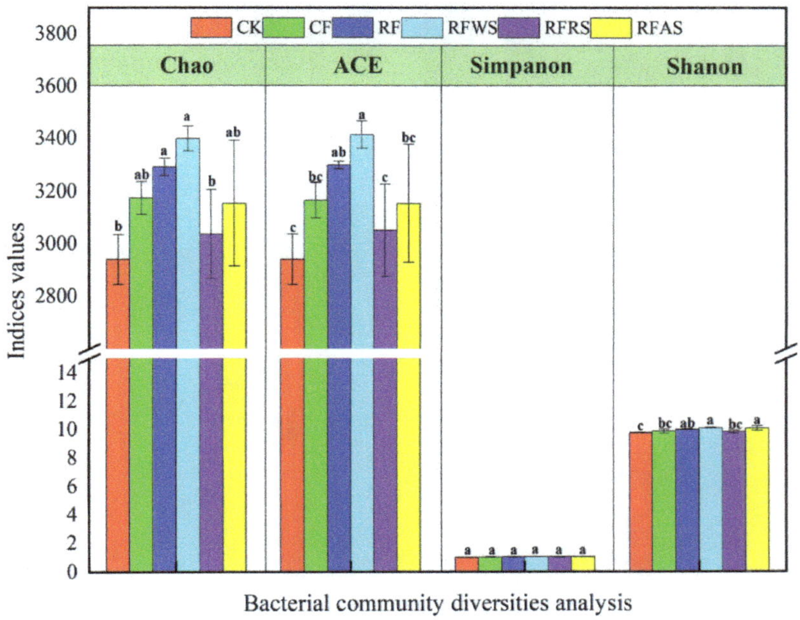

Figure 4. Analysis of bacterial community diversity in various treatments (i.e., CK: control treatment without organic material and fertilizer; CF: conventional fertilization; RF: 20% fertilizer reduction compared with CF; RFWS: RF plus wheat straw amendment; RFRS: RF plus rapeseed shell amendment; RFAS: RF plus astragalus smicus amendment). Data (means ± SD, $n = 3$) followed by different letters indicate significant differences among treatments at $p < 0.05$.

The top ten dominant bacterial communities were identified in Figure 5A,B at the phylum and class levels, respectively. In this study, *Proteobacteria*, *Actinobacteria*, and *Acidobacteria* were regarded as the three major bacterial communities at the phylum level, with average proportions of 35.2%, 12.3%, and 15.3%, respectively. Among them, the relative distribution of *Proteobacteria* was significantly higher than that in all treatments, but no difference existed among the different treatments. For *Firmicutes*, the proportion of RFRS (11%) was higher than that in the CK (8%), RF (8%), RFWS (7%), RFAS (6%), and CF (5%) treatments. In addition, *Gammaproteobacteria*, *Alphaproteobacteria*, *Deltaproteobacteria*, and *Acidoproteobacteria* were considered to be the main bacterial communities at the class level, with average contributions of 12.8%, 12.2%, 10.0%, and 8.8%, respectively. Compared to CK (11%), the distribution of *Gammaproteobacteria* in the CF, RF, RFWS, RFRS, and RFAS amendments increased to 14%, 12%, 12%, 12%, and 14%, respectively. The contribution of *Alphaproteobacteria* was the highest in the RFWS treatments among all bacterial relative abundances, accounting for 15%. The abundance of *Deltaproteobacteria* ranged from 9% to 11%, which was higher than the distribution of *Acidoproteobacteria* in all treatments, ranging from 7% to 11%.

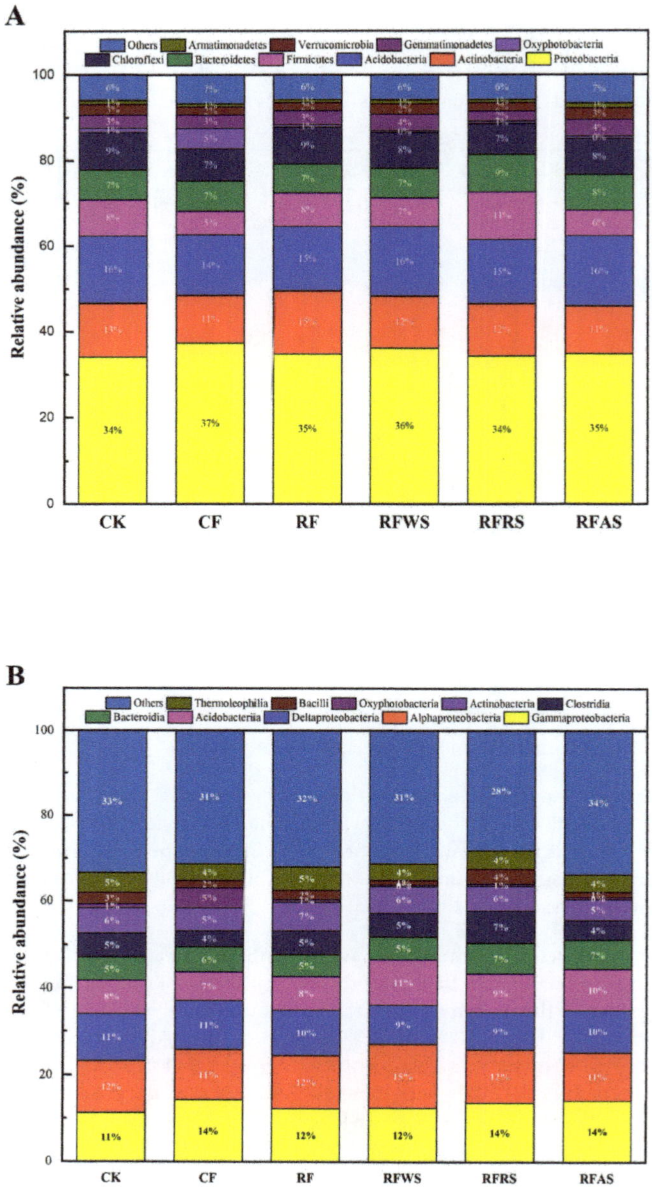

Figure 5. Top ten dominant bacterial communities in soil (0–10 cm) at phylum (**A**) and class (**B**) levels after 85-day leaching experiment in various treatments (i.e., CK: control treatment without organic material and fertilizer; CF: conventional fertilization; RF:20% fertilizer reduction compared with CF; RFWS: RF plus wheat straw amendment; RFRS: RF plus rapeseed shell amendment; RFAS: RF plus astragalus smicus amendment).

3.4. Correlation of the Bacterial Communities to Environmental Factors

To understand the effect of fertilizer reduction combined with organic materials on the structural composition of soil microorganisms, the differences in soil bacterial community structures were analyzed by PCoA (Figure 6A) and PCA (Figure 6B). The loadings for all variables involved in the PCA in different treatments are shown in the Supplementary Materials. PC1 and PC2 were obtained, with contributions of 14.7% on the first axis and 8.75% on the second axis in PCoA, which indicated a similar OTU composition between the RFWS and RFAS treatments. However, the CF and RFRS treatments showed similar variations in OUT compositions in PCoA, with the dispersion of data points in the four quadrants.

Conversely, the OTU compositions of the CK and RF treatments were similar, which demonstrated the fertilizer application rate was the main factor in bacterial community structures. Additionally, PC1 and PC2 were acquired by PCA, accounting for 11.8% and 9.67% of the differences in bacterial structure. The points of the RFWS, RFRS and RFAS treatments were located in the same quadrant above the main axis in PCA, which shows the semblable structural composition. However, compared to CF, the samples of the CK and RF treatments were located in the same quadrant below the main axis, which indicates the importance of the fertilizer reduction regime on soil microbial variations.

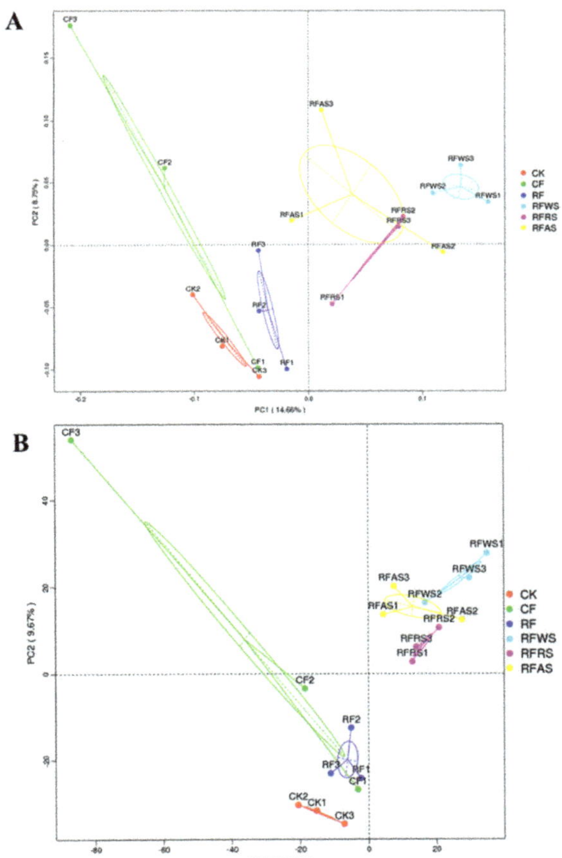

Figure 6. PCoA (**A**) and PCA (**B**) analysis of soil bacterial community structures (i.e., CK: control treatment without organic material and fertilizer; CF: conventional fertilization; RF:20% fertilizer reduction compared with CF; RFWS: RF plus wheat straw amendment; RFRS: RF plus rapeseed shell amendment; RFAS: RF plus astragalus smicus amendment). The chromatic dot represents the distribution of community composition of each treatment and the chromatic dotted box represents samples with similar microbial community composition.

To explore the relationship between bacterial communities and environmental factors (Table S4), the changes of the top ten dominant soil bacterial communities and soil relative indices are shown by heatmap analysis at the phylum and class levels (Figure 7A,B). The correlation of soil bacterial communities and soil indices was analyzed using Origin Pro (Origin 2021, USA) and a *p*-value at 0.05, 0.01 and 0.001 levels, respectively. Overall, the indices of FI (Fluorescence intensity), HIX (Humification index) and α:β (ratio of new and old carbon), which represented soil dissolved organic carbon (DOC) spectral characteristics, had a highly significant correlation ($p < 0.001$). At the phylum level, *Verrucomicrobia* had a significant association with *Armatimonadetes* ($p < 0.01$), and the contribution of the *Bacteroidetes* community was significantly correlated with soil NH_4^+ ($p < 0.01$), suggesting the crucial effect of soil characteristics on soil microbial structures. Regarding the class level, the relative abundance of *Bacteroidia* was strongly correlated with soil NH_4^+ ($p < 0.01$). Additionally, the relative distribution of *Actinobacteria* showed a significant relationship with soil DOC and microbial biomass nitrogen (MBN) ($p < 0.05$), and the Bacilli community was significantly related to soil AP concentration.

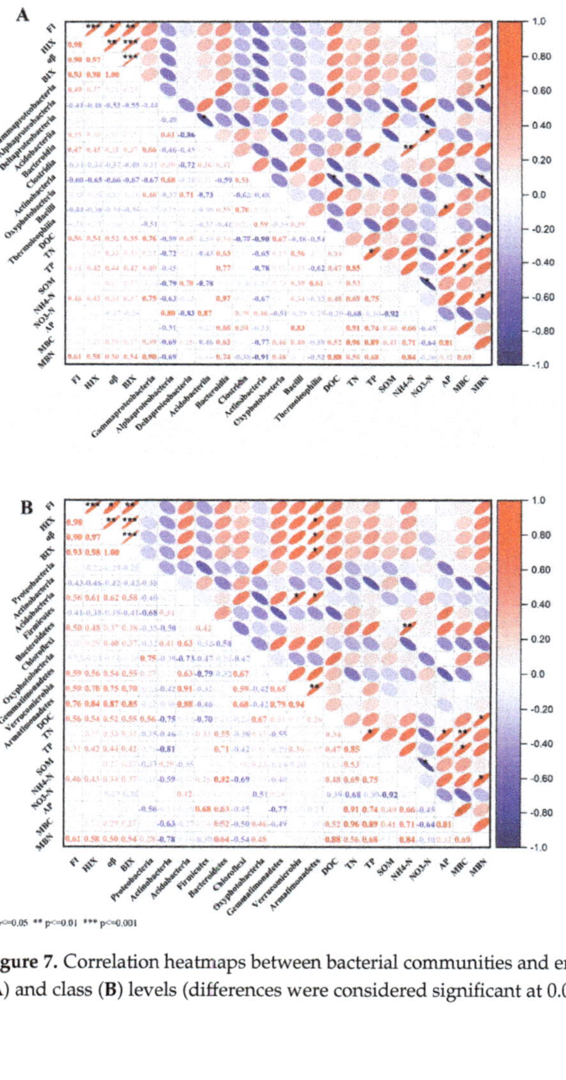

Figure 7. Correlation heatmaps between bacterial communities and environmental factors at phylum (**A**) and class (**B**) levels (differences were considered significant at 0.05, 0.01 and 0.001 levels).

4. Discussion

4.1. Fertilizer Reduction Plus Organic Materials Accelerate the CH_4 Emissions from the Flooded Paddy Soil

Based on the soil's basic properties, the indices of AK, AP, CEC, and TN exhibit a similar variation trend where the upper-layer content is higher than that in the lower layer, which is attributed to the application of organic or inorganic fertilizers during long-term cultivation. The organic materials (WS, RS, and AS) used in this study are rich in carbon (C) and nitrogen (N), providing ample energy sources for soil microorganisms involved in CH_4 production. Previous studies have also demonstrated that variations in soil depth can significantly influence the distribution and availability of essential nutrients [35]. The high C and N contents in the organic materials could potentially stimulate specific microbial metabolic pathways involved in CH_4 production, warranting further investigation into the underlying mechanisms. The FITR analysis showed that WS contained more plentiful C=C and C=O groups than in RS and AS, which is consistent with the results of elements analysis.

Generally, the contents of carbon (C) and nitrogen (N) have significant influences on microbial metabolic pathways and CH_4 production. However, the effects of reduced chemical fertilizer combined with crop residues on soil bacterial community composition over the course of long-term leaching experiments need to be explored deeply. Previous studies reported that reduced chemical fertilizer combined with straw retention can enhance CH_4 emission by regulating the abundance of soil methanogens and methane-oxidising bacteria, which is consistent with the present results [36,37]. The exogenous organic matter can be biodegraded under both aerobic and anaerobic conditions, especially in the drying and wetting cycling process [38]. Moreover, it is widely agreed that straw amendments supply additional organic C substrate for *methanogens* and increase plentiful CH_4 emissions [39,40]. On the other hand, fertilization improves photosynthetic production and therefore provides more rhizodeposition C for *methanogenesis*, as well as the oxidization of rhizosphere soil for *mathanotrophs* [41]. The relative strategy can also facilitate soil oxygen transportation and enhance its available concentration [42]. The current results also indicated that the CH_4 emissions flux and the abundance of bacterial diversity significantly rose for the RFWS, RFRS, and RFAS treatments; however, there was no significant difference in the emission flux among the CK, CF, and RF treatments, which is in line with the published studies [39,43]. This phenomenon could be mainly attributed to the 85-day soil column leaching experiment that incorporated fresh straw into the waterlogged soil when the temperature was relatively high, as the CH_4 emissions flux peaked shortly after fertilization for 28 days (Figure 3) [44,45]. This was also likely due to the fact that the application of fresh crop residues to paddy soil accelerates CH_4 emission by regulating soil pH or enhancing anaerobic oxidation processes as an electron acceptor [46]. Therefore, to the best of our knowledge, the RFRS treatment should be recommended as the moderate strategy that mitigates CH_4 emission in rice paddies.

4.2. Potential Mechanism of CH_4 Emission under Fertilizer Reduction Plus Organic Materials

As is known, microbial communities play a crucial role in the decomposition of organic matter. Different microorganisms have specific enzymes and metabolic pathways for breaking down complex carbon compounds. This process can release substrates that are subsequently utilized for CH_4 production. On the other hand, certain microorganisms such as *Methanotrophic* bacteria can in return anaerobically oxidize CH_4, influencing the net emissions. Understanding these complex interactions is crucial for developing sustainable agricultural practices that balance crop productivity and environmental protection. It is well-known that N fertilization has a positive influence on paddy soil microbial communities, particularly in a waterlogged environment [47,48]. Nevertheless, the research on the response of soil bacterial community composition to fertilizer reduction combined with organic materials during the leaching simulation period is lacking. In this study, the top ten bacterial community compositions were identified among various treatments (Figure 5A,B), indicating that reducing chemical fertilizer plus using organic matter amendments can significantly regulate the main bacterial communities of methanogenesis after a long-term leaching experiment. As the biodegradable organic compounds decompose, the living conditions of the related organisms may deteriorate, and the favorable habitat of the related microorganisms may be destroyed due to the lack of fresh substrates or favorable soil conditions. The abundant carbohydrates, including starch and glucose, which may be released from organic matter, provide an energy substrate for the metabolism of most microorganisms [49]. Additionally, the redox potential decreases with the consumption of oxygen during straw decomposition, which leads to the growth of *methanogens* and inhibits the activity of *methanotrophs* [50]. The increase in degradable C causing the increase in CH_4 emissions due to straw incorporation is associated with reduced fertilizer [51]. With the leaching experiment process, the soil C/N, pH and redox significantly changed because of the reduction of water and nutrient contents, which may directly affect the activity and abundance of *methanogens* and *methanotrophs.* Future research could focus on unraveling the specific molecular mechanisms underlying the observed changes in microbial communities and their functional implications for soil processes.

Previous studies showed that organic amendments could significantly increase the soil microbial activity and microbial biomass [24,52]. In this study, fertilizer reduction plus organic material treatments increased the soil microbial C substrate, which could supply more C and N sources for methanogenic growth. In general, *methanomicrobia* and *methanobacteria* were considered to be the main compositions of the methanogenic archaeal community, shifting the basal metabolic activities in paddy soils [53]. The results were consistent with a previous study on the functional and structural responses of methanogenic archaeal communities with rice straw addition, which demonstrated that the hydrogenotrophic methanogenic pathway was in particular controlled by the availability of degradable OC [54]. PCoA and PCA analyses showed that the RFWS, RFRS and RFAS treatments had similar bacterial communities and OTU structural composition; however, relatively lower bacterial abundances were found in the CK, CF and RF treatments, which suggests the important role of organic C and moderate N sources on microorganism growth [55]. If there is an excess of one without a corresponding amount of the other, it can lead to nutrient imbalances that affect microbial growth efficiency [43]. Additionally, the form in which these nutrients are present (e.g., whether N is in an ammonium or nitrate form) can also modulate the impact on microbial growth. The correlation heatmap revealed that the distributions of *Bacteroidetes* and *Bacteroidia* were dramatically close to soil NH_4^+ concentration, which was devoted to the degradation of crop residue or the growth of *Methanogens* [56]. Reporter indicated that hydrogenotrophic *Methanocella* and acetoclastic *Methanosaeta,* which can generate methane by using hydrogen as an electron donor and carbon dioxide as a carbon source, were considered the major methanogenic communities associated with CH_4 production and emission [57]. The major bacterial communities including *Bacteroidia, Acidobacteriia, Actinobacteria* and *Proteobacteri* obtained in this study were related to *Methanocella,* whose variations may mediate the CH_4 emission rate during the period of the leaching experiment. Moreover, factors such as the availability of hydrogen and carbon dioxide, as well as the overall environmental conditions (e.g., temperature, pH, and redox potential), can affect the growth and metabolic rate of the main bacterial communities relative to methanogenesis, thereby impacting CH_4 net emission. Altogether, our results showed that fertilizer reduction plus organic materials generated great changes in the CH_4 emission and the abundance of relative bacterial community composition in paddy soils during a long-term leaching experiment. However, the present experiment was isolated from field trials, and no plants participated in the leaching experiment process, which scarcely reflects the actual field conditions. Hence, further investigations of the fertilization regime's impact on the CH_4 emission molecular mechanism should be conducted in a long-term field test.

4.3. Implication

The effects of fertilizer reduction combined with various crop residues on the variations of CH_4 emission and bacterial communities were investigated via an 85-day soil column leaching experiment.

It was demonstrated that the RFRS treatment significantly decreased and the RFWS regime treatment significantly increased the CH_4 emission from paddy soil. Simultaneously, this was primarily associated with the proliferation of *Bacteroidetes* and *Bacteroidia*, which were capable of better degrading various organics or mediating the composition of microbial substrates in waterlogged anaerobic conditions. Nevertheless, the biochemical processes of *Methanomicrobia* and *Methanobacteria* genes, which regulate CH_4 production, remain ambiguous. In future studies, the potential influence of the redox process on altering the CH_4 emissions mechanism by modifying soil-dissolved organic matter or the electron transport process should also be considered [58,59]. To reduce errors in experimental data and sampling, an automatic collection device should be adopted instead of manual operation, which can improve the consistency of the results in deep study. The release of ebullition fluxes that occurred during the duration of gas sampling should also be considered in the follow-up study. Further studies are also necessary to elucidate how different microbial communities are formed with different types of organic material amendments. Moreover, the present study only observed a slight variation in CH_4 emission with fertilizer reduction plus organic materials during the leaching process; therefore, more long-term field experiments should be carried out to verify the changes more accurately.

5. Conclusions

In this study, the ability of organic materials to balance soil fertility and regulate CH_4 emissions in paddy soil was demonstrated. The results showed that the CH_4 flux ranged from 0.001 $\mu g\ m^{-2}\ h^{-1}$ to 320.51 $\mu g\ m^{-2}\ h^{-1}$ using various plus organic material treatments over the course of an 85-day leaching experiment. RFWS, RFRS and RFAS showed higher CH_4 flux than that of the CK, RF and CF treatments, which illustrated the importance of N stimulation on the degradation of organic C from crop residue during the leaching experiment. Additionally, *Bacteroidia*, *Acidobacteriia*, *Actinobacteria* and *Proteobacteria* were confirmed to be the major bacterial community compositions at the phylum and the class analysis levels. Notably, *Bacteroidetes* and *Bacteroidia* were correlated with soil MBC, NH_4^+-N, suggesting that the vital role of modest N should not be ignored when considering the CH_4 emission from waterlogged paddy soils. The results indicated that the application rates of chemical fertilizers and the choices of crop residues were equally important in rice cultivation.

Supplementary Materials: The following supporting information can be downloaded at: https://www.mdpi.com/article/10.3390/agronomy14091972/s1.

Author Contributions: Methodology, J.G. and Z.S.; investigation, J.G. and J.L.; data curation, J.G. and Z.M.; writing—original draft preparation, J.G.; writing—review and editing, J.G. and J.L.; supervision, J.G. and J.L.; project administration, J.G., L.L. and Z.S.; and funding acquisition, L.L. and Z.S. All authors have read and agreed to the published version of the manuscript.

Funding: This study was funded by the Natural Science Foundation of Henan Province (222300420146, 242300420598), the Henan Province Key Research and Development and Promotion Project (212102110286), and the State Key Laboratory Fund (F2010121002-202321).

Data Availability Statement: The data presented in this study are openly available in NCBI Sequence Read Archive (SRA) at https://submit.ncbi.nlm.nih.gov/subs/bioproject/SUB14156894/overview (accessed on 25 March 2023), reference number PRJNA1065809.

Conflicts of Interest: We declare that the authors have no conflicts of interest.

References

1. Haque, M.M.; Sang, Y.K.; Ali, M.A.; Kim, P.J. Contribution of greenhouse gas emissions during cropping and fallow seasons on total global warming potential in mono-rice paddy soils. *Plant Soil* **2015**, *387*, 251–264. [CrossRef]
2. Xu, Y.; Zhang, L.; Ou, S.; Wang, R.; Wang, Y.; Chu, C.; Yao, S. Natural variations of SLG1 confer high-temperature tolerance in indica rice. *Nat. Commun.* **2020**, *11*, 5441. [CrossRef] [PubMed]
3. Neubauer, S.C.; Megonigal, J.P. Moving beyond global warming potentials to quantify the climatic role of ecosystems. *Ecosystems* **2015**, *18*, 1000–1013. [CrossRef]
4. Ministry of Ecology and Environment of the People's Republic of China. Available online: https://www.mee.gov.cn/ywgz/ydqhbh/wsqtkz/201907/P020190701765971866571.pdf (accessed on 25 March 2023).
5. Wang, W.; Lai, D.Y.F.; Wang, C.; Pan, T.; Zeng, C. Effects of rice straw incorporation on active soil organic carbon pools in a subtropical paddy field. *Soil Tillage Res.* **2015**, *152*, 8–16. [CrossRef]
6. Yao, Z.; Zheng, X.; Dong, H.; Wang, R.; Mei, B.; Zhu, J. A 3-year record of N_2O and CH_4 emissions from a sandy loam paddy during rice seasons as affected by different nitrogen application rates. *Agric. Ecosyst. Environ.* **2012**, *152*, 1–9. [CrossRef]
7. Wang, Y.; Bai, R.; Di, H.J.; Mo, L.Y.; Han, B.; Zhang, L.M.; He, J.Z. Differentiated mechanisms of biochar mitigating straw-induced greenhouse gas emissions in two contrasting paddy soils. *Front. Microbiol.* **2018**, *9*, 2566. [CrossRef]
8. Zhou, J.; Chen, H.; Tao, Y.; Thring, R.W.; Mao, J. Biochar amendment of chromium polluted paddy soil suppresses greenhouse gas emissions and decreases chromium uptake by rice grain. *J. Soils Sediments* **2019**, *19*, 1756–1766. [CrossRef]
9. Pan, F.F.; Yu, W.T.; Ma, Q.; Zhou, H.; Jiang, C.M.; Xu, Y.G.; Ren, J.F. Influence of 15N-labeled ammonium sulfate and straw on nitrogen retention and supply in different fertility soils. *Biol. Fertil. Soils* **2017**, *53*, 303–313. [CrossRef]
10. Sun, J.; Peng, H.; Chen, J.; Wang, X.; Wei, M.; Li, W.; Yang, L.; Zhang, Q.; Wang, W.; Mellouki, A. An estimation of CO_2 emission via agricultural crop residue open field burning in China from 1996 to 2013. *J. Clean. Prod.* **2016**, *112*, 2625–2631. [CrossRef]
11. Zhang, J.W.; Tang, H.Y.; Zhu, J.G.; Lin, X.G.; Feng, Y.Z. Divergent responses of methanogenic archaeal communities in two rice cultivars to elevated ground-level O_3. *Environ. Pollut.* **2016**, *213*, 127–134. [CrossRef] [PubMed]
12. Wu, X.H.; Wang, W.; Xie, K.J.; Yin, C.M.; Hou, H.J.; Xie, X.L. Combined effects of straw and water management on CH_4 emissions from rice fields. *J. Environ. Manag.* **2019**, *231*, 1257–1262. [CrossRef] [PubMed]
13. Ye, R.Z.; Horwath, W.R. Influence of rice straw on priming of soil C for dissolved organic C and CH_4 production. *Plant Soil* **2017**, *417*, 231–241. [CrossRef]
14. Wang, H.; Li, X.; Li, X.; Wang, J.; Li, X.; Guo, Q.; Yu, Z.; Yang, T.; Zhang, H. Long-term no-tillage and different residue amounts alter soil microbial community composition and increase the risk of maize root rot in northeast China. *Soil Till Res.* **2020**, *196*, 104452. [CrossRef]
15. Liu, B.; Xia, H.; Jiang, C.; Riaz, M.; Yang, L.; Chen, Y.; Fan, X.; Xia, X. 14 year applications of chemical fertilizers and crop straw effects on soil labile organic carbon fractions, enzyme activities and microbial community in rice-wheat rotation of middle China. *Sci. Total Environ.* **2022**, *841*, 156608. [CrossRef]
16. Wang, X.; Sun, B.; Mao, J.; Sui, Y.; Cao, X. Structural convergence of maize and wheat straw during two-year decomposition under different climate conditions. *Environ. Sci. Technol.* **2012**, *46*, 7159–7165. [CrossRef]
17. Jiang, Y.; van Groenigen, K.J.; Huang, S.; Hungate, B.A.; van Kessel, C.; Hu Liechty, Z.; Santos-Medellín, C.; Edwards, J.; Nguyen, B.; Mikhail, D.; et al. Comparative analysis of root microbiomes of rice cultivars with high and low methane emissions reveals differences in abundance of methanogenic archaea and putative upstream fermenters. *mSystems* **2020**, *5*, e819–e897.
18. Pump, J.; Pratscher, J.; Conrad, R. Colonization of rice roots with methanogenic archaea controls photosynthesis-derived methane emission. *Environ. Microbiol.* **2015**, *17*, 2254–2260. [CrossRef]
19. Hofmann, K.; Praeg, N.; Mutschlechner, M.; Wagner, A.O.; Illmer, P. Abundance and potential metabolic activity of methanogens in well-aerated forest and grassland soils of an alpine region. *FEMS Microbiol. Ecol.* **2016**, *92*, fiv171. [CrossRef]
20. Yang, Y.-L.; Shen, L.-D.; Zhao, X.; Shan, J.; Wang, S.-W.; Zhou, W.; Liu, J.-Q.; Liu, X.; Tian, M.-H.; Yang, W.-T.; et al. Long-term incorporation of wheat straw changes the methane oxidation potential, abundance and community composition of methanotrophs in a paddy ecosystem. *Appl. Soil Ecol.* **2022**, *173*, 1043. [CrossRef]
21. Gütlein, A.; Gerschlauer, F.; Kikoti, I.; Kiese, R. Impacts of climate and land use on N_2O and CH_4 fluxes from tropical ecosystems in the Mt. Kilimanjaro region, Tanzania. *Glob. Chang. Biol.* **2018**, *24*, 1239–1255. [CrossRef]
22. Qi, L.; Pokharel, P.; Huang, R.; Chang, S.X.; Gong, X.Q.; Sun, T.; Wu, Y.Q.; Wang, Z.F.; Gao, M. Crop residue retention increases greenhouse gas emissions but reduces chemical fertilizer requirement in a vegetable-rice rotation. *Agric. For. Meteorol.* **2023**, *342*, 109723. [CrossRef]
23. Jiang, Y.; Qian, H.Y.; Huang, S.; Zhang, X.Y.; Wang, L.; Zhang, L.; Shen, M.X.; Xiao, X.P.; Chen, F.; Zhang, H.L.; et al. Acclimation of methane emissions from rice paddy fields to straw addition. *Sci. Adv.* **2019**, *5*, eaau9038. [CrossRef] [PubMed]
24. Blanca Pascual, M.; Sánchez-Monedero, M.A.; Chacón, F.J.; Sánchez-García, M.; Cayuela, M.L. Linking biochars properties to their capacity to modify aerobic CH4 oxidation in an upland agricultural soil. *Geoderma* **2020**, *363*, 114179. [CrossRef]
25. Luo, D.; Li, Y.Y.; Yao, H.Y.; Chapman, S.J. Effect of different carbon sources on methane production and the methanogenic communities in iron rich flooded paddy soil. *Sci. Total Environ.* **2022**, *823*, 153636. [CrossRef] [PubMed]
26. He, C.; Yang, Z.L.; Huang, G.Q.; Chen, L.J.; Han, L.J. Qualitative and quantitative analysis of straw content in straw-coal blends using vis/nir spectroscopy. *Trans. Chin. Soc. Agric. Eng.* **2013**, *29*, 188–195.

27. Gao, J.K.; Liang, C.L.; Shen, G.Z.; Lv, J.L.; Wu, H.M. Spectral characteristics of dissolved organic matter in various agricultural soils throughout China. *Chemosphere* **2017**, *176*, 108–116. [CrossRef]
28. Fellman, J.B.; Hood, E.; Spencer, R.G. Fluorescence spectroscopy opens new windows into dissolved organic matter dynamics in freshwater ecosystems: A review. *Limnol. Oceanogr.* **2010**, *55*, 2452–2462. [CrossRef]
29. Zsolnay, Á. Dissolved organic matter: Artefacts, definitions, and functions. *Geoderma* **2003**, *113*, 187–209. [CrossRef]
30. Li, W.; Li, X.; Han, C.; Gao, L.; Wu, H.; Li, M. A new view into three−dimensional excitation−emission matrix fluorescence spectroscopy for dissolved organic matter. *Sci. Total Environ.* **2023**, *855*, 158963. [CrossRef]
31. Gao, J.; Liu, L.; Shi, Z.; Lv, J. Biochar Amendments Facilitate Methane Production by Regulating the Abundances of Methanogens and Methanotrophs in Flooded Paddy Soil. *Front. Soil Sci.* **2022**, *2*, 801227. [CrossRef]
32. Wu, J.; Zhang, J.; Jia, W. Impact of COD/N ratio on nitrous oxide emission from microcosm wetlands and their performance in removing nitrogen from wastewater. *Bioresour. Technol.* **2009**, *100*, 2910–2917. [CrossRef] [PubMed]
33. Munyaka, P.M.; Khafipour, A.; Wang, H.; Eissa, N.; Khafipour, E.; Ghia, J.E. Mo1774 Prenatal Antibiotic Treatment Increases Offspring's Susceptibility to Experimental Colitis: A Role of the Gut Microbiota. *PLoS ONE* **2015**, *10*, e0142536. [CrossRef]
34. Jami, E.; Israel, A.; Kotser, A.; Mizrahi, I. Exploring the bovine rumen bacterial community from birth to adulthood. *ISME J.* **2013**, *7*, 1069–1079. [CrossRef]
35. Liu, D.; Huang, Y.M.; An, S.S.; Sun, H.Y.; Bhople, P.; Chen, Z.W. Soil physicochemical and microbial characteristics of contrasting land-use types along soil depth gradients. *Catena* **2017**, *162*, 345–353. [CrossRef]
36. Qin, H.L.; Tang, Y.F.; Shen, J.L.; Wang, C.; Chen, C.L.; Yang, J.; Liu, Y.; Chen, X.B.; Li, Y.; Hou, H.J. Abundance of transcripts of functional gene reflects the inverse relationship between CH_4 and N_2O emissions during mid-season drainage in acidic paddy soil. *Biol. Fertil. Soils* **2018**, *54*, 885–895. [CrossRef]
37. Wang, W.; Chen, C.L.; Wu, X.H.; Xie, K.J.; Yin, C.M.; Hou, H.J.; Xie, X.L. Effects of reduced chemical fertilizer combined with straw retention on greenhouse gas budget and crop production in double rice fields. *Biol. Fertil. Soils* **2019**, *55*, 89–96. [CrossRef]
38. Liu, H.; Hu, Z.; Zhang, J.; Ngo, H.H.; Guo, W.; Liang, S. Optimizations on supply and distribution of dissolved oxygen in constructed wetlands: A review. *Bioresour. Technol.* **2016**, *214*, 797–805. [CrossRef] [PubMed]
39. Hu, N.J.; Wang, B.J.; Gu, Z.H.; Tao, B.R.; Zhang, Z.W.; Hu, S.J.; Zhu, L.Q.; Meng, Y.L. Effects of different straw returning modes on greenhouse gas emissions and crop yields in a rice-wheat rotation system. *Agric. Ecosyst. Environ.* **2016**, *223*, 115–122. [CrossRef]
40. Wang, C.; Liu, J.; Shen, J.; Chen, D.; Li, Y.; Jiang, B.; Wu, J. Effects of bio-char amendment on net greenhouse gas emissions and soil fertility in a double rice cropping system: A 4-year field experiment. *Agric. Ecosyst. Environ.* **2018**, *262*, 83–96. [CrossRef]
41. Wu, Z.; Sun, L.; Dong, Y.; Xu, X.; Xiong, Z. Contrasting effects of different field-aged biochars on potential methane oxidation between acidic and saline paddy soils. *Sci. Total Environ.* **2022**, *853*, 158643. [CrossRef]
42. Jiang, Y.; van Groenigen, K.J.; Huang, S.; Hungate, B.A.; van Kessel, C.; Hu, S.J.; Zhang, J.; Wu, L.H.; Yan, X.J.; Wang, L.L.; et al. Higher yields and lower methane emissions with new rice cultivars. *Glob. Chang. Biol.* **2017**, *23*, 4728–4738. [CrossRef] [PubMed]
43. Wu, X.; Wang, W.; Xie, X.; Yin, C.; Hou, H.; Yan, W.; Wang, G. Net global warming potential and greenhouse gas intensity as affected by different water management strategies in Chinese double rice cropping systems. *Sci. Rep.* **2018**, *8*, 779. [CrossRef] [PubMed]
44. Huang, S.; Sun, Y.; Yu, X.; Zhang, W. Interactive effects of temperature and moisture on CO_2 and CH_4 production in a paddy soil under long-term different fertilization regimes. *Biol Fertil Soils* **2016**, *52*, 285–294. [CrossRef]
45. Zhou, M.; Zhu, B.; Wang, X.; Wang, Y. Long-term field measurements of annual methane and nitrous oxide emissions from a Chinese subtropical wheat-rice rotation system. *Soil Biol. Biochem.* **2017**, *115*, 21–34. [CrossRef]
46. Zhang, Y.; Xu, X.; Zhang, P.; Ling, Z.; Qiu, H.; Cao, X. Pyrolysis-temperature depended quinone and carbonyl groups as the electron accepting sites in barley grass derived biochar. *Chemosphere* **2019**, *232*, 273–280. [CrossRef]
47. Li, X.M.; Ding, L.J.; Li, X.M.; Zhu, Y.G. Abundance, diversity, and structure of geobacteraceae community in paddy soil under long-term fertilization practices. *Appl. Soil Ecol.* **2020**, *153*, 103577. [CrossRef]
48. Li, D.D.; Zhao, B.Z.; Zhang, J.B.; Liu, K.L.; Huang, Q.H. Different traits from the paddy soil and upland soil regulate bacterial community and molecular composition under long-term fertilization regimes. *Appl. Soil Ecol.* **2021**, *165*, 103982. [CrossRef]
49. Meier, K.K.; Jones, S.M.; Kaper, T.; Hansson, H.; Koetsier, M.J.; Karkehabadi, S.; Solomon, E.I.; Sandgren, M.; Kelemen, B. Oxygen activation by Cu LPMOs in recal-citrant carbohydrate polysaccharide conversion to monomer sugars. *Chem. Rev.* **2018**, *118*, 2593–2635. [CrossRef]
50. Björn, B.; Martin, B.B.; Melanie, K.; Ralf, C. Croprotation of flooded rice with upland maize impacts the resident and active methanogenic microbial community. *Environ. Microbiol.* **2015**, *18*, 2868–2885.
51. Conrad, R. Microbial ecology of methanogens and methanotrophs. *Adv. Agron.* **2007**, *96*, 1–63.
52. Azeem, M.; Hayat, R.; Hussain, Q.; Ahmed, M.; Imran, M.; Crowley, D.E. Effect of biochar amendment on soil microbial biomass abundance enzyme activity in the mash bean field. *J. Biod. Environ. Sci.* **2016**, *8*, 2222–3045.
53. Watanabe, T.; Kimura, M.; Asakawa, S. Community structure of methanogenic archaea in paddy field soil under double cropping (rice–wheat). *Soil Biol. Biochem.* **2006**, *38*, 1264–1274. [CrossRef]
54. Ji, Y.; Liu, P.; Conrad, R. Change of the pathway of methane production with progressing anoxic incubation of paddy soil. *Soil Biol. Biochem.* **2018**, *121*, 177–184. [CrossRef]
55. Rui, J.; Peng, J.; Lu, Y. Succession of Bacterial Populations during Plant Residue De-composition in Rice Field Soil. *Appl. Environ. Microbiol.* **2009**, *75*, 4879–4886. [CrossRef] [PubMed]

56. Hernández, M.; Conrad, R.; Klose, M.; Ma, K.; Lu, Y. Structure and function of methanogenic microbial communities in soils from flooded rice and upland soybean fields from Sanjiang Plain, NE China. *Soil Biol. Biochem.* **2017**, *105*, 81–91. [CrossRef]
57. Yuan, J.; Yuan, Y.; Zhu, Y.; Cao, L. Effects of different fertilizers on methane emissions and methanogenic community structures in paddy rhizosphere soil. *Sci. Total Environ.* **2018**, *627*, 770–781. [CrossRef] [PubMed]
58. Fan, L.; Dippold, M.A.; Ge, T.; Wu, J.; Thiel, V.; Kuzyakov, Y.; Dorodnikov, M. Anaerobic oxidation of methane in paddy soil: Role of electron acceptors and fertilization in mitigating ch4 fluxes. *Soil Biol. Biochem.* **2020**, *141*, 107685. [CrossRef]
59. Zhang, Y.; Wang, F.; Jia, Z. Electron shuttles facilitate anaerobic methane oxidation coupled to nitrous oxide reduction in paddy soil. *Soil Biol. Biochem.* **2021**, *153*, 108091. [CrossRef]

MDPI AG
Grosspeteranlage 5
4052 Basel
Switzerland
Tel.: +41 61 683 77 34

Agronomy Editorial Office
E-mail: agronomy@mdpi.com
www.mdpi.com/journal/agronomy

www.ingramcontent.com/pod-product-compliance
Lightning Source LLC
Chambersburg PA
CBHW061842210926
51850CB00006B/947